JEE Main 2025

January Session
(22-30 January)

10 Solved Papers

with Important Concept Marker

PHYSICS
CHEMISTRY
MATHS

Authors

- **Deepak Paliwal** (Physics)
- **S. Raychaudhuri** (Chemistry)
- **DK Upadhyay** (Mathematics)

arihant
SUCCESS AND NOTHING LESS

arihant
SUCCESS AND NOTHING LESS

ARIHANT PRAKASHAN (SERIES), MEERUT

ADMINISTRATIVE & PRODUCTION OFFICES
Regd. Office
'Ramchhaya' 4577/15, Agarwal Road, Darya Ganj, New Delhi -110002
Tele: 011- 47630600, 43518550; Fax: 011- 23280316

Head Office
Kalindi, TP Nagar, Meerut (UP) - 250002, Tel: 0121-7156203, 7156204

SALES & SUPPORT OFFICES
Agra, Ahmedabad, Bengaluru, Bareilly, Chennai, Delhi, Guwahati, Hyderabad, Jaipur, Jhansi, Kolkata, Lucknow, Nagpur & Pune.

PRICE ₹235.00

PO No : TXT-XX-XXXXXXX-X-XX

Published By Arihant Publications (India) Ltd.

For further information about the books published by Arihant, log on to www.arihantbooks.com or e-mail at info@arihantbooks.com

Follow us on

Class-wise Snapshot

JEE Main 2025, Session 1

All Papers Shift 1

- In the overall analysis of class-wise weightage in the JEE Main 2025 January session, the major portion was from the syllabus of Class 12. As per the data, around 44.44% of the questions were from the Class 11 syllabus, and 55.55% were from Class 12.
- Cumulative subjectwise questions distribution of all shift-1 papers is given below

Subject	Total Percentage	
	Class 11	Class 12
Mathematics	56.00%	44.00%
Chemistry	49.2%	50.8%
Physics	44.8%	55.2%
Grand Total	**50%**	**50%**

All Papers Shift 2

- In the overall analysis of class-wise weightage in the JEE Main 2025 January session, the major portion was from the syllabus of Class 12. As per the data, around 42.22% of the questions were from the Class 11 syllabus, and 57.78% were from Class 12.
- Cumulative subjectwise questions distribution of all shift-2 papers is given below

Subject	Total Percentage	
	Class 11	Class 12
Mathematics	30.00%	70.00%
Chemistry	46.66%	53.33%
Physics	50.00%	50.00%
Grand Total	**42.22%**	**57.78%**

Subjectwise Overall Difficulty Level

- As per the analysis we can say that the subjectwise difficulty was in the following order: Mathematics> Chemistry>Physics
- Subjectwise overall difficulty level on the scale of 1 to 3 is given below

Subject	Difficulty Level
Mathematics	2.20
Chemistry	1.90
Physics	2.15

Subject-wise Overall Analysis

JEE Main 2025, Session 1

Mathematics

- Math papers in JEE Main 2025, session 1 were more or less balanced certainly with lengthy calculations. Class-wise, topics from Class 11th had more & class 12th had less weightage.
- Topics like Complex, Numbers Sequence & Series, Co-ordinate Geometry (straight line, conic section, circle), Binomial Theorem, Function & Inverse Trigonometric Functions, Limit, Continuity & Differentiation, Indefinite & Definite Integration, Vector-3D, Probability had higher weightage. At the same time, Trigonometric Equations, Determinants, and Quadratic Equations had lower weightage.
- Even in some papers questions were framed on concepts outside the syllabus like properties of determinants and straight line in space. Overall, the level of the papers were moderate to difficult type.

Physics

- The Physics papers in JEE Main 2025, session 1 were surprisingly different from the expectations. Overall, in all the shifts the papers were unbalanced. Class 12 topical weightage was higher than the Class 11 topical weightage.
- As expected, chapters like Unit and Measurements, Work Energy and Power, System of Particles and Rotational Motion, Mechanical Properties of Fluids, Thermodynamics, Electrostatic Potential, Current Electricity, Moving Charges and Magnetism EMW, Ray Optice, Wave Optics, Dual Nature, and Semiconductor had higher weightage. As a usual trend the question distribution was uneven and unpredictable. In every paper, some portion of syllabus was completely missing.
- Difficulty level wise the papers were easy to moderate and contained many fact-based questions as well from NCERT to check students' knowledge.

Chemistry

- The overall papers of Chemistry were also had some surprising elements. Branchwise analysis of chemistry is given below.
- In all the papers, Organic Chemistry was mostly NCERT-based. In general, its weightage in the paper of every shift was higher than the weightage of inorganic chemistry but lower than the weightage of physical chemistry.
- Questions were asked from GOC, Amines, Aldehydes & Ketones, Biomolecules, Phenol & Ethers, Aryl & Alkyl Halides alongwith mixed concept questions. The nature of questions was tricky and conceptual. Overall, the paper level was easy to moderate covering all the important concepts.
- There were more questions from the Periodic Table, Some Basic Concept of Chemistry, Thermodynamics, Equilibrium, Coordination Compounds, Chemical Bonding and Amines. Fact-based questions were directly asked from the NCERT.
- As compared to organic and inorganic chemistry, this year in session 1 questions from Physical Chemistry more less in all the shifts.

JEE Main 2025 Session 1

Subjectwise and Unitwise Questions Average

Physics	Average	Chemistry	Average	Mathematics	Average
Physics & Measurements	1-2	Some Basic Concepts in Chemistry	1-2	Sets	0-1
Motion in a Straight Line	0-1	Atomic Structure	0-1	Relations & Functions	1-2
Motion in a Plane	0-1	Classification of Elements and Periodicity in Properties	1-2	Trigonometry	1-2
Laws of Motion	0-1	Chemical Bonding and Molecular Structure	0-1	Complex Number & Quadratic Equations	2-3
Work, Energy & Power	1-2	Thermodynamics	1-2	Permutations and Combinations	1-2
System of Particles & Rotational Motion	1-2	Equilibrium	1-2	Binomial Theorem and its Application	0-1
Gravitation	0-1	Redox Reactions	0-1	Sequence and Series	1-2
Mechanical Properties of Solids	0-1	*p*-block Elements-1	0-1	Straight Lines and Pair of Straight Lines	1-2
Mechanical Properties of Fluids	1-2	Purification & Characterisation of Organic Compounds	0-1	Circle	0-1
Thermal Properties of Matter	0-1	Some Basic Principles of Organic Chemistry	2-3	Conic Sections	2-3
Thermodynamics	1-2	Hydrocarbons	1-2	Matrices and Deferminants	0-1
Kinetic Theory of Gases	0-1	Solutions	0-1	Limit, Continuity and Differentiability	1-2
Oscillations	1-2	Electrochemistry	1-2	Differentiation	0-1
Waves	0-1	Chemical Kinetics	1-2	Application of Derivatives	0-1
Electric Charges and Fields	1-2	*p*-block Elements - 2	0-1	Integrals	1-2
Electrostatics Potential and Capacitance	1-2	*d* & *f*-block Elements	1-2	Application of Intergrals	0-1
Current Electricity	1-2	Coordination Compounds	1-2	Vector Algebra	1-2
Moving Charges and Magnetism	1-2	Organic Compounds Containing Halogens	0-1	Three Dimentional Geometry	0-1
Magnetism and Matter	0-1	Alcohols, Phenols & Ethers	0-1	Statistics	0-1
Electromagnetic Induction	0-1	Aldehydes, Ketones and Carboxylic Acids	1-2	Probability	0-1
Alternating Current	0-1	Organic Compounds Containing Nitrogen	1-2	Differential Equation	0-1
Electromagnetic Waves	0-1	Biomolecules	1-2		
Ray Optics & Optical Instruments	2-3	Principles Related to Practical Chemistry	0-1		
Wave Optics	1-2				
Dual Nature of Matter & Radiation	1-2				
Atoms	0-1				
Nuclei	0-1				
Semiconductor Electronics	1-2				
Experimental Physics	0-1				

Shiftwise Percentile *vs* Score and Difficulty Level
JEE Main 2025, Session 1

Papers Shift	Total Score at 99.9 Percentile	Subjectwise scores at 99.9 Percentile			Overall Difficulty Level
		Physics	Chemistry	Maths	
Jan 22, Shift-I	279	95	92	92	Easy
Jan 22, Shift-II	233	82	81	70	Moderate
Jan 23, Shift-I	255	91	95	69	Moderate
Jan 23, Shift-II	228	86	82	60	Moderate
Jan 24, Shift-I	233	83	80	70	Moderate
Jan 24, Shift-II	233	86	85	62	Moderate
Jan 28, Shift-I	224	89	79	56	Difficult
Jan 28, Shift-II	224	84	80	60	Moderate to Difficult
Jan 29, Shift-I	233	85	85	63	Moderate to Difficult
Jan 29, Shift-II	240	87	89	64	Easey to Moderate

JEE Main Cutoff - Previous Years'

Category	2024	2023	2022
General	93.2362181	90.7788642	88.4121
EWS	81.3266412	75.6229025	63.1114
OBC-NCL	79.6757881	73.6114227	67.0090
SC	60.0923182	51.9776027	43.0820
ST	46.6975840	37.2348772	26.7771
PWD	0.0018	0.0013527	0.00310

Branchwise Opening and Closing Rank of Top NITs in JEE Main 2024

Name of NIT	Course	Opening Rank	Closing Rank
NIT Trichy	Computer Science and Engineering	1196	4178
	Electronics and Communications Engineering	2712	6799
	Electronics and Electronics Engineering	6688	9043
	Instrumentation and Control Engineering	5571	11050
	Mechanical Engineering	5681	8125
	Chemical Engineering	12706	13950
	Metallurgical and Materials Engineering	6251	15520
	Civil Engineering	19131	30998
NIT Rourkela	Computer Science and Engineering	2379	7801
	Electronics and Communication Engineering	9521	9936
	Electrical Engineering	7421	18080
	Electronics and Instrumentation Engineering	13084	18305
	Mechanical Engineering	9600	21690
	Civil Engineering	25338	32946
NIT Karnataka, Surathkal	Computer Science and Engineering	1865	3406
	Artificial Intelligence	2836	3505
	Information Technology	2911	3877
	Electronics and Communication Engineering	2763	4948
	Computational and Data Science	1190	2696
	Electrical and Electronics Engineering	8120	9921
	Mechanical Engineering	3340	12856
	Civil Engineering	10933	27416
NIT Calicut	Computer Science and Engineering	4488	4965
	Engineering Physics	14312	78121
	Electrical and Electronics Engineering	54211	11991
	Mechanical Engineering	15470	21214
	Civil Engineering	25853	35519
MNNIT Allahabad	Computer Science and Engineering	1125	2322
	Electrical Engineering	8821	10111
	Electronics and Communication Engineering	1572	1750
	Mechanical Engineering	5680	6395
	Civil Engineering	16831	23584

Branch-wise JEE Advanced Cutoff 2024 for Top 5 IITs

In Indian Institutes of Technology (IITs) the CSE branch is the first choice among Engineering aspirants. Apart from Computer Science and Engineering, Electrical Engineering, Mathematics and Computing Engineering and Mechanical Engineering are the most popular branches but they differ from one IIT to another. As per the IIT JEE Advanced 2024 cutoff, IIT Bombay has the most competitive cutoff rank for the CSE branch as compared with other IITs. Given below are the top 5 IITs branch-wise cutoff 2024 for the General All India category candidates.

1. IIT Bombay	B.Tech. in Computer Science and Engineering	68
	B.Tech. in Electrical Engineering	496
	Integrated B.Tech. in Electrical Engineering + M.Tech. in Microelectronics	992
	B.Tech. in Mechanical Engineering	666
	B.Tech. in Engineering Physics	1875
2. IIT Delhi	B.Tech. in Computer Science and Engineering	116
	Integrated B.Tech. + M.Tech. in Computer Science and Engineering	189
	B.Tech. in Mathematics and Computing Engineering	355
	Integrated B.Tech. + M.Tech. in Mathematics and Computing	433
	B.Tech. in Electrical Engineering	592
	B.Tech. in Electrical Engineering (Power and Automation)	625
	B.Tech. in Mechanical Engineering	1774
	B.Tech.	3412
3. IIT Madras	B.Tech. in Civil Engineering	7212
	B.Tech. in Computer Science and Engineering	159
	B.Tech. in Electrical Engineering	838
	B.Tech. in Mechanical Engineering	2712
	B.Tech. in Engineering Physics	1922
4. IIT Kanpur	B.Tech. in Computer Science and Engineering	252
	B.Tech. in Electrical Engineering	1290
	B.Tech. in Mechanical Engineering	2141
	B.Tech. in Chemical Engineering	4214
	B.Tech. in Civil Engineering	4295
5. IIT Roorkee	B.Tech. in Computer Science and Engineering	481
	B.Tech. in Data Science and Artificial Intelligence	825
	B.Tech. in Electronics and Communication Engineering	2033
	B.Tech. in Electrical Engineering	2172
	B.Tech. in Engineering Physics	4590
	B.Tech. in Mechanical Engineering	10517

CONTENTS

JEE Main 2025 January Session-I, 10 Solved Papers (Online)

Scan & Attempt Online Mock Tests

Test 1 Test 2 Test 3

SYLLABUS

Physics

UNIT 1 Physics and Measurement

Units of measurements, System of Units, S I Units, fundamental and derived units, least count, significant figures, Errors in measurements, Dimensions of Physics quantities, dimensional analysis, and its applications.

UNIT 2 Kinematics

The frame of reference, motion in a straight line, Position- time graph, speed and velocity; Uniform and non-uniform motion, average speed and instantaneous velocity, uniformly accelerated motion, velocity-time, position-time graph, relations for uniformly accelerated motion, Scalars and Vectors, Vector. Addition and subtraction, scalar and vector products, Unit Vector, Resolution of a Vector. Relative Velocity, Motion in a plane, Projectile Motion, Uniform Circular Motion.

UNIT 3 Laws of Motion

Force and inertia, Newton's First law of motion; Momentum, Newton's Second law of motion, Impulses; Newton's Third law of motion. Law of conservation of linear momentum and its applications. Equilibrium of concurrent forces.

Static and Kinetic friction, laws of friction, rolling friction. Dynamics of uniform circular motion: centripetal force and its applications: vehicle on a level circular road, vehicle on a banked road.

UNIT 4 Work, Energy and Power

Work done by a constant force and a variable force; kinetic and potential energies, work-energy theorem, power.

The potential energy of spring conservation of mechanical energy, conservative and non-conservative forces; motion in a vertical circle: Elastic and inelastic collisions in one and two dimensions.

UNIT 5 Rotational Motion

Centre of the mass of a two-particles system, Centre of the mass of a rigid body; Basic concepts of rotational motion; moment of a force; torque, angular momentum, conservation of angular momentum and its applications;

The moment of inertia, the radius of gyration, values of moments of inertia for simple geometrical objects, parallel and perpendicular axes theorems and their applications.

Equilibrium of rigid bodies, rigid body rotation and equations of rotational motion, comparison of linear and rotational motions.

UNIT 6 Gravitation

The universal law of gravitation. Acceleration due to gravity and its variation with altitude and depth. Kepler's law of planetary motion. Gravitational potential energy; gravitational potential. Escape velocity, Motion of a satellite, orbital velocity, time period and energy of satellite.

UNIT 7 Properties of Solids and Liquids

Elastic behaviour, Stress-strain relationship, Hooke's Law. Young's modulus, bulk modulus, and modulus of rigidity. Pressure due to a fluid column; Pascal's law and its applications. Effect of gravity on fluid pressure.

Viscosity, Stokes' law, terminal velocity, streamline, and turbulent flow, critical velocity. Bernoulli's principle and its applications. Surface energy and surface tension, angle of contact, excess of pressure across a curved surface, application of surface tension - drops, bubbles, and capillary rise. Heat, temperature, thermal expansion; specific heat capacity, calorimetry; change of state, latent heat. Heat transfer conduction, convection and radiation.

UNIT 8 Thermodynamics

Thermal equilibrium, zeroth law of thermodynamics, the concept of temperature. Heat, work and internal energy. The first law of thermodynamics, isothermal and adiabatic processes. The second law of thermodynamics: reversible and irreversible processes.

UNIT 9 Kinetic Theory of Gases

Equation of state of a perfect gas, work done on compressing a gas, Kinetic theory of gases -assumptions, the concept of pressure. Kinetic interpretation of temperature: RMS speed of gas molecules: Degrees of freedom. Law of equipartition of energy and applications to specific heat capacities of gases; Mean free path. Avogadro's number.

UNIT 10 Oscillations and Waves

Oscillations and periodic motion – time period, frequency, displacement as a function of time. Periodic functions. Simple harmonic motion (SHM) and its equation; phase: oscillations of a spring -restoring force and force constant: energy in SHM - Kinetic and potential energies; Simple pendulum - derivation of expression for its time period:

Wave motion. Longitudinal and transverse waves, speed of the travelling wave. Displacement relation for a progressive wave. Principle of superposition of waves, reflection of waves. Standing waves in strings and organ pipes, fundamental mode and harmonics. Beats.

UNIT 11 Electrostatics

Electric charges: Conservation of charge. Coulomb's law forces between two point charges, forces between multiple charges: superposition principle and continuous charge distribution.

Electric field: Electric field due to a point charge, Electric field lines. Electric dipole, Electric field due to a dipole. Torque on a dipole in a uniform electric field Electric flux. Gauss's law and its applications to find field due to infinitely long uniformly charged straight wire uniformly charged infinite plane sheet and uniformly charged thin spherical shell. Electric potential and its calculation for a point charge, electric dipole and system of charges; potential difference, Equipotential surfaces, Electrical potential energy of a system of two point charges and of electric dipole in an electrostatic field.

Conductors and insulators. Dielectrics and electric polarization, capacitors and capacitances, the combination of capacitors in series and parallel, and capacitance of a parallel plate capacitor with and without dielectric medium between the plates. Energy stored in a capacitor.

UNIT 12 Current Electricity

Electric current. Drift velocity, mobility, and their relation with electric current. Ohm's law, Electrical resistance. V-I characteristics of Ohmic and non-ohmic conductors. Electrical energy and power. Electrical resistivity and conductivity. Series and parallel combinations of resistors; Temperature dependence of resistance.

Internal resistance, potential difference, and emf of a cell, a combination of cells in series and parallel. Kirchhoff's laws and their applications. Wheatstone bridge. Metre bridge.

UNIT 13 Magnetic Effects of Current and Magnetism

Biot - Savart' law and its application to the current-carrying circular loop. Ampere's law and its applications to infinitely long current carrying straight wire and solenoid. Force on a moving charge in uniform magnetic and electric fields.

Force on a current-carrying conductor in a uniform magnetic field. The force between two parallel currents carrying conductors-definition of ampere. Torque experienced by a current loop in a uniform magnetic field: Moving coil galvanometer, its sensitivity, and conversion to ammeter and voltmeter.

Current loop as a magnetic dipole and its magnetic dipole moment. Bar magnet as an equivalent solenoid, magnetic field lines; Magnetic field due to a magnetic dipole (bar magnet) along its axis and perpendicular to its axis. Torque on a magnetic dipole in a uniform magnetic field. Para, dia and ferromagnetic substances with examples, the effect of temperature on magnetic properties.

UNIT 14 Electromagnetic Induction and Alternating Currents

Electromagnetic induction: Faraday's law. Induced emf and current: Lenz's Law, Eddy currents. Self and mutual inductance. Alternating currents, peak and RMS value of alternating current/ voltage: reactance and impedance: LCR series circuit, resonance: power in AC circuits, wattless current. AC generator and transformer.

UNIT 15 Electromagnetic Waves

Displacement current. Electromagnetic waves and their characteristics, Transverse nature of electromagnetic waves, Electromagnetic spectrum (radio waves, microwaves, infrared, visible, ultraviolet, X-rays, Gamma rays), Applications of EM waves.

UNIT 16 Optics

Reflection of light, spherical mirrors, mirror formula. Refraction of light at plane and spherical surfaces, thin lens formula, and lens maker formula. Total internal reflection and its applications. Magnification. Power of a Lens. Combination of thin lenses in contact. Refraction of light through a prism. Microscope and Astronomical Telescope (reflecting and refracting) and their magnifying powers.

Wave optics: wavefront and Huygens' principle. Laws of reflection and refraction using Huygens principle. Interference, Young's double-slit experiment, and expression for fringe width, coherent sources, and sustained interference of light. Diffraction due to a single slit, width of central maximum. Polarization, plane-polarized light: Brewster's law, uses of plane-polarized light and Polaroid.

UNIT 17 Dual Nature of Matter and Radiation

Dual nature of radiation. Photoelectric effect. Hertz and Lenard's observations; Einstein's photoelectric equation: particle nature of light. Matter waves-wave nature of particle, de Broglie relation.

UNIT 18 Atoms and Nuclei

Alpha-particle scattering experiment; Rutherford's model of atom; Bohr model, energy levels, hydrogen spectrum. Composition and size of nucleus, atomic masses, Mass-energy relation, mass defect; binding energy per nucleon and its variation with mass number, nuclear fission, and fusion.

UNIT 19 Electronic Devices

Semiconductors; semiconductor diode: I-V characteristics in forward and reverse bias; diode as a rectifier; I-V characteristics of LED, the photodiode, solar cell, and Zener diode; Zener diode as a voltage regulator. Logic gates (OR, AND, NOT, NAND and NOR).

UNIT 20 EXPERIMENTAL SKILLS

Familiarity with the basic approach and observations of the experiments and activities:

1. Vernier calipers -its use to measure the internal and external diameter and depth of a vessel.
2. Screw gauge - its use to determine the thickness/ diameter of thin sheet/wire.
3. Simple Pendulum-dissipation of energy by plotting a graph between the square of amplitude and time.
4. Metre Scale - the mass of a given object by the principle of moments.
5. Young's modulus of elasticity of the material of a metallic wire.
6. Surface tension of water by capillary rise and effect of detergents.
7. Co-efficient of Viscosity of a given viscous liquid by measuring the terminal velocity of a given spherical body.
8. Speed of sound in air at room temperature using a resonance tube.
9. Specific heat capacity of a given (i) solid and (ii) liquid by method of mixtures.
10. The resistivity of the material of a given wire using a metre bridge.
11. The resistance of a given wire using Ohm's law.
12. Resistance and figure of merit of a galvanometer by half deflection method.
13. The focal length of;
    (i) Convex mirror
    (ii) Concave mirror
    (iii) Convex lens, using the parallax method.
14. The plot of the angle of deviation vs angle of incidence for a triangular prism.
15. The refractive index of a glass slab using a travelling microscope.
16. Characteristic curves of a p-n junction diode in forward and reverse bias.
17. Characteristic curves of a Zener diode and finding reverse breakdown voltage.
18. Identification of Diode. LED, Resistor. A capacitor from a mixed collection of such items.

Chemistry

Physical Chemistry

UNIT 1 Some Basic Concepts in Chemistry

Matter and its nature, Dalton's atomic theory: Concept of atom, molecule, element, and compound:: Laws of chemical combination; Atomic and molecular masses, mole concept, molar mass, percentage composition, empirical and molecular formulae: Chemical equations and stoichiometry.

UNIT 2 Atomic Structure

Nature of electromagnetic radiation, photoelectric effect; Spectrum of the hydrogen atom. Bohr model of a hydrogen atom - its postulates, derivation of the relations for the energy of the electron and radii of the different orbits, limitations of Bohr's model; Dual nature of matter, de Broglie's relationship. Heisenberg uncertainty principle. Elementary ideas of quantum mechanics, quantum mechanics, the quantum mechanical model of the atom, and its important features. Concept of atomic orbitals as one-electron wave functions: Variation of y and y^2 with r for 1s and 2s orbitals; various quantum numbers (principal, angular momentum, and magnetic quantum numbers) and their significance; shapes of s, p, and d - orbitals, electron spin and spin quantum number: Rules for filling electrons in orbitals – Aufbau principle. Pauli's exclusion principle and Hund's rule, electronic configuration of elements, and extra stability of half-filled and completely filled orbitals.

UNIT 3 Chemical Bonding and Molecular Structure

Kossel-Lewis approach to chemical bond formation, the concept of ionic and covalent bonds.

Ionic Bonding: Formation of ionic bonds, factors affecting the formation of ionic bonds; calculation of lattice enthalpy.

Covalent Bonding: Concept of electronegativity. Fajan's rule, dipole moment: Valence Shell Electron Pair Repulsion (VSEPR) theory and shapes of simple molecules.

Quantum mechanical approach to covalent bonding: Valence bond theory - its important features, the concept of hybridization involving s, p, and d orbitals; Resonance.

Molecular Orbital Theory - Its important features. LCAOs, types of molecular orbitals (bonding, antibonding), sigma and pi-bonds, molecular orbital electronic configurations of homonuclear diatomic molecules, the concept of bond order, bond length, and bond energy. Elementary idea of metallic bonding. Hydrogen bonding and its applications.

UNIT 4 Chemical Thermodynamics

Fundamentals of thermodynamics: System and surroundings, extensive and intensive properties, state functions, entropy, types of processes.

The first law of thermodynamics - Concept of work, heat internal energy and enthalpy, heat capacity, molar heat capacity; Hess's law of constant heat summation; Enthalpies of bond dissociation, combustion, formation, atomization, sublimation, phase transition, hydration, ionization, and solution.

The second law of thermodynamics - Spontaneity of processes; DS of the universe and DG of the system as criteria for spontaneity. DG° (Standard Gibbs energy change) and equilibrium constant.

UNIT 5 Solutions

Different methods for expressing the concentration of solution - molality, molarity, mole fraction, percentage (by volume and mass both), the vapour pressure of solutions and Raoult's Law - Ideal and non-ideal solutions, vapour pressure - composition, plots for ideal and non-ideal solutions; Colligative properties of dilute solutions - a relative lowering of vapour pressure, depression of freezing point, the elevation of boiling point and osmotic pressure; Determination of molecular mass using colligative properties; Abnormal value of molar mass, Van't Hoff factor and its significance.

UNIT 6 Equilibrium

Meaning of equilibrium is the concept of dynamic equilibrium.

Equilibria involving physical processes: Solid-liquid, liquid-gas - gas and solid-gas equilibria, Henry's law. General characteristics of equilibrium involving physical processes.

Equilibrium involving chemical processes: Law of chemical equilibrium, equilibrium constants (K_p and K_c) and their significance, the significance of DG and DG° in chemical equilibrium, factors affecting equilibrium concentration, pressure, temperature, the effect of catalyst; Le Chatelier's principle.

Ionic equilibrium: Weak and strong electrolytes, ionization of electrolytes, various concepts of acids and bases (Arrhenius. Bronsted - Lowry and Lewis) and their ionization, acid-base equilibria (including multistage ionization) and ionization constants, ionization of water. pH scale, common ion effect, hydrolysis of salts and pH of their solutions, the solubility of sparingly soluble salts and solubility products, and buffer solutions.

UNIT 7 Redox Reactions and Electrochemistry

Electronic concepts of oxidation and reduction, redox reactions, oxidation number, rules for assigning oxidation number and balancing of redox reactions.

Electrolytic and metallic conduction, conductance in electrolytic solutions, molar conductivities and their variation with concentration: Kohlrausch's law and its applications.

Electrochemical cells - Electrolytic and Galvanic cells, different types of electrodes, electrode potentials including standard electrode potential, half-cell and cell reactions, emf of a Galvanic cell and its measurement: Nernst equation and its applications; Relationship between cell potential and Gibbs' energy change: Dry cell and lead accumulator; Fuel cells.

UNIT 8 Chemical Kinetics

Rate of a chemical reaction, factors affecting the rate of reactions: concentration, temperature, pressure, and catalyst; elementary and complex reactions, order and molecularity of reactions, rate law, rate constant and its units, differential and integral forms of zero and first-order reactions, their characteristics and half-lives, the effect of temperature on the rate of reactions, Arrhenius theory, activation energy and its calculation, collision theory of bimolecular gaseous reactions (no derivation).

Inorganic Chemistry

UNIT 9 Classification of Elements and Periodicity in Properties

Modem periodic law and present form of the periodic table, s, p. d and f block elements, periodic trends in properties of elements, atomic and ionic radii, ionisation enthalpy, electron gain enthalpy, valence, oxidation states, and chemical reactivity.

UNIT 10 *p*-block Elements

Group-13 to Group 18 Elements

General Introduction: Electronic configuration and general trends in physical and chemical properties of elements across the periods and down the groups; unique behaviour of the first element in each group.

UNIT 11 *d*- and *f*- Block Elements

Transition Elements

General introduction, electronic configuration, occurrence and characteristics, general trends in properties of the first-row transition elements - physical properties, ionization enthalpy, oxidation states, atomic radii, colour, catalytic behaviour, magnetic properties, complex formation, interstitial compounds, alloy formation; Preparation, properties, and uses of $K_2Cr_2O_7$, and $KMnO_4$.

Inner Transition Elements

Lanthanoids - Electronic configuration, oxidation states, and lanthanoid contraction.

Actinoids - Electronic configuration and oxidation states.

UNIT 12 Co-ordination Compounds

Introduction to coordination compounds. Werner's theory; ligands, coordination number, denticity. chelation; IUPAC nomenclature of mononuclear co-ordination compounds, isomerism; Bonding-Valence bond approach and basic ideas of Crystal field theory, colour and magnetic properties; Importance of co-ordination compounds (in qualitative analysis, extraction of metals, and in biological systems).

Organic Chemistry

UNIT 13 Purification and Characterisation of Organic Compounds

Purification - Crystallization, sublimation, distillation, differential extraction, and chromatography - principles and their applications.

Quantitative analysis Detection of nitrogen, sulphur, phosphorus, and halogens.

Quantitative analysis (basic principles only) - Estimation of carbon, hydrogen, nitrogen, halogens, sulphur, and phosphorus. Calculations of empirical formulae and molecular formulae: Numerical problems in organic quantitative analysis.

UNIT 14 Some Basic Principles of Organic Chemistry

Tetravalency of carbon: Shapes of simple molecules - hybridization (s and p): Classification of organic compounds based on functional groups: and those containing halogens, oxygen, nitrogen, and sulphur; Homologous series: Isomerism - structural and stereoisomerism.

Nomenclature (Trivial and IUPAC)

Covalent bond fission - Homolytic and heterolytic: free radicals, carbocations and carbanions; stability of carbocations and free radicals, electrophiles and nucleophiles.

Electronic displacement in a covalent bond - Inductive effect, electromeric effect, resonance and hyperconjugation.

Common types of organic reactions - Substitution, addition, elimination, and rearrangement

UNIT 15 Hydrocarbons

Classification, isomerism, IUPAC nomenclature, general methods of preparation, properties, and reactions.

Alkanes - Conformations: Sawhorse and Newman projections (of ethane): Mechanism of halogenation of alkanes.

Alkenes - Geometrical isomerism: Mechanism of electrophilic addition: addition of hydrogen, halogens, water, hydrogen halides (Markownikoffs and peroxide effect): Ozonolysis and polymerization.

Alkynes - Acidic character: Addition of hydrogen, halogens, water, and hydrogen halides: Polymerisation.

Aromatic hydrocarbons - Nomenclature, benzene - structure and aromaticity: Mechanism of electrophilic substitution: halogenation, nitration.

Friedel-Craft's alkylation and acylation, directive influence of the functional group in monosubstituted benzene.

UNIT 16 Organic Compounds Containing Halogens

General methods of preparation, properties, and reactions: Nature of C—X bond; Mechanisms of substitution reactions. Uses: Environmental effects of chloroform, iodoform, freons and DDT.

UNIT 17 Organic Compounds Containing Oxygen

General methods of preparation, properties, reactions, and uses.

ALCOHOLS, PHENOLS, AND ETHERS

Alcohols: Identification of primary, secondary, and tertiary alcohols: mechanism of dehydration.

Phenols: Acidic nature, electrophilic substitution reactions: halogenation. nitration and sulphonation. Reimer - Tiemann reaction.

Ethers: Structure.

Aldehyde and Ketones: Nature of carbonyl group; Nucleophilic addition to $C>O=$ group, relative reactivities of aldehydes and ketones; Important reactions such as - Nucleophilic addition reactions (addition of HCN. NH_3, and its derivatives), Grignard reagent; oxidation: reduction (Wolf Kishner and Clemmensen); the acidity of a-hydrogen. aldol condensation, Cannizzaro reaction. Haloform reaction, Chemical tests to distinguish between aldehydes and ketones.

Carboxylic Acids

Acidic strength and factors affecting it.

UNIT 18 Organic Compounds Containing Nitrogen

General methods of preparation. Properties, reactions, and uses.

Amines: Nomenclature, classification structure, basic character, and identification of primary, secondary, and tertiary amines and their basic character.

Diazonium Salts: Importance in synthetic organic chemistry.

UNIT 19 Biomolecules

General introduction and importance of biomolecules.

CARBOHYDRATES - Classification; aldoses and ketoses: monosaccharides (glucose and fructose) and constituent monosaccharides of oligosaccharides (sucrose, lactose, and maltose).

PROTEINS - Elementary Idea of a-amino acids, peptide bond, polypeptides. Proteins: primary, secondary, tertiary, and quaternary structure (qualitative idea only), denaturation of proteins, enzymes.

VITAMINS – Classification and functions.

NUCLEIC ACIDS – Chemical constitution of DNA and RNA.

Biological functions of nucleic acids.

Hormones (General introduction)

UNIT 20 Principles Related To Practical Chemistry

Detection of extra elements (Nitrogen, Sulphur, halogens) in organic compounds; Detection of the following functional groups; hydroxyl (alcoholic and phenolic), carbonyl (aldehyde and ketones) carboxyl, and amino groups in organic compounds.

- The chemistry involved in the preparation of the following:
  Inorganic compounds; Mohr's salt, potash alum.
  Organic compounds: Acetanilide, p-nitro acetanilide, aniline yellow, iodoform.
- The chemistry involved in the titrimetric exercises – Acids, bases, and the use of indicators, oxalic-acid vs $KMnO_4$, Mohr's salt vs $KMnO_4$.
- Chemical principles involved in the qualitative salt analysis:

Cations — Pb^{2+}, Cu^{2+}, Al^{3+}, Fe^{3+}, Zn^{2+}, Ni^{2+}, Ca^{2+}, Ba^{2+}, Mg^{2+} NH^{4+}.
Anions – CO_3^{2-}, S^{2-}, SO_4^{2-}, NO_2, NO_3, Cl^-, Br^-, I^- (Insoluble salts excluded).

Chemical principles involved in the following experiments:

1. Enthalpy of solution of $CuSO_4$.
2. Enthalpy of neutralization of strong acid and strong base.
3. Preparation of lyophilic and lyophobic sols.
4. Kinetic study of the reaction of iodide ions with hydrogen peroxide at room temperature.

MATHEMATICS

UNIT 1 Sets, Relations And Functions

Sets and their representation: Union, intersection, and complement of sets and their algebraic properties; Power set; Relation, Type of relations, equivalence relations, functions; one-one, into and onto functions, the composition of functions

UNIT 2 Complex Numbers And Quadratic Equations

Complex numbers as ordered pairs of reals, Representation of complex numbers in the form a + ib and their representation in a plane, Argand diagram, algebra of complex number, modulus, and argument (or amplitude) of a complex number, Quadratic equations in real and complex number system and their solutions, Relations between roots and co-efficient, nature of roots, the formation of quadratic equations with given roots

UNIT 3 Matrices And Determinants

Matrices, algebra of matrices, type of matrices, determinants, and matrices of order two and three, evaluation of determinants, area of triangles using determinants, Adjoint and evaluation of inverse of a square matrix using determinates and Test of consistency and solution of simultaneous linear equations in two or three variables using matrices

UNIT 4 Permutations And Combinations

The fundamental principle of counting, permutation as an arrangement and combination as section, Meaning of P (n,r) and C (n,r), simple applications

UNIT 5 Binomial Theorem And Its Simple Applications

Binomial theorem for a positive integral index, general term and middle term, and simple applications

UNIT 6 Sequence And Series

Arithmetic and Geometric progressions, insertion of arithmetic, geometric means between two given numbers, Relation between A.M and G.M

UNIT 7 Limit, Continuity and Differentiability

Real–valued functions, algebra of functions, polynomials, rational, trigonometric, logarithmic, and exponential functions, inverse function. Graphs of simple functions. Limits, continuity and differentiability. Differentiation of the sum, difference, product, and quotient of two functions. Differentiation of trigonometric, inverse trigonometric, logarithmic, exponential, composite, and implicit functions; derivatives of order up to two, Applications of derivatives:
Rate of change of quantities, monotonic-Increasing and decreasing functions, Maxima and minima of functions of one variable,

UNIT 8 Integral Calculus

Integral as an anti-derivative, Fundamental integral involving algebraic, trigonometric, exponential, and logarithmic functions. Integrations by substitution, by parts and by partial functions. Integration using trigonometricidentities. Evaluation of simple integrals of the type

$$\int \frac{dx}{x^2 \pm a^2},\ \int \frac{dx}{\sqrt{x^2 \pm a^2}},\ \int \frac{dx}{a^2 - x^2},\ \int \frac{dx}{\sqrt{a^2 - x^2}},\ \int \frac{dx}{ax^2+bx+c},\ \int \frac{dx}{\sqrt{ax^2+bx+c}},\ \int \frac{(px+q)dx}{ax^2+bx+c},\ \int \frac{(px+q)dx}{\sqrt{ax^2+bx+c}},\ \int \sqrt{a^2 \pm x^2}\,dx,\ \int \sqrt{x^2 - a^2}\,dx$$

The fundamental theorem of calculus, properties of definite integrals. Evaluation of definite integrals, determining areas of the regions bounded by simple curves in standard form

UNIT 9 Differential Equations

Ordinary differential equations, their order, and degree, the solution of differential equation by the method of separation of variables, solution of a homogeneous and linear differential equation of the type $\frac{dy}{dx} + p(x)y = q(x)$

UNIT 10 Co-ordinate Geometry

Cartesian system of rectangular coordinates in a plane, distance formula, sections formula, locus and its equation, the slope of a line, parallel and perpendicular lines, intercepts of a line on the co-ordinate axis.
Straight line - Various forms of equations of a line, intersection of lines, angles between two lines, conditions for concurrence of three lines, the distance of a point form a line, co-ordinate of the centroid, orthocentre, and circumcentre of a triangle,
Circle, conic sections - A standard form of equations of a circle, the general form of the equation of a circle, its radius and central, equation of a circle when the endpoints of a diameter are given, points of intersection of a line and a circle with the centre at the origin and sections of conics, equations of conic sections (parabola, ellipse, and hyperbola) in standard forms,

UNIT 11 Three Dimensional Geometry

Coordinates of a point in space, the distance between two points, section formula, directions ratios, and direction cosines, and the angle between two intersecting lines. Skew lines, the shortest distance between them, and its equation. Equations of a line

UNIT 12 Vector Algebra

Vectors and scalars, the addition of vectors, components of a vector in two dimensions and three-dimensional space, scalar and vector products

UNIT 13 Statistics and Probability

Measures of discretion; calculation of mean, median, mode of grouped and ungrouped data calculation of standard deviation, variance, and mean deviation for grouped and ungrouped data.
Probability: Probability of an event, addition and multiplication theorems of probability, Baye's theorem, probability distribution of a random variate,

UNIT 14 Trigonometry

Trigonometrical identities and trigonometrical functions, inverse trigonometrical functions, and their properties,

Scan & Attempt Online Mock Tests

Test 1　　Test 2　　Test 3

Time 9:00-12:00 **MM** : 300 Marks

22 January
Shift-I

Online Question Paper

JEE Main 2025

Mathematics

Section A
Objective Type Questions

01/ A circle C of radius 2 lies in the second quadrant and touches both the coordinate axes. Let r be the radius of a circle that has centre at the point (2, 5) and intersects the circle C at exactly two points.

If the set of all possible values of r is the interval (α, β), then $3\beta - 2\alpha$ is equal to

(a) 14 (b) 12
(c) 15 (d) 10

02/ Let the foci of a hyperbola be (1, 14) and (1, – 12). If it passes through the point (1, 6), then the length of its latus-rectum is

(a) $\frac{144}{5}$ (b) $\frac{288}{5}$
(c) $\frac{24}{5}$ (d) $\frac{25}{6}$

03/ The product of all solutions of the equation $e^{5(\log_e x)^2+3} = x^8, x > 0$ is

(a) $e^{\frac{6}{5}}$ (b) e
(c) $e^{\frac{8}{5}}$ (d) e^2

04/ Let $A = \{1, 2, 3, \ldots\ldots, 10\}$ and $B = \left\{\frac{m}{n} : m, n \in A, m < n \text{ and } \gcd(m, n) = 1\right\}$.

Then, $n(B)$ is equal to

(a) 29 (b) 36 (c) 31 (d) 37

Imp.

05/ Two balls are selected at random one by one without replacement from a bag containing 4 white and 6 black balls. If the probability that the first selected ball is black given that the second selected ball is also black is m / n, where $\gcd(m, n) = 1$, then $m + n$ is equal to

(a) 14 (b) 4
(c) 13 (d) 11

Baye's Theorem

06/ Let $a_1, a_2, a_3, \ldots\ldots$ be a GP of increasing positive terms.
If $a_1 a_5 = 28$ and $a_2 + a_4 = 29$, then a_6 is equal to

(a) 526 (b) 628 (c) 812 (d) 784

07/ The area of the region, inside the circle $(x - 2\sqrt{3})^2 + y^2 = 12$ and outside the parabola $y^2 = 2\sqrt{3}x$ is

(a) $6\pi - 8$ (b) $3\pi + 8$
(c) $6\pi - 16$ (d) $3\pi - 8$

08/ Let $x = x(y)$ be the solution of the differential equation
$y^2 dx + \left(x - \frac{1}{y}\right) dy = 0$. If $x(1) = 1$, then $x(1/2)$ is

(a) $\frac{1}{2} + e$ (b) $3 + e$ (c) $\frac{3}{2} + e$ (d) $3 - e$

09/ Let $f(x)$ be a real differentiable function such that $f(0)=1$ and $f(x+y)=f(x)f'(y)+f'(x)f(y)\ \forall$ $x, y \in R$. Then, $\sum_{n=1}^{100} \log_e f(n)$ is equal to

(a) 2525 (b) 2384

(c) 5220 (d) 2406

10/ The number of non-empty equivalence relations on the set $\{1, 2, 3\}$ is

(a) 5 (b) 7

(c) 4 (d) 6

11/ Let $f: R \to R$ be a twice differentiable function such that $f(x+y)=f(x)f(y), \forall x, y \in R$. If $f'(0)=4a$ and f satisfies $f''(x)-3af'(x)-f(x)=0$, $a>0$, then the area of the region $R=\{|(x, y)|, 0 \le y \le f(ax), 0 \le x \le 2\}$ is

(a) e^2-1 (b) e^4+1

(c) e^4-1 (d) e^2+1

12/ From all the English alphabets, five letters are chosen and are arranged in alphabetical order. The total number of ways, in which the middle letter is '*M*' is

(a) 6084 (b) 14950

(c) 4356 (d) 5148

13/ If $\sum_{r=1}^{n} T_r = \frac{(2n-1)(2n+1)(2n+3)(2n+5)}{64}$, then $\lim_{n\to\infty} \sum_{r=1}^{n}\left(\frac{1}{T_r}\right)$ is equal to

(a) 0 (b) 1/3

(c) 1 (d) 2/3

14/ Let $L_1: \frac{x-1}{2}=\frac{y-2}{3}=\frac{z-3}{4}$ and $L_2: \frac{x-2}{3}=\frac{y-4}{4}=\frac{z-5}{5}$ be two lines. Then, which of the following points lies on the line of the shortest distance between L_1 and L_2?

(a) $\left(\frac{8}{3}, -1, \frac{1}{3}\right)$ (b) $\left(-\frac{5}{3}, -7, 1\right)$

(c) $\left(\frac{14}{3}, -3, \frac{22}{3}\right)$ (d) $\left(2, 3, \frac{1}{3}\right)$

15/ Let the parabola $y=x^2+px-3$ meet the coordinate axes at the points P, Q and R. If the circle C with centre at $(-1, -1)$ passes through the points P, Q and R, then the area of ΔPQR is

(a) 5 (b) 6

(c) 7 (d) 4

16/ Let the ΔPQR be the image of the triangle with vertices (1, 3), (3, 1) and (2, 4) in the line $x+2y=2$. If the centroid of ΔPQR is the point (α, β), then $15(\alpha-\beta)$ is equal to

(a) 21 (b) 22 (c) 24 (d) 19

17/ Let for $f(x)=7\tan^8 x+7\tan^6 x-3\tan^4 x-3\tan^2 x$, $I_1=\int_0^{\frac{\pi}{4}} f(x)\,dx$ and $I_2=\int_0^{\frac{\pi}{4}} xf(x)\,dx$. Then, $7I_1+12I_2$ is equal to

(a) 1 (b) 2π

(c) 2 (d) π

18/ Using the principal values of the inverse trigonometric functions, the sum of the maximum and the minimum values of $16[(\sec^{-1} x)^2+(\text{cosec}^{-1} x)^2]$ is

(a) $24\pi^2$ (b) $22\pi^2$

(c) $18\pi^2$ (d) $31\pi^2$

19/ A coin is tossed three times. Let X denote the number of times a tail follows a head. If μ and σ^2 denote the mean and variance of X, then the value of $64(\mu+\sigma^2)$ is

(a) 48 (b) 32

(c) 64 (d) 51

20/ Let z_1, z_2 and z_3 be three complex numbers on the circle $|z|=1$ with $\arg(z_1)=\frac{-\pi}{4}$, $\arg(z_2)=0$ and $\arg(z_3)=\frac{\pi}{4}$. If $|z_1\bar{z}_2+z_2\bar{z}_3+z_3\bar{z}_1|^2=\alpha+\beta\sqrt{2}$, $\alpha, \beta \in Z$, then the value of $\alpha^2+\beta^2$ is

(a) 41 (b) 29 (c) 31 (d) 24

Section B

Numerical Type Questions

21/ Let **c** be the projection vector of $\mathbf{b}=\lambda\hat{\mathbf{i}}+4\hat{\mathbf{k}}$, $\lambda>0$, on the vector $\mathbf{a}=\hat{\mathbf{i}}+2\hat{\mathbf{j}}+2\hat{\mathbf{k}}$. If $|\mathbf{a}+\mathbf{c}|=7$, then the area of the parallelogram formed by the vectors **b** and **c** is ....

Imp.

22/ Let A be a square matric of order 3 such that $\det(A)=-2$ and $\det(3\text{adj}(-6\text{adj}(3A)))=2^{m+n}$, $m>n$. Then, $4m+2n$ is equal to ......

Properties of Adjoint of a Matrix

23/ Let the function $f(x) = \begin{cases} -3ax^2 - 2, & x < 1 \\ a^2 + bx, & x \geq 1 \end{cases}$ be differentiable, $\forall x \in R$, where $a > 1$, $b \in R$. If the area of the region enclosed by $y = f(x)$ and the line $y = -20$ is $\alpha + \beta\sqrt{3}$, $\alpha, \beta \in Z$, then the value of $\alpha + \beta$ is ....

24/ If $\sum_{r=0}^{5} \frac{^{11}C_{2r+1}}{2r+2} = \frac{m}{n}$, gcd $(m, n) = 1$, then $m - n$ is equal to ....

25/ Let $L_1 : \frac{x-1}{3} = \frac{y-1}{-1} = \frac{z+1}{0}$ and $L_2 : \frac{x-2}{2} = \frac{y}{0} = \frac{z+4}{\alpha}$, $\alpha \in R$ be two lines, which intersect at the point B.
If P is the foot of perpendicular from the point $A(1, 1, -1)$ on L_2, then the value of $26\alpha(PB)^2$ is ....

Physics

Section A
Objective Type Questions

Imp.

26/ Which of the following circuits represents a forward biased diode?

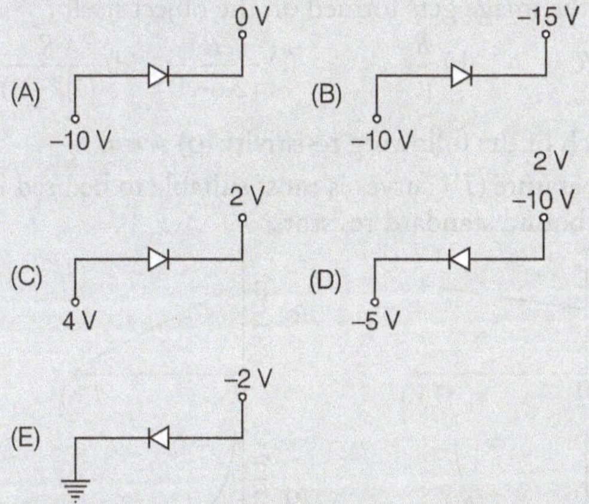

Choose the correct answer from the options given below:
Options
(a) (B), (C) and (E) only
(b) (C) and (E) only
(c) (B), (D) and (E) only
(d) (A) and (D) only

Biasing of *p-n* Junction Diode

27/ A closed organ and an open organ tube are filled by two different gases having same bulk modulus but different densities ρ_1 and ρ_2, respectively. The frequency of 9th harmonic of closed tube is identical with 4th harmonic of open tube. If the length of the closed tube is 10 cm and the density ratio of the gases is $\rho_1 : \rho_2 = 1 : 16$, then the length of the open tube is

(a) $\frac{15}{7}$ cm
(b) $\frac{20}{9}$ cm
(c) $\frac{20}{7}$ cm
(d) $\frac{15}{9}$ cm

28/ Given below are two statements.
Statement I The equivalent emf of two non-ideal batteries connected in parallel is smaller than either of the two emfs.
Statement II The equivalent internal resistance of two non-ideal batteries connected in parallel is smaller than the internal resistance of either of the two batteries.
In the light of the above statements, choose the correct answer from the options given below.
(a) Both Statement I and Statement II are true.
(b) Both Statement I and Statement II are false.
(c) Statement I is true but Statement II is false.
(d) Statement I is false but Statement II is true.

29/ The work functions of Cesium (Cs) and Lithium (Li) metals are 1.9 eV and 2.5 eV respectively. If we incident a light of wavelength 550 nm on these two metal surfaces, then photoelectric effect is possible for the case of
(a) Both Cs and Li
(b) Neither Cs nor Li
(c) Cs only
(d) Li only

30/ Two spherical bodies of same materials having radii 0.2 m and 0.8 m are placed in same atmosphere. The temperature of the smaller body is 800 K and temperature of the bigger body is 400 K. If the energy radiated from the smaller body is E, the energy radiated from the bigger body is (assume, effect of the surrounding temperature to be negligible),
(a) 256 E
(b) 16 E
(c) 64 E
(d) E

31/ Given below are two statements.

Statement I In a vernier callipers, one vernier scale division is always smaller than one main scale division.

Statement II The vernier constant is given by one main scale division multiplied by the number of vernier scale divisions.

In the light of the above statements, choose the correct answer from the options given below.

(a) Statement I is true but Statement II is false.
(b) Both Statement I and Statement II are true.
(c) Statement I is false but Statement II is true.
(d) Both Statement I and Statement II are false.

32/ A small point of mass m is placed at a distance $2R$ from the centre O of a big uniform solid sphere of mass M and radius R. The gravitational force on m due to M is F_1. A spherical part of radius $\frac{R}{3}$ is removed from the big sphere as shown in the figure and the gravitational force on m due to remaining part of M is found to be F_2. The value of ratio $F_1 : F_2$ is

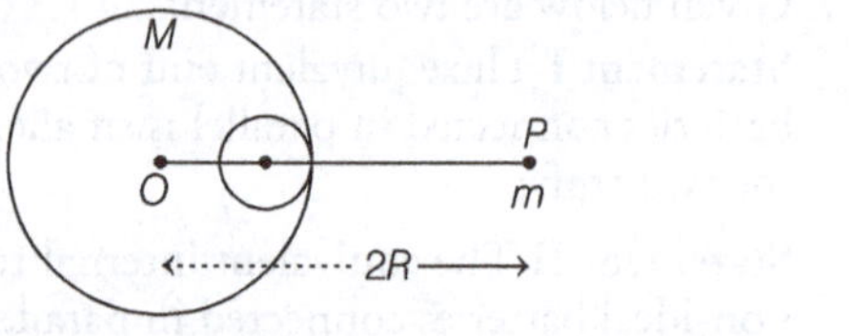

(a) 16 : 9 (b) 12 : 11 (c) 11 : 10 (d) 12 : 9

33/ In the diagram given below, there are three lenses formed. Considering negligible thickness of each of them as compared to $|R_1|$ and $|R_2|$, i.e., the radii of curvature for upper and lower surface of the glass lens, the power of the combinations is

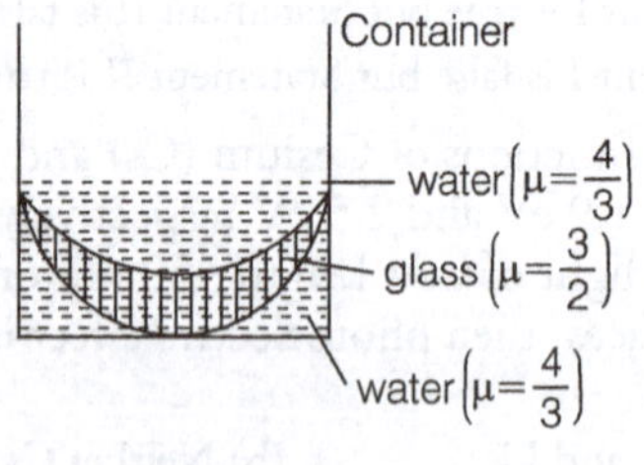

(a) $-\frac{1}{6}\left(\frac{1}{|R_1|}-\frac{1}{|R_2|}\right)$

(b) $\frac{1}{6}\left(\frac{1}{|R_1|}-\frac{1}{|R_2|}\right)$

(c) $\frac{1}{6}\left(\frac{1}{|R_1|}+\frac{1}{|R_2|}\right)$

(d) $-\frac{1}{6}\left(\frac{1}{|R_1|}+\frac{1}{|R_2|}\right)$

34/ A uniform circular disc of radius R and mass M is rotating about an axis perpendicular to its plane and passing through its centre. A small circular part of radius $\frac{R}{2}$ is removed from the original disc as shown in the figure. Find the moment of inertia of the remaining part of the original disc about the axis as given above.

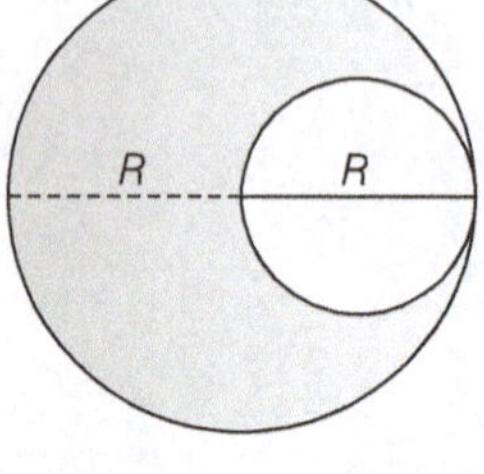

(a) $\frac{13}{32}MR^2$ (b) $\frac{7}{32}MR^2$ (c) $\frac{9}{32}MR^2$ (d) $\frac{17}{32}MR^2$

35/ Given is a thin convex lens of glass (refractive index μ) and each side having radius of curvature R. One side is polished for complete reflection. At what distance from the lens an object be placed on the optic axis, so that the image gets formed on the object itself?

(a) μR (b) $\frac{R}{\mu}$ (c) $\frac{R}{(2\mu-1)}$ (d) $\frac{R}{(2\mu-3)}$

36/ Which of the following resistivity (ρ) *versus* temperature (T) curves is most suitable to be used in wire bound standard resistors?

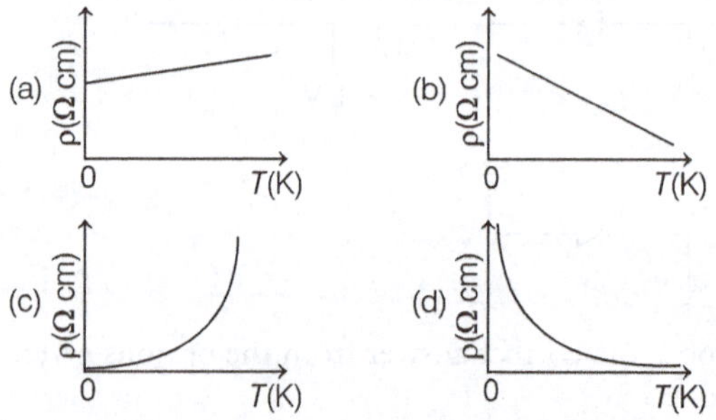

37/ An electron is made to enter symmetrically between two parallel and equally but oppositely charged metal plates, each of 10 cm length. The electron emerges out of the electric field region with a horizontal component of velocity 10^6 m/s. If the magnitude of the electric field between the plates is 9.1 V/cm, then the vertical component of velocity of electrons is (mass of electron $= 9.1 \times 10^{-31}$ kg and charge of electron $= 1.6 \times 10^{-19}$ C)

(a) 16×10^4 m/s (b) 0
(c) 1×10^6 m/s (d) 16×10^6 m/s

38/ If B magnetic field and μ_0 is permeability of free space, then the dimensions of $\left(\frac{B}{\mu_0}\right)$ is

(a) $[ML^2T^{-2}A^{-1}]$ (b) $[L^{-1}A]$
(c) $[MT^{-2}A^{-1}]$ (d) $[LT^{-2}A^{-1}]$

39/ A bob of mass m is suspended at a point O by a light string of length l and left to perform vertical motion (circular) as shown in figure. Initially, by applying horizontal velocity v_0 at the point A, the string becomes slack when, the bob reaches at the point D. The ratio of the kinetic energy of the bob at the points B and C is ....

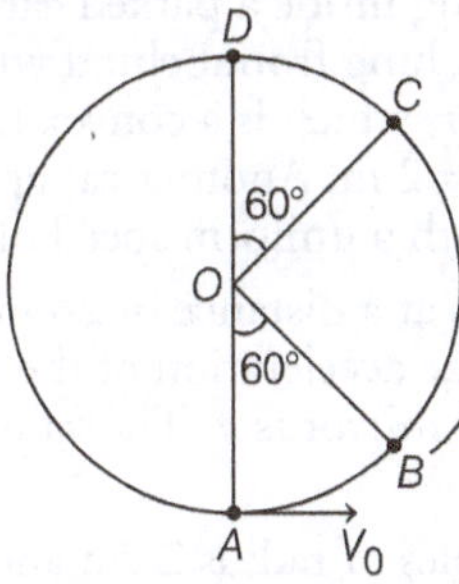

(a) 3 (b) 1 (c) 4 (d) 2

40/

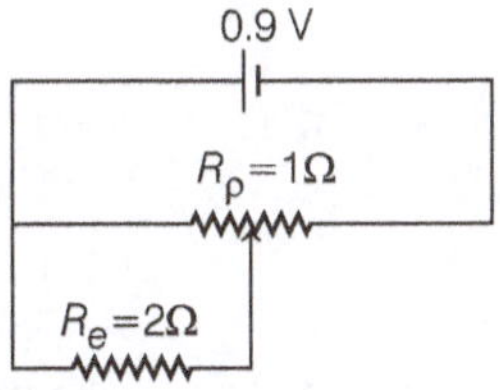

Sliding contact of a potentiometer is in the middle of the potentiometer wire having resistance $R_p = 1\Omega$ as shown in the figure. An external resistance of $R_e = 2\Omega$ is connected *via* the sliding contact. The electric current in the circuit is

(a) 1.35 A (b) 1.0 A (c) 0.9 A (d) 0.3 A

41/ Given below are two Statements. One is labelled as **Assertion** (A) and the other is labelled as **Reason** (R).

Assertion (A) If Young's double slit experiment is performed in an optically denser medium than air, then the consecutive fringes come closer.

Reason (R) The speed of light reduces in an optically denser medium than air while its frequency does not change.

In the light of the above statements, choose the most appropriate answer from the options given below.

(a) (A) is true but (R) is false.
(b) Both (A) and (R) are true but (R) is not the correct explanation of (A).
(c) Both (A) and (R) are true and (R) is the correct explanation of (A).
(d) (A) is false but (R) is true.

42/ An electron in the ground state of the hydrogen atom has the orbital radius of 53×10^{-11} m while that for the electron in third excited state is 8.48×10^{-10} m. The ratio of the de-Broglie wavelengths of electron in the ground state to that in the excited state is

(a) 3 (b) 4 (c) 16 (d) 9

43/ A line charge of length $a/2$ is kept at the centre of an edge BC of a cube $ABCDEFGH$ having edge length a as shown in the figure. If the density of line charge is λC per unit length, then the total electric flux through all the faces of the cube will be .... (Take, ε_0 as the free space permittivity)

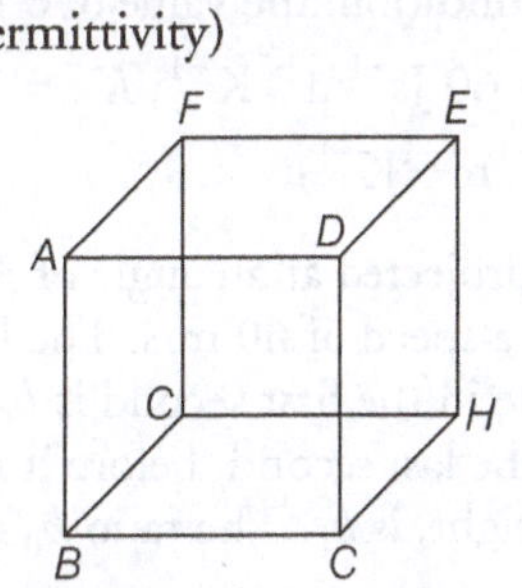

(a) $\frac{\lambda a}{4\varepsilon_0}$ (b) $\frac{\lambda a}{2\varepsilon_0}$ (c) $\frac{\lambda a}{16\varepsilon_0}$ (d) $\frac{\lambda a}{8\varepsilon_0}$

Imp.

44/ An amount of ice of mass 10^{-3} kg and temperature $-10°$ C is transformed to vapour of temperature 110°C by applying heat. The total amount of work required for this conversion is, (Take, specific heat of ice = 2100 J $kg^{-1}K^{-1}$, specific heat of water = 4180 J $kg^{-1}K^{-1}$, specific heat of steam = 1920 J $kg^{-1}K^{-1}$, latent heat of ice = 3.35×10^5 Jkg^{-1} and latent heat of steam = 2.25×10^6 Jkg^{-1}).

(a) 3043 J (b) 3003 J
(c) 3022 J (d) 3024 J

Calorimetry

45/ A parallel plate capacitor of capacitance 40 μF is connected to a 100 V power supply. Now, the intermediate space between the plates is filled with a dielectric material of dielectric constant $K = 2$.

Due to the introduction of dielectric material, the extra charge and the change in the electrostatic energy in the capacitor, respectively, are

(a) 4 mC and 0.2 J (b) 2 mC and 0.4 J
(c) 8mC and 2.0 J (d) 2 mC and 0.2 J

Section B

Numerical Type Questions

46/ Three conductors of same length having thermal conductivity K_1, K_2 and K_3 are connected as shown in figure.

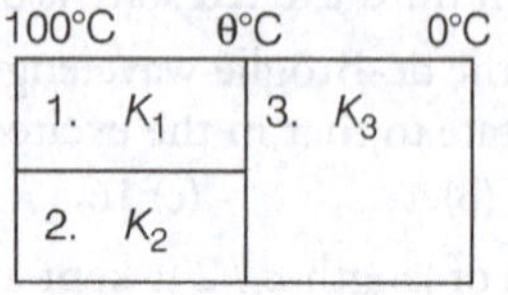

Area of cross-sections of 1st and 2nd conductor are same and for 3nd conductor it is double of the 1st conductor. The temperature are given in the figure. In steady state condition, the value of θ is ...... °C. (Given, $K_1 = 60 \text{ Js}^{-1}\text{m}^{-1}\text{K}^{-1}$, $K_2 = 120 \text{ Js}^{-1}\text{m}^{-1}\text{K}^{-1}$, $K_3 = 135 \text{ Js}^{-1}\text{m}^{-1}\text{K}^{-1}$).

47/ A particle is projected at an angle of 30° from horizontal at a speed of 60 m/s. The height traversed by the particle in the first second is h_0 and height traversed in the last second, before it reaches the maximum height, is h_1. The ratio $h_0 : h_1$ is .... [Take, $g = 10 \text{ m/s}^2$]

48/ The position vectors of two 1 kg particles , A and B, are given by

$\mathbf{r}_A = (\alpha_1 t^2 \hat{\mathbf{i}} + \alpha_2 t \hat{\mathbf{j}} + \alpha_3 t \hat{\mathbf{k}})\text{m}$ and

$\mathbf{r}_B = (\beta_1 t \hat{\mathbf{i}} + \beta_2 t^2 \hat{\mathbf{j}} + \beta_3 t \hat{\mathbf{k}})\text{m}$, respectively; ($\alpha_1 = 1 \text{ m/s}^2$. $\alpha_2 = 3n$ m/s, $\alpha_3 = 2$ m/s, $\beta_1 = 2$ m/s, $\beta_2 = -1 \text{ m/s}^2$, $\beta_3 = 4p$ m/s), where t is time, n and p are constants. At $t = 1$ s, $|\mathbf{v}_A| = |\mathbf{v}_B|$ and velocities $\mathbf{v}_A$ and $\mathbf{v}_B$ of the particles are orthogonal to each other. At $t = 1$ s, the magnitude of angular momentum of particle A with respect to the position of particle B is $\sqrt{L} \text{ kgm}^2\text{s}^{-1}$. The value of L is ....

49/ The driver sitting inside a parked car is watching vehicles approaching from behind with the help of his side view mirror, which is a convex mirror with radius of curvature $R = 2$ m. Another car approaches him from behind with a uniform speed of 90 km/h. When the car is at a distance of 24 m from him, the magnitude of the acceleration of the image of the car in the side view mirror is a. The value of $100a$ is ....m/s^2.

50/ Two soap bubbles of radius 2 cm and 4 cm, respectively, are in contact with each other. The radius of curvature of the common surface, in cm, is ......

Chemistry

Section A

Objective Type Questions

51/ Which of the following statement is not true for radioactive decay?

(a) Amount of radioactive substance remained after three Half-lifes is 1/8th of original amount.

(b) Half-life is ln2 times of $\frac{1}{\text{rate constant}}$.

(c) Decay constant does not depend upon temperature.

(d) Decay constant increases with increase in temperature.

Imp.

52/ Lanthanoid ions with $4f^7$ configuration are

(A) Eu^{2+} (B) Gd^{3+} (C) Eu^{3+} (D) Tb^{3+} (E) Sm^{2+}

Choose the correct answer from the options given below

(a) (*A*) and (*D*) only (b) (*A*) and (*B*) only

(c) (*B*) and (*E*) only (d) (*B*) and (*C*) only

Electronic Configuration of Inner-transition Element

53/ From the magnetic behaviour of $[NiCl_4]^{2-}$ (paramagnetic) and $[Ni(CO)_4]$ (diamagnetic), choose the correct geometry and oxidation state.

(a) $[NiCl_4]^{2-}$: Ni^{II}, square planar
$[Ni(CO)_4]$: Ni(0), square planar

(b) $[NiCl_4]^{2-}$: Ni^{II}, tetrahedral
$[Ni(CO)_4]$: Ni^{II}, square planar

(c) $[NiCl_4]^{2-}$: Ni(0), tetrahedral
$[Ni(CO)_4]$: Ni(0), square planar

(d) $[NiCl_4]^{2-}$: Ni^{II}, tetrahedral
$[Ni(CO)_4]$: Ni(0), tetrahedral

54/ Which of the following electrolyte can be used to obtain $H_2S_2O_8$ by the process of electrolysis?

(a) Acidified dilute solution of sodium sulphate.

(b) Dilute solution of sulphuric acid.

(c) Concentrated solution of sulphuric acid.

(d) Dilute solution of sodium sulphate.

55/ Match List-I with List-II.

	List-I		List-II
(A)	$Al^{3+} < Mg^{2+} < Na^{+} < F^{-}$	(I)	Ionisation enthalpy
(B)	B < C < O < N	(II)	Metallic character
(C)	B < Al < Mg < K	(III)	Electronegativity
(D)	Si < P < S < Cl	(IV)	Ionic radii

Choose the correct answer from the options given below.

(a) (A)-(IV), (B)-(I), (C)-(III), (D)-(II)
(b) (A)-(II), (B)-(III), (C)-(IV), (D)-(I)
(c) (A)-(III), (B)-(IV), (C)-(II), (D)-(I)
(d) (A)-(IV), (B)-(I), (C)-(II), (D)-(III)

56/ Given below are two statements:

Statement I One mole of propyne reacts with excess of sodium to liberate half a mole of H_2 gas.

Statement II Four grams of propyne reacts with $NaNH_2$ to liberate NH_3 gas which occupies 224 mL at STP.

In the light of the above statements, choose the most appropriate answer from the options given below:

(a) Both Statement I and Statement II are correct.
(b) Both Statement I and Statement II are incorrect.
(c) Statement I is correct but Statement II is incorrect.
(d) Statement I is incorrect but Statement II is correct.

57/ Arrange the following solutions in order of their increasing boiling points.

(i) 10^{-4} M NaCl (ii) 10^{-4} M Urea
(iii) 10^{-3} M NaCl (iv) 10^{-2} M NaCl

(a) (ii) < (i) < (iii) < (iv)
(b) (ii) < (i) < (iii) < (iv)
(c) (iv) < (iii) < (i) < (ii)
(d) (i) < (ii) < (iii) < (iv)

Imp.

58/ Which of the following electronegativity order is incorrect ?

(a) Al < Mg < B < N (b) S < Cl < O < F
(c) Mg < Be < B < N (d) Al < Si < C < N

Order of Electronegativity

59/ The products formed in the following reaction sequence are

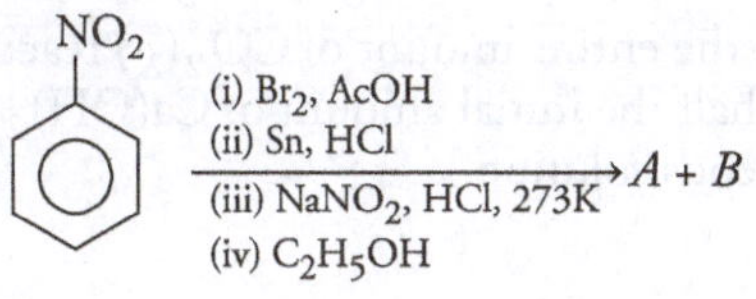

(a) 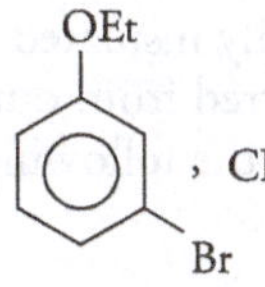

, $CH_3—COOH$

(b) 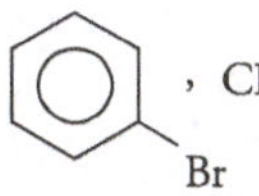

, $CH_3—CHO$

(c) 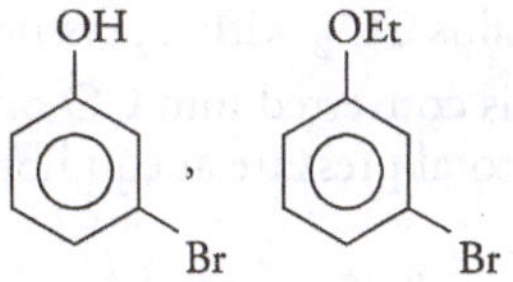

(d) 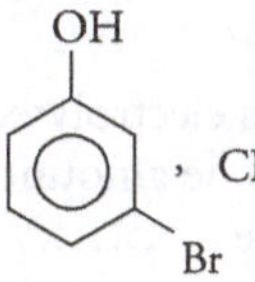

, $CH_3—CHO$

60/ The IUPAC name of the following compound is

$$CH_3—\underset{}{CH}(COOH)—CH_2—CH_2—CH(COOCH_3)—CH_3$$

(a) Methyl-6-carboxy-2, 5-dimethylhexanoate
(b) Methyl-5-carboxy-2-methylhexanoate
(c) 2-carboxy-5-methoxycarbonylhexane
(d) 6-methoxycarbonyl-2, 5-dimethylhexanoic acid

61/ How many different stereoisomers are possible for the given molecule?

$$CH_3—CH(OH)—CH=CH—CH_3$$

(a) 3 (b) 4 (c) 2 (d) 1

62/ The compounds which give positive Fehling's test are

(A) 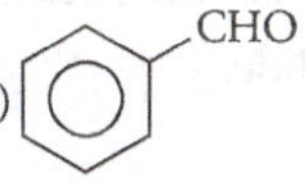

(B) 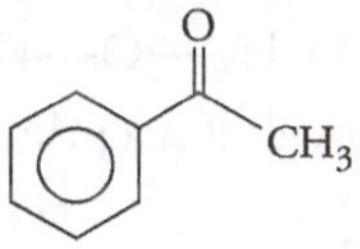

(C) $HOCH_2—CO—(CHOH)_3—CH_2—OH$

(D) $CH_3—\overset{O}{\overset{\|}{C}}—H$

(E) 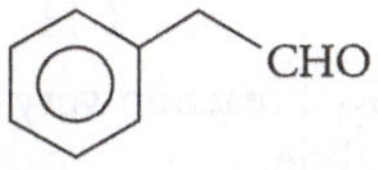

Choose the correct answer from the options given below.

(a) (A), (B) and (C) only
(b) (C), (D) and (E) only
(c) (A), (C) and (D) only
(d) (A), (D) and (E) only

63/ A liquid when kept inside a thermally insulated closed vessel at 25°C was mechanically stirred from outside. What will be the correct option for the following thermodynamic parameters?

(a) $\Delta U = 0, q < 0, w > 0$
(b) $\Delta U = 0, q = 0, w = 0$
(c) $\Delta U > 0, q = 0, w > 0$
(d) $\Delta U < 0, q = 0, w > 0$

64/ A vessel at 1000 K contains CO_2 with a pressure of 0.5 atm. Some of CO_2 is converted into CO on addition of graphite. If total pressure at equilibrium is 0.8 atm, then K_p is

(a) 0.18 atm (b) 0.3 atm
(c) 3 atm (d) 1.8 atm

65/ A solution of aluminium chloride is electrolysed for 30 minutes using a current of 2A. The amount of the aluminium deposited at the cathode is ....

[Given, molar mass of aluminium and chlorine are 27 g mol^{-1} and 35.5 g mol^{-1} respectively. Faraday constant = 96500 C mol^{-1}]

(a) 1.007 g (b) 0.336 g (c) 0.441 g (d) 1.660 g

66/ Radius of the first excited state of helium ion is given as: $a_0 \rightarrow$ radius of first stationary state of hydrogen atom.

(a) $r = \frac{a_0}{2}$ (b) $r = 2a_0$
(c) $r = \frac{a_0}{4}$ (d) $r = 4a_0$

67/ Which of the following acids is a vitamin?

(a) Aspartic acid (b) Ascorbic acid
(c) Saccharic acid (d) Adipic acid

68/ Given below are two statements.

Statement I $CH_3—O—CH_2—Cl$ will undergo S_N1 reaction though it is a primary halide.

Statement II $CH_3—C(CH_3)_2—CH_2—Cl$ will not undergo S_N2 reaction very easily though it is a primary halide.

In the light of the above statements, choose the most appropriate answer from the options given below

(a) Both Statement I and Statement II are correct.
(b) Statement I is incorrect but Statement II is correct.
(c) Statement I is correct but Statement II is incorrect.
(d) Both Statement I and Statement II are incorrect.

69/ The incorrect statements regarding geometrical isomerism are

(A) Propene shows geometrical isomerism.
(B) *trans* isomer has identical atoms/groups on the opposite sides of the double bond.
(C) *cis*-but-2-ene has higher dipole moment than *trans*-but-2-ene.
(D) 2-methylbut-2-ene shows two geometrical isomers.
(E) *trans*-isomer has lower melting point than *cis* isomer.

Choose the correct answer from the options given below.

(a) (A), (D) and (E) Only
(b) (B) and (C) Only
(c) (A) and (E) Only
(d) (C), (D) and (E) Only

70/ In which of the following complexes the CFSE, Δ_0 will be equal to zero?

(a) $[Fe(NH_3)_6]Br_2$ (b) $K_3[Fe(SCN)_6]$
(c) $[Fe(en)_3]Cl_3$ (d) $K_4[Fe(CN)_6]$

Section B
Numerical Type Questions

71/ In Carius method for estimation of halogens, 180 mg of an organic compound produced 143.5 mg of AgCl. The percentage composition of chlorine in the compound is ...... %.

(Given, molar mass in g mol^{-1} of Ag : 108, Cl : 35.5)

72/ Consider the following sequence of reactions,

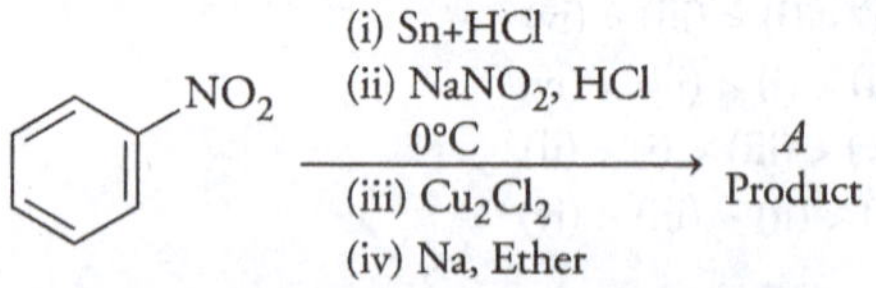

Molar mass of the product formed (*A*) is ...... g mol^{-1}.

73/ Some CO_2 gas was kept in a sealed container at a pressure of 1 atm and at 273 K. This entire amount of CO_2 gas was later passed through an aqueous solution of $Ca(OH)_2$. The excess unreacted $Ca(OH)_2$ was later neutralised with 0.1 M of 40 mL HCl. If the volume of the sealed container of CO_2 was *x*, then *x* is .... cm^3 (nearest integer).

[Given, the entire amount of $CO_2(g)$ reacted with exactly half the initial amount of $Ca(OH)_2$ present in the aqueous solution.]

$A \rightarrow B$

74/ The molecule A changes into its isomeric from B by following a first order kinetics at a temperature of 1000 K.

If the energy barrier with respect to reactant energy for such isomeric transformation is 191.48 kJ mol^{-1} and the frequency factor is 10^{20}, the time required for 50% molecules of A to become B is .... picoseconds (nearest integer).

$[R = 8.314 \text{ J K}^{-1}\text{mol}^{-1}]$

Imp.

75/ The number of molecules/ions that show linear geometry among the following is ....

SO_2, $BeCl_2$, CO_2, N_3^-, NO_2, F_2O, XeF_2, NO_2^+, I_3^-, O_3

Geometry of Molecules/Ions

Answer Key

Mathematics

1 (c)	2 (b)	3 (c)	4 (c)	5 (a)	6 (d)	7 (c)	8 (d)	9 (a)	10 (a)
11 (a)	12 (d)	13 (d)	14 (c)	15 (b)	16 (b)	17 (a)	18 (b)	19 (a)	20 (b)
21 (16)	22 (34)	23 (34)	24 (2035)	25 (216)					

Physics

26 (a)	27 (b)	28 (d)	29 (c)	30 (d)	31 (a)	32 (b)	33 (a)	34 (a)	35 (c)
36 (a)	37 (d)	38 (b)	39 (d)	40 (b)	41 (c)	42 (*)	43 (d)	44 (a)	45 (a)
46 (40)	47 (5)	48 (90)	49 (8)	50 (4)					

Chemistry

51 (d)	52 (b)	53 (d)	54 (c)	55 (d)	56 (c)	57 (b)	58 (a)	59 (b)	60 (d)
61 (b)	62 (b)	63 (c)	64 (d)	65 (b)	66 (b)	67 (b)	68 (a)	69 (a)	70 (b)
71 (20)	72 (154)	73 (45)	74 (69)	75 (6)					

***Note** (*) None of the option is correct.*

Detailed Explanations

22 January Shift-I

Mathematics

1. (c)

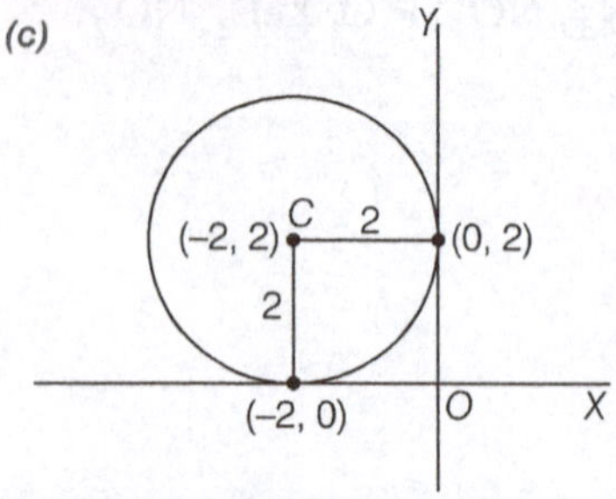

Equation of circle C :

$$(x+2)^2+(y-2)^2=2^2$$

Centre $C_1(-2, 2)$, $r_1 = 2$

$\because$ An another circle having centre $C_2(2, 5)$ and radius r intersects the circle C at exactly 2 points.

$\therefore \ |r_1 - r_2| < C_1C_2 < r_1 + r_2$

$|r-2|<5<r+2$

$-5<r-2<5$ and $r+2>5$

$-3<r<7$ and $r>3$

$\therefore \quad r\in(3, 7)$

$\Rightarrow \alpha=3, \beta=7$

$\therefore \ 3\beta-2\alpha=21-6=15$

2. (b) Foci of hyperbola (1, 14), (1, – 12)

$\therefore \ x=1$ be the transverse axis.

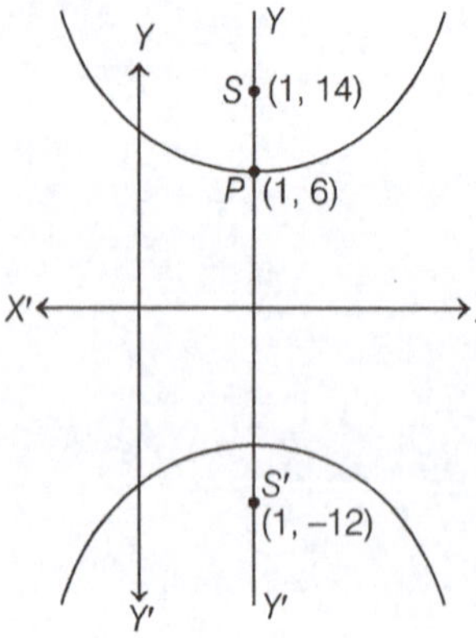

$$|SP-S'P|=2b$$

$$\Rightarrow \quad 2b=|8-18|$$

$\Rightarrow b=5$ and $SS'=2be$

$26=2be$

$be=13$

$$e=\frac{13}{5}$$

$\therefore$ Length of latus-rectum

$$=\frac{2a^2}{b}=\frac{2b^2(e^2-1)}{b}$$

$$=2(5)\left(\frac{144}{25}\right)=\frac{288}{5}$$

3. (c) $e^{5(\ln x)^2+3}=x^8$

On taking ln both sides, we get

$$5(\ln x)^2+3=8\ln x$$

$$\Rightarrow 5(\ln x)^2-8\ln x+3=0$$

$$\Rightarrow 5(\ln x)^2-5\ln x-3\ln x+3=0$$

$$\Rightarrow \quad (5\ln x-3)(\ln x-1)=0$$

$$\Rightarrow \quad \ln x=\frac{3}{5}, 1 \Rightarrow x=e^{\frac{3}{5}}, e^1$$

$$\therefore \quad \text{Product}=e^{3/5}\cdot e^1=e^{8/5}$$

4. (c) $A=\{1, 2, 3, ..., 10\}$

$$B=\left\{\frac{m}{n}: m, n\in A, m<n \text{ and } gcd(m, n)=1\right\}$$

$m=1\Rightarrow n=2, 3, 4, ..., 10$ $\quad\rightarrow 9$ elements

$m=2\Rightarrow n=3, 5, 7, 9$ $\quad\rightarrow 4$ elements

$m=3\Rightarrow n=4, 5, 7, 8, 10$ $\quad\rightarrow 5$ elements

$m=4\Rightarrow n=5, 7, 9$ $\quad\rightarrow 3$ elements

$m=5\Rightarrow n=6, 7, 8, 9$ $\quad\rightarrow 4$ elements

$m=6\Rightarrow n=7$ $\quad\rightarrow 1$ element

$m=7\Rightarrow n=8, 9, 10$ $\quad\rightarrow 3$ elements

$m=8\Rightarrow n=9$ $\quad\rightarrow 1$ element

$m=9\Rightarrow n=10$ $\quad\rightarrow 1$ element

$\therefore \ n(B)=31$

5. (a)

4 White balls
6 Black balls

E_1 = First selected ball is white.

E_2 = First selected ball is black.

A = Second selected ball is black.

$$P\left(\frac{E_2}{A}\right)=\frac{P(E_2)\cdot P\left(\frac{A}{E_2}\right)}{P(E_1)\cdot P\left(\frac{A}{E_1}\right)+P(E_2)\,P\left(\frac{A}{E_2}\right)}$$

$$=\frac{\frac{6}{10}\cdot\frac{5}{9}}{\frac{4}{10}\cdot\frac{6}{9}+\frac{6}{10}\cdot\frac{5}{9}}$$

$$=\frac{30}{24+30}=\frac{30}{54}=\frac{5}{9}=\frac{m}{n}$$

$\therefore \ m+n=5+9=14$

6. (d)

Key idea

General term of GP.

$T_n=ar^{n-1}$

$a_1, a_2, a_3, ...$ are in GP.

$a_1a_5=28$ and $a_2+a_4=29$

Let a be the first term and r be the common ratio.

$a(ar^4)=28$ and $ar+ar^3=29$

$a^2r^4=28$

$\Rightarrow \ ar^2=2\sqrt{7}$

$$\Rightarrow \ \frac{2\sqrt{7}}{r}+2\sqrt{7}r=29$$

$\Rightarrow \ 2\sqrt{7}r^2-29r+2\sqrt{7}=0$

$\Rightarrow \ 2\sqrt{7}r^2-28r-r+2\sqrt{7}=0$

$\Rightarrow \ 2\sqrt{7}r(r-2\sqrt{7})-1(r-2\sqrt{7})=0$

$\Rightarrow \ (r-2\sqrt{7})(2\sqrt{7}r-1)=0$

$$\Rightarrow \ r=2\sqrt{7}, r=\frac{1}{2\sqrt{7}}$$

$\because$ Increasing GP $\Rightarrow r=2\sqrt{7}$

$$\therefore \quad a=\frac{1}{2\sqrt{7}}$$

$\therefore \ a_6=ar^5=(2\sqrt{7})^4=(28)^2=784$

7. (c) Circle : $(x-2\sqrt{3})^2+y^2=12$

$\Rightarrow \quad x=2\sqrt{3}\pm\sqrt{12-y^2}$

Parabola : $y^2=2\sqrt{3}x$

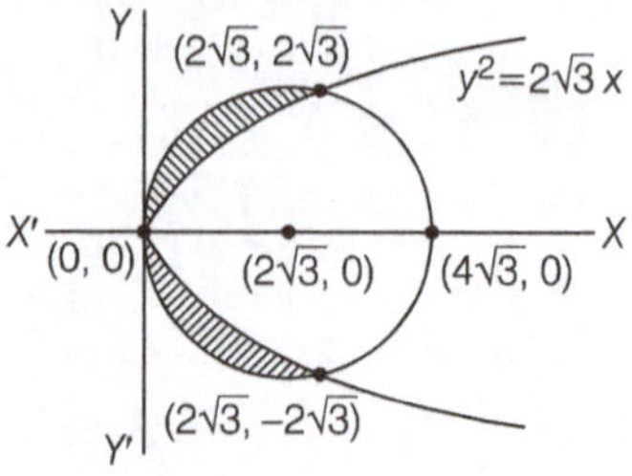

Point of intersection $\Rightarrow y^2 = 12$

$\Rightarrow \quad y = \pm 2\sqrt{3}$

$\therefore \quad x = 2\sqrt{3}$

Area of required region

$$= 2\int_0^{2\sqrt{3}} (x_1 - x_2)\,dy$$

$$= 2\int_0^{2\sqrt{3}} \left(\frac{y^2}{2\sqrt{3}} - 2\sqrt{3} + \sqrt{12 - y^2}\right) dy$$

$$= 2\left[\frac{y^3}{3\sqrt{3}} - 2\sqrt{3}y + \frac{y}{2}\sqrt{12 - y^2} + \frac{12}{2}\sin^{-1}\frac{y}{2\sqrt{3}}\right]_0^{2\sqrt{3}}$$

$$= 2\left[4 - 12 + 0 + 6\times\frac{\pi}{2}\right] = 6\pi - 16$$

8. *(d)* $y^2\,dx + \left(x - \frac{1}{y}\right)dy = 0$

$$\Rightarrow \quad y^2\frac{dx}{dy} + x = \frac{1}{y}$$

$$\Rightarrow \quad \frac{dx}{dy} + \left(\frac{1}{y^2}\right)x = \frac{1}{y^3}$$

$$P = \frac{1}{y^2} \text{ and } Q = \frac{1}{y^3}$$

$$\text{IF} = e^{\int P dy} = e^{\int \frac{1}{y^2}dy} = e^{-\frac{1}{y}}$$

Solution of differential equation

$$x(\text{IF}) = \int Q\cdot(\text{IF})dy + C$$

$$xe^{-\frac{1}{y}} = \int \frac{1}{y^3}e^{-\frac{1}{y}}\,dy + C \qquad \ldots(\text{i})$$

Put $\quad e^{-\frac{1}{y}} = t \Rightarrow -\frac{1}{y} = \log t$

$$e^{-\frac{1}{y}}\cdot\frac{1}{y^2}\,dy = dt$$

From Eq. (i), we get

$$\therefore \quad xe^{-\frac{1}{y}} = \int -\ln t\,dt + C$$

$$xe^{-\frac{1}{y}} = -t(\ln t - 1) + C$$

$$xe^{-\frac{1}{y}} = -e^{-\frac{1}{y}}\left(-\frac{1}{y} - 1\right) + C$$

$$x = \frac{1}{y} + 1 + Ce^{\frac{1}{y}}$$

Put $x = y = 1$

$$1 = 1 + 1 + Ce \Rightarrow C = -\frac{1}{e}$$

$$\therefore \quad x\left(\frac{1}{2}\right) = 2 + 1 - \left(\frac{1}{e}\right)\cdot e^2 = 3 - e$$

> **Concept Enhancer**
>
> Solution of linear differential equation
>
> $\frac{dy}{dx} + py = Q$ is $y(\text{IF}) = \int Q(\text{IF})dx + C$,
>
> where $\text{IF} = e^{\int P dx}$

9. *(a)* Given, $f(x+y) = f(x)\cdot f'(y) + f'(x)\cdot f(y)$

Put $y = 0$

$$f(x) = f(x)\cdot f'(0) + f'(x)f(0)$$

$$f(x) = f(x)\cdot f'(0) + f'(x) \qquad \ldots(\text{i})$$

$[\because f(0) = 1]$

Put $x = y = 0$

$$f(0) = f'(0) + f'(0) \Rightarrow f'(0) = \frac{1}{2}$$

Hence, equation (i) becomes

$$f(x) = \frac{f(x)}{2} + f'(x)$$

$$f'(x) = \frac{f(x)}{2} \Rightarrow \frac{f'(x)}{f(x)} = \frac{1}{2}$$

On integrating both sides, we get

$$\ln f(x) = \frac{x}{2} + C$$

Put $x = 0 \Rightarrow C = 0$

$$\Rightarrow \quad \ln f(x) = \frac{x}{2} \Rightarrow f(x) = e^{\frac{x}{2}}$$

$$\therefore \quad f(n) = e^{\frac{n}{2}}$$

$$\log_e f(n) = \frac{n}{2}$$

$$\therefore \quad \sum_{n=1}^{100}\log_e f(n) = \frac{1}{2}\sum_{n=1}^{100} n$$

$$= \frac{1}{2}\sum_{n=1}^{100} n = \frac{1}{2}\left(\frac{100}{2}(100+1)\right)$$

$$= 25\times 101 = 2525$$

10. *(a)*

> **Key idea**
>
> A relation is equivalence when
>
> (i) $(a, a) \in R \ \forall\ a \in A$
>
> (ii) $(a, b) \in R \Rightarrow (b, a) \in R \ \forall\ a, b \in A$
>
> (iii) $(a, b) \in R$, $(b, c) \in R \Rightarrow (a, c) \in R \ \forall\ a, b, c \in A$

$$A = \{1, 2, 3\}$$

$\therefore$ Non-empty equivalence relation

$R_1 = \{(1, 1), (2, 2), (3, 3)\}$

$R_2 = \{(1, 1), (2, 2), (3, 3), (1, 2), (2, 1)\}$

$R_3 = \{(1, 1), (2, 2), (3, 3), (2, 3), (3, 2)\}$

$R_4 = \{(1, 1), (2, 2), (3, 3), (1, 3), (3, 1)\}$

$R_5 = \{(1, 1), (2, 2), (3, 3), (1, 2), (2, 1), (1, 3), (3, 1), (2, 3), (3, 2)\}$

$\therefore$ Number of relation $= 5$

11. *(a)* $f(x+y) = f(x)\cdot f(y),\ \forall\ x, y \in R$

$$f(x) = k^x$$

$$f'(x) = k^x \ln k$$

$$f'(0) = \ln k = 4a$$

$$k = e^{4a}$$

$$f(x) = (e^{4a})^x = e^{4ax}$$

$$f'(x) = 4a\,e^{4ax}$$

$$f''(x) = 16a^2\,e^{4ax}$$

Now, $f''(x) - 3af'(x) - f(x) = 0$

$$16a^2e^{4ax} - 12a^2e^{4ax} - e^{4ax} = 0$$

$$(4a^2 - 1)e^{4ax} = 0$$

$$4a^2 = 1$$

$$a = \frac{1}{2} \qquad [\because a > 0]$$

$$\therefore \quad y = e^{2x} \Rightarrow f(ax) = e^x$$

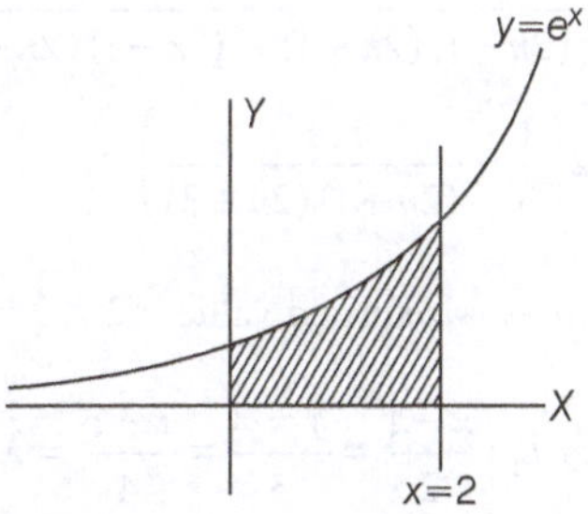

Area of region

$$R = \{(x, y); 0 \le y \le f(ax), 0 \le x \le 2\}$$

$$A = \int_0^2 e^x\,dx = [e^x]_0^2 = e^2 - 1$$

12. *(d)* Total alphabet = 26

(12 alphabets) (13 alphabets)

M

Remaining alphabet = 25

Now, arrangement in alphabetical order

$$= {}^{12}C_2 \times {}^{13}C_2 \times 1$$

$$= \frac{12\times 11}{2\times 1} \times \frac{13\times 12}{2\times 1} = 5148$$

13. *(d)*

$$\sum_{r=1}^{n} T_r = \frac{(2n-1)(2n+1)(2n+3)(2n+5)}{64}$$

$$T_n = S_n - S_{n-1} = \frac{(2n-1)(2n+1)(2n+3)(2n+5)}{64} - \frac{(2n-3)(2n-1)(2n+1)(2n+3)}{64}$$

$$= \frac{(2n-1)(2n+1)(2n+3)(2n+5-2n+3)}{64}$$

$$= \frac{(2n-1)(2n+1)(2n+3)}{8}$$

$$\therefore \sum_{r=1}^{n} \frac{1}{T_r} = 8\sum_{r=1}^{n} \frac{1}{(2r-1)(2r+1)(2r+3)}$$

$$= \frac{8}{4}\left(\sum_{r=1}^{n} \frac{1}{(2r-1)(2r+1)} - \frac{1}{(2r+1)(2r+3)}\right)$$

$$= 2\left[\left(\frac{1}{1\cdot 3} - \frac{1}{3\cdot 5}\right) + \left(\frac{1}{3\cdot 5} - \frac{1}{5\cdot 7}\right) \ldots \left(\frac{1}{(2n-1)(2n+1)} - \frac{1}{(2n+1)(2n+3)}\right)\right]$$

$$= 2\left(\frac{1}{3} - \frac{1}{(2n+1)(2n+3)}\right)$$

At $n \to \infty$ required value $= 2 \times \frac{1}{3} = \frac{2}{3}$

14. *(c)* $L_1 : \frac{x-1}{2} = \frac{y-2}{3} = \frac{z-3}{4} = \lambda$ (say)

$L_2 : \frac{x-2}{3} = \frac{y-4}{4} = \frac{z-5}{5} = \mu$ (say)

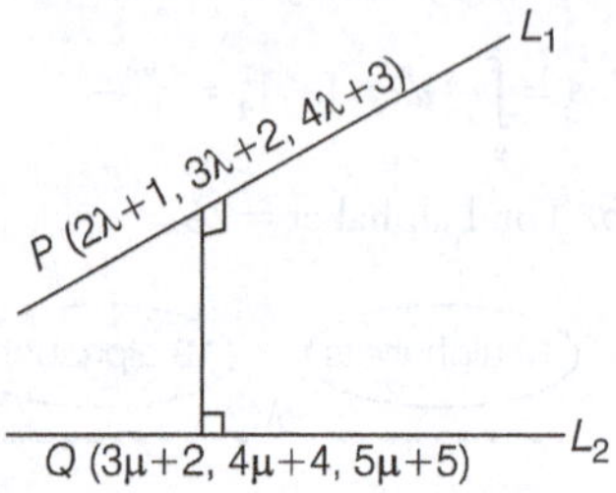

$$\text{PQ} = (3\mu - 2\lambda + 1)\hat{i} + (4\mu - 3\lambda + 2)\hat{j} + (5\mu - 4\lambda + 2)\hat{k}$$

$\because$ $\text{PQ} \perp \mathbf{b}_1$ and $\text{PQ} \perp \mathbf{b}_2$

$\therefore$ PQ is parallel to $\mathbf{b}_1 \times \mathbf{b}_2$

$$\mathbf{b}_1 \times \mathbf{b}_2 = \begin{vmatrix} \hat{i} & \hat{j} & \hat{k} \\ 2 & 3 & 4 \\ 3 & 4 & 5 \end{vmatrix} = -\hat{i} + 2\hat{j} - \hat{k}$$

$\text{PQ} \parallel \mathbf{b}_1 \times \mathbf{b}_2$

$$\frac{3\mu - 2\lambda + 1}{-1} = \frac{4\mu - 3\lambda + 2}{2} = \frac{5\mu - 4\lambda + 2}{-1}$$

$$6\mu - 4\lambda + 2 = -4\mu + 3\lambda - 2$$

$$\Rightarrow \quad 10\mu - 7\lambda + 4 = 0 \times 7$$

$$-4\mu + 3\lambda - 2 = 10\mu - 8\lambda + 4$$

$$\Rightarrow \quad -14\mu + 11\lambda - 6 = 0 \times 5$$

$$70\mu - 49\lambda + 28 = 0$$

$$-70\mu + 55\lambda - 30 = 0$$

$$6\lambda - 2 = 0$$

$$\lambda = \frac{1}{3}, \mu = \left(\frac{7}{3} - 4\right)\frac{1}{10} = -\frac{5}{3\times 10} = -\frac{1}{6}$$

$$P = \left(2\left(\frac{1}{3}\right) + 1, 3\left(\frac{1}{3}\right) + 2, 4\left(\frac{1}{3}\right) + 3\right)$$

$$= \left(\frac{5}{3}, 3, \frac{13}{3}\right)$$

Equation of PQ

$$\frac{x - \frac{5}{3}}{-1} = \frac{y-3}{2} = \frac{z - \frac{13}{3}}{-1}$$

For $\left(\frac{14}{3}, -3, \frac{22}{3}\right)$ satisfies the equation.

15. *(b)* $y = x^2 + px - 3$

$\because$ $x = 0, y = -3$

$\therefore (0, -3)$ be point where parabola cuts Y-axis.

$\therefore$ Equation of circle C,

$$(x+1)^2 + (y+1)^2 = (-1-0)^2 + (-1+3)^2$$

$$\Rightarrow \quad (x+1)^2 + (y+1)^2 = 5$$

Circle cuts X-axis at $y = 0$.

$\therefore$ $(x+1)^2 = 4 \Rightarrow x + 1 = \pm 2$

$\Rightarrow$ $x = 1, -3$

$\therefore$ Points are (1, 0), (– 3, 0) and (0, – 3).

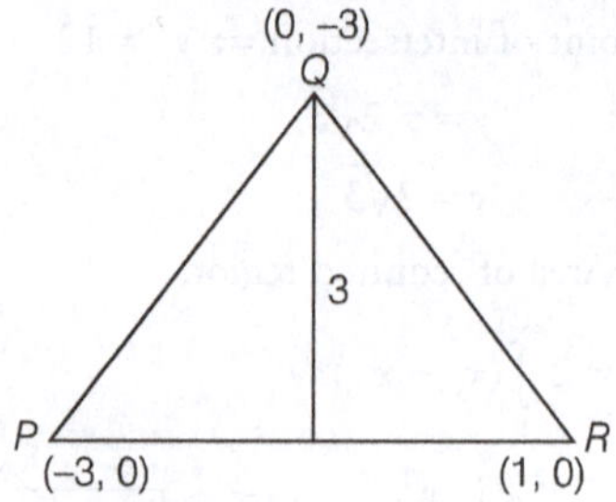

Area of $\Delta PQR = \frac{1}{2} \times 4 \times 3 = 6$

16. *(b)* $\because$ ΔPQR is image of ΔABC w.r.t. line $x + 2y = 2$

Centroid of ΔABC

$$G = \left(2, \frac{8}{3}\right)$$

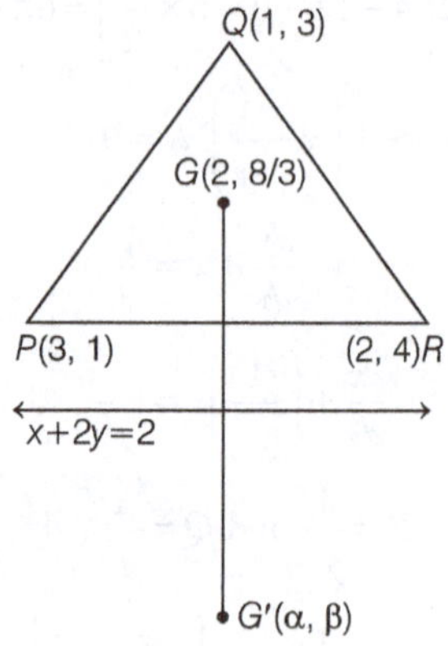

$\therefore$ Centroid of ΔPQR is $G'(\alpha, \beta)$

That is image of G w.r.t. $x + 2y - 2 = 0$

$$\frac{\alpha - 2}{1} = \frac{\beta - \frac{8}{3}}{2} = \frac{-2\left(2 + \frac{16}{3} - 2\right)}{1+4}$$

$$\Rightarrow \quad \frac{\alpha - 2}{1} = \frac{\beta - \frac{8}{3}}{2} = \frac{-32}{15}$$

$$\Rightarrow \quad \alpha = 2 - \frac{32}{15} = -\frac{2}{15}$$

$$\Rightarrow \quad \beta = \frac{8}{3} - \frac{64}{15} = -\frac{24}{15}$$

$\therefore$ $15(\alpha - \beta) = 22$

Time Saver Tech

Let the image of a point (x_1, y_1) with respect to $ax + by + c = 0$ be (x_2, y_2), then

$$\frac{x_2 - x_1}{a} = \frac{y_2 - y_1}{b} = \frac{-2(ax_1 + by_1 + c)}{a^2 + b^2}$$

17. (a) $f(x) = 7\tan^8 x + 7\tan^6 x - 3\tan^4 x - 3\tan^2 x$

$= 7\tan^6 x(\tan^2 x + 1) - 3\tan^2 x (\tan^2 x + 1)$

$= \sec^2 x(7\tan^6 x - 3\tan^2 x)$

$= \sec^2 x \tan^2 x (7\tan^4 x - 3)$

$$I_1 = \int_0^{\frac{\pi}{4}} f(x)dx$$

$$= \int_0^{\frac{\pi}{4}} \sec^2 x(7\tan^6 x - 3\tan^2 x)dx$$

Let $\tan x = t$

$\Rightarrow \sec^2 x\, dx = dt$

$$= \int_0^1 (7t^6 - 3t^2)dt$$

$$= [t^7 - t^3]_0^1 = 0$$

$$I_2 = \int_0^{\frac{\pi}{4}} x f(x)dx$$

$$= x\int_0^{\frac{\pi}{4}} f(x) - \int_0^{\frac{\pi}{4}} (1 \cdot \int f(x)dx)\, dx$$

$$= 0 - \int_0^{\frac{\pi}{4}} \tan^7 x - \tan^3 x$$

$$I_2 = -\int_0^{\frac{\pi}{4}} \tan^3 x(\tan^4 x - 1)dx$$

$$= -\int_0^{\frac{\pi}{4}} \tan^3 x(\tan^2 x - 1)(\tan^2 x + 1)dx$$

$$= -\int_0^{\frac{\pi}{4}} \tan^3 x(\tan^2 x - 1)\sec^2 x\, dx$$

Let $\tan x = t$

$\sec^2 x dx = dt$

when $x = 0, t = 0$

and $x = \frac{\pi}{4}, t = 1$

$$= \int_0^1 t^3(1 - t^2)dt = \int_0^1 (t^3 - t^5)dt$$

$$= \left[\frac{t^4}{4} - \frac{t^6}{6}\right]_0^1 = \frac{1}{4} - \frac{1}{6} = \frac{1}{12}$$

$\therefore \quad 7I_1 + 12I_2 = 1$

18. (b) Let $y = (\sec^{-1} x)^2 + (\text{cosec}^{-1} x)^2$

$= (\sec^{-1} x + \text{cosec}^{-1} x)^2 - 2\sec^{-1} x \,\text{cosec}^{-1} x$

$$= \left(\frac{\pi}{2}\right)^2 - 2(\sec^{-1} x)\left(\frac{\pi}{2} - \sec^{-1} x\right)$$

$$= \frac{\pi^2}{4} + 2(\sec^{-1} x)^2 - \pi \sec^{-1} x$$

$$\Rightarrow 2\left((\sec^{-1} x)^2 - \frac{\pi}{2}(\sec^{-1} x) + \frac{\pi^2}{8}\right)$$

$$\Rightarrow 2\left[\left(\sec^{-1} x - \frac{\pi}{4}\right)^2 + \frac{\pi^2}{8} - \frac{\pi^2}{16}\right]$$

$$\Rightarrow 2\left(\sec^{-1} x - \frac{\pi}{4}\right)^2 + \frac{\pi^2}{8}$$

$\because \quad \sec^{-1} x \in [0, \pi] - \left\{\frac{\pi}{2}\right\}$

$\therefore$ Minimum value of $y = 0 + \frac{\pi^2}{8}$

Maximum value of

$$y = 2\left(\frac{9\pi^2}{16}\right) + \frac{\pi^2}{8}$$

$$= \frac{10\pi^2}{8}$$

$\therefore$ Required sum

$$= 16\left(\frac{\pi^2}{8} + \frac{10\pi^2}{8}\right)$$

$$= 22\pi^2$$

19. (a)

Key idea
Mean $(\mu) = \Sigma P_i x_i$,
Variance $(\sigma^2) = \Sigma P_i x_i^2 - (\Sigma P_i x_i)^2$

X = Number of times a tail follows a head.

Outcomes	X_i	P_i
HHH	0	1/8
HHT	1	1/8
HTH	1	1/8
THH	0	1/8
HTT	1	1/8
THT	1	1/8
TTH	0	1/8
TTT	0	1/8

$$P(X = 0) = \frac{4}{8} = \frac{1}{2}$$

$$P(X = 1) = \frac{4}{8} = \frac{1}{2}$$

$$\mu = \sum P_i x_i = 0 + \frac{1}{2} = \frac{1}{2}$$

$$\sigma^2 = \sum P_i x_i^2 - \mu^2 = 0 + \frac{1}{2} - \frac{1}{4} = \frac{1}{4}$$

$$\therefore \quad 64(\mu + \sigma^2) = 64\left(\frac{3}{4}\right) = 48$$

20. (b) Given, z_1, z_2, z_3 are three complex numbers lie on circle $|z| = 1$

$\arg(z_1) = -\frac{\pi}{4}$, $\arg(z_2) = 0$

and $\arg(z_3) = \frac{\pi}{4}$

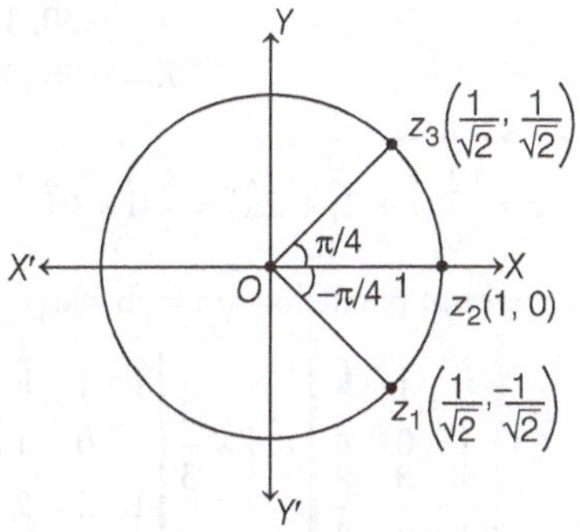

$$\therefore \quad z_1 = \frac{1}{\sqrt{2}}(1 - i), \bar{z}_1 = \frac{1}{\sqrt{2}}(1 + i)$$

$$z_2 = 1 + 0i, \bar{z}_2 = 1$$

$$z_3 = \frac{1}{\sqrt{2}}(1 + i), \bar{z}_3 = \frac{1}{\sqrt{2}}(1 - i)$$

$\therefore \quad |z_1\bar{z}_2 + z_2\bar{z}_3 + z_3\bar{z}_1|^2$

$$= \left|\frac{1}{\sqrt{2}}(1 - i) + \frac{1}{\sqrt{2}}(1 - i) + \frac{1}{2}(1 + i)^2\right|^2$$

$$= \left|\sqrt{2} - \sqrt{2}i + \frac{1}{2}(2i)\right|^2$$

$= |\sqrt{2} + (1 - \sqrt{2})i|^2 = 2 + (1 - \sqrt{2})^2$

$= 2 + 1 + 2 - 2\sqrt{2} = 5 - 2\sqrt{2}$

$\therefore \quad \alpha = 5, \beta = -2$

$\Rightarrow \alpha^2 + \beta^2 = 25 + 4 = 29$

21. (16) $\mathbf{b} = \lambda\hat{\mathbf{i}} + 4\hat{\mathbf{k}}, \lambda > 0$

$\mathbf{c}$ = Projection of $\mathbf{b}$ on vector

$\mathbf{a} = \hat{\mathbf{i}} + 2\hat{\mathbf{j}} + 2\hat{\mathbf{k}}$

$= (|\mathbf{b}|\cos\theta)\hat{\mathbf{a}}$

$$= \left(\frac{\mathbf{b} \cdot \mathbf{a}}{|a|}\right)\hat{\mathbf{a}}$$

$\Rightarrow \; \mathbf{c} = \dfrac{(\lambda+8)}{9} \cdot (\hat{\mathbf{i}} + 2\hat{\mathbf{j}} + 2\hat{\mathbf{k}})$

$\because \; |\mathbf{a} + \mathbf{c}| = 7$

$|\mathbf{a}|^2 + |\mathbf{c}|^2 + 2\mathbf{a}\cdot\mathbf{c} = 49$

$\Rightarrow \; 9 + \left(\dfrac{\lambda+8}{9}\right)^2 (9) + 2\dfrac{(\lambda+8)}{9}(9) = 49$

$\Rightarrow \; 9 + \dfrac{(\lambda+8)^2}{9} + 2(\lambda+8) = 49$

Let $\lambda + 8 = t$

$t^2 + 81 + 18t = 49 \times 9$

$\Rightarrow \; t^2 + 18t - 360 = 0$

$\Rightarrow \; t^2 + 30t - 12t - 360 = 0$

$\Rightarrow \; (t+30)(t-12) = 0$

$t = -30, 12$

$\lambda = -38, 4$

$\because \; \lambda > 0, \lambda = 4$

$\mathbf{c} = \dfrac{12}{9}(\hat{\mathbf{i}} + 2\hat{\mathbf{j}} + 2\hat{\mathbf{k}}) = \dfrac{4}{3}(\hat{\mathbf{i}} + 2\hat{\mathbf{j}} + 2\hat{\mathbf{k}})$

$\therefore$ Area of parallelogram $= |\mathbf{b} \times \mathbf{c}|$

$$= \left\| \begin{vmatrix} \hat{\mathbf{i}} & \hat{\mathbf{j}} & \hat{\mathbf{k}} \\ 4 & 0 & 4 \\ \frac{4}{3} & \frac{8}{3} & \frac{8}{3} \end{vmatrix} \right\| = 4 \times \frac{4}{3} \left\| \begin{vmatrix} \hat{\mathbf{i}} & \hat{\mathbf{j}} & \hat{\mathbf{k}} \\ 1 & 0 & 1 \\ 1 & 2 & 2 \end{vmatrix} \right\|$$

$$= \frac{16}{3}\sqrt{(2)^2 + (1)^2 + (2)^2} = 16$$

22. *(34)* $|A| = -2$

Order of matrix $A = 3(n)$

$\therefore \; |3\,\text{adj}\,(-6\,\text{adj}\,3A)|$

$|3\,\text{adj}(-6 \cdot 3^{3-1}\,\text{adj}\,A|$

$= |3(-6 \cdot 3^2)^{3-1}\,\text{adj adj}\,A|$

$= |3 \cdot 6^2 \cdot 3^4\,\text{adj adj}\,A|$

$= 3^3 \cdot (6^2)^3\,(3^4)^3\,|\text{adj adj}\,A|$

$= 3^3 \cdot 2^6 \cdot 3^6 \cdot 3^{12}\,|A|^{(n-1)^2}$

$= 2^6 \cdot 3^{21} \cdot (-2)^4 = 2^{10}\,3^{21}$

$\therefore \; m + n = 10$

$mn = 21$ and $m > n$

$m = 7$ and $n = 3$

$\Rightarrow \; 4m + 2n = 28 + 6 = 34$

> **Concept Enhancer**
> $\text{adj}\,(KA) = k^{n-1}\text{adj}(A)$ and $|KA| = K^n|A|$

23. *(34)* $f(x) = \begin{cases} -3ax^2 - 2, & x < 1 \\ a^2 + bx, & x \ge 1 \end{cases}$

$\because \; f(x)$ is differentiable.

$f'(x) = \begin{cases} -6ax, & x < 1 \\ b, & x > 1 \end{cases}$

$f(1^+) = f(1^-) = f(1)$

$\therefore \; -3a - 2 = a^2 + b$

$a^2 + 3a + b = -2 \quad \ldots\text{(i)}$

and $\; f'(1^+) = f'(1^-)$

$-6a = b \quad \ldots\text{(ii)}$

Solving Eqs. (i) and (ii), we get

$a^2 - 3a + 2 = 0$

$(a-1)(a-2) = 0$

$\Rightarrow \; a = 1, 2 \quad [\because a > 1]$

$\Rightarrow \; a = 2$

$b = -12$

Now, $y = f(x) = \begin{cases} -6x^2 - 2, & x < 1 \\ -12x + 4, & x \ge 1 \end{cases}$

$-6x^2 - 2 = -20$

$6x^2 = 18$

$x^2 = 3 \;\Rightarrow x = -\sqrt{3}$

or $\; -12x + 4 = -20$

$12x = 24$

$x = 2$

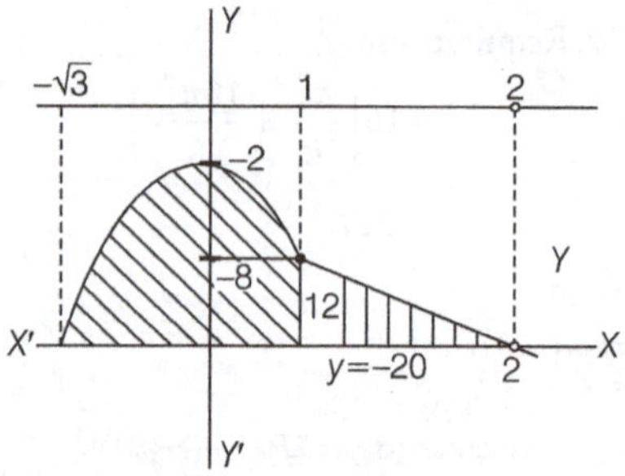

$\therefore$ Required area $= 20(\sqrt{3} + 1)$

$$-\left|\int_{-\sqrt{3}}^{1} y\,dx\right| + \frac{1}{2} \times 1 \times 12$$

$$= 20(\sqrt{3}+1) - \left|\int_{-\sqrt{3}}^{1} (-6x^2 - 2)\,dx\right| + 6$$

$= 20\sqrt{3} + 26 - |[2x^3 + 2x]_{-\sqrt{3}}^{1}|$

$= 20\sqrt{3} + 26 - |4 + 8\sqrt{3}|$

$= 20\sqrt{3} + 26 - 8\sqrt{3} - 4 = 22 + 12\sqrt{3}$

$\alpha = 22$ and $\beta = 12$

$\therefore \; \alpha + \beta = 34$

24. *(2035)* $\displaystyle\sum_{r=0}^{5} \frac{{}^{11}C_{2r+1}}{2r+2} = \frac{1}{12}\sum_{r=0}^{5} {}^{12}C_{2r+2}$

$= \dfrac{1}{12}({}^{12}C_2 + {}^{12}C_4 + {}^{12}C_6 + {}^{12}C_8 + {}^{12}C_{10} + {}^{12}C_{12})$

$= \dfrac{1}{12}(2^{11} - 1) = \dfrac{2^{11} - 1}{12} = \dfrac{2047}{12} = \dfrac{m}{n}$

$\Rightarrow \; m - n = 2047 - 12 = 2035$

> **Concept Enhancer**
> $\dfrac{{}^{n+1}C_{r+1}}{n+1} = \dfrac{{}^{n}C_{r}}{r+1}$

25. *(216)* $L_1 : \dfrac{x-1}{3} = \dfrac{y-1}{-1} = \dfrac{z+1}{0} = \lambda$ (say)

$\Rightarrow \; P(3\lambda + 1, -\lambda + 1, -1)$

$L_2 : \dfrac{x-2}{2} = \dfrac{y}{0} = \dfrac{z+4}{\alpha} = \mu$ (say)

$\Rightarrow \; Q(2\mu + 2, 0, \alpha\mu - 4)$

$\because \; L_1$ and L_2 intersect.

$\therefore \; 3\lambda + 1 = 2\mu + 2$

and $\; -\lambda + 1 = 0$

$\lambda = 1$

and $\; -1 = \alpha\mu - 4$

$\Rightarrow \; -1 = \alpha - 4 \;\Rightarrow\; \alpha = 3$

$4 = 2\mu + 2$

$\Rightarrow \; \mu = 1 \;\Rightarrow\; B = (4, 0, -1)$

$\therefore \; L_2 : \dfrac{x-2}{2} = \dfrac{y}{0} = \dfrac{z+4}{3}$

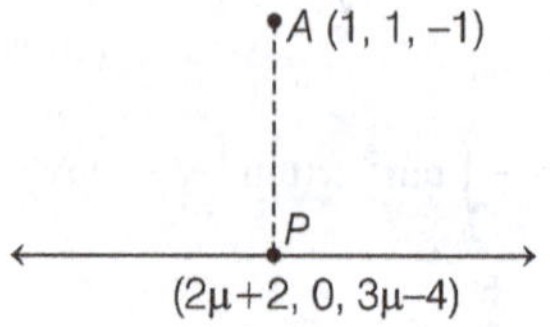

$\text{AM} \perp (2\hat{\mathbf{i}} + 0\hat{\mathbf{j}} + 3\hat{\mathbf{k}})$

$\Rightarrow \; \text{AM} \cdot (2\hat{\mathbf{i}} + 3\hat{\mathbf{k}}) = 0$

$\Rightarrow \; 2(2\mu + 1) + 3(3\mu - 3) = 0$

$\Rightarrow \; 4\mu + 2 + 9\mu - 9 = 0$

$\Rightarrow \; 13\mu = 7$

$\Rightarrow \; \mu = \dfrac{7}{13}$

$PB = \sqrt{(2\mu - 2)^2 + 0^2 + (3\mu - 3)^2}$

$= \sqrt{(\mu - 1)^2\,(4 + 9)} = \sqrt{\left(\dfrac{6}{13}\right)^2 (13)}$

$= \dfrac{6}{\sqrt{13}}$

$\therefore \; 26\alpha\,(PB)^2 = 26(3)\dfrac{36}{13}$

$6 \times 36 = 216$

Physics

26. *(a)*

Key idea

To check weather a diode is in forward bias or in reverse bias, check sign of potentials of p and n side,

If $V_p - V_n = \begin{cases} +, \text{forward bias} \\ -, \text{Reverse bias} \end{cases}$

For forward bias $V_p - V_n$ is positive. Among the given circuits, only B, C and E are forward bias.

In forward biasing, positive or high concentration part is connect to high potential and N part of semiconductor is connect to low potential.

27. *(b)* Harmonic of closed pipe is given by $= \dfrac{nv_1}{4l_1}$

Frequency of 9th harmonic of closed pipe $= 4 \cdot \dfrac{v_2}{2l_2} = \dfrac{9v_1}{4l_1}$ ...(i)

Frequency of 4th harmonic of open pipe $= \dfrac{2v_2}{l_2}$ ...(ii)

As frequency of 9th and 4th harmonics are equal.

$$\frac{9v_1}{4l_1} = \frac{2v_2}{l_2}$$

$$\therefore \quad \frac{9}{4l_1}\sqrt{\frac{B}{\rho_1}} = \frac{2}{l_2}\sqrt{\frac{B}{\rho_2}} \quad \text{...(iii)}$$

$$l_1 = 10 \text{ cm}, \rho_1 : \rho_2 = 1:16$$

Substituting all values in Eq.(iii)

$$l_2 = \frac{20}{9} \text{ cm}$$

28. *(d)* Statement I is false but statement II is correct. The correct form of statement I is

When two non-ideal batteries are connected in parallel, then equivalent emf is lies between E_1 and E_2

$$E_{eq} = \frac{\dfrac{E_1}{r_1} + \dfrac{E_2}{r_2}}{\dfrac{1}{r_1} + \dfrac{1}{r_2}} \Rightarrow \frac{E_1 r_2 + E_2 r_1}{r_1 + r_2}$$

29. *(c)* For photoelectric effect, incident energy must be equal to or greater than the work function.

For wavelength, $\lambda = 550$ nm

incident energy

$$E = \frac{1240}{\lambda} = \frac{1240}{550} \simeq 2.25 \text{ eV}$$

Hence, only cesium will show photoelectric effect.

30. *(d)* According to Stefan law, rate of heat radiated is given by

$$\frac{dQ}{dt} = \sigma e A T^4 \Rightarrow P \propto AT^4$$

$$P \propto R^2 T^4$$

$$\frac{P_{\text{smaller}}}{P_{\text{larger}}} = \frac{(0.2)^2 \times (800)^4}{(0.8)^2 \times (400)^4} = 1$$

$$\therefore \quad P_{\text{larger}} = P_{\text{smaller}}$$

31. *(a)* Statement I is true and Statement II is false.

In vernier scale, one VSD is smaller than one MSD. Also least count is given by one main scale division/number of vernier scale division for normal vernier calliper.

32. *(b)*

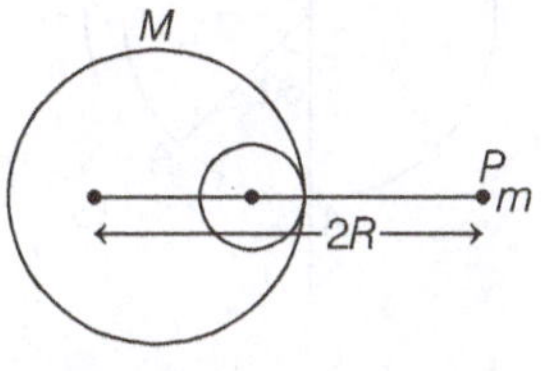

According to Newton's gravitational law,

When whole mass is accounted.

$$F_1 = \frac{GMm}{(2R)^2} \quad \text{...(i)}$$

When section of mass (of radius $\frac{R}{3}$) is removed;

$$F_2 = \frac{GMm}{(2R)^2} - \left[\frac{G\left(\frac{M}{27}\right)m}{\left(\frac{4R}{3}\right)^2}\right]$$

$$F_2 = \frac{11}{48}\frac{GMm}{R^2} \quad \text{...(ii)}$$

From Eqs. (i) and (ii)

$$F_1 : F_2 = 12:11$$

33. *(a)* There are three lens in the given diagram,

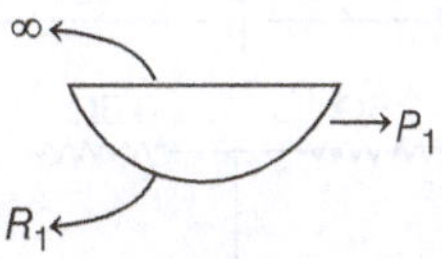

Where R is radius of curvature and P is power of lens

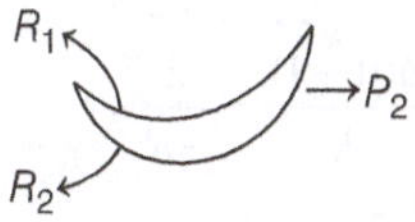

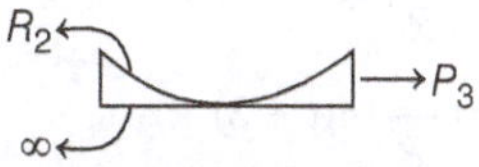

Total power is given by

$$P_{\text{total}} = P_1 + P_2 + P_3$$

$$\Rightarrow P_1 = \left[\frac{4}{3} - 1\right]\left[\frac{1}{\infty} - \frac{1}{|R_1|}\right]$$

$$\Rightarrow P_1 = \left[\frac{1}{3|R_1|}\right]$$

$$P_2 = \frac{1}{2}\left[\frac{1}{-|R_1|} - \frac{1}{-|R_2|}\right]$$

$$P_3 = \frac{1}{3}\left[\frac{1}{-|R_2|} - \frac{1}{\infty}\right] \Rightarrow \frac{-1}{3|R_2|}$$

$$P_{\text{total}} = P_1 + P_2 + P_3 = \frac{-1}{6}\left[\frac{1}{|R_1|} - \frac{1}{|R_2|}\right]$$

34. *(a)* Moment of inertia, when the section is taken out is

$$I = \frac{MR^2}{2} - \left[\frac{\frac{M}{4}\left(\frac{R}{2}\right)^2}{2} + \frac{M}{4}\left(\frac{R}{2}\right)^2\right]$$

$$I = \frac{13}{32}MR^2$$

Hence, option (a) is correct.

35. *(c)*

Key idea

On silvering surface of a lens it behaves like a mixture. As light ray passes twice through lens in process of image formation from a silvered lens, we have

$$\frac{1}{F} = \frac{2}{f_l} + \frac{1}{f_m}$$

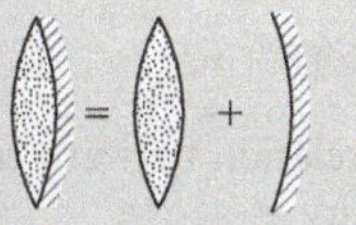

Here, $f_l = \dfrac{R}{2(\mu - 1)}$ and $f_m = \dfrac{R}{2}$

When object's image is lies on object itself that means, the combination of two lens acts as mirror.

$$P_{eq} = 2P_l + P_m$$

$$\frac{-1}{f_{eq}} = \frac{2}{f_l} - \frac{1}{f_m}$$
$$= \frac{4(\mu - 1)}{R} - \frac{2}{-R}$$
$$= \frac{1}{R}[4\mu - 4 + 2]$$
$$\frac{1}{f_{eq}} = \frac{-1}{R}(4\mu - 2) \quad \ldots(i)$$
$$\Rightarrow f_{eq} = \frac{R}{2}$$
$$R = 2f_{eq} = \frac{-2R}{4\mu - 2} \text{ or } \frac{-R}{(2\mu - 1)}$$

36. *(a)* For wire bound standard resistors, the value of resistivity changes very minimally when there is increase in temperature.

Hence, the correct graph is

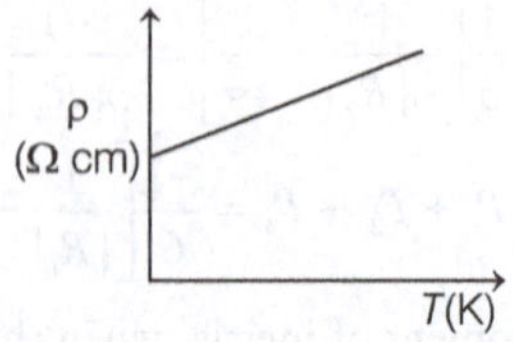

37. *(d)* The diagram for the situation is

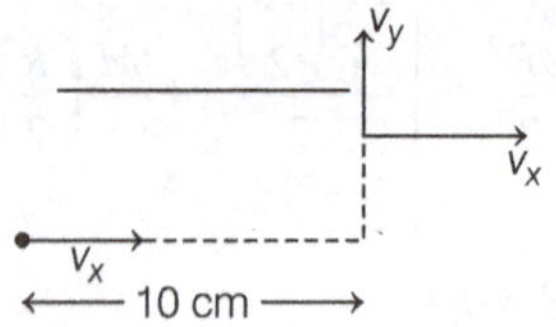

Time to travel 10 cm,
$$t = \frac{L}{v_x} = \frac{10 \times 10^{-2}}{10^6} = 10^{-7} \text{ s}$$

From 1st equation of motion
$$v_y = u_y + a_y t$$
$$v_y = 0 + \frac{eE}{m} \times 10^{-7}$$
$$v_y = \frac{1.6 \times 10^{-19}}{9.1 \times 10^{-31}} \times 9.1 \times 10^{-2} \times 10^{-7}$$
$$v_y = 16 \times 10^6 \text{ m/s}$$

38. *(b)* The relation between field and permeability is
$$B = \mu_0 ni$$
$$\frac{B}{\mu_0} = ni$$

For $n = \frac{N}{L}$, for $i = \text{A}$
$$\frac{B}{\mu_0} = [\text{L}^{-1}\text{A}]$$

39. *(d)*

Key idea

As particle rises in the vertical plane, it's kinetic energy is converted into potential energy. Also as no loss of energy is given, so total initial energy (at *A*) and total final energy at some point (*B* or *C*) remains same.

For slacking at Point *D*
$$v_A = \sqrt{5gl}$$
$$v_D = \sqrt{gl}$$

Applying conservation of energy at point *A*,

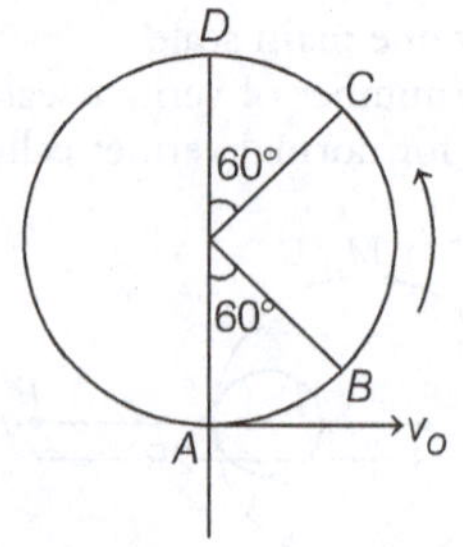

$$\frac{1}{2}mv_A^2 = \frac{1}{2}mv_B^2 + mgh$$

As string becomes slack at top,
At lowest point $v = \sqrt{5gl}$

$$\therefore \quad \frac{1}{2}m(5gl) = \frac{1}{2}mv_B^2 + mg\frac{l}{2}$$
$$\Rightarrow \quad \frac{1}{2}mv_B^2 = \frac{5mgl}{2} - \frac{mgl}{2}$$
$$\Rightarrow \quad (\text{KE})_B = 2mgl \quad \ldots(i)$$

Now at point *C*
$$\frac{1}{2}mv_C^2 = \frac{1}{2}mv_D^2 + mg\frac{l}{2}$$
$$(\text{KE})_C = \frac{1}{2}mgl + mg\frac{1}{2} = mgl \quad \ldots(ii)$$
$$\frac{(\text{KE})_B}{(\text{KE})_C} = \frac{2}{1}$$

40. *(b)* R_{eq} of the given circuit is,

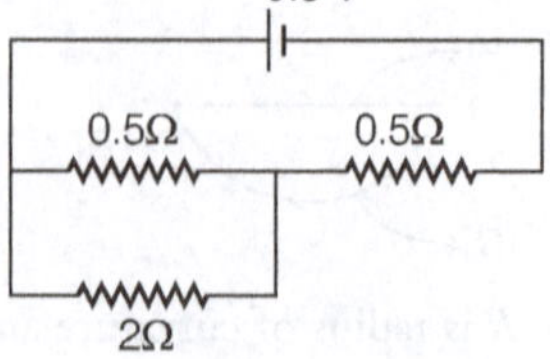

$$R_{eq} = 0.5 + \frac{0.5 \times 2}{2 + 0.5}$$

As for series $R_{eq} = R_1 + R_2$ and for parallel connection $R_{eq} = \frac{R_1 \times R_2}{R_1 + R_2}$

$$\therefore \quad R_{eq} = \frac{45}{50} = \frac{9}{10} = 0.9\Omega$$

Current is given by Ohm's law
$$i = \frac{V}{R} = \frac{0.9}{0.9} = 1\text{A}$$

41. *(c)* Both A and R is true and R is the correct explanation of A.

$$\beta \cdot (\text{fringe width}) = \frac{\lambda D}{d}$$

In denser medium, λ decrease therefore β will also decreases.

∴ Fringe will come closer.

Also $\mu = \frac{c}{v}$ or $v = \frac{c}{\mu}$

Frequency remains same.
$$\Rightarrow \mu = \frac{\lambda_{\text{Vaccum}} \times f}{v_{\text{medium}} \times f} \Rightarrow \lambda_{\text{medium}} = \frac{\lambda_{\text{vaccum}}}{\mu}$$

42. *(*)* According to de - Broglie's equation,
$$\lambda = \frac{h}{p} \Rightarrow \frac{h}{mv} \quad \ldots(i)$$

From Bohr's 2nd postulate,
$$mvr = \frac{nh}{2\pi}$$
$$mv = \frac{nh}{2\pi r} \quad \ldots(ii)$$

Substitute the value of *mv* from Eq (i)
$$\lambda = \frac{2\pi rh}{nh} \Rightarrow \lambda \propto \frac{r}{n}$$
$$\frac{\lambda_1}{\lambda_4} = \frac{r_1 n_4}{r_4 n_1} \quad \ldots(ii)$$

Substituting all the values
$$\therefore \quad \frac{\lambda_1}{\lambda_4} = \frac{5.3 \times 10^{-11} \times 4}{1 \times 8.48 \times 10^{-10}} = \frac{1}{4}$$

43. *(d)* As the charge is kept at edge's centre, then total electric flux is given by

$\phi = \frac{q}{\varepsilon_0}$ and $q = \frac{a\lambda}{8}$, where λ is linear change density.

$$\therefore \quad \phi = \frac{\lambda a}{8\varepsilon_0}$$

44. *(a)* Ice from −10°C changes to vapour with temperature.

The different stage of heat energy required is

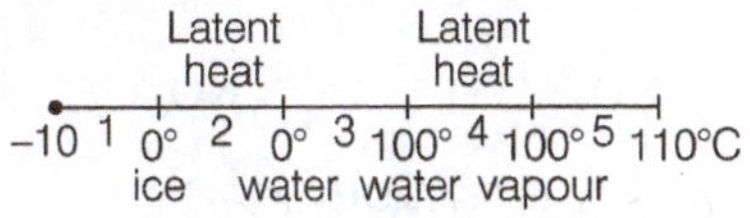

Heat required in each step is,

$$\Delta Q_1 = ms_i \Delta T = 10^{-3} \times 2100 \times 10$$
$$= 21\,\text{J}$$
$$\Delta Q_2 = mL_f = 10^{-3} \times 3.35 \times 10^5 = 335\,\text{J}$$
$$\Delta Q_3 = m \times s_w \times \Delta T$$
$$= 10^{-3} \times 4180 \times 100 = 418\,\text{J}$$
$$\Delta Q_4 = m \times L_v = 10^{-3} \times 2.25 \times 10^6$$
$$= 2250\,\text{J}$$
$$\Delta Q_5 = m \times s_v \times \Delta T = 10^{-3} \times 1920 \times 10$$
$$= 19.2\,\text{J}$$
$$\Delta Q_{\text{total}} = \Delta Q_1 + \Delta Q_2 + \Delta Q_3 + \Delta Q_4 + \Delta Q_5$$
$$\Delta Q_{\text{total}} = 3043.25\,\text{J}$$

45. *(a)* The charge store in capacitor is

$$\Delta q = (KC - C)V$$
$$= 40 \times 10^{-6} \times 100$$
$$= 4 \times 10^{-3} = 4\,\text{mC}$$

Also, $\Delta U = \frac{1}{2}C'V^2 - \frac{1}{2}CV^2$

$$= \frac{1}{2}CV^2(2-1)$$
$$= \frac{1}{2} \times 40 \times 10^{-6} \times 10000$$
$$= 0.2\,\text{J}$$

46. *(40)* The diagram for the given data is

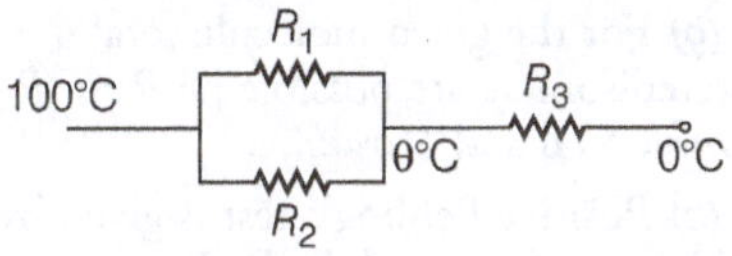

Resistance $R_1 = \frac{2L}{K_1 A}$

$$R_2 = \frac{2L}{K_2 A}$$
$$R_3 = \frac{L}{K_3 A}$$

$$\frac{\theta - 100}{\frac{R_1 R_2}{R_1 + R_2}} + \frac{\theta - 0}{R_3} = 0$$
$$\theta = 40°\text{C}$$

47. *(5)*

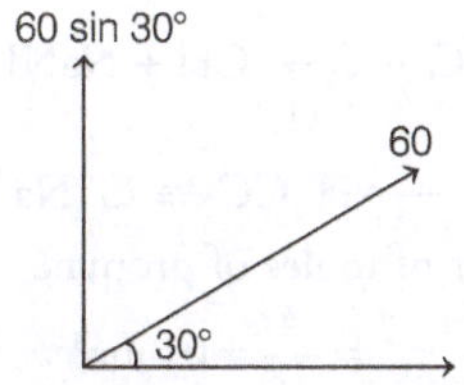

Using 2nd equation of motion for first second,

$$s = ut + \frac{1}{2}at^2$$
$$h_0 = 30 \times 1 - \frac{1}{2} \times 10 \times 1 = 25$$

Time to reach maximum height

$$t = \frac{T}{2} = \frac{u \sin\theta}{g} = 3\text{s}$$

For last second

$$h_1 = 30 + \left(\frac{-10}{2}\right) \times (2 \times 3 - 1) = 5$$
$$\frac{h_0}{h_1} = \frac{25}{5} = 5$$

48. *(90)* Vector $\mathbf{v}_A = (2t\hat{i} + 3n\hat{j} + 2\hat{k})$

$$\mathbf{v}_B = (2\hat{i} - 2t\hat{j} + 4p\hat{k})$$
$$\mathbf{v}_A \cdot \mathbf{v}_B = 0$$
$$4 - 6n + 8p = 0$$
$$2 - 3n + 4p = 0$$
$$3n = 2 + 4p$$
$$|\mathbf{v}_A| = |\mathbf{v}_B|$$
$$4 + 9n^2 + 4 = 4 + 4 + 16p^2$$
$$p = \frac{-1}{4} \Rightarrow n = \frac{1}{3}$$
$$\mathbf{L} = m_A(\mathbf{r}_{A/B} \times \mathbf{v}_A)$$
$$\mathbf{r}_{A/B} = (\alpha_1 - \beta_1)\hat{i} + (\alpha_2 - \beta_2)\hat{j} + (\alpha_3 - \beta_3)\hat{k}$$
$$= (1-2)\hat{i} + (1+1)\hat{j} + 3\hat{k}$$

$$\therefore \mathbf{L} = \begin{vmatrix} \hat{i} & \hat{j} & \hat{k} \\ -1 & 2 & 3 \\ 2 & 1 & 2 \end{vmatrix} = \hat{i} + 8\hat{j} - 5\hat{k}$$

$$|\mathbf{L}| = \sqrt{1 + 64 + 25} = \sqrt{90}$$

49. *(8)* Using mirror formula

$$v = \frac{uf}{u - f} = \frac{-24 \times 1}{-24 - 1} = \frac{24}{25}$$
$$m = \frac{-v}{u} = \frac{24}{25(-24)} = \frac{1}{25}$$

Using the formula

$$\frac{1}{v} - \frac{1}{u} = \frac{1}{f} \quad \ldots\text{(i)}$$
$$v_I = -m^2 v_0 = \frac{-1}{(25)^2} \times 25 = \frac{-1}{25}$$

Differentiating Eq. (i)

$$\frac{-1}{v^2}\left(\frac{dv}{dt}\right) + \frac{1}{u^2}\left(\frac{du}{dt}\right) = 0$$
$$\left[\frac{dv}{dt} = v_I, \frac{du}{dt} = v_o\right]$$
$$\frac{+2}{v^3}(v_I)^2 - \frac{1}{v^2} a_1 - \frac{2}{u^3}(v_o)^2 + \frac{1}{u^2} a_o = 0$$
$$\frac{a_1}{v^2} = \frac{2}{v^3} \cdot v_1^2 - \frac{2}{u^3} \cdot v_o^2$$

Substituting the value

$$a_1 = \frac{2}{24.25} - \frac{2}{24}$$
$$100 a_1 = \frac{2}{25} \times 100 = 8$$

50. *(4)* Radius, $r_1 = 2$ cm

Radius, $r_2 = 4$ cm

The radius of curvature of common surface.

$$r = \frac{r_1 \cdot r_2}{r_1 - r_2}$$
$$= \frac{2 \times 4}{4 - 2} = 4\,\text{cm}$$

Chemistry

51. *(d)* Statement given in option (d) is not true for radioactive decay. Its correct form is decay constant is independent of temperature.

52. *(b)* The electronic configuration of given species are

(A) $Eu^{2+}: [Xe] 4f^7 6s^0$

(B) $Gd^{3+}: [Xe]\ 4f^7 5d^0 6s^0$

(C) $Eu^{3+}: [Xe] 4f^6 6s^0$

(D) $Tb^{3+}: [Xe] 4f^8 6s^0$

(E) $Sm^{2+}: [Xe] 4f^6 6s^0$

Thus, Eu^{2+} and Gd^{3+} have $4f^7$ configuration i.e. *A* and *B* only.

53. *(d)* In complex $[NiCl_4]^{2-}$, Ni is in + 2 oxidation state. So, configuration of $Ni^{2+} = [Ar]3d^8$. Also, since Cl is a weak field ligand. So, the geometry of compound is tetrahedral with sp^3 hybridisation.

Similarly in complex $[Ni(CO)_4]$, Ni is in zero oxidation state. So, configuration of $Ni^0 = [Ar]3d^8 4s^2$. Also, since CO is a strong field ligand. So, after rearrangement the configuration of Ni $= [Ar]3d^{10} 4s^0$ So, the geometry of the compound is tetrahedral with sp^3 hybridisation.

So, option (d) is the correct answer.

54. *(c)* Concentrated solution of sulphuric acid is used as an electrolyte to obtain $H_2S_2O_8$ by the process of electrolysis.

55. *(d)* The correct match is A-IV, B-I, C-II, D-III.

- For isoelectronic species the cation with the greater positive charge will have smaller radius while anion with greater negative charge will have larger radius. So, the order is

$$Al^{3+} < Mg^{2+} < Na^+ < F^-.$$

- Ionisation enthalpy order is B < C < O < N. As it increases along the period and decreases down the group. Further it increases with half or fully-filled orbital.
- Metallic character increases down the group and decreases along the period. So, the correct order is B < Al < Mg < K
- The correct order of electronegativity is

$$\underset{(1.9)}{Si} < \underset{(2.1)}{P} < \underset{(2.5)}{S} < \underset{(3.0)}{Cl}$$

56. (c) $H_3C-C\equiv CH + \underset{(4g)}{NaNH_2}$

$$\longrightarrow H_3CC\equiv C^-Na^+ + NH_3$$

Number of moles of propyne

$$= \frac{4}{40} = 0.1 \text{ mol}$$

Number of moles of NH_3 produced = 0.1

Volume occupied by 0.1 mol of NH_3 at STP = 2240 mL

Thus, Statement I is correct but Statement II is incorrect.

57. *(b)*

Key idea To solve this problem, first of all deduce the relation between ΔT_b, i and concentration. After that substitute the given values to obtain the order of boiling points.

We know that,

$$\Delta T_b = i \times m \times k_b$$

$$\Rightarrow \quad \Delta T_b \propto iM$$

(Where, M = concentration)

For NaCl, the value of i is 2 while for urea the value of i is 1.

(i) For 10^{-4} M NaCl, the value of

$$\Delta T_b \propto 10^{-4} \times 2$$

(ii) For 10^{-4} M urea, the value of

$$\Delta T_b \propto 10^{-4} \times 1$$

(iii) For 10^{-3} M NaCl, the value of

$$\Delta T_b \propto 10^{-3} \times 2$$

(iv) For 10^{-2} M NaCl, the value of

$$\Delta T_b \propto 10^{-2} \times 2$$

So, the correct increasing order of boiling points is

(ii) < (i) < (iii) < (iv)

58. *(a)* The incorrect order of electronegativity is given in option (a). Its correct order is

$$\underset{(1.2)}{Mg} < \underset{(1.5)}{Al} < \underset{(2.0)}{B} < \underset{(3.0)}{N}$$

59. *(b)* The complete reaction sequence involved is as follows.

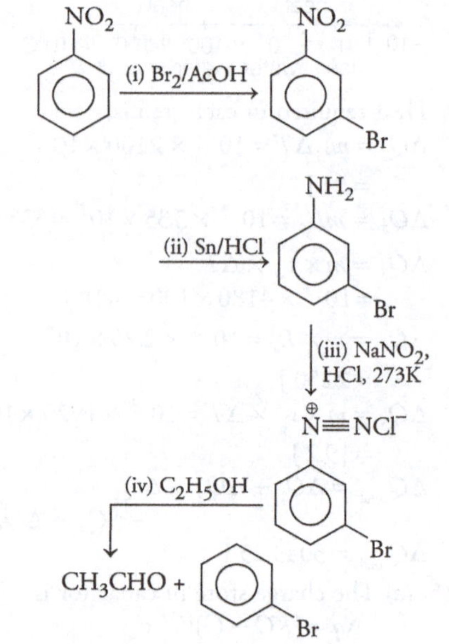

60. *(d)*

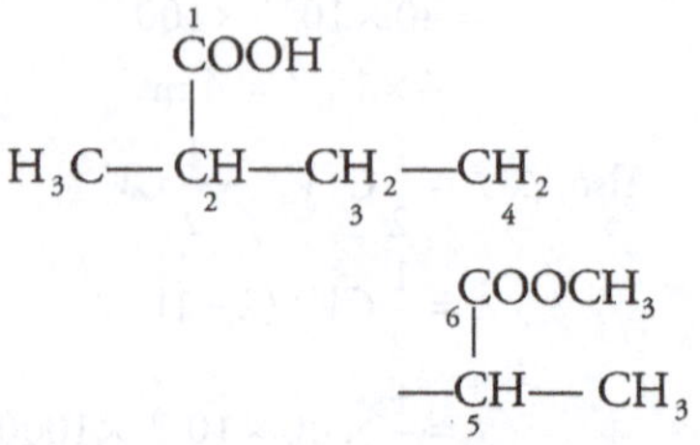

So, the IUPAC name of the given compound is 6-methoxycarbonyl -2, 5-dimethylhexanoic acid.

61. *(b)* For the given molecule total 4 stereoisomers are possible i.e *R cis*, *R trans*, *S cis* and *S trans*.

62. *(b)* Positive Fehling's test is given by aldehydes having alpha hydrogen. Also, ketones generally do not react until they are α-hydroxy ketones.

So, among the given options, compounds given in C, D and E i.e. $HOCH_2-CO-(CHOH)_3-CH_2-OH$,

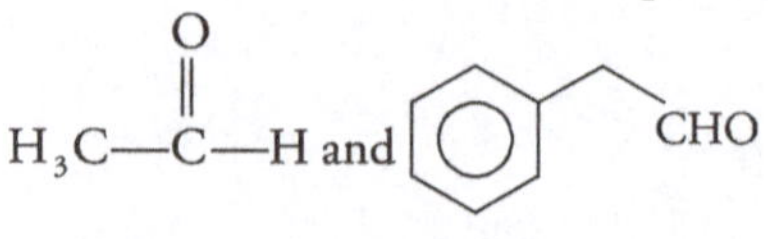

will give positive Fehling's test.

63. *(c)* For thermally insulated system

$$q = 0$$

Also from 1st law of thermodynamics

$$\Delta U = q + W$$

So, $\Delta U = W$

$\Rightarrow \quad W > 0, \Delta U > 0$

So, the correct thermodynamic parameter is

$$\Delta U > 0, q = 0, W > 0$$

64. *(d)* The reaction involved in conversion of CO_2 into CO is as follows

$$CO_2(g) + C(s) \rightleftharpoons 2CO(g)$$

$$\begin{matrix} 0.5 & & - \\ 0.5 - x & & 2x \end{matrix}$$

$$p_{total} = 0.5 + x = 0.8$$

$$x = 0.3$$

So, $k_p = \dfrac{(0.6)^2}{0.2} = 1.8$ atm

So, the value of k_p is 1.8 atm

65. *(b)* From Faraday's law,

gram equivalent of Al deposited

$$= \frac{It}{96500}$$

$$\frac{w}{27} \times 3 = \frac{2 \times 30 \times 60}{96500}$$

so $w = 0.366$ g

Thus, the amount of aluminium deposited at cathode is 0.336 g.

66. *(b)* For first excited state, $n = 2$

So, substituting the value of n in equation $r = \dfrac{a_0 n^2}{Z}$ we get

$$r = \frac{(2)^2 a_0}{2} = 2a_0 \quad (\because Z = 2)$$

So, $r = 2a_0$

67. *(b)* Ascorbic acid is vitamin C.

Concept Enhancer

Vitamin A	Retinol
Vitamin B_1	Thiamine
Vitamin B_2	Niacin
Vitamin B_5	Pantothenic acid
Vitamin B_6	Pyridoxine
Vitamin B_{12}	Cyanocobalamin
Vitamin D	Calciferol

68. *(a)* Both statement I and statement II are correct.

69. *(a)* Statements given in A, D and E are incorrect while B and C are correct.

The correct form of incorrect statements are

(A) Propene does not show geometrical isomerism.

(D) 2-methylbut-2-ene does not show geometrical isomerism.

(E) *trans* isomer has higher melting point than *cis* isomer.

70. *(b)* Electronic configuration of Fe^{3+} ion in complex $K_3[Fe(SCN)_6]$ is given by

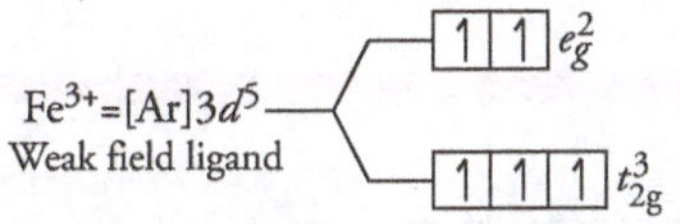

So, using the formula,

$$= (-0.4 \times t_{2g}e^- + 0.6 \times e_g e^-)\Delta_0$$

$$= (-0.4 \times 3 + 0.6 \times 2)\Delta_0$$

$$= 0\Delta_0$$

71. *(20)* Given,

mass of organic compound = 180 mg

Mass of AgCl = 143.5 g/mol

% Cl = ?

$$n_{Cl} = n_{AgCl} = \frac{143.5 \times 10^{-3}}{143.5} = 10^{-3}$$

$$\text{Now, \% Cl} = \frac{10^{-3} \times 35.5}{180 \times 10^{-3}} \times 100 = 19.72$$

$$\approx 20$$

72. *(154)* Complete reaction sequence involved is as follows

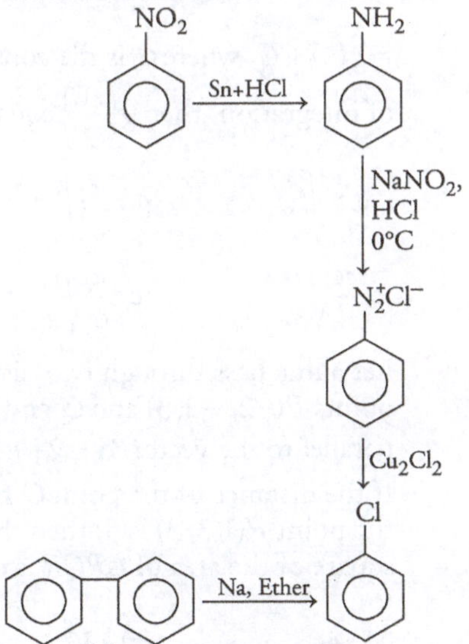

(A)

So, the molar mass of the product (A) is 154 g/mol.

73. *(45)* Let the number of moles of $CO_2 = y$.

Total initial moles of $Ca(OH)_2 = 2y$

Excess $Ca(OH)_2 = y$

Gram equivalent of $Ca(OH)_2$

= Gram equivalent of HCl

$$y \times 2 = \frac{0.1 \times 40}{1000} \times 1$$

$$y = 2 \times 10^{-3}$$

Volume of CO_2

$$= x = 2 \times 10^{-3} \times 22400$$

$$= 44.8 \text{ cm}^3$$

$$\approx 45 \text{ cm}^3.$$

So, the value of x is 45.

74. *(69)* Given, $A = 10^{20}$

$E_a = 191.48$ kJ/mol

$T = 1000$ K $R = 8.314$ JK^{-1} mol^{-1}

$t_{1/2} = ?$

We know that, $t_{1/2} = \dfrac{0.693}{k}$

and k can be calculated by using,

$$= Ae^{\frac{-E_a}{RT}}$$

$$= 10^{20} \times e^{-\frac{191.48 \times 10^3}{8.314 \times 1000}}$$

$$= 10^{20} \times e^{-23.031}$$

$$= \frac{10^{20}}{10^{10}} = 10^{10} s^{-1}$$

Now, substitute this value in half-life equation

$$t_{1/2} = \frac{0.693}{10^{10}} = 6.93 \times 10^{-11} \text{ s}$$

or $t_{1/2} = 69.3 \times 10^{-12}$ s

So, time required for 50% molecule of A to become molecule B is 69 picoseconds.

75. *(6)* Among the given species the geometry of given molecules are linear i.e

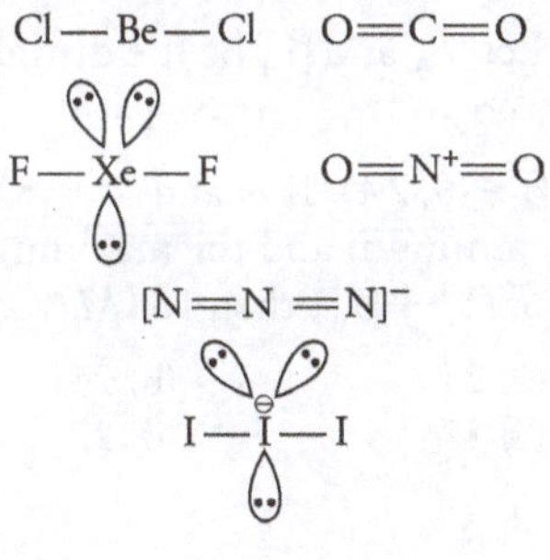

Time 3:00-6:00 | MM : 300 Marks

22 January
Shift-II

Online Question Paper

JEE Main 2025

Mathematics

Section A
Objective Type Questions

01/ Suppose that the number of terms in an AP is $2k, k \in N$. If the sum of all odd terms of the AP is 40, then sum of all even terms is 55 and the last term of the AP exceeds the first term by 27, then k is equal to

(a) 4 (b) 6
(c) 8 (d) 5

02/ Let $f(x) = \int_0^{x^2} \frac{t^2 - 8t + 15}{e^t} dt, x \in R$.

Then, the number of local maximum and local minimum points of f, respectively are

(a) 1 and 3 (b) 3 and 2
(c) 2 and 3 (d) 2 and 2

03/ If $\lim_{x \to \infty} \left(\left(\frac{e}{1-e}\right)\left(\frac{1}{e} - \frac{x}{1+x}\right)\right)^x = \alpha$,

then the value of $\frac{\log_e \alpha}{1 + \log_e \alpha}$ equals

(a) e (b) e^{-2} (c) e^2 (d) e^{-1}

04/ Let α_θ and β_θ be the distinct roots of $2x^2 + (\cos\theta)x - 1 = 0$, $\theta \in (0, 2\pi)$. If m and M are the minimum and the maximum values $\alpha_\theta^4 + \beta_\theta^4$, then $16(M + m)$ equals

(a) 24 (b) 25
(c) 17 (d) 27

05/ For a 3×3 matrix M, let trace (M) denote the sum of all the diagonal elements of M. Let A be a 3×3 matrix such that $|A| = \frac{1}{2}$ and trace $(A) = 3$. If $B = \text{adj}(\text{adj}(2A))$, then the value of $|B| +$ trace (B) equals

(a) 174 (b) 280
(c) 132 (d) 56

06/ If

$$\int e^x \left(\frac{x \sin^{-1} x}{\sqrt{1-x^2}} + \frac{\sin^{-1} x}{(1-x^2)^{\frac{3}{2}}} + \frac{x}{1-x^2}\right) dx$$

$= g(x) + C$, where C is the constant of integration, then $g\left(\frac{1}{2}\right)$ equals

(a) $\frac{\pi}{6}\sqrt{\frac{e}{3}}$ (b) $\frac{\pi}{4}\sqrt{\frac{e}{2}}$
(c) $\frac{\pi}{4}\sqrt{\frac{e}{3}}$ (d) $\frac{\pi}{6}\sqrt{\frac{e}{2}}$

07/ Let a line pass through two distinct points $P(-2, -1, 3)$ and Q and be parallel to the vector $3\hat{i} + 2\hat{j} + 2\hat{k}$. If the distance of the point Q from the point $R(1, 3, 3)$ is 5, then the square of the area of ΔPQR is equal to

(a) 144 (b) 136
(c) 148 (d) 140

08/ Let **a** and **b** be two unit vectors such that the angle between them is $\frac{\pi}{3}$. If $\lambda\mathbf{a}+2\mathbf{b}$ and $3\mathbf{a}-\lambda\mathbf{b}$ are perpendicular to each other, then the number of values of λ in $[-1, 3]$ is

(a) 1 (b) 2
(c) 0 (d) 3

Imp.

09/ If the system of linear equations

$$x+y+2z=6,$$
$$2x+3y+az=a+1,$$
$$-x-3y+bz=2b,$$

where $a, b \in R$ has infinitely many solutions, then $7a+3b$ is equal to

(a) 9 (b) 22
(c) 16 (d) 12

Solution of System of Linear Equations

10/ Let the curve $z(1+i)+\bar{z}(1-i)=4$, $z \in C$ divide the region $|z-3| \le 1$ into two parts of areas α and β. Then, $|\alpha-\beta|$ equals

(a) $1+\frac{\pi}{4}$ (b) $1+\frac{\pi}{2}$
(c) $1+\frac{\pi}{3}$ (d) $1+\frac{\pi}{6}$

11/ Let $A=\{1, 2, 3, 4\}$ and $B=\{1, 4, 9, 16\}$. Then, the number of many-one functions $f: A \to B$ such that $1 \in f(A)$ is equal to

(a) 151 (b) 139
(c) 163 (d) 127

12/ If $x=f(y)$ is the solution of the differential equation $(1+y^2)+(x-2e^{\tan^{-1}y})\frac{dy}{dx}=0$, $y \in \left(-\frac{\pi}{2}, \frac{\pi}{2}\right)$ with $f(0)=1$, then $f\left(\frac{1}{\sqrt{3}}\right)$ is equal to

(a) $e^{\pi/4}$ (b) $e^{\pi/6}$
(c) $e^{\pi/12}$ (d) $e^{\pi/3}$

13/ If A and B are two events such that $P(A \cap B)=0.1$ and $P(A|B)$ and $P(B|A)$ are the roots of the equation $12x^2-7x+1=0$, then the value of $\frac{P(\bar{A}\cup\bar{B})}{P(\bar{A}\cap\bar{B})}$ is

(a) 5/3 (b) 7/4
(c) 4/3 (d) 9/4

14/ Let α, β, γ and δ be the coefficients of x^7, x^5, x^3 and x respectively in the expansion of $(x+\sqrt{x^3-1})^5+(x-\sqrt{x^3-1})^5$, $x>1$. If u and v satisfy the equations $\alpha u+\beta v=18$, $\gamma u+\delta v=20$, then $u+v$ equals

(a) 4 (b) 8 (c) 3 (d) 5

15/ Let $E: \frac{x^2}{a^2}+\frac{y^2}{b^2}=1$, $a>b$ and $H: \frac{x^2}{A^2}-\frac{y^2}{B^2}=1$. Let the distance between the foci of E and the foci of H be $2\sqrt{3}$. If $a-A=2$ and the ratio of the eccentricities of E and H is $\frac{1}{3}$, then the sum of the lengths of their latus rectums is equal to

(a) 8 (b) 10 (c) 7 (d) 9

16/ Let $P(4, 4\sqrt{3})$ be a point on the parabola $y^2=4ax$ and PQ be a focal chord of the parabola. If M and N are the foot of perpendiculars drawn from P and Q, respectively on the directrix of the parabola, then the area of the quadrilateral $PQMN$ is equal to

(a) $\frac{343\sqrt{3}}{8}$ (b) $17\sqrt{3}$ (c) $\frac{263\sqrt{3}}{8}$ (d) $\frac{34\sqrt{3}}{3}$

17/ The perpendicular distance of the line $\frac{x-1}{2}=\frac{y+2}{-1}=\frac{z+3}{2}$ from the point $P(2, -10, 1)$ is

(a) $5\sqrt{2}$ (b) $3\sqrt{5}$
(c) $4\sqrt{3}$ (d) 6

18/ The sum of all values of $\theta \in [0, 2\pi]$ satisfying $2\sin^2\theta=\cos 2\theta$ and $2\cos^2\theta=3\sin\theta$ is

(a) 4π (b) $\frac{5\pi}{6}$
(c) $\frac{\pi}{2}$ (d) π

19/ The area of the region enclosed by the curves $y=x^2-4x+4$ and $y^2=16-8x$ is

(a) 5 (b) 8
(c) 4/3 (d) 8/3

20/ In a group of 3 girls and 4 boys, there are two boys B_1 and B_2. The number of ways, in which these girls and boys can stand in a queue such that all the girls stand together, all the boys stand together but B_1 and B_2 are not adjacent to each other is

(a) 144 (b) 96
(c) 72 (d) 120

Section B

Numerical Type Questions

21/ Let $y = f(x)$ be the solution of the differential equation $\frac{dy}{dx} + \frac{xy}{x^2 - 1} = \frac{x^6 + 4x}{\sqrt{1 - x^2}}$, $-1 < x < 1$ such that $f(0) = 0$.

If $6 \int_{-\frac{1}{2}}^{1/2} f(x)dx = 2\pi - \alpha$, then α^2 is equal to ……

22/ Let the distance between two parallel lines be 5 units and a point P lie between the lines at a unit distance from one of them. An equilateral ΔPQR is formed such that Q lies on one of the parallel lines, while R lies on the other. Then, $(QR)^2$ is equal to ……… .

23/ If $\sum_{r=1}^{30} \frac{r^2 ({}^{30}C_r)^2}{{}^{30}C_{r-1}} = \alpha \times 2^{29}$, then α is equal to …… .

Imp.

24/ Let $A(6, 8)$, $B(10\cos\alpha, -10\sin\alpha)$ and $C(-10\sin\alpha, 10\cos\alpha)$ be the vertices of a triangle. If $L(a, 9)$ and $G(h, k)$ be its orthocenter and centroid respectively, then $(5a - 3h + 6k + 100\sin 2\alpha)$ is equal to ……… .

Orthocentre and Centroid of a Triangle

25/ Let $A = \{1, 2, 3\}$. The number of relations on A, containing (1, 2) and (2, 3), which are reflexive and transitive but not symmetric is ……… .

Physics

Section A

Objective Type Questions

26/ A rectangular metallic loop is moving out of a uniform magnetic field region to a field free region with a constant speed. When the loop is partially inside the magnetic field, the plot of magnitude of induced emf (ε) with time (t) is given by

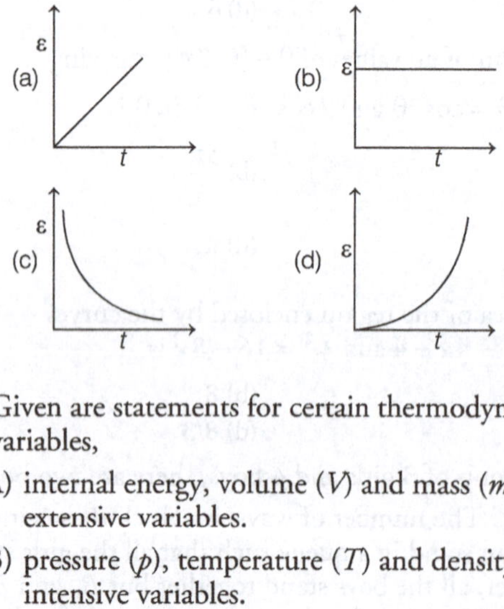

27/ Given are statements for certain thermodynamic variables,

(A) internal energy, volume (V) and mass (m) are extensive variables.

(B) pressure (p), temperature (T) and density (ρ) are intensive variables.

(C) volume (V), temperature (T) and density (ρ) are intensive variables.

(D) mass (m), temperature (T) and internal energy are extensive variables.

Choose the correct answer from the options given below.

(a) (C) and (D) Only
(b) (B) and (C) Only
(c) (D) and (A) Only
(d) (A) and (B) Only

28/ A small rigid spherical ball of mass M is dropped in a long vertical tube containing glycerine. The velocity of the ball becomes constant after some time. If the density of glycerine is half of the density of the ball, then the viscous force acting on the ball will be (consider g as acceleration due to gravity)

(a) $\frac{Mg}{2}$
(b) $\frac{3}{2}Mg$
(c) $2\,Mg$
(d) Mg

29/ A symmetric thin biconvex lens is cut into four equal parts by two planes AB and CD as shown in figure. If the power of original lens is 4D, then the power of a part of the divided lens is

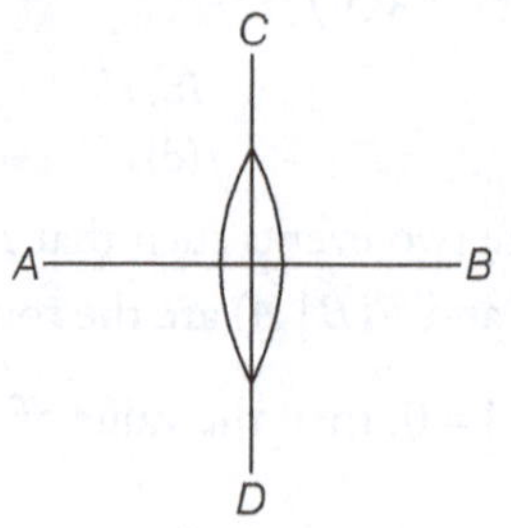

(a) D
(b) 2D
(c) 4D
(d) 8D

30/ A force $\mathbf{F} = 2\hat{\mathbf{i}} + b\hat{\mathbf{j}} + \hat{\mathbf{k}}$ is applied on a particle and it undergoes a displacement $\hat{\mathbf{i}} - 2\hat{\mathbf{j}} - \hat{\mathbf{k}}$. What will be the value of b, if work done on the particle is zero?

(a) 2 (b) 0 (c) 1/3 (d) 1/2

31/

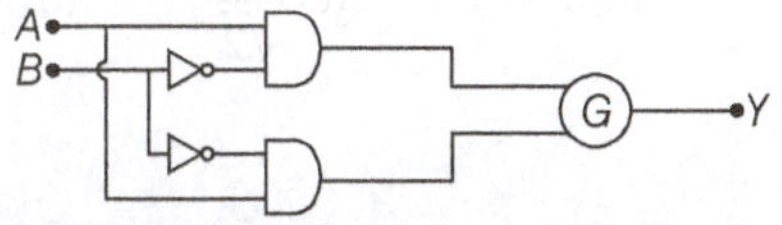

A	*B*	*Y*
0	0	1
0	1	0
1	0	0
1	1	1

To obtain the given truth table, following logic gate should be placed at *G*.

(a) AND gate (b) OR gate (c) NAND gate (d) NOR gate

32/ Given below are two statements. One is labelled as Assertion (A) and the other is labelled as Reason (R).

Assertion (A) In Young's double slit experiment, the fringes produced by red light are closer as compared to those produced by blue light.

Reason (R) The fringe width is directly proportional to the wavelength of light.

In the light of the above statements, choose the correct answer from the options given below.

(a) A is false but R is true.
(b) A is true but R is false.
(c) Both A and R are true and R is the correct explanation of A.
(d) Both A and R are true but R is not the correct explanation of A.

33/ A transparent film of refractive index, 2.0 is coated on a glass slab of refractive index, 1.45. What is the minimum thickness of transparent film to be coated for the maximum transmission of green light of wavelength 550 nm? [Assume that the light is incident nearly perpendicular to the glass surface.]

(a) 94.8 nm (b) 275 nm (c) 68.7 nm (d) 137.5 nm

34/

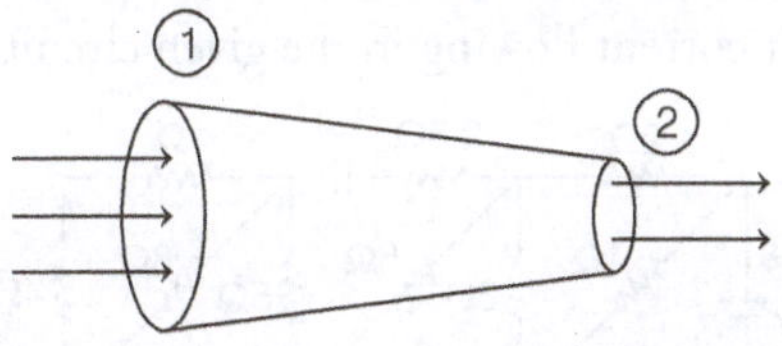

A tube of length *L* is shown in the figure. The radius of cross-section at the point 1 is 2 cm and at the point 2 is 1 cm, respectively. If the velocity of water entering at point 1 is 2 m/s, then velocity of water leaving the point 2 will be

(a) 4 m/s (b) 8 m/s (c) 6 m/s (d) 2 m/s

35/ The torque due to the force $(2\hat{\mathbf{i}} + \hat{\mathbf{j}} + 2\hat{\mathbf{k}})$ about the origin, acting on a particle whose position vector is $(\hat{\mathbf{i}} + \hat{\mathbf{j}} + \hat{\mathbf{k}})$, would be

(a) $\hat{\mathbf{i}} + \hat{\mathbf{k}}$ (b) $\hat{\mathbf{i}} - \hat{\mathbf{j}} + \hat{\mathbf{k}}$ (c) $\hat{\mathbf{j}} + \hat{\mathbf{k}}$ (d) $\hat{\mathbf{i}} - \hat{\mathbf{k}}$

36/ Given below are two statements. One is labelled as Assertion (A) and the other is labelled as Reason (R).

Assertion (A) A simple pendulum is taken to a planet of mass and radius, 4 times and 2 times, respectively, than the Earth. The time period of the pendulum remains same on Earth and the planet.

Reason (R) The mass of the pendulum remains unchanged at Earth and the other planet.

In the light of the above statements, choose the correct answer from the options given below.

(a) Both A and R are true but R is not the correct explanation of A.
(b) Both A and R are true and R is the correct explanation of A.
(c) A is false but R is true.
(d) A is true but R is false.

Imp.

37/ A light source of wavelength λ illuminates a metal surface and electrons are ejected with maximum kinetic energy of 2 eV. If the same surface is illuminated by a light source of wavelength λ/2, then the maximum kinetic energy of ejected electrons will be (the work function of metal is 1 eV)

(a) 3 eV (b) 6 eV (c) 2 eV (d) 5 eV

Einstein Equation of Photoelectric Effect

38/ For a diatomic gas, if $\gamma_1 = \left(\frac{C_p}{C_V}\right)$ for rigid molecules and $\gamma_2 = \left(\frac{C_p}{C_V}\right)$ for another diatomic molecules, but also having vibrational modes. Then, which one of the following options is correct? (C_p and C_V are specific heats of the gas at constant pressure and volume)

(a) $\gamma_2 = \gamma_1$ (b) $2\gamma_2 = \gamma_1$ (c) $\gamma_2 < \gamma_1$ (d) $\gamma_2 > \gamma_1$

39/ An electron projected perpendicular to a uniform magnetic field *B* moves in a circle. If Bohr's quantisation is applicable, then the radius of the electronic orbit in the first excited state is

(a) $\sqrt{\frac{h}{2\pi eB}}$ (b) $\sqrt{\frac{2h}{\pi eB}}$ (c) $\sqrt{\frac{h}{\pi eB}}$ (d) $\sqrt{\frac{4h}{\pi eB}}$

40/ For a short dipole at origin O, the dipole moment p is along X-axis, as shown in the figure. If the electric potential and electric field at A are V_0 and E_0, respectively, then the correct combination of the electric potential and electric field, respectively, at point B on the Y-axis is given by

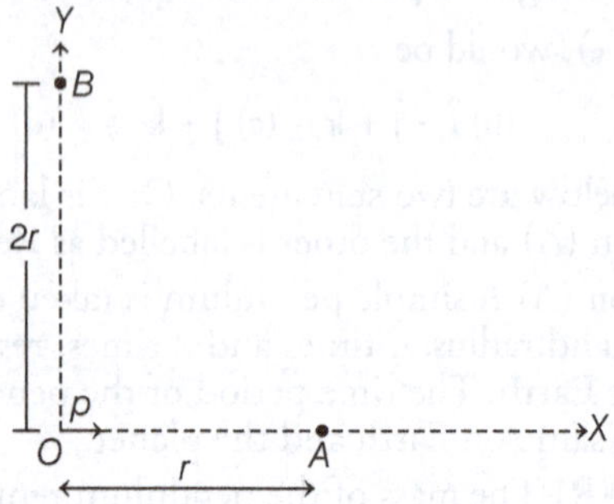

(a) $\frac{V_0}{2}$ and $\frac{E_0}{16}$ (b) zero and $\frac{E_0}{16}$

(c) zero and $\frac{E_0}{8}$ (d) V_0 and $\frac{E_0}{4}$

41/ The maximum percentage error in the measurement of density of a wire is

[Given, mass of wire = (0.60 ± 0.003) g
radius of wire = (0.50 ± 0.01) cm
length of wire = (10.00 ± 0.05) cm]

(a) 5 (b) 8
(c) 4 (d) 7

42/ A body of mass 100 g is moving in circular path of radius 2 m on vertical plane as shown in figure. The velocity of the body at point A is 10 m/s. The ratio of its kinetic energies at point B and C is

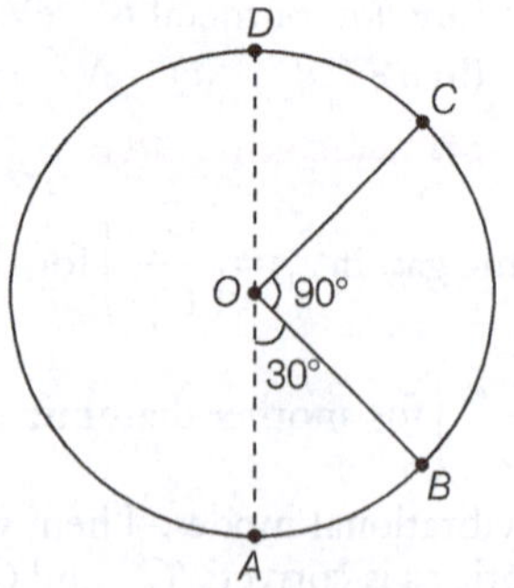

(Take acceleration due to gravity as 10 m/s^2)

(a) $\frac{2+\sqrt{2}}{3}$ (b) $\frac{3+\sqrt{3}}{2}$

(c) $\frac{2+\sqrt{3}}{3}$ (d) $\frac{3-\sqrt{2}}{2}$

43/ A series L-C-R circuit is connected to an alternating source of emf E. The current amplitude at resonant frequency is I_0. If the value of resistance R becomes twice of its initial value, then amplitude of current at resonance will be

(a) $2I_0$ (b) I_0 (c) $\frac{I_0}{\sqrt{2}}$ (d) $\frac{I_0}{2}$

Imp.

44/ A ball of mass 100 g is projected with velocity 20 m/s at 60° with horizontal. The decrease in kinetic energy of the ball during the motion from point of projection to highest point is

(a) 20 J (b) 5 J
(c) zero (d) 15 J

Work Energy Theorem

45/ Which one of the following is the correct dimensional formula for the capacitance in F ? M, L, T and C stand for unit of mass, length, time and charge.

(a) $[F] = [C^2M^{-1}L^{-2}T^2]$
(b) $[F] = [CM^{-2}L^{-2}T^{-2}]$
(c) $[F] = [CM^{-1}L^{-2}T^2]$
(d) $[F] = [C^2M^{-2}L^2T^2]$

Section B

Numerical Type Questions

46/ A tube of length 1 m is filled completely with an ideal liquid of mass $2M$ and closed at both ends. The tube is rotated uniformly in horizontal plane about one of its ends. If the force exerted by the liquid at the other end is F, then angular velocity of the tube is $\sqrt{F/\alpha M}$ in SI unit. The value of α is ......

47/ A proton is moving undeflected in a region of crossed electric and magnetic fields at a constant speed of 2×10^5 ms^{-1}. When the electric field is switched off, the proton moves along a circular path of radius 2 cm. The magnitude of electric field is $x \times 10^4$ N/C. The value of x is ....
Take the mass of the proton = 1.6×10^{-27} kg.

48/ The net current flowing in the given circuit is ...... A.

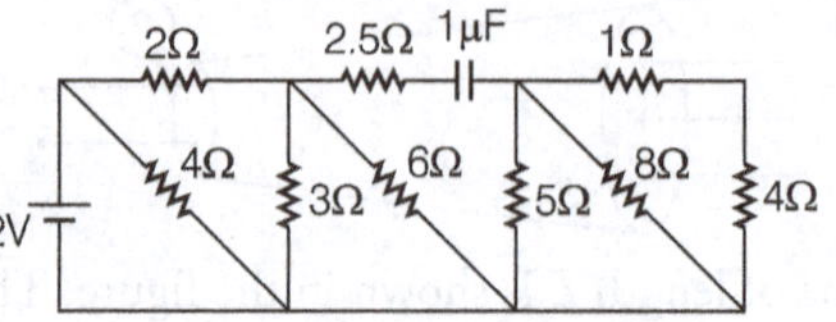

49/ Two long parallel wires X and Y, separated by a distance of 6 cm, carry currents of 5 A and 4 A, respectively, in opposite directions as shown in the figure. Magnitude of the resultant magnetic field at point P at a distance of 4 cm from wire Y is $x \times 10^{-5}$ T.

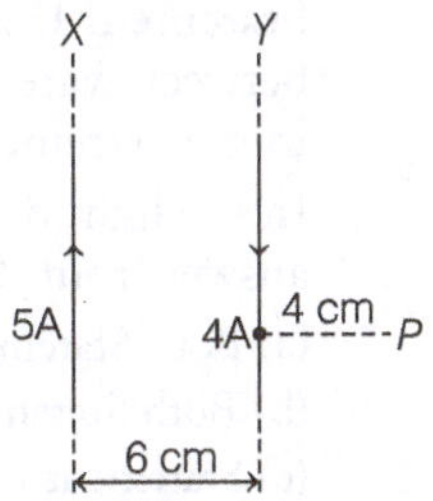

The value of x is .... Take permeability of free space as $\mu_0 = 4\pi \times 10^{-7}$ SI units.

50/ A parallel plate capacitor of area $A = 16\ \text{cm}^2$ and separation between the plates 10 cm, is charged by a DC current. Consider a hypothetical plane surface of area $A_0 = 3.2\ \text{cm}^2$ inside the capacitor and parallel to the plates. At an instant, the current through the circuit is 6A. At the same instant the displacement current through A_0 is .... mA.

Chemistry

Section A

Objective Type Questions

51/ Given below are two statements.

Statement I Nitrogen, sulphur, halogen and phosphorus present in an organic compound are detected by Lassaigne's test.

Statement II The elements present in the compound are converted from covalent form into ionic form by fusing the compound with magnesium in Lassaigne's test.

In the light of the above statements, choose the correct answer from the options given below:

(a) Statement I is false but Statement II is true.
(b) Both Statement I and Statement II are true.
(c) Statement I is true but Statement II is false.
(d) Both Statement I and Statement II are false.

52/ Consider the given figure and choose the correct option

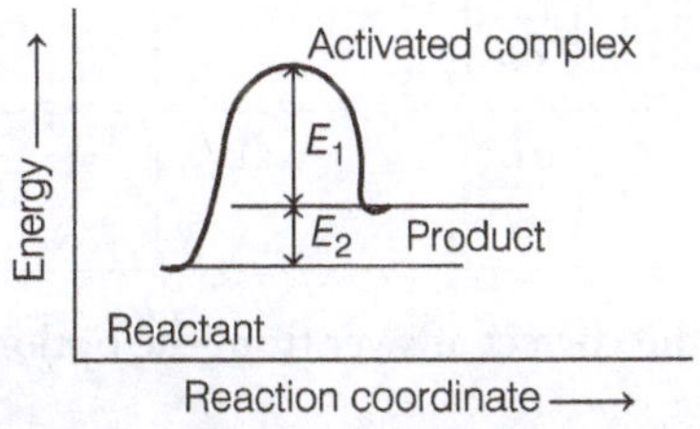

(a) Activation energy of forward reaction is $E_1 + E_2$ and product is less stable than reactant.
(b) Activation energy of both forward and backward reaction is $E_1 + E_2$ and reactant is more stable than product.
(c) Activation energy of forward reaction is $E_1 + E_2$ and product is more stable than reactant.
(d) Activation energy of backward reaction is E_1 and product is more stable than reactant.

53/ The molar solubility(S) of zirconium phosphate with molecular formula $(Zr^{4+})_3\ (PO_4^{3-})_4$ is given by relation

(a) $\left(\dfrac{K_{sp}}{6912}\right)^{\frac{1}{7}}$ (b) $\left(\dfrac{K_{sp}}{9612}\right)^{\frac{1}{3}}$

(c) $\left(\dfrac{K_{sp}}{8435}\right)^{\frac{1}{7}}$ (d) $\left(\dfrac{K_{sp}}{5348}\right)^{\frac{1}{6}}$

54/ Identify the homoleptic complex(es) that is/are low spin.

(A) $[Fe(CN)_5NO]^{2-}$ (B) $[CoF_6]^{3-}$
(C) $[Fe(CN)_6]^{4-}$ (D) $[Co(NH_3)_6]^{3+}$
(E) $[Cr(H_2O)_6]^{2+}$

Choose the correct answer from the options given below:

(a) (C) and (D) only (b) (A) and (C) only
(c) (C) only (d) (B) and (E) only

55/ Identify the number of structure/s from the following which can be correlated to D-glyceraldehyde.

(A) 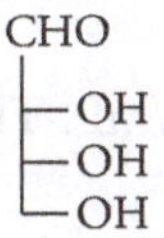

(B) 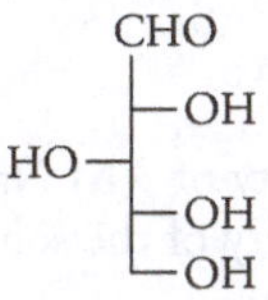

(C) 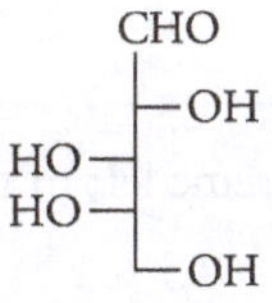

(D) 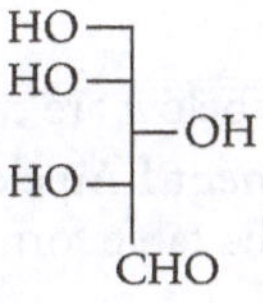

(a) four (b) one
(c) two (d) three

56/ The maximum covalency of a non-metallic group 15 element 'E' with weakest E—E bond is
(a) 4 (b) 3 (c) 6 (d) 5

57/ The alkane from below having two secondary hydrogens is
(a) 4-ethyl-3,4-dimethyloctane
(b) 2,2,3,3-tetramethylpentane
(c) 2,2,4,4-tetramethylhexane
(d) 2,2,4,5-tetramethylheptane

58/ Match the compounds (List-I) with the appropriate Catalyst/Reagents (List-II) for their reduction into corresponding amines.

	List-I (Compounds)		List-II (Catalyst/Reagents)
(A)	$R-\overset{\overset{O}{\|}}{C}-NH_2$	(I)	NaOH (aqueous)
(B)	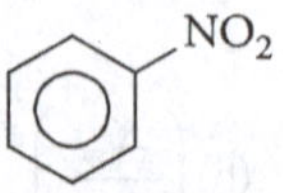	(II)	H_2 / Ni
(C)	$R-C\equiv N$	(III)	$LiAlH_4$, H_2O
(D)	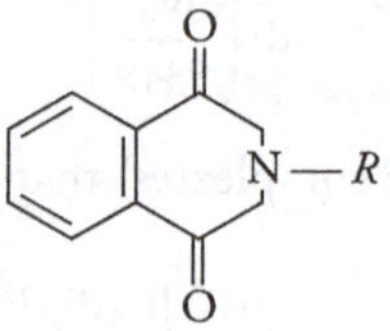	(IV)	Sn, HCl

Choose the correct answer from the options given below:
(a) (A)-(II), (B)-(I), (C)-(III), (D)-(IV)
(b) (A)-(III), (B)-(II), (C)-(IV), (D)-(I)
(c) (A)-(III), (B)-(IV), (C)-(II), (D)-(I)
(d) (A)-(II), (B)-(IV), (C)-(III), (D)-(I)

59/ The species which does not undergo disproportionation reaction is
(a) ClO_3^- (b) ClO_2^-
(c) ClO^- (d) ClO_4^-

60/ Density of 3 M NaCl solution is 1.25 g/mL. The molality of the solution is
(a) 2.79 m (b) 1.79 m
(c) 3 m (d) 2 m

61/ Given below are two statements.
Statement I An element in the extreme left of the periodic table forms acidic oxides.
Statement II Acid is formed during the reaction between water and oxide of a reactive element present in the extreme right of the periodic table.
In the light of the above statements, choose the correct answer from the options given below:
(a) Both Statement I and Statement II are true.
(b) Both Statement I and Statement II are false.
(c) Statement I is false but Statement II is true.
(d) Statement I is true but Statement II is false.

62/ When *sec*-butylcyclohexane reacts with bromine in the presence of sunlight, the major product is

(a) 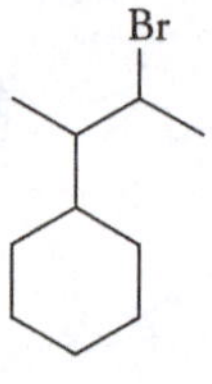

(b) 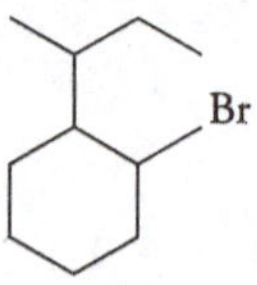

(c) 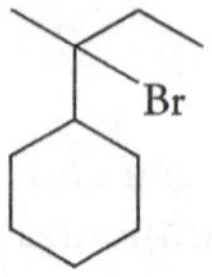

(d) 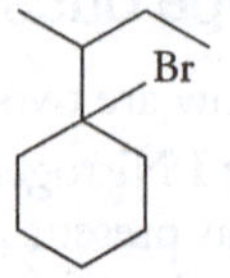

Imp.

63/ Match List-I with List-II.

	List-I (Partial derivatives)		List-II (Thermodynamic quantity)
(A)	$\left(\frac{\partial G}{\partial T}\right)_p$	(I)	C_p
(B)	$\left(\frac{\partial H}{\partial T}\right)_p$	(II)	$-S$
(C)	$\left(\frac{\partial G}{\partial p}\right)_T$	(III)	C_V
(D)	$\left(\frac{\partial U}{\partial T}\right)_V$	(IV)	V

Choose the correct answer from the options given below
(a) (A)-(I), (B)-(II), (C)-(IV), (D)-(III)
(b) (A)-(II), (B)-(III), (C)-(I), (D)-(IV)
(c) (A)-(II), (B)-(I), (C)-(IV), (D)-(III)
(d) (A)-(II), (B)-(I), (C)-(III), (D)-(IV)

Partial Derivative of Thermodynamic Quantity

64/ The most stable carbocation from the following is

(a) 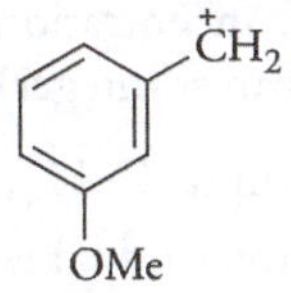

(b) 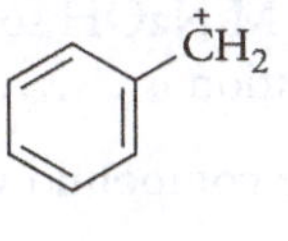

(c) 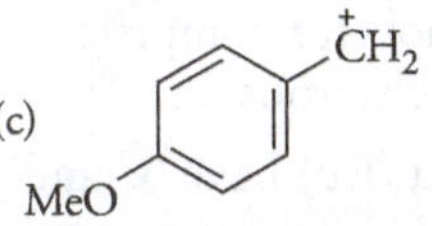

(d) 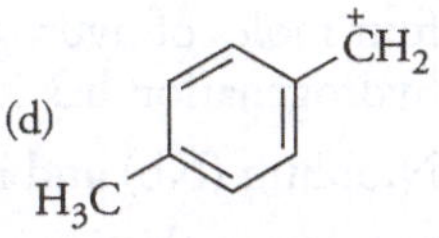

65/ Given below are two statements:

Statement I Corrosion is an electrochemical phenomenon in which pure metal acts as an anode and impure metal as a cathode.

Statement II The rate of corrosion is more in alkaline medium than in acidic medium.

In the light of the above statements, choose the correct answer from the options given below:

(a) Statement I is true but Statement II is false.
(b) Both Statement I and Statement II are true.
(c) Statement I is false but Statement II is true.
(d) Both Statement I and Statement II are false.

66/

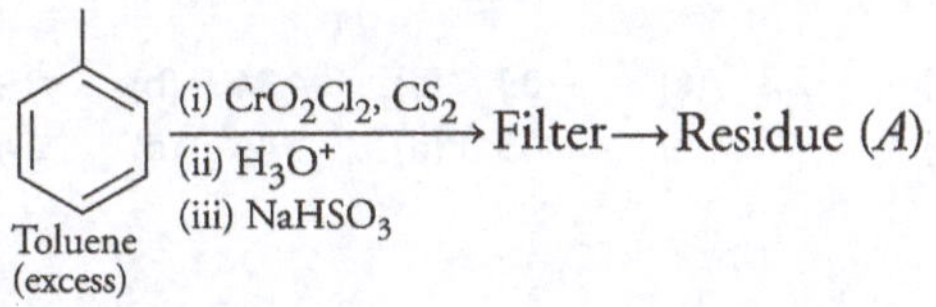

Residue (*A*) + HCl(dil.) $\longrightarrow$ Compound (*B*)

Structure of residue (*A*) and compound (*B*) formed respectively is

	[*A*]	[*B*]
(a)	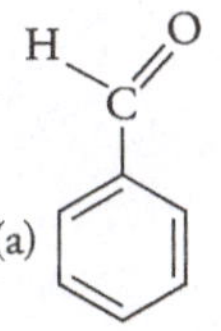	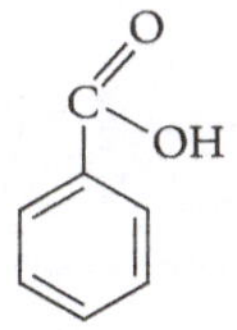
(b)	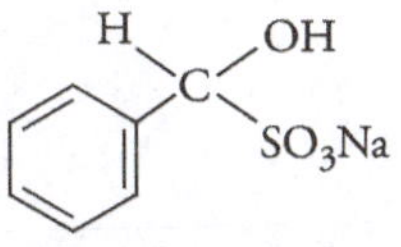	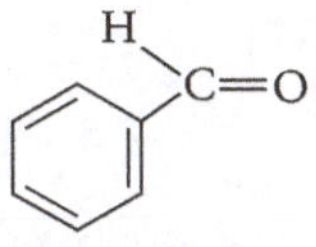
(c)	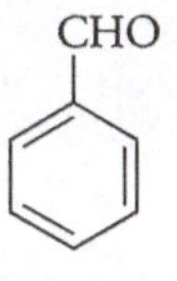	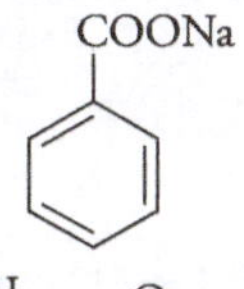
(d)	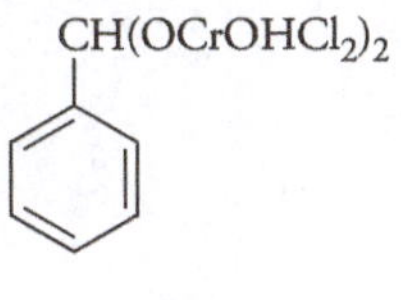	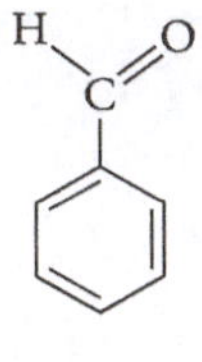

Imp.

67/ The correct order of the following complexes in terms of their crystal field stabilisation energies is

(a) $[Co(NH_3)_4]^{2+} < [Co(NH_3)_6]^{2+} < [Co(NH_3)_6]^{3+} < [Co(en)_3]^{3+}$
(b) $[Co(en)_3]^{3+} < [Co(NH_3)_6]^{3+} < [Co(NH_3)_6]^{2+} < [Co(NH_3)_4]^{2+}$
(c) $[Co(NH_3)_6]^{2+} < [Co(NH_3)_6]^{3+} < [Co(NH_3)_4]^{2+} < [Co(en)_3]^{3+}$
(d) $[Co(NH_3)_4]^{2+} < [Co(NH_3)_6]^{2+} < [Co(en)_3]^{3+} < [Co(NH_3)_6]^{3+}$

CFSE of Coordination Compounds

68/ Given below are two statements.

Statement I A spectral line will be observed for a $2p_x \rightarrow 2p_y$ transition.

Statement II $2p_x$ and $2p_y$ are degenerate orbitals.

In the light of the above statements, choose the correct answer from the options given below:

(a) Both Statement I and Statement II are true.
(b) Statement I is false but Statement II is true.
(c) Both Statement I and Statement II are false.
(d) Statement I is true but Statement II is false.

69/ Arrange the following compounds in increasing order of their dipole moment

HBr, H_2S, NF_3 and $CHCl_3$

(a) $NF_3 < HBr < H_2S < CHCl_3$
(b) $CHCl_3 < NF_3 < HBr < H_2S$
(c) $H_2S < HBr < NF_3 < CHCl_3$
(d) $HBr < H_2S < NF_3 < CHCl_3$

70/

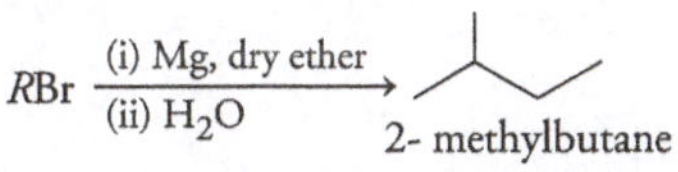

The maximum number of *R*Br producing 2-methylbutane by above sequence of reactions is .... (Consider the structural isomers only)

(a) 5 (b) 3
(c) 4 (d) 1

Section B

Numerical Type Questions

71/ The complex of Ni^{2+} ion and dimethyl glyoxime contains .... number of hydrogen (H) atoms.

72/ Consider the following cases of standard enthalpy of reaction (ΔH_r° in kJ mol^{-1})

$$C_2H_6(g) + \frac{7}{2}O_2(g) \longrightarrow 2CO_2(g) + 3H_2O(l); \Delta H_1^\circ = -1550$$

$$C\,(graphite) + O_2(g) \longrightarrow CO_2(g); \Delta H_2^\circ = -393.5$$

$$H_2(g) + \frac{1}{2}O_2(g) \longrightarrow H_2O(l); \Delta H_3^\circ = -286$$

The magnitude of $\Delta H_f^\circ(C_2H_6(g))$ is .... kJ mol^{-1} (Nearest integer).

73/ 20 mL of 2 M NaOH solution is added to 400 mL of 0.5 M NaOH solution. The final concentration of the solution is .... $\times 10^{-2}$ M. (Nearest integer)

74/ The compound with molecular formula C_6H_6, which gives only one monobromo derivative and takes up four moles of hydrogen per mole for complete hydrogenation has .... π-electrons.

75/ Niobium (Nb) and ruthenium (Ru) have 'x' and 'y' number of electrons in their respective $4d$-orbitals. The value of $x + y$ is ....

Answer Key

Mathematics

1 (d)	2 (c)	3 (a)	4 (b)	5 (b)	6 (a)	7 (b)	8 (c)	9 (c)	10 (b)
11 (a)	12 (b)	13 (d)	14 (d)	15 (a)	16 (a)	17 (b)	18 (d)	19 (d)	20 (a)
21 (27)	22 (28)	23 (465)	24 (145)	25 (3)					

Physics

26 (b)	27 (d)	28 (a)	29 (b)	30 (d)	31 (*)	32 (a)	33 (d)	34 (b)	35 (d)
36 (a)	37 (d)	38 (c)	39 (c)	40 (b)	41 (a)	42 (b)	43 (d)	44 (d)	45 (a)
46 (1)	47 (2)	48 (1)	49 (1)	50 (1200)					

Chemistry

51 (c)	52 (a)	53 (a)	54 (a)	55 (d)	56 (a)	57 (b)	58 (c)	59 (d)	60 (a)
61 (c)	62 (a)	63 (c)	64 (c)	65 (a)	66 (b)	67 (a)	68 (b)	69 (a)	70 (c)
71 (14)	72 (95)	73 (57)	74 (8)	75 (11)					

***Note** (*) None of the option is correct.*

Detailed Explanations

22 January
Shift-II

Mathematics

1. *(d)*

> **Key idea**
> For AP $S_n = \frac{n}{2}[2a+(n-1)d]$
> $a_n = a+(n-1)d$

Let AP $a_1, a_2, a_3, \ldots, a_{2k}$

Given, $\sum_{r=1}^{k} a_{2r-1} = 40$

and $\sum_{r=1}^{k} a_{2r} = 55$

$$\frac{k}{2}[2a_1+(k-1)\,2d] = 40 \quad \ldots\text{(i)}$$

and $$\frac{k}{2}[2a_2+(k-1)\,2d] = 55 \quad \ldots\text{(ii)}$$

Dividing Eq. (i) by Eq. (ii)

$$\frac{2a_1+(k-1)\,2d}{2a_2+(k-1)\,2d} = \frac{40}{55}$$

$$\Rightarrow \quad \frac{a_1+(k-1)\,d}{a_1+d+(k-1)\,d} = \frac{8}{11}$$

$$\Rightarrow \quad 11a_1 + 11(k-1)\,d = 8a_1 + 8d + 8(k-1)\,d$$

$$3a_1 = 8d - 3(k-1)\,d \quad \ldots\text{(iii)}$$

$$\because \quad a_{2k} = a_1 + 27$$

$$a_1 + (2k-1)d = a_1 + 27$$

$$d = \frac{27}{2k-1} \quad \ldots\text{(iv)}$$

From Eq. (i), we get

$$a_1 + (k-1)\,d = \frac{40}{k}$$

$$\Rightarrow \quad a_1 = \frac{40}{k} - (k-1)\,d$$

From Eq. (ii), we get

$$a_2 + (k-1)\,d = \frac{55}{k}$$

$$\Rightarrow \quad a_1 = \frac{55}{k} - (k-1)\,d - d$$

We get,

$$\frac{40}{k} - (k-1)\,d = \frac{55}{k} - (k-1)\,d - d$$

$$\Rightarrow \quad d = \frac{15}{k}$$

$$\Rightarrow \quad \frac{27}{2k-1} = \frac{15}{k} \Rightarrow 27k = 30k - 15$$

$$\Rightarrow \quad 3k = 15$$

$$\therefore \quad k = 5$$

2. *(c)* $f(x) = \int_0^{x^2} \frac{t^2-8t+15}{e^t}\,dt$

$$f'(x) = \frac{2x\,(x^4 - 8x^2 + 15)}{e^{x^2}}$$

For extreme points

$$f'(x) = 0$$

$$2x\,(x^2-5)\,(x^2-3) = 0$$

$$x = 0, \pm\sqrt{5}, \pm\sqrt{3}$$

Sign. of $f'(x)$

$$\underset{}{-}\;\; -\sqrt{5} \;\; + \;\; -\sqrt{3} \;\; - \;\; 0 \;\; + \;\; \sqrt{3} \;\; - \;\; \sqrt{5} \;\; +$$

$\therefore$ Local maximum at $x = -\sqrt{3}, \sqrt{3}$

Local minimum at $x = -\sqrt{5}, 0, \sqrt{5}$

> **Concept Enhancer**
> $f(x) = \int_{g(x)}^{h(x)} F(t)dt$
> $\Rightarrow f'(x) = h'(x)F(h(x)) - g'(x)F(g(x))$

3. *(a)* $\alpha = \lim_{x\to\infty}\left(\left(\frac{e}{1-e}\right)\left(\frac{1}{e} - \frac{x}{1+x}\right)\right)^x$

(1^∞ form)

$$= e^{\lim_{x\to\infty}\left[\left(\frac{e}{1-e}\right)\left(\frac{1}{e}-\frac{x}{1+x}\right)-1\right]x}$$

$$= e^{\lim_{x\to\infty} x\left(\frac{e}{1-e}\right)\left(\frac{1}{e}-\frac{x}{1+x}-\frac{1-e}{e}\right)}$$

$$= e^{\lim_{x\to\infty} x\left(\frac{e}{1-e}\right)\left(1-\frac{x}{1+x}\right)}$$

$$= e^{\lim_{x\to\infty} x\left(\frac{e}{1-e}\right)\left(\frac{1}{1+x}\right)}$$

$$= e^{\lim_{x\to\infty} \frac{e}{1-e}\left(\frac{1}{1+1/x}\right)}$$

$$\alpha = e^{\frac{e}{1-e}}$$

$$\Rightarrow \quad \log_e \alpha = \frac{e}{1-e}$$

$$\therefore \quad \frac{\log_e \alpha}{1+\log_e \alpha} = \frac{\frac{e}{1-e}}{1+\frac{e}{1-e}} = \frac{e}{1-e+e} = e$$

4. *(b)*

$$2x^2 + (\cos\theta)\,x - 1 = 0 \begin{array}{l}\nearrow \alpha_\theta \\ \searrow \beta_\theta\end{array} \quad \ldots\text{(i)}$$

$$\theta \in (0, 2\pi)$$

α_θ and β_θ be roots of Eq. (i), we get

$$\alpha_\theta + \beta_\theta = -\frac{\cos\theta}{2}$$

$$\alpha_\theta\,\beta_\theta = -\frac{1}{2}$$

$$\therefore \alpha_\theta^4 + \beta_\theta^4 = (\alpha_\theta^2 + \beta_\theta^2)^2 - 2\alpha_\theta^2\beta_\theta^2$$

$$= ((\alpha_\theta + \beta_\theta)^2 - 2\alpha_\theta\beta_\theta)^2 - 2\left(-\frac{1}{2}\right)^2$$

$$= \left(\frac{\cos^2\theta}{4} + 1\right)^2 - \frac{1}{2}$$

$$\therefore \quad m = 1 - \frac{1}{2} = \frac{1}{2}$$

$$M = \frac{25}{16} - \frac{1}{2}$$

$$\therefore \quad 16\,(M+m) = 25$$

5. *(b)* $|A| = \frac{1}{2}\,tr(A) = 3, n = 3$

$$B = \text{adj}\,(\text{adj}\,2A)$$

$$= |2A|^{n-2}\,(2A)$$

$$= |2A|(2A) = 2^3\,|A| \cdot 2A$$

$$= 2^4 \times \frac{1}{2} \times A = 2^3\,A$$

$$|B| = |2^3\,A| = (2^3)^3\,|A| = 2^8 = 256$$

$$trace\,(B) = 2^3\;trace\,(A) = 2^3 \times 3 = 24$$

$$\therefore \; |B| + trace\,(B) = 256 + 24 = 280$$

6. *(a)* $I = \int e^x \left(\frac{x\sin^{-1}x}{\sqrt{1-x^2}} + \frac{\sin^{-1}x}{(1-x^2)^{3/2}} + \frac{x}{1-x^2}\right)dx$

Let $f(x) = \dfrac{x \sin^{-1} x}{\sqrt{1-x^2}}$

Differentiating w.r.t. x, we get

$$f'(x) = \frac{\sqrt{1-x^2}\left(x \times \dfrac{1}{\sqrt{1-x^2}} + \sin^{-1} x\right) - x \sin^{-1} x \dfrac{1}{2\sqrt{1-x^2}}(-2x)}{(1-x^2)}$$

$$= \frac{x}{1-x^2} + \frac{\sin^{-1} x}{(1-x^2)^{3/2}}$$

$\therefore \quad I = \int e^x (f(x) + f'(x))\, dx$

$$= e^x f(x) + c$$

$\therefore \quad I = e^x \left(\dfrac{x \sin^{-1} x}{\sqrt{1-x^2}}\right) + c$

$$g(x) = e^x \left(\frac{x \sin^{-1} x}{\sqrt{1-x^2}}\right)$$

$$g\left(\frac{1}{2}\right) = \frac{e^{1/2} \dfrac{1}{2}\left(\dfrac{\pi}{6}\right)}{\dfrac{\sqrt{3}}{2}} = \frac{\pi}{6}\sqrt{\frac{e}{3}}$$

Concept Enhancer

$\int e^x (f(x) + f'(x)) dx = e^x f(x) + C$

7. *(b)* $\mathbf{PQ} \,||\, 3\hat{\mathbf{i}} + 2\hat{\mathbf{j}} + 2\hat{\mathbf{k}}$

Equation of line PQ

$$\frac{x+2}{3} = \frac{y+1}{2} = \frac{z-3}{2} = \lambda \quad \text{(say)}$$

Let $Q\,(3\lambda - 2,\, 2\lambda - 1,\, 2\lambda + 3)$

$R\,(1, 3, 3)$

$\because \quad QR = 5$

$\Rightarrow \quad QR^2 = 25$

$\Rightarrow (3\lambda - 3)^2 + (2\lambda - 4)^2 + (2\lambda)^2 = 25$

$\Rightarrow \quad 17\lambda^2 - 34\lambda = 0$

$17\lambda(\lambda - 2) = 0$

$\lambda = 0, 2$

Area of $PQR = \dfrac{1}{2}|\mathbf{PQ} \times \mathbf{QR}|$

$$A = \frac{1}{2}\begin{vmatrix} \hat{\mathbf{i}} & \hat{\mathbf{j}} & \hat{\mathbf{k}} \\ 6 & 4 & 4 \\ 3 & 4 & 0 \end{vmatrix}$$

$$A = \sqrt{64 + 36 + 36} = \sqrt{136}$$

$$A^2 = 136$$

8. *(c)*

Key idea
If two vectors **a** and **b** are perpendicular to each other than $\mathbf{a} \cdot \mathbf{b} = 0$

$$|\mathbf{a}| = |\mathbf{b}| = 1 \quad \theta = \frac{\pi}{3}$$

$$(\lambda\mathbf{a} + 2\mathbf{b}) \perp^r 3\mathbf{a} - \lambda\mathbf{b}$$

$\therefore \quad (\lambda\mathbf{a} + 2\mathbf{b}) \cdot (3\mathbf{a} - \lambda\mathbf{b}) = 0$

$3\lambda|a|^2 - \lambda^2 \mathbf{a} \cdot \mathbf{b} + 6\mathbf{b} \cdot \mathbf{a} - 2\lambda|\mathbf{b}|^2 = 0$

$\Rightarrow \; 3\lambda - \lambda^2 (\cos\theta) + 6\cos\theta - 2\lambda = 0$

$\Rightarrow \; \lambda - \lambda^2\left(\cos\dfrac{\pi}{3}\right) + 6\cos\dfrac{\pi}{3} = 0$

$\Rightarrow \lambda - \dfrac{\lambda^2}{2} + 3 = 0 \Rightarrow \lambda^2 - 2\lambda - 6 = 0$

$$\lambda = 1 \pm \sqrt{7}$$

$\therefore$ Number of values of λ in $[-1, 3]$ is 0.

9. *(c)* $x + y + 2z = 6$

$$2x + 3y + az = a + 1$$
$$-x - 3y + bz = 2b$$

For infinite many solution

$$\Delta = \Delta_x = \Delta_y = \Delta_z = 0$$

$$\Delta = \begin{vmatrix} 1 & 1 & 2 \\ 2 & 3 & a \\ -1 & -3 & b \end{vmatrix} = 0$$

$\Rightarrow \quad 1(3b + 3a) - 1(2b + a) - 2(-6 + 3) = 0$

$\Rightarrow \quad 2a + b - 6 = 0 \qquad \ldots\text{(i)}$

$$\Delta_z = \begin{vmatrix} 1 & 1 & 6 \\ 2 & 3 & a+1 \\ -1 & -3 & 2b \end{vmatrix} = 0$$

$= 1(6b + 3a + 3) - 1(4b + a + 1) + 6(-6 + 3) = 0$

$\Rightarrow \quad a + b - 8 = 0 \qquad \ldots\text{(ii)}$

Solving Eqs. (i) and (ii), we get

$$a = -2 \text{ and } b = 10$$

$$7a + 3b = 7(-2) + 3(10) = 16$$

10. *(b)* Let $z = x + iy$

$$z(1 + i) + \bar{z}(1 - i) = 4, \quad z \in C$$

$\Rightarrow (x + iy)(1 + i) + (x - iy)(1 - i) = 4$

$\Rightarrow x + ix + iy - y + x - ix - iy - y = 4$

$\Rightarrow \quad x - y = 2$

and $\quad |z - 3| \le 1$

$\Rightarrow \quad (x - 3)^2 + y^2 \le 1$

Solving $x - y = 2$ and $(x - 3)^2 + y^2 = 1$

we get $x = 2$ or 3

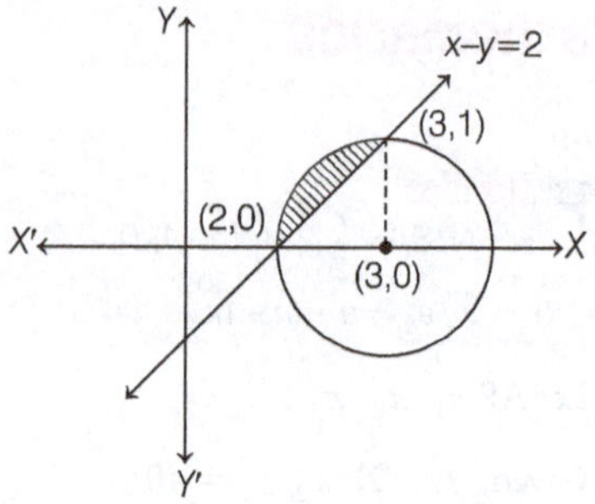

Shaded area $= \pi(1)^2\left(\dfrac{90^\circ}{360^\circ}\right) - \dfrac{1}{2} \times 1 \times 1$

$$\alpha = \frac{\pi}{4} - \frac{1}{2}$$

Unshaded area

$$= \beta = \pi - \left(\frac{\pi}{4} - \frac{1}{2}\right) = \frac{3\pi}{4} + \frac{1}{2}$$

$$|\alpha - \beta| = \frac{\pi}{2} + 1$$

11. *(a)* Total function $= 4^4 = 256$

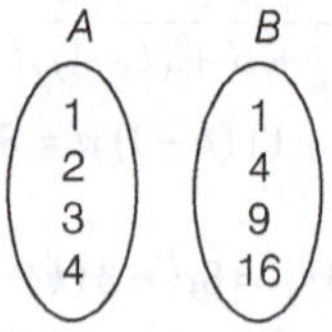

Total possible one-one function

$$= 4! = 24$$

$\therefore$ Many one function $= 256 - 24 = 232$

Many one function where $1 \notin f(A)$

$$= 3 \cdot 3 \cdot 3 \cdot 3 = 81$$

$\therefore$ Required number of many one function

$$= 232 - 81 = 151$$

12. *(b)* $(1 + y^2) + (x - 2e^{\tan^{-1} y})\dfrac{dy}{dx} = 0$

$\Rightarrow \quad (1 + y^2)\dfrac{dx}{dy} + x = 2e^{\tan^{-1} y}$

$\Rightarrow \quad \dfrac{dx}{dy} + \dfrac{x}{1 + y^2} = \dfrac{2e^{\tan^{-1} y}}{1 + y^2}$

$$P = \frac{1}{1 + y^2}, \; Q = \frac{2e^{\tan^{-1} y}}{1 + y^2}$$

If $= e^{\int P\, dy} = e^{\int \frac{1}{1+y^2} dy} = e^{\tan^{-1} y}$

Solution of differential equation

$$x\,(\text{IF}) = \int Q\cdot(\text{IF})\,dy + C$$

$$\Rightarrow xe^{\tan^{-1} y} = 2\int \frac{(e^{\tan^{-1} y})^2}{1+y^2}\,dy + C$$

Put $e^{\tan^{-1} y} = t \Rightarrow e^{\tan^{-1} y}\dfrac{1}{1+y^2}\,dy = dt$

$$\Rightarrow \quad xe^{\tan^{-1} y} = 2\int t\,dt + C$$

$$\Rightarrow \quad xe^{\tan^{-1} y} = t^2 + C$$

$$\Rightarrow \quad xe^{\tan^{-1} y} = e^{2\tan^{-1} y} + C$$

$$\Rightarrow \quad x = e^{\tan^{-1} y} + Ce^{-\tan^{-1} y}$$

$$x = 1,\ y = 0$$

$$1 = e^0 + Ce^0$$

$$C = 0$$

$$\therefore \quad x = e^{\tan^{-1} y}$$

$$\text{at } y = \frac{1}{\sqrt{3}},\ x = e^{\tan^{-1}\frac{1}{\sqrt{3}}}$$

$$= e^{\pi/6}$$

13. *(d)* Given, $P(A \cap B) = 0.1$

$$12x^2 - 7x + 1 = 0$$

$$\Rightarrow \quad (3x-1)(4x-1) = 0$$

$$x = \frac{1}{3},\ x = \frac{1}{4}$$

$$P\left(\frac{A}{B}\right) = \frac{1}{3}, \qquad P\left(\frac{B}{A}\right) = \frac{1}{4}$$

$$\frac{P(A\cap B)}{P(B)} = \frac{1}{3} \quad \text{or} \quad \frac{P(B\cap A)}{P(A)} = \frac{1}{4}$$

$$P(B) = 0.3 \quad \text{or} \quad P(A) = 0.4$$

Now, $P(\bar{A}\cup\bar{B}) = P(\overline{A\cap B})$

$$= 1 - P(A\cap B)$$

$$= 1 - 0.1 = 0.9$$

and $\quad P(\bar{A}\cap\bar{B}) = P(\overline{A\cup B})$

$$= 1 - P(A\cup B)$$

$$= 1 - [P(A) + P(B) - P(A\cap B)]$$

$$= 1 - (0.3 + 0.4 - 0.1) = 0.4$$

$$\therefore \quad \frac{P(\bar{A}\cup\bar{B})}{P(\bar{A}\cap\bar{B})} = \frac{0.9}{0.4} = \frac{9}{4}$$

14. *(d)* $\because (a+b)^n + (a-b)^n$

$$= 2(\text{sum of all odd terms})$$

$$\therefore \ (x + \sqrt{x^3-1})^5 + (x - \sqrt{x^3-1})^5$$

$$= 2\,({}^5C_0 x^5 + {}^5C_2 x^3 (\sqrt{x^3-1})^2$$

$$+ {}^5C_4 x^1 (\sqrt{x^3-1})^4]$$

$$= 2\,[x^5 + 10x^3(x^3-1)$$

$$+ 5x(x^6 - 2x^3 + 1)]$$

$$\alpha = \text{Cofficient of } x^7 = 2\times 5 = 10$$

$$\beta = \text{Cofficient of } x^5 = 2\times 1 = 2$$

$$\gamma = \text{Cofficient of } x^3$$

$$= 2(-10) + 2(5) = -20$$

$$\delta = \text{Cofficient of } x = 2\times 5 = 10$$

$$\because \quad \alpha u + \beta v = 18 \Rightarrow 10u + 2v = 18$$

$$\Rightarrow \quad 5u + v = 9 \qquad \ldots\text{(i)}$$

$$\text{and} \quad \gamma u + \delta v = 20$$

$$\Rightarrow \quad -20u + 10v = 20$$

$$\Rightarrow \quad -2u + v = 2 \qquad \ldots\text{(ii)}$$

By solving Eqs. (i) and (ii)

$$7u = 7$$

$$\Rightarrow \quad u = 1$$

$$\text{and} \quad v = 2 + 2 = 4$$

$$\therefore \quad u + v = 5$$

15. *(a)* $E: \dfrac{x^2}{a^2} + \dfrac{y^2}{b^2} = 1$

$$a > b \Rightarrow \text{focii } (ae, 0), (-ae, 0)$$

$$H: \frac{x^2}{A^2} - \frac{y^2}{B^2} = 1$$

$$\Rightarrow \quad \text{focii } (Ae', 0), (-Ae', 0)$$

$\because$ Given $2ae = 2\sqrt{3}$ and $2Ae' = 2\sqrt{3}$

$$ae = \sqrt{3} \quad Ae' = \sqrt{3}$$

$$ae = Ae'$$

$$\frac{a}{A} = \frac{e'}{e} \Rightarrow 3$$

$$\Rightarrow \quad a = 3A \text{ and } a - A = 2$$

$$2A = 2 \Rightarrow A = 1 \Rightarrow a = 3$$

$$Ae' = \sqrt{3} \quad \text{and} \quad ae = \sqrt{3}$$

$$e' = \sqrt{3}, \quad e = \frac{1}{\sqrt{3}}$$

Sum of lengths of their latus-rectum

$$= \frac{2b^2}{a} + \frac{2B^2}{A}$$

$$= \frac{2a^2(1-e^2)}{a} + \frac{2A^2[(e')^2 - 1]}{A}$$

$$= 2(3)\left(1 - \frac{1}{3}\right) + 2(1)(3-1)$$

$$= 4 + 4 = 8$$

16. *(a)* $\because P(4, 4\sqrt{3})$ lies on parabola

$$y^2 = 4ax$$

$$\therefore \quad (4\sqrt{3})^2 = 4a\times 4$$

$$a = 3$$

$\therefore$ Parabola is $y^2 = 12x$

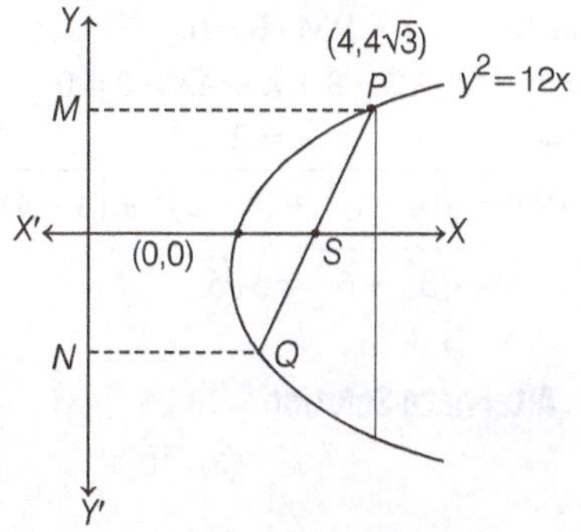

$$P(4, 4\sqrt{3})$$

$$2at = 4\sqrt{3}$$

$$t = \frac{4\sqrt{3}}{6} = \frac{2}{\sqrt{3}}$$

Parametric form of $Q = -\dfrac{1}{t} = -\dfrac{\sqrt{3}}{2}$

$$\therefore \quad Q = \left(3\left(-\frac{\sqrt{3}}{2}\right)^2, 2(3)\left(-\frac{\sqrt{3}}{2}\right)\right)$$

$$= \left(\frac{9}{4}, -3\sqrt{3}\right)$$

$\therefore$ Area of trapezium $PQNM$

$$= \frac{1}{2}(PM + QN)\times MN$$

$$= \frac{1}{2}(PS + SQ)\,MN$$

$$= \frac{1}{2}\,PQ\cdot MN$$

$$= \frac{1}{2}\sqrt{\left(\frac{9}{4} - 4\right)^2 + (-3\sqrt{3} - 4\sqrt{3})^2}\cdot 7\sqrt{3}$$

$$= \frac{1}{2}\times 7\times\frac{7}{4}\times 7\sqrt{3} = \frac{343\sqrt{3}}{8}$$

17. *(b)* $L: \dfrac{x-1}{2} = \dfrac{y+2}{-1} = \dfrac{z+3}{2} = \lambda$ (say)

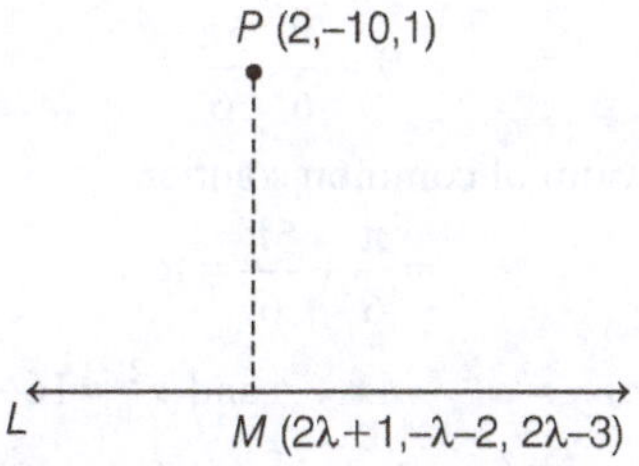

Let $M = (2\lambda + 1, -\lambda - 2, 2\lambda - 3)$

$$\text{PM} = (2\lambda - 1)\,\hat{i} + (8 - \lambda)\,\hat{j} + (2\lambda - 4)\,\hat{k}$$

$$\mathbf{b} = 2\hat{\mathbf{i}} - \hat{\mathbf{j}} + 2\hat{\mathbf{k}}$$

$\because \quad \mathbf{PM} \perp \mathbf{b}$

$\Rightarrow \quad \mathbf{PM} \cdot \mathbf{b} = 0$

$\Rightarrow \quad 4\lambda - 2 - 8 + \lambda + 4\lambda - 8 = 0$

$\Rightarrow \quad \lambda = 2$

$$PM = \sqrt{(4-1)^2 + (8-2)^2 + (4-4)^2}$$

$$= \sqrt{3^2 + 6^2} = 3\sqrt{5}$$

Alternate Solution

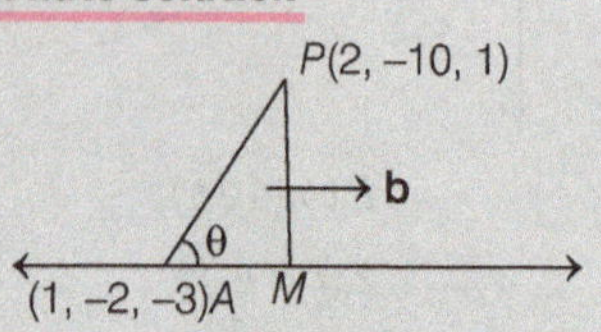

$$PM = PA \sin\theta = \frac{|\mathbf{PA} \times \mathbf{b}|}{|\mathbf{b}|}$$

$$= \frac{|(-\hat{\mathbf{i}} + 8\hat{\mathbf{j}} - 4\hat{\mathbf{k}}) \times (2\hat{\mathbf{i}} - \hat{\mathbf{j}} + 2\hat{\mathbf{k}})|}{\sqrt{4+1+4}}$$

$$= \frac{\sqrt{(12)^2 + (6)^2 + (15)^2}}{3} = 3\sqrt{5}$$

18. *(d)* $2\sin^2\theta = \cos 2\theta$

$\Rightarrow \quad 1 - \cos 2\theta = \cos 2\theta$

$\Rightarrow \quad 2\cos 2\theta = 1$

$$\cos 2\theta = \frac{1}{2}$$

$$2\theta = \frac{\pi}{3}, 2\pi - \frac{\pi}{3}, 2\pi + \frac{\pi}{3}, 4\pi - \frac{\pi}{3}$$

$$\theta = \frac{\pi}{6}, \frac{5\pi}{6}, \frac{7\pi}{6}, \frac{11\pi}{6}$$

and $\quad 2\cos^2\theta = 3\sin\theta$

$\Rightarrow \quad 2(1 - \sin^2\theta) = 3\sin\theta$

$\Rightarrow \quad 2\sin^2\theta + 3\sin\theta - 2 = 0$

$\Rightarrow 2\sin^2\theta + 4\sin\theta - \sin\theta - 2 = 0$

$\Rightarrow \quad (2\sin\theta - 1)(\sin\theta + 2) = 0$

$$\sin\theta = \frac{1}{2} \quad (\because \sin\theta \neq -2)$$

$$\theta = \frac{\pi}{6}, \frac{5\pi}{6}$$

$\therefore$ Sum of common solutions is

$$= \frac{\pi}{6} + \frac{5\pi}{6} = \pi$$

19. *(d)* $y = x^2 - 4x + 4$ and $y^2 = 16 - 8x$

$$y = (x-2)^2, \; y^2 = -8(x-2)$$

Let $x - 2 = X$

$\therefore \quad y = X^2$

and $\quad y^2 = -8X$

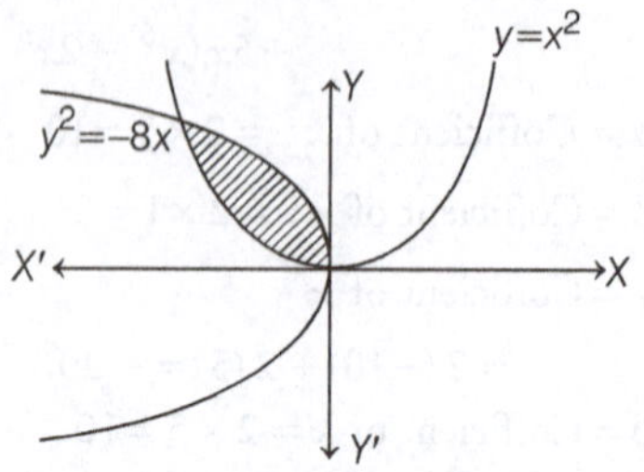

$$\therefore \text{ Area enclosed} = \frac{16ab}{3}$$

$$= \frac{16 \times \frac{1}{4} \times \frac{8}{4}}{3} = \frac{8}{3}$$

20. *(a)* Group : 3 Girls + 4 boys $\begin{cases} B_1 \\ B_2 \end{cases}$

Number of required arrangements

= Total – When B_1 and B_2 are together

$= 2!(3!)\,4! - 2!(3!)(3!\cdot 2!)$

$= 144$

21. *(27)* $\dfrac{dy}{dx} + \left(\dfrac{x}{x^2 - 1}\right) y$

$$= \frac{x^6 + 4x}{\sqrt{1-x^2}} \quad -1 < x < 1$$

$$P = \frac{-x}{1-x^2}, \quad Q = \frac{x^6 + 4x}{\sqrt{1-x^2}}$$

$$\text{IF} = e^{\int P\,dx} = e^{\int \frac{-x}{1-x^2}}$$

$$= e^{-\frac{1}{2}\ln(1-x^2)} = \sqrt{1-x^2}$$

Solution of differential equation

$$y\sqrt{1-x^2} = \int \frac{x^6 + 4x}{\sqrt{1-x^2}} \cdot \sqrt{1-x^2}\, dx + C$$

$$y\sqrt{1-x^2} = \frac{x^7}{7} + 2x^2 + C$$

$$x = 0, \quad y = 0$$

$\Rightarrow \quad C = 0$

$$y = \frac{\frac{x^7}{7} + 2x^2}{\sqrt{1-x^2}} = f(x)$$

$$\therefore \int_{-1/2}^{1/2} f(x)\,dx = \int_{-1/2}^{1/2} \frac{\frac{x^7}{7} + 2x^2}{\sqrt{1-x^2}}\, dx$$

$$= \int_{-1/2}^{1/2} \frac{x^7}{7\sqrt{1-x^2}}\,dx + 2\int_{-1/2}^{1/2} \frac{x^2}{\sqrt{1-x^2}}\,dx$$

$\downarrow$ odd function $\qquad \downarrow$ even function

$$\therefore \quad = 4\int_0^{1/2} \frac{x^2}{\sqrt{1-x^2}}\,dx$$

$$= -4\int_0^{1/2} \frac{1 - x^2 - 1}{\sqrt{1-x^2}}\,dx$$

$$= -4\int_0^{1/2} \sqrt{1-x^2}\,dx + 4\int_0^{1/2} \frac{1}{\sqrt{1-x^2}}\,dx$$

$$= -4\left[\frac{x}{2}\sqrt{1-x^2} + \frac{1}{2}\sin^{-1} x\right]_0^{1/2} + 4(\sin^{-1} x)_0^{1/2}$$

$$= -4\left(\frac{\sqrt{3}}{8} + \frac{1}{2} \times \frac{\pi}{6}\right) + 4\left(\frac{\pi}{6}\right)$$

$$= -\frac{\sqrt{3}}{2} + \frac{4\pi}{6}\left(-\frac{1}{2} + 1\right) = -\frac{\sqrt{3}}{2} + \frac{\pi}{3}$$

$\therefore$ Required value $= 2\pi - 3\sqrt{3}$

$\therefore \quad \alpha = 3\sqrt{3}$

$$\alpha^2 = 27$$

Concept Enhancer

The solution of the differential equation $\dfrac{dy}{dx} + Py = Q$ is

$y(\text{IF}) = \int Q \cdot (\text{IF})dx + C$, where $\text{IF} = e^{\int P dx}$

22. *(28)* Let side of equilateral ΔPQR be a.

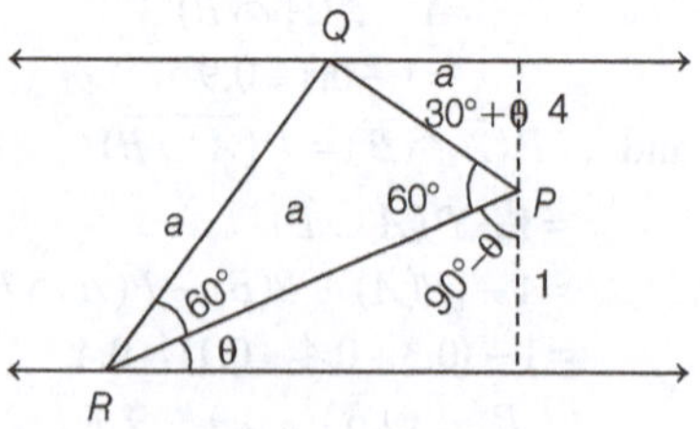

$PR = 1\operatorname{cosec}\theta$ and $PQ = 4\sec(30° + \theta)$

$\because \quad PR = PQ$

$$\operatorname{cosec}\theta = 4\sec(30° + \theta)$$

$\Rightarrow \quad 4\sin\theta = \cos(30° + \theta)$

$\Rightarrow 4\sin\theta = \cos 30° \cos\theta - \sin 30° \sin\theta$

$\Rightarrow \quad 4\sin\theta = \dfrac{\sqrt{3}}{2}\cos\theta - \dfrac{1}{2}\sin\theta$

$$\Rightarrow \frac{9}{2}\sin\theta = \sqrt{3}\cos\theta$$

$$\Rightarrow \tan\theta = \frac{\sqrt{3}}{9} = \frac{1}{3\sqrt{3}}$$

$$\therefore (QR)^2 = (PR)^2$$

$$= \frac{(3\sqrt{3})^2 + 1}{1} = 28$$

23. *(465)* Given, $\sum_{r=1}^{30} \frac{r^2\,({}^{30}C_r)^2}{{}^{30}C_{r-1}} = \alpha\,(2^{29})$

$$\because \frac{{}^{30}C_r}{{}^{30}C_{r-1}} = \frac{30-r+1}{r} = \frac{31-r}{r}$$

$$\therefore \text{LHS} = \sum_{r=1}^{30} r^2\cdot{}^{30}C_r\cdot\frac{31-r}{r}$$

$$= \sum_{r=1}^{30} r\cdot{}^{30}C_r\,(31-r)$$

$$\Rightarrow \sum_{r=1}^{30} \frac{30!\,(31-r)}{(r-1)!\,(30-r)!}$$

$$\Rightarrow 30\sum_{r=1}^{30} \frac{(31-r)\,29!}{(r-1)!\,(30-r)!}$$

$$= 30\sum_{r=1}^{30}(31-r)\,{}^{29}C_{r-1}$$

$$= 30\left(31\sum_{r=1}^{30}{}^{29}C_{r-1} - \sum_{r=1}^{30} r\cdot{}^{29}C_{r-1}\right)$$

$$= 30\left((31)(2^{29}) - \sum_{r=1}^{30}(r-1)\,{}^{29}C_{r-1} - \sum_{r=1}^{30}{}^{29}C_{r-1}\right)$$

$$= 30\left(31\times 2^{29} - \sum_{r'=0}^{29} r'\cdot{}^{29}C_{r'} - 2^{29}\right)$$

$$= 30\,(31\times 2^{29} - 29\cdot 2^{28} - 2^{29})$$

$$= 30\times 2^{28}\,(31\times 2 - 29 - 2)$$

$$= 30\times 2^{28}\times 31$$

$$= 15\times 31\times 2^{29}$$

$$\alpha = 15\times 31 = 465$$

Concept Enhancer

$${}^nC_r = \frac{n}{r}\cdot{}^{n-1}C_{r-1}$$

$$\sum_{r=0}^{n}{}^nC_r = 2^n,\ \sum_{r=1}^{n} r\cdot{}^nC_r = n\cdot 2^{n-1}$$

24. *(145)* Given $A\,(6, 8)$,

$B\,(10\cos\alpha, -10\sin\alpha)$

and $C\,(-10\sin\alpha, 10\cos\alpha)$

$\because$ All three points A, B and C lies on the circle

$$x^2 + y^2 = 100$$

$\therefore$ Circumcentre of $\Delta\,ABC$ is $O\,(0, 0)$

orthocenter $H\,(a, 9)$

centroid $G\,(h, k)$

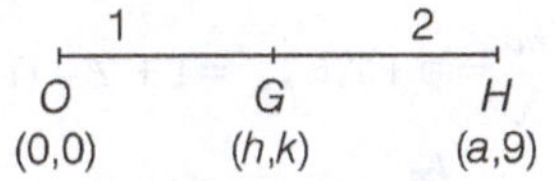

$$h = \frac{a}{3} \text{ and } k = \frac{9}{3} = 3$$

$$\frac{6 + 10\cos\alpha - 10\sin\alpha}{3} = \frac{a}{3}$$

and $\frac{8 - 10\sin\alpha + 10\cos\alpha}{3} = 3$

$$10\,(\cos\alpha - \sin\alpha) = 1$$

$$\cos\alpha - \sin\alpha = \frac{1}{10}$$

$$\therefore \frac{a}{3} = \frac{6 + 10\left(\frac{1}{10}\right)}{3} \Rightarrow a = 7$$

Now, $5a - 3h + 6k + 100\sin 2\alpha$

$$= 5\,(7) - 3\left(\frac{7}{3}\right) + 6(3) + 100\,[1 - (\cos\alpha - \sin\alpha)^2]$$

$$= 35 - 7 + 18 + 100\left(1 - \frac{1}{100}\right)$$

$$= 28 + 18 + 99$$

$$= 145$$

25. *(3)* $A = \{1, 2, 3\}$

$R = \{(1, 2), (2, 3)\}$

for reflexive $(1, 1), (2, 2), (3, 3) \in R$

for transitive $(1, 2), (2, 3) \in R$

$\Rightarrow (1, 3) \in R$

Relation which are reflexive, transitive but not symmetric

$R_1 = \{(1, 1)\,(2, 2)\,(3, 3)\,(1, 2)\,(2, 3)\,(1, 3)\}$

$R_2 = \{(1, 1)\,(2, 2)\,(3, 3)\,(1, 2)\,(2, 1)\,(2, 3)\,(1, 3)\}$

$R_3 = \{(1, 1)\,(2, 2)\,(3, 3)\,(1, 2)\,(2, 3)\,(3, 2)\,(1, 3)\}$

Physics

26. *(b)*

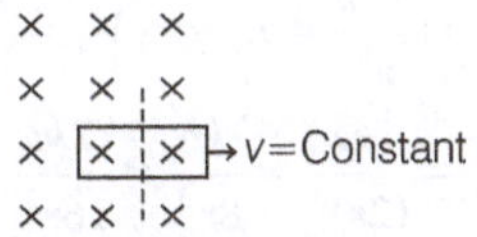

B = Constant

Motional emf $\Rightarrow \varepsilon = Blv$ = Constant

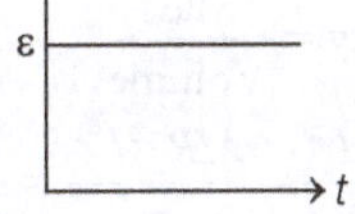

27. *(d)* Extensive variable depends on size or mass of system

Example - Internal energy (ΔU)

Volume (V)

Mass (m)

Intensive variables are

Pressure (p), Temperature (T), Density (ρ)

A and B are correct.

28. *(a)*

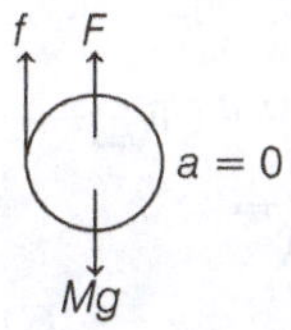

$$Mg - F_B = f \Rightarrow Mg - \frac{Mg}{2} = f$$

$$f = \frac{2Mg - Mg}{2}$$

$$f = \frac{Mg}{2}$$

29. *(b)*

$$\frac{1}{f_1} = (\mu - 1)\frac{2}{R} \Rightarrow P = 4\text{D}$$

$\Rightarrow$  $\frac{1}{f_2} = (\mu - 1)\frac{1}{R} = \frac{P}{2}$

$$= \frac{4\text{D}}{2} = 2\text{D}$$

$$P' = 2\text{D}$$

30. *(d)* $\mathbf{F} = 2\hat{i} + b\hat{j} + \hat{k},\ W = 0$ (given)

$$\mathbf{s} = \hat{i} - 2\hat{j} - \hat{k}$$

$$W = \mathbf{F}\cdot\mathbf{s} = (2\hat{i} + b\hat{j} + \hat{k})\cdot(\hat{i} - 2\hat{j} - \hat{k})$$

$$0 = 2 - 2b - 1$$

$$2b = 1 \Rightarrow b = 1/2$$

31. (*)

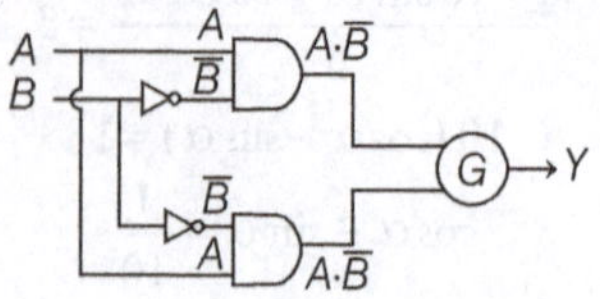

A	B	Y
0	0	1
0	1	0
1	0	0
1	1	1

$$Y=\overline{A\overline{B}+\overline{A}B}=\overline{A\overline{B}}\cdot\overline{\overline{A}B}$$
$$=(\overline{A}+B)(A+\overline{B})$$

Thus, output G is NOT gate.
Hence, no option is correct.
(Bonus marks)

32. (a) In YDSE fringe width

$$\beta=\frac{\lambda D}{d} \text{ and } \lambda_R>\lambda_B$$
$$\beta_R>\beta_B$$

Concept Enhancer

On immersing the apparatus in liquid (denser medium), then wavelength decreases and hence fringe width also decreases.

i.e. $\beta_{med}=\frac{\beta_{air}}{\mu}$

33. (d)

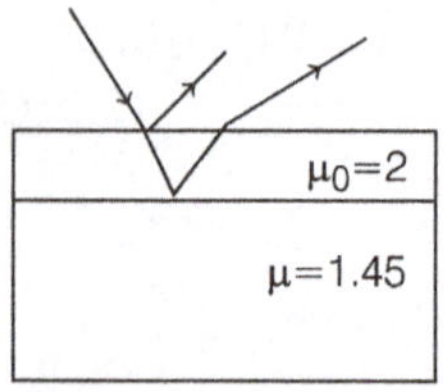

$$\Delta p=2\mu_0 t=n\lambda$$
$$t=\frac{n\lambda}{2\mu_0}\quad\therefore\quad t_{min}=\frac{\lambda}{2\mu_0}$$
$$t_{min}=\frac{550}{2\times2}=137.5\text{ nm}$$

34. (b)

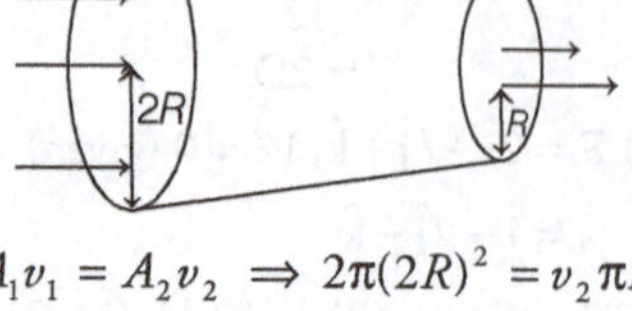

$$A_1v_1=A_2v_2\Rightarrow 2\pi(2R)^2=v_2\pi R^2$$
$$8\pi R^2=v_2\pi R^2\Rightarrow v_2=8\text{ m/s}$$

35. (d) $\mathbf{r}=\hat{i}+\hat{j}+\hat{k}$

$$\mathbf{F}=2\hat{i}+\hat{j}+2\hat{k}$$
$$\tau=\mathbf{r}\times\mathbf{F}=\begin{vmatrix}\hat{i}&\hat{j}&\hat{k}\\1&1&1\\2&1&2\end{vmatrix}$$
$$=\hat{i}-0\hat{j}-\hat{k}=\hat{i}-\hat{k}$$

36. (a) $g=\frac{GM}{R_e^2}$

Now, $M\to4M$, $R\to2R$

$$g'=\frac{G(4M)}{(2R_e)^2}=\frac{G(4M)}{4R_e^2}$$
$$g'=\frac{GM}{R_e^2}=g\Rightarrow T=2\pi\sqrt{\frac{l}{g}}$$

A and R are correct.

but $T=2\pi\sqrt{\frac{l}{g}}$ does not depend on mass ; R does not explain A.

37. (d) $\frac{hc}{\lambda}=\phi+KE_{max}$

Given $KE_{max}=2\text{eV}$

Work function $(\phi)=1\text{eV}$

$$\frac{hc}{\lambda}=\phi+KE_{max}=1+2=(1+2)\text{ eV}$$
$$\frac{hc}{\lambda}=3\text{ eV}\qquad\ldots(i)$$

According to 2nd condition

$$\lambda\to\frac{\lambda}{2}$$

Then, $\frac{hc}{\frac{\lambda}{2}}=\phi+KE_{max}$

$$2\frac{hc}{\lambda}=1\text{ eV}+KE_{max}$$

As, $\frac{hc}{\lambda}=3\text{eV}$

$$2(3)\text{eV}=1\text{ eV}+KE_{max}$$
$$6-1=KE_{max}$$
$$KE_{max}=5\text{eV}$$

38. (c) $\gamma_1=\frac{C_p}{C_V}\Rightarrow\gamma_2=\frac{C_p}{C_V}$

$\gamma=\frac{2}{f}+1$, where $f\to$ degree of freedom without vibration : $f=5$ (for rigid molecules)

$$\gamma_1=\frac{2}{5}+1=0.4+1=1.40$$

For diatomic molecules with vibration $f=7$

$$\gamma_2=\frac{2}{f}+1=\frac{2}{7}+1=1.28$$
$$\therefore\quad\gamma_1>\gamma_2$$
$$\gamma_2<\gamma_1$$

39. (c) Radius $R=\frac{mv}{qB}$ ...(i)

and $mvR=\frac{nh}{2\pi}$ ...(ii)

Put the value of mv from Eq.(i) in Eq. (ii)

$$mv=ReB\Rightarrow ReB(R)=\frac{nh}{2\pi}$$
$$R^2eB=\frac{nh}{2\pi}\Rightarrow R^2=\frac{nh}{2\pi eB}$$
$$R=\sqrt{\frac{nh}{2\pi eB}}$$

For first excited state $n=2$

$$\therefore\quad R=\sqrt{\frac{2h}{2\pi eB}}=\sqrt{\frac{h}{\pi eB}}$$

40. (b)

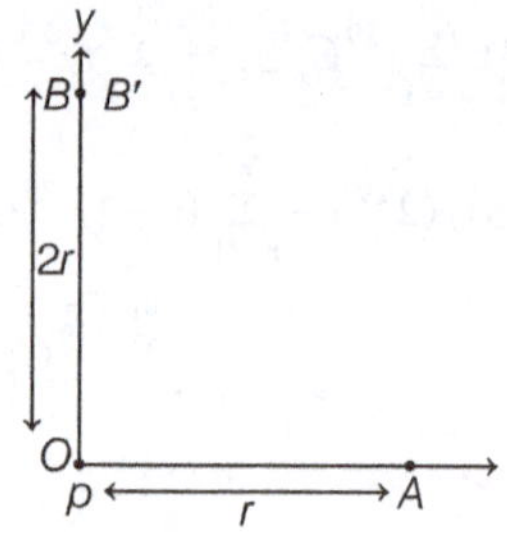

$$E_A=\frac{2kp}{r^3}=E_0\Rightarrow V_A=\frac{kp}{r^2}=V_0$$
$$E_B=\frac{kp}{(2r)^3}=\frac{kp}{8r^3}=\frac{2kp}{16r^3}$$
$$E_B=\frac{E_0}{16}\Rightarrow V_B=0$$

41. (a) Density $=\frac{\text{Mass}}{\text{Volume}}$

$$\rho=\frac{M}{\pi R^2 l}=\frac{MR^{-2}l^{-1}}{\pi}=\frac{1}{\pi}MR^{-2}l^{-1}$$
$$\ln\rho=\ln\left(\frac{1}{\pi}\right)+\ln M+\ln R^{-2}+\ln l^{-1}$$

After differentiation -

$$\frac{d\rho}{\rho}=0+\frac{dM}{M}+\frac{2dR}{R}+\frac{dl}{l}$$

$$\frac{d\rho}{\rho}\times 100=\frac{dM}{M}\times 100+\frac{2dR}{R}\times 100+\frac{dl}{l}\times 100$$

$$\frac{d\rho}{\rho}\times 100=\left(\frac{dM}{M}+\frac{2dR}{R}+\frac{dl}{l}\right)\times 100$$

$$=\left(\frac{0.003}{0.6}+\frac{2\times 0.01}{0.5}+\frac{0.05}{10}\right)\times 100$$

$$\frac{d\rho}{\rho}\times 100=5\%$$

42. *(b)*

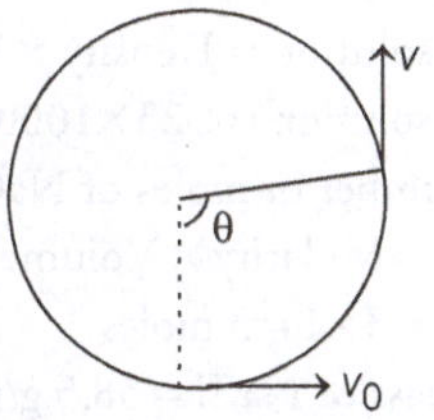

$$v=\sqrt{v_0^2-2gr(1-\cos\theta)}$$

For $\theta=30^\circ$

$$v_b=\sqrt{10^2-2\times 10\times 2(1-\cos 30^\circ)}$$

$$v_b=\sqrt{100-40\left(1-\frac{\sqrt{3}}{2}\right)}$$

For $\theta=120^\circ$

$$v_c=\sqrt{10^2-2\times 10\times 2(1-\cos 120^\circ)}$$

$$=\sqrt{100-\frac{40\times 3}{2}}$$

As per question

$$\frac{KB_B}{KE_C}=\frac{\frac{1}{2}mv_B^2}{\frac{1}{2}mv_C^2}$$

$$=\frac{100-40\left(1-\frac{\sqrt{3}}{2}\right)}{40}$$

$$=\frac{60+20\sqrt{3}}{40}=\frac{3+\sqrt{3}}{2}$$

43. *(d)* Initially $I_0=\frac{\varepsilon_m}{R}$

R become twice of initial value

$R\to 2R$

Finally $I_0'=\frac{\varepsilon_m}{2R}=\frac{I_0}{2}$

44. *(d)* Given, $m=100\text{ g}=0.1\text{ kg}$

$u=20\text{ m/s}$

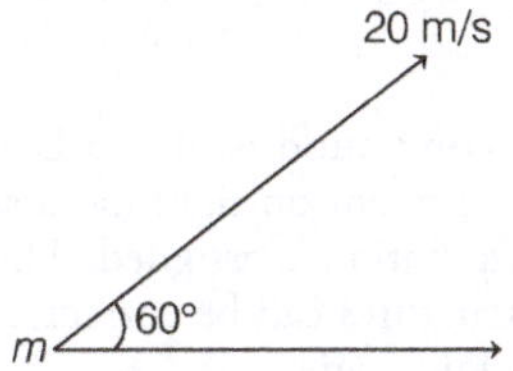

$$KE_i=\frac{1}{2}mv^2$$

$$KE_f=\frac{1}{2}m(v\cos 60^\circ)^2=\frac{1}{2}mv^2\left(\frac{1}{2}\right)^2$$

$$=\frac{1}{2}mv^2\times\frac{1}{4}=\frac{1}{8}mv^2$$

$$\Delta K=K_i-K_f=\frac{1}{2}mv^2-\frac{1}{8}mv^2$$

$$=\frac{3}{8}mv^2=\frac{3}{8}\times 0.1\times(20)^2=15\text{ J}$$

45. *(a)* $C=\frac{q}{V}=\frac{q\cdot q}{V\cdot q}=\frac{q^2}{W}\quad\left[V=\frac{W}{q}\right]$

$$=\frac{[C^2]}{[ML^2T^{-2}]}=[M^{-1}L^{-2}T^2C^2]$$

46. *(1)*

$$F=2M\omega^2\frac{l}{2}=M\omega^2 l\Rightarrow\omega^2=\frac{F}{Ml}$$

$$\omega=\sqrt{\frac{F}{Ml}}\Rightarrow\omega=\sqrt{\frac{F}{Ml}}\quad[l=1\text{ m}]$$

$$\omega=\sqrt{\frac{F}{M(1)}}\Rightarrow\alpha=1$$

47. *(2)* For uniform speed, $v=E/B$

Radius of path when electric field is switched off ;

$$R=\frac{mv}{eB}\Rightarrow R=\frac{mv^2}{eE}$$

$$E=\frac{mv^2}{eR}$$

$$=\frac{1.6\times 10^{-27}\times 4\times 10^{10}}{1.6\times 10^{-19}\times 2\times 10^{-2}}$$

$$=2\times 10^4\text{ N/C}$$

$x=2$

48. *(1)*

Key idea

A capacitor acts an open switch in a DC circuit.

$$R_{CB}=2\Omega+2\Omega=4\Omega$$

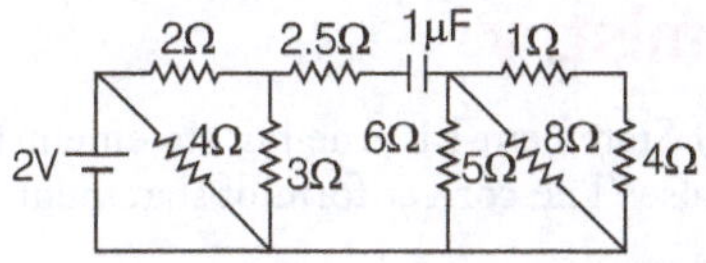

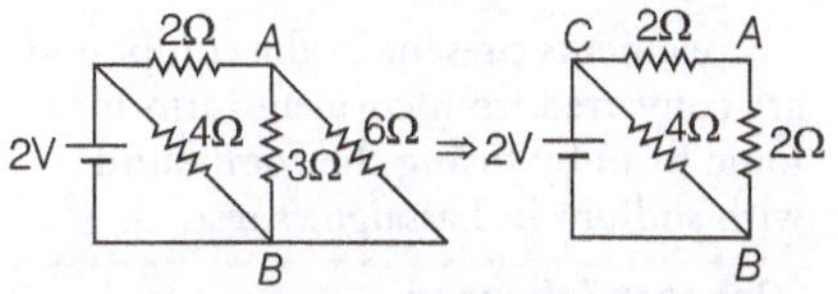

2Ω and 2Ω are in series

$$\frac{1}{R_{AB}}=\frac{1}{6}+\frac{1}{3}\Rightarrow\frac{1}{R_{AB}}=\frac{3}{6}\Rightarrow R_{AB}=2$$

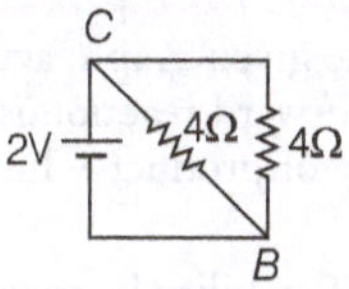

$$\frac{1}{R_{eq}}=\frac{1}{4}+\frac{1}{4}=\frac{2}{4}\Rightarrow R_{eq}=2$$

$$I=V/R=2/2=1\text{ A}$$

49. *(1)*

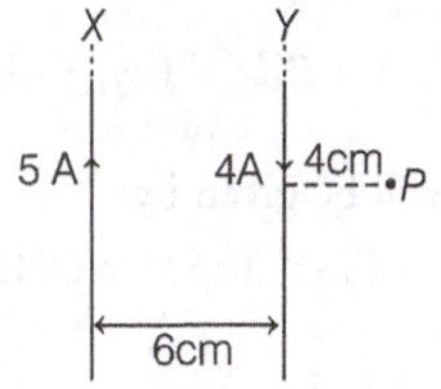

$$B_1=\frac{\mu_0 I_1}{2\pi r_1}$$

$$r_1=6\text{cm}+4\text{cm}=10\text{cm}=0.1\text{ m}$$

$$B_2=\frac{\mu_0 I_2}{2\pi r_2}\Rightarrow r_2=4\text{ cm}=0.04\text{ cm}$$

$$B_1=\frac{\mu_0(5)}{2\pi(0.1)}\otimes\Rightarrow B_2=\frac{\mu_0(4)}{2\pi\times 0.04}\odot$$

$$\mathbf{B}_{net}=\mathbf{B}_1-\mathbf{B}_2$$

$$=\frac{\mu_0(5)}{2\pi(0.1)}-\frac{\mu_0(4)}{2\pi(0.04)}=\frac{-100\mu_0}{4\pi}$$

$$=-100\times 10^{-7}=-1\times 10^{-5}\text{ T}$$

$x=1$

50. *(1200)* Given, $A_0=3.2\text{ cm}^2$

$I=6\text{A}$

Displacement current = ?

$J_d=I/A$

Current density $=\frac{\text{Current}}{\text{Area}}=\frac{6}{16}$

Current through small area

$$=J_d\times A_0=\frac{6}{16}\times 3.2$$

$$=1.2\text{ A}=1200\text{ mA}$$

Chemistry

51. *(c)* Statement I is true but statement II is false. The correct form of statement II is

The elements present in the compound are converted from covalent form into ionic form by fusing the compound with sodium in Lassaigne's test.

> **Concept Enhancer**
> Lassaigne's extract is a chemical extract used to detect the presence of nitrogen, sulphur and halogens in organic compounds. It's also known as the sodium fusion test.

52. *(a)* For the given graph, activation energy of forward reaction is $E_1 + E_2$. Also energy of product > Energy of reactant.

So, order of stability is reactant > product.

53. *(a)* The reaction involving dissociation of zirconium phosphate is as follows

$$Zr_3(PO_4)_4(s) \rightleftharpoons \underset{3S}{3Zr^{4+}(aq)} + \underset{4S}{4PO_4^{3-}(aq)}$$

So, K_{sp} will be given by

$$K_{sp} = (3S)^3\,(4S)^4 = 6912\,S^7$$

$$S = \left(\frac{K_{sp}}{6912}\right)^{\frac{1}{7}}$$

54. *(a)*

(A) $[Fe(CN)_5NO]^{2-}$ is a heteroleptic with d^2sp^3 hybridisation, low spin (as strong field ligand).

(B) $[CoF_6]^{3-}$ is a homoleptic complex with sp^3d^2 hybridisation. It is high spin (as weak field ligand).

(C) $[Fe(CN)_6]^{4-}$ is a homoleptic complex, with d^2sp^3 hybridisation low spin (as strong field ligand).

(D) $[Co(NH_3)_6]^{3+}$ is a homoleptic complex with, d^2sp^3 hybridisation low spin (as Co is in +3 oxidation state. Thus, NH_3 will act (as strong field ligand).

(E) $[Cr(H_2O)_6]^{2+}$ is a homoleptic complex with d^2sp^3 hybridisation high spin (as H_2O is a weak field ligand). Thus, option a is correct.

55. *(d)* In compound A, B and D —OH group is present on right hand side, so D-configuration is assigned. Thus, total three structures can be corrected to D-glyceraldehyde.

56. *(a)* In group 15 elements, weakest $E—E$ bond is present in nitrogen. Its maximum possible covalency is 4.

57. *(b)* The structure of given alkanes are as follows.

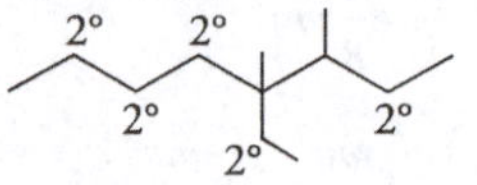

⇒ 10 secondary hydrogens are present.

4-ethyl-3, 4-dimethyloctane

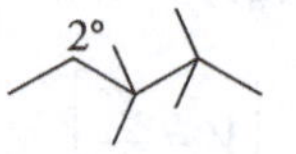

⇒ 2 secondary hydrogens are present.

2, 2, 3, 3- tetramethylpentane

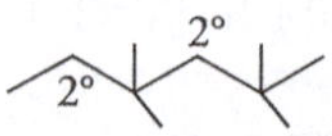

⇒ 4 secondary hydrogens are present.

2, 2, 4, 4- tetramethylhexane

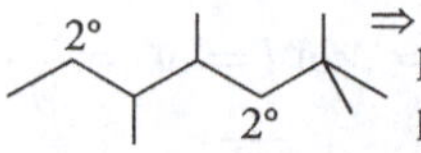

⇒ 4 secondary hydrogens are present.

2, 2, 4, 5- tetramethylheptane

So, 2, 2, 3, 3-tetramethylpentane has 2 secondary hydrogen.

58. *(c)* The correct match is A-III, B-IV, C-II, D-I.

(A) $R—\overset{\overset{O}{\|}}{C}—NH_2 \xrightarrow[H_2O]{LiAlH_4} R—CH_2—NH_2$

(B) 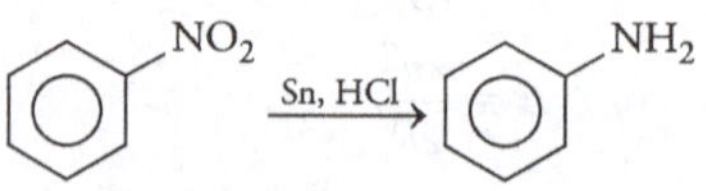

(C) $R—C\equiv N \xrightarrow{H_2/Ni} R—CH_2NH_2$

(D) 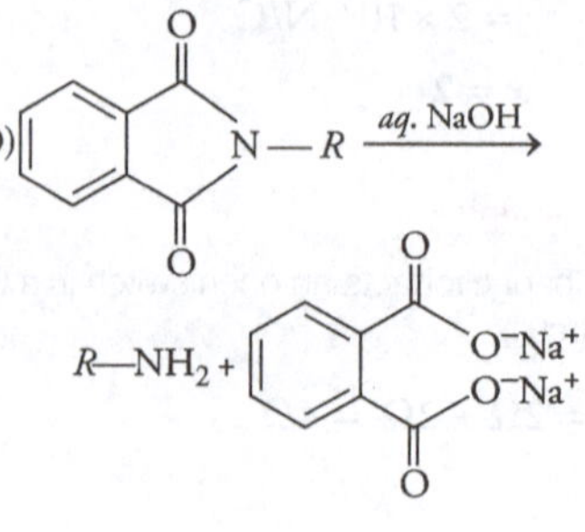

59. *(d)* Compound containing chlorine in maximum oxidation state will not undergo disproportionation reaction. So, among the given options, ClO_4^- has chlorine in maximum possible oxidation state i.e. 7. Thus, it will not undergo disproportionation.

60. *(a)* Given, Molarity = 3M

Density = 1.25 g/mL

Molality = ?

Mass of solution = Density × Volume

Mass of solution $= 1.25 \times 1000 = 1250$ g

Now, number of moles of NaCl

= Molarity × Volume

$= 3 \times 1 = 3$ moles

Molar mass of NaCl = 58.5 g/mol

Mass of NaCl $= 3 \times 58.5 = 175.5$ g

Now mass of solvent $= 1250 - 175.5$

$= 1074.5$ g

So, molality $= \dfrac{\text{Number of moles} \times 1000}{\text{Mass of solvent (in g)}}$

$= \dfrac{3 \times 1000}{1074.5} = 2.79$ m

> **Alternate Solution**
>
> $\text{Molality} = \dfrac{M \times 1000}{1000d - M \times Mw}$
>
> $= \dfrac{3000}{1250 - 175.5} = 2.79$ m

61. *(c)* Statement I is false but Statement II is true. It is because elements in extreme left are metallic in nature which generally forms basic oxides.

62. *(a)* The reaction involved is as follows.

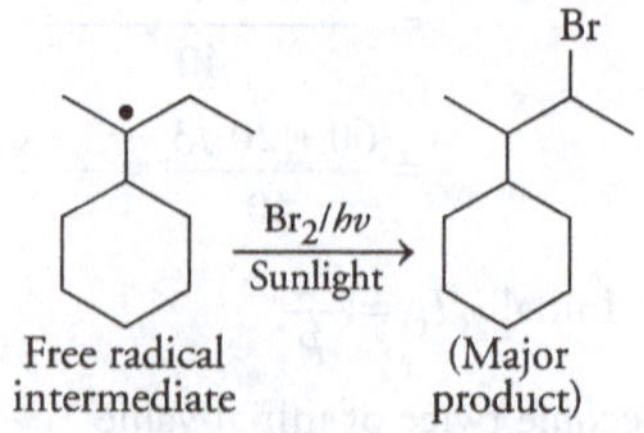

63. *(c)* The correct match is A-II, B-I, C-IV, D-III.

- We know that,

$$dG = Vdp - SdT$$

At constant pressure $Vdp = 0$

So, $\quad dG = -SdT$

or $\left(\frac{\partial G}{\partial T}\right)_p = -S$

- $dH = (dq)_p = C_p dT$

$\left(\frac{\partial H}{\partial T}\right)_p = C_p$

- $dG = Vdp - SdT$

At constant temperature, $dG = Vdp$

$\left(\frac{\partial G}{\partial p}\right)_T = V$

- $dU = C_V dT = (q)_V$

$\left(\frac{\partial U}{\partial T}\right)_V = C_V$

64. *(c)* The order of stability of given carbocations are as follows.

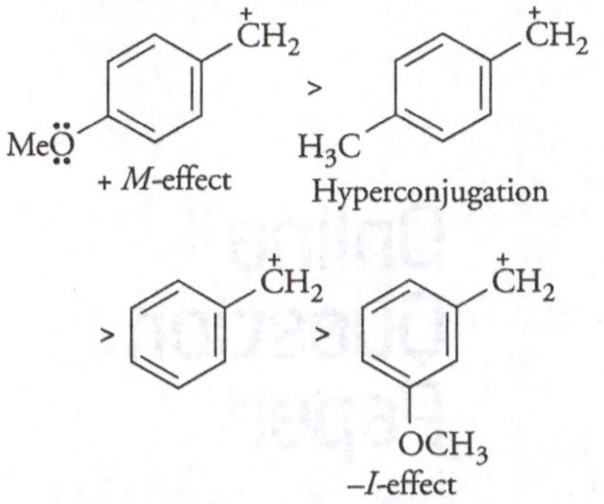

So, carbocation given in option (c) is most stable.

65. *(a)* Statement I is true but Statement II is false. It is because corrosion is more favourable in acidic medium than alkaline medium. So, the rate of corrosion is high in acidic medium than alkaline.

66. *(b)* The complete reaction sequence involved is as follows.

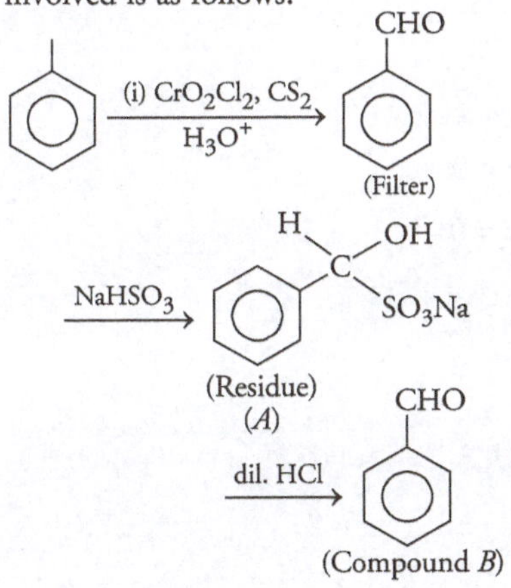

Thus option (b) is the correct answer.

67. *(a)* As the charge on complex increases, the value of CFSE increases. Also, as the strength of ligand increases the value of CFSE increases.

Thus, the correct order of CFSE is

$[Co(NH_3)_4]^{2+} < [Co(NH_3)_6]^{2+} < [Co(NH_3)_6]^{3+} < [Co(en)_3]^{3+}$

68. *(b)* Statement I is false but Statement II is true.

The correct form of statement I is :

No spectral line will be observed for a $2p_x \longrightarrow 2p_y$ transition because $2p_x$ and $2p_y$ orbitals are degenerate orbitals.

69. *(a)* The correct order of increasing dipole moment of given compounds is

NF_3	<	HBr	<	H_2S	<	$CHCl_3$
(0.24 D)		(0.79 D)		(0.95 D)		(1.04 D)

70. *(c)* For the given reaction

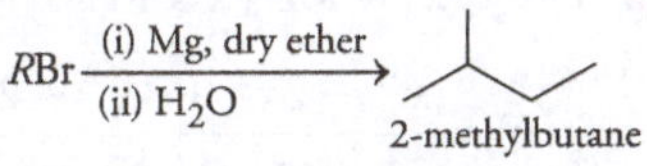

The maximum number of *R*Br are

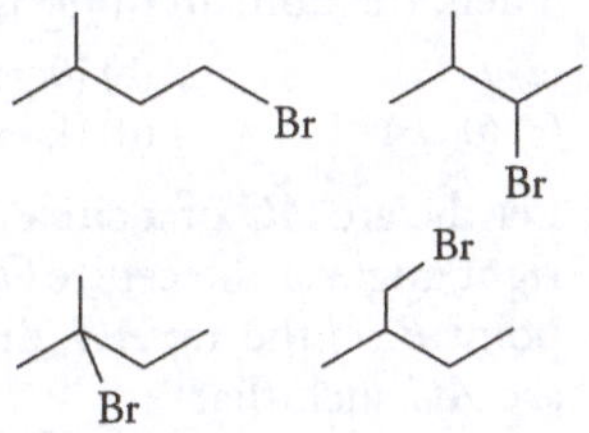

Thus, total 4 possible structures are possible.

71. *(14)* The structure of complex Ni^{2+} ion and dimethyl glyoxime is

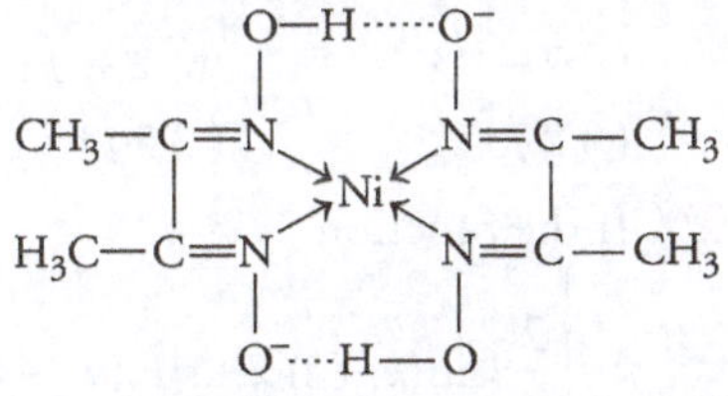

So, the number of H-atom = 14

72. *(95)*

Key idea

This question is based upon the concept of Hess's law to calculate the required ΔH value for the given reaction. First obtain the required equation by multiplying Eq. (ii) by 2, (iii) by 3 then from this value subtract Eq. (i). After that calculate the value by using

$2\Delta H_2 + 3\Delta H_3 - \Delta H_1$

Given,

$C_2H_6(g) + \frac{7}{2}O_2(g) \longrightarrow 2CO_2(g) + 3H_2O(l);$

$\Delta H_1^\circ = -1550$ kJ/mol ...(i)

$C(\text{Graphite}) + O_2(g) \longrightarrow CO_2(g)$ ...(ii)

$\Delta H_2^\circ = -393.5$ kJ/mol

$H_2(g) + \frac{1}{2}O_2(g) \longrightarrow H_2O(l)$ ...(iii)

$\Delta H_3^\circ = -286$ kJ/mol

$2C(\text{graphite}) + 3H_2(g) \longrightarrow C_2H_6(g);$

$\Delta H_f(C_2H_6(g)) = ?$

$\Delta H_f = 2\Delta H_2^\circ + 3\Delta H_3^\circ - \Delta H_1^\circ$

$= 95$ kJ/mole

73. *(57)* Given $M_1 = 2$

$V_1 = 20$

$M_2 = 0.5$

$V_2 = 400$

$M_F = ?$

We know that, $M_F(V_1 + V_2) = M_1V_1 + M_2V_2$

On substituting the given values in above equation, we get

$M_F = \frac{2 \times 20 + 0.5 \times 400}{420}$

$= 0.571$ M

$M_F = 57.1 \times 10^{-2}$ M

$M_F = 57$

74. *(8)* The reaction involved is as follows.

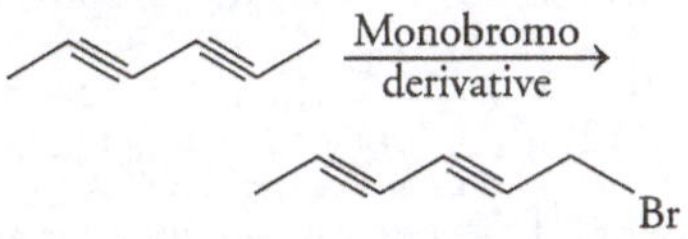

Number of $\pi e^- = 8$

75. *(11)* The electronic configuration of niobium (Nb) with atomic number 41 $= [Kr]4d^4 5s^1$

Number of electrons in $4d$ orbitals $= 4 = x$

Similarly the electronic configuration of ruthenium (Ru) with atomic number 44 $= [Kr]\, 4d^7 5s^1$

Number of electrons in $4d$ orbitals is $= 7 = y$

So, $x + y = 11$

Time 9:00-12:00 | **MM** : 300 Marks

23 January
Shift-I

Online Question Paper

—

JEE Main 2025

Mathematics

Section A
Objective Type Questions

01/ Let $R = \{(1, 2), (2, 3) (3, 3)\}$ be a relation defined on the set $\{1, 2, 3, 4\}$. Then, the minimum number of elements, needed to be added in R so that R becomes an equivalence relation, is

(a) 10 (b) 9
(c) 8 (d) 7

02/ If the first term of an AP is 3 and the sum of its first four terms is equal to one-fifth of the sum of the next four terms, then the sum of the first 20 terms is equal to

(a) -120 (b) -1080
(c) -1020 (d) -1200

03/ If $\frac{\pi}{2} \le x \le \frac{3\pi}{4}$, then $\cos^{-1}\left(\frac{12}{13}\cos x + \frac{5}{13}\sin x\right)$ is equal to

(a) $x + \tan^{-1}\frac{5}{12}$ (b) $x - \tan^{-1}\frac{5}{12}$
(c) $x - \tan^{-1}\frac{4}{3}$ (d) $x + \tan^{-1}\frac{4}{5}$

04/ Let a curve $y = f(x)$ passes through the point $(0, 5)$ and $(\log_e 2, k)$. If the curve satisfies the differential equation $2(3 + y)e^{2x}\,dx - (7 + e^{2x})\,dy = 0$, then k is equal to

(a) 8 (b) 16
(c) 4 (d) 32

05/ Let $f(x) = \log_e x$ and

$$g(x) = \frac{x^4 - 2x^3 + 3x^2 - 2x + 2}{2x^2 - 2x + 1}$$

Then, the domain of fog is

(a) R (b) $[0, \infty)$
(c) $(0, \infty)$ (d) $[1, \infty)$

06/ Let the arc AC of a circle subtend a right angle at the centre O. If the point B on the arc AC, divides the arc AC such that

$$\frac{\text{length of arc } AB}{\text{length of arc } BC} = \frac{1}{5}$$

and $\mathbf{OC} = \alpha\,\mathbf{OA} + \beta\,\mathbf{OB}$, then $\alpha + \sqrt{2}\,(\sqrt{3} - 1)\beta$ is equal to

(a) $2 - \sqrt{3}$ (b) $2 + \sqrt{3}$
(c) $2\sqrt{3}$ (d) $5\sqrt{3}$

07/ If the function

$$f(x) = \begin{cases} \frac{2}{x}\{\sin(k_1 + 1)\,x + \sin(k_2 - 1)x\}, & x < 0 \\ 4, & x = 0 \\ \frac{2}{x}\log_e\left(\frac{2 + k_1 x}{2 + k_2 x}\right), & x > 0 \end{cases}$$

is continuous at $x = 0$, then $k_1^2 + k_2^2$ is equal to

(a) 5 (b) 10
(c) 8 (d) 20

08/ The number of words, which can be formed using all the letters of the word "DAUGHTER", so that all the vowels never come together is

(a) 34000 (b) 37000
(c) 36000 (d) 35000

09/ Let the area of ΔPQR with vertices $P(5,4), Q(-2,4)$ and $R(a,b)$ be 35 square units. If its orthocenter and centroid are $O\left(2, \frac{14}{5}\right)$ and $C(c,d)$ respectively, then $c+2d$ is equal to

(a) $\frac{7}{3}$ (b) $\frac{8}{3}$
(c) 2 (d) 3

10/ Let $I(x) = \int \frac{dx}{(x-11)^{\frac{11}{13}}(x+15)^{\frac{15}{13}}}$.

If $I(37) - I(24) = \frac{1}{4}\left(\frac{1}{b^{1/13}} - \frac{1}{c^{1/13}}\right), b, c, \in N$, then $3(b+c)$ is equal to

(a) 22 (b) 26
(c) 39 (d) 40

11/ The value of $(\sin 70°)(\cot 10° \cot 70° - 1)$ is

(a) $\frac{2}{3}$ (b) 0
(c) $\frac{3}{2}$ (d) 1

12/ Marks obtained by all the students of class 12 are presented in a frequency distribution with classes of equal width. Let the median of this grouped data be 14 with median class interval 12-18 and median class frequency 12. If the number of students whose marks are less than 12 is 18, then the total number of students is

(a) 52 (b) 44 (c) 48 (d) 40

13/ One die has two faces marked 1, two faces marked 2, one face marked 3 and one face marked 4.

Another die has one face marked 1, two faces marked 2, two faces marked 3 and one face marked 4. The probability of getting the sum of numbers to be 4 or 5, when both the dice are thrown together, is

(a) $\frac{1}{2}$ (b) $\frac{2}{3}$ (c) $\frac{4}{9}$ (d) $\frac{3}{5}$

14/ Let the position vectors of the vertices A, B and C of a tetrahedron $ABCD$ be $\hat{i}+2\hat{j}+\hat{k}$, $\hat{i}+3\hat{j}-2\hat{k}$ and $2\hat{i}+\hat{j}-\hat{k}$, respectively. The altitude from the vertex D to the opposite face ABC meets the median line segment through A of the ΔABC at the point E. If the length of AD is $\frac{\sqrt{110}}{3}$ and the volume of the tetrahedron is $\frac{\sqrt{805}}{6\sqrt{2}}$, then the position vector of E is

(a) $\frac{1}{2}(\hat{i}+4\hat{j}+7\hat{k})$ (b) $\frac{1}{12}(7\hat{i}+4\hat{j}+3\hat{k})$
(c) $\frac{1}{6}(7\hat{i}+12\hat{j}+\hat{k})$ (d) $\frac{1}{6}(12\hat{i}+12\hat{j}+\hat{k})$

15/ Let $\left|\frac{\bar{z}-i}{2\bar{z}+i}\right| = \frac{1}{3}, z \in C$ be the equation of a circle with centre at C. If the area of the triangle, whose vertices are at the point $(0, 0)$, C and $(\alpha, 0)$ is 11 sq. units, then α^2 equals :

(a) 81/25 (b) 121/25 (c) 100 (d) 50

16/ If A, B and $(\text{adj}(A^{-1}) + \text{adj}(B^{-1}))$ are non-singular matrices of same order, then the inverse of $A(\text{adj}(A^{-1}) + \text{adj}(B^{-1}))^{-1} B$, is equal to

(a) $\frac{1}{|AB|}(\text{adj}(B) + \text{adj}(A))$

(b) $\text{adj}(B^{-1}) + \text{adj}(A^{-1})$

(c) $AB^{-1} + A^{-1}B$

(d) $\frac{AB^{-1}}{|A|} + \frac{BA^{-1}}{|B|}$

17/ The value of

$$\int_{e^2}^{e^4} \frac{1}{x}\left(\frac{e^{((\log_e x)^2+1)^{-1}}}{e^{((\log_e x)^2+1)^{-1}} + e^{((6-\log_e x)^2+1)^{-1}}}\right) dx \text{ is}$$

(a) $\log_e 2$ (b) e^2 (c) 2 (d) 1

Imp.

18/ Let P be the foot of the perpendicular from the point $Q(10, -3, -1)$ on the line $\frac{x-3}{7} = \frac{y-2}{-1} = \frac{z+1}{-2}$.

Then, the area of the right angled ΔPQR, where R is the point $(3, -2, 1)$ is

(a) $\sqrt{30}$ (b) $9\sqrt{15}$ (c) $8\sqrt{15}$ (d) $3\sqrt{30}$

Foot of the Perpendicular

19/ If the system of equations

$$(\lambda - 1)x + (\lambda - 4)y + \lambda z = 5$$
$$\lambda x + (\lambda - 1)y + (\lambda - 4)z = 7$$
$$(\lambda + 1)x + (\lambda + 2)y - (\lambda + 2)z = 9$$

has infinitely many solutions, then $\lambda^2 + \lambda$ is equal to

(a) 20 (b) 10 (c) 12 (d) 6

20/ If the line $3x - 2y + 12 = 0$ intersects the parabola $4y = 3x^2$ at the points A and B, then at the vertex of the parabola, the line segment AB subtends an angle equal to

(a) $\frac{\pi}{2} - \tan^{-1}\left(\frac{3}{2}\right)$ (b) $\tan^{-1}\left(\frac{11}{9}\right)$

(c) $\tan^{-1}\left(\frac{4}{5}\right)$ (d) $\tan^{-1}\left(\frac{9}{7}\right)$

Section B

Numerical Type Questions

21/ Let the circle C touches the line $x - y + 1 = 0$ have the centre on the positive X-axis and cut off a chord of length $\frac{4}{\sqrt{13}}$ along the line $-3x + 2y = 1$. Let H be the hyperbola $\frac{x^2}{\alpha^2} - \frac{y^2}{\beta^2} = 1$, whose one of the foci is the centre of C and the length of the transverse axis is the diameter of C. Then, $2\alpha^2 + 3\beta^2$ is equal to ……… .

22/ If the set of all values of a, for which the equation $5x^3 - 15x - a = 0$ has three distinct real roots, the interval (α, β), then $\beta - 2\alpha$ is equal to ……….. .

23/ If the equation $a(b-c)x^2 + b(c-a)x + c(a-b) = 0$ has equal roots, where $a + c = 15$ and $b = \frac{36}{5}$, then $a^2 + c^2$ is equal to ………… .

Imp.

24/ If the area of the larger portion bounded between the curves $x^2 + y^2 = 25$ and $y = |x - 1|$ is $\frac{1}{4}(b\pi + c)$, $b, c \in N$, then $b + c$ is equal to ……… .

Area Bounded by Two Curves

25/ The sum of all rational terms in the expansion of $(1 + 2^{\frac{1}{3}} + 3^{\frac{1}{2}})^6$ is equal to ……….. .

Physics

Section A

Objective Type Questions

26/ The electric flux is $\phi = \alpha\sigma + \beta\lambda$,

where λ and σ are linear and surface charge density, respectively, $\left(\frac{\alpha}{\beta}\right)$ represents

(a) electric field (b) area

(c) charge (d) displacement

27/ Given below are two statements.

Statement I The hot water flows faster than cold water.

Statement II Soap water has higher surface tension as compared to fresh water.

In the light above statements, choose the correct answer from the options given below.

(a) Statement I is true but Statement II is false.

(b) Both Statement I and Statement II is false.

(c) Statement I is false but Statement II is true.

(d) Both Statement I and Statement II are true.

28/ A light hollow cube of side length 10 cm and mass 10g, is floating in water. It is pushed down and released to execute simple harmonic oscillations. The time period of oscillations is $y\pi \times 10^{-2}$s, where the value of y is

(Acceleration due to gravity, $g = 10\text{ m/s}^2$, density of water $= 10^3\text{ kg/m}^3$)

(a) 2 (b) 4 (c) 6 (d) 1

29/ Identify the valid statements relevant to the given circuit at the instant when the key is closed.

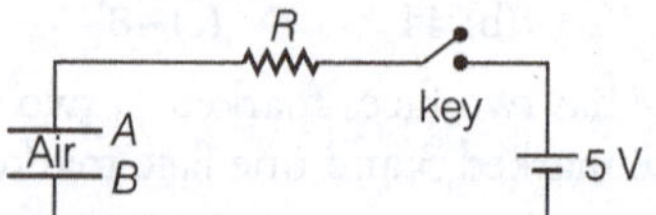

A. There will be no current through resistor R.

B. There will be maximum current in the connecting wires.

C. Potential difference between the capacitor plates A and B is minimum.

D. Charge on the capacitor plates is minimum.

Choose the correct answer from the options given below.

(a) C, D Only (b) A, B, D Only

(c) A, C Only (d) B, C, D Only

30/ A radioactive nucleus n_2 has 3 times the decay constant as compared to the decay constant of another radioactive nucleus n_1. If initial number of both nuclei are the same, what is the ratio of number of nuclei of n_2 to the number of nuclei of n_1, after one half-life of n_1?

(a) 4 (b) 1/4 (c) 8 (d) 1/8

31/ Refer to the circuit diagram given in the figure, which of the following observations are correct?

A. Total resistance of circuit is $6\,\Omega$.
B. Current in ammeter is 1 A.
C. Potential across AB is 4 V.
D. Potential across CD is 4 V.
E. Total resistance of the circuit is $8\,\Omega$.

Choose the correct answer from the options given below.

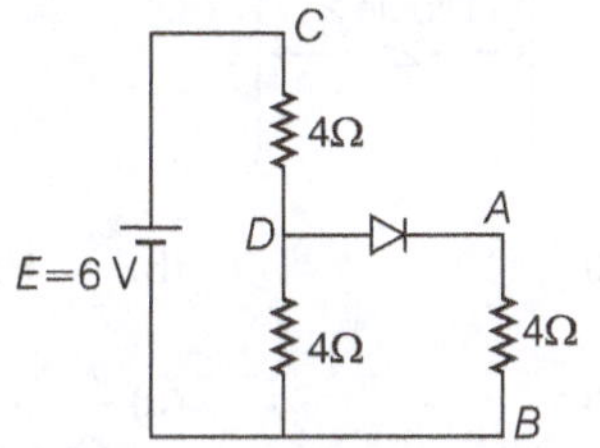

(a) B, C and E Only (b) A, B and D Only
(c) A, C and D Only (d) A, B and C Only

32/ What is the lateral shift of a ray refracted through a parallel sided glass slab of thickness h in terms of the angle of incidence i and angle of refraction r, if the glass slab is placed in air medium?

(a) $\dfrac{h\tan(i-r)}{\tan r}$ (b) $\dfrac{h\cos(i-r)}{\sin r}$

(c) $\dfrac{h\sin(i-r)}{\cos r}$ (d) h

33/ A gun fires a lead bullet of temperature 300 K into a wooden block. The bullet having melting temperature of 600 K penetrates into the block and melts down. If the heat required for the process is 625 J, then the mass of the bullet is .... grams.

(Latent heat of fusion of lead $= 2.5\times10^4$ Jkg^{-1} and specific heat capacity of lead $= 125$ Jkg^{-1} K^{-1})

(a) 15 (b) 10 (c) 5 (d) 20

34/ Regarding self-inductance.

A. The self-inductance of the coil depends on its geometry.
B. Self-inductance does not depend on the permeability of the medium.
C. Self-induced emf opposes any change in the current in a circuit.
D. Self-inductance is electromagnetic analogue of mass in mechanics.
E. Work needs to be done against self-induced emf in establishing the current.

Choose the correct answer from the options given below.

(a) A, B, C, E only (b) A, B, C, D only
(c) A, C, D, E only (d) B, C, D, E only

35/ The electric field of an electromagnetic wave in free space is

$$E = 57\cos[7.5\times10^6 t - 5\times10^{-3}(3x+4y)]\,(4\hat{i}-3\hat{j})\text{ N/C}$$

The associated magnetic field in Tesla is

(a) $\mathbf{B} = \dfrac{57}{3\times10^8}\cos[7.5\times10^6 t - 5\times10^{-3}(3x+4y)]\,(5\hat{k})$

(b) $\mathbf{B} = -\dfrac{57}{3\times10^8}\cos[7.5\times10^6 t - 5\times10^{-3}(3x+4y)]\,(5\hat{k})$

(c) $\mathbf{B} = -\dfrac{57}{3\times10^8}\cos[7.5\times10^6 t - 5\times10^{-3}(3x+4y)]\,(\hat{k})$

(d) $\mathbf{B} = \dfrac{57}{3\times10^8}\cos[7.5\times10^6 t - 5\times10^{-3}(3x+4y)]\,(\hat{k})$

36/ Given a thin convex lens (refractive index μ_2), kept in a liquid (refractive index $\mu_1, \mu_1 < \mu_2$) having radii of curvatures $|R_1|$ and $|R_2|$. Its second surface is silver polished. Where should an object be placed on the optic axis, so that a real and inverted image is formed at the same place?

(a) $\dfrac{\mu_1|R_1|\cdot|R_2|}{\mu_2(|R_1|+|R_2|)-\mu_1|R_2|}$

(b) $\dfrac{\mu_1|R_1|\cdot|R_2|}{\mu_1(|R_1|+|R_2|)-\mu_2|R_2|}$

(c) $\dfrac{\mu_1|R_1|\cdot|R_2|}{\mu_1(2|R_1|+|R_2|)-\mu_2\sqrt{|R_1|\cdot|R_2|}}$

(d) $\dfrac{(\mu_2+\mu_1)|R_1|}{(\mu_2-\mu_1)}$

37/ A spherical surface of radius of curvature R, separates air from glass (refractive index = 1.5). The centre of curvature is in the glass medium. A point object O placed in air on the optic axis of the surface. So, that its real image is formed at I inside glass. The line OI intersects the spherical surface at P and $PO = PI$. The distance PO equals to

(a) $5R$ (b) $1.5\,R$
(c) $3\,R$ (d) $2\,R$

38/ A solid sphere of mass m and radius r is allowed to roll without slipping from the highest point of an inclined plane of length L and makes an angle 30° with the horizontal. The speed of the particle at the bottom of the plane is v_1. If the angle of inclination is increased to 45° while keeping L constant. Then the new speed of the sphere at the bottom of the plane is v_2. The ratio $v_1^2 : v_2^2$ is

(a) 1 : 2 (b) $1:\sqrt{3}$

(c) 1 : 3 (d) $1:\sqrt{2}$

Imp.

39/ The motion of an airplane is represented by velocity-time graph as shown below. The distance covered by airplane in the first 30.5 second is …… km.

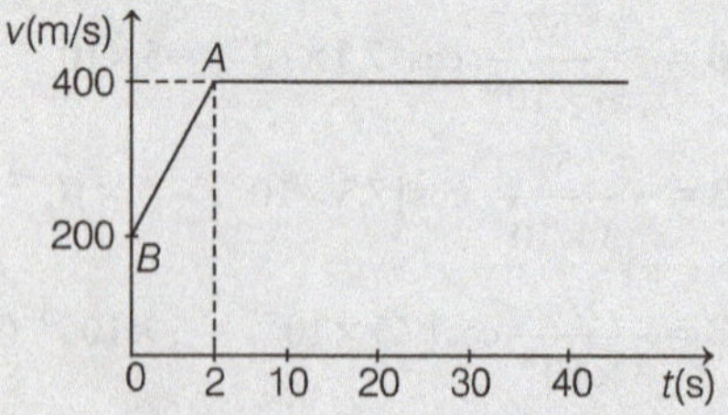

(a) 3 (b) 6 (c) 9 (d) 12

Velocity-Time Graph

40/ The position of a particle moving on X-axis is given by $x(t) = A\sin t + B\cos^2 t + Ct^2 + D$, where t is time.

The dimension of $\frac{ABC}{D}$ is

(a) $[L^3T^{-2}]$ (b) $[L^2]$ (c) $[L^2T^{-2}]$ (d) [L]

41/ Match the List-I with List-II.

	List-I		List-II
A.	Pressure varies inversely with volume of an ideal gas.	I.	Adiabatic process
B.	Heat absorbed goes partly to increase internal energy and partly to do work.	II.	Isochoric process
C.	Heat is neither absorbed nor released by a system	III.	Isothermal process
D.	No work is done on or by a gas.	IV.	Isobaric process

Choose the correct answer from the options given below.

(a) A-I, B-IV, C-II, D-III (b) A-III, B-IV, C-I, D-II

(c) A-I, B-III, C-II, D-IV (d) A-III, B-I, C-IV, D-II

42/ A point particle of charge Q is located at P along the axis of an electric dipole 1 at a distance r as shown in the figure. The point P is also on the equatorial plane of a second electric dipole 2 at a distance r. The dipoles are made of opposite charge q separated by a distance $2a$. For the charge particle at P not to experience any net force, which of the following correctly describes the situation?

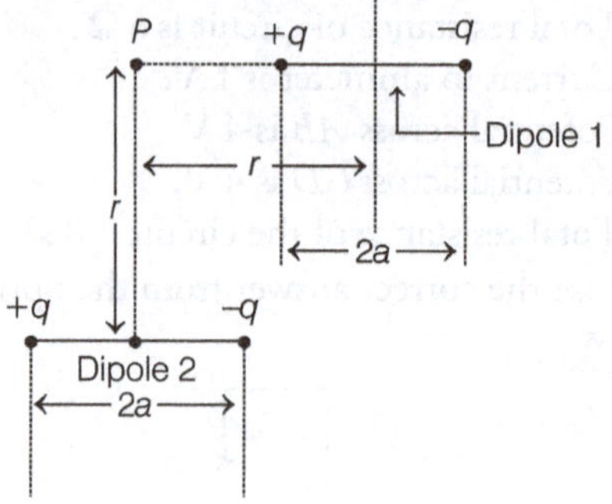

(a) $\frac{a}{r} \sim 20$ (b) $\frac{a}{r} \sim 10$

(c) $\frac{a}{r} \sim 0.5$ (d) $\frac{a}{r} \sim 3$

43/ Consider a moving coil galvanometer (MCG).

A. The torsional constant in moving coil galvanometer has dimentions $[ML^2T^{-2}]$.

B. Increasing the current sensitivity may not necessarily increase the voltage sensitivity.

C. If we increase number of turns (N) to its double ($2N$), then the voltage sensitivity doubles.

D. MCG can be converted into an ammeter by introducing a shunt resistance of large value in parallel with galvanometer.

E. Current sensitivity of MCG depends inversely on number of turns of coil.

Choose the correct answer from the options given below.

(a) B, D, E Only (b) A, D, Only

(c) A, B Only (d) A, B, E Only

44/ A sub-atomic particle of mass 10^{-30} kg is moving with a velocity 2.21×10^6 m/s. Under the matter wave consideration, the particle will behave closely like ………

($h = 6.63 \times 10^{-34}$ J-s)

(a) Gamma rays

(b) X-rays

(c) Visible radiation

(d) Infrared radiation

Imp.

45/ Consider a circular disc of radius 20 cm with centre located at the origin. A circular hole of radius 5 cm is cut from this disc in such a way that the edge of the hole touches the edge of the disc. The distance of centre of residual or remaining disc from the origin will be

(a) 0.5 cm (b) 2.0 cm (c) 1.0 cm (d) 1.5 cm

Center of Mass

Section B

Numerical Type Questions

46/

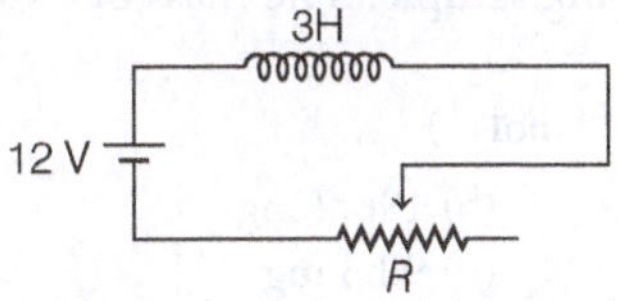

In the given circuit the sliding contact is pulled outwards such that electric current in the circuit changes at the rate of 8 A/s. At an instant when R is 12 Ω, the value of the current in the circuit will be .... A.

47/ Two particles are located at equal distance from origin. The position vectors of those are represented by $\mathbf{A} = 2\hat{\mathbf{i}} + 3n\hat{\mathbf{j}} + 2\hat{\mathbf{k}}$ and $\mathbf{B} = 2\hat{\mathbf{i}} - 2\hat{\mathbf{j}} + 4p\hat{\mathbf{k}}$, respectively. If both the vectors are at right angle to each other, the value of n^{-1} is ....

48/ An ideal gas initially at 0°C temperature, is compressed suddenly to one-fourth of its volume. If the ratio of specific heat at constant pressure to that at constant volume is $\frac{3}{2}$, the change in temperature due to the thermodynamic process is .... K.

49/ A force $F = x^2 y\hat{i} + y^2\hat{j}$ acts on a particle in a plane $x + y = 10$. The work done by this force during a displacement from (0, 0) to (4 m, 2 m) is .... J (round off to the nearest integer)

50/ A positive ion A and a negative ion B has charges 6.67×10^{-19} C and 9.6×10^{-10} C and masses 19.2×10^{-27} kg and 9×10^{-27} kg respectively. At an instant, the ions are separated by a certain distance r. At that instant the ratio of the magnitudes of electrostatic force to gravitational force is $P \times 10^{-13}$, where the value of P is ....

(Take $\frac{1}{4\pi\varepsilon_0} = 9 \times 10^9$ Nm^{2}C^{-2} and universal gravitational constant as 6.67×10^{-11} Nm^{2}kg^{-2})

Chemistry

Section A

Objective Type Questions

51/ Match the List-I with List-II.

	List-I (Classification of molecules based on octet rule)		List-II (Example)
A.	Molecules obeying octet rule	I.	NO, NO_2
B.	Molecules with incomplete octet	II.	BCl_3, $AlCl_3$
C.	Molecules with incomplete octet with odd electron	III.	H_2SO_4, PCl_5
D.	Molecules with expanded octet	IV.	CCl_4, CO_2

Choose the correct answer from the options given below:

(a) A-II, B-IV, C-III, D-I (b) A-III, B-II, C-I, D-IV
(c) A-IV, B-II, C-I, D-III (d) A-IV, B-I, C-III, D-II

52/ Given below are two statements:

Statement I Fructose does not contain an aldehydic group but still reduces Tollen's reagent.

Statement II In the presence of base, fructose undergoes rearrangement to given glucose.

In the light of the above statements, choose the correct answer from the options give below

(a) Both Statement I and Statement II are true.
(b) Statement I is false but Statement II is true.
(c) Both Statement I and Statement II are false.
(d) Statement I is true but Statement II is false.

53/ The *d*-electronic configuration of an octahedral Co(II) complex having magnetic moment of 3.95 BM is

(a) $e^4 t_2^3$ (b) $t_{2g}^6 e_g^1$

(c) $t_{2g}^3 e_g^0$ (d) $t_{2g}^5 e_g^2$

54/ The complex that shows *facial - meridional* isomerism is

(a) $[Co(NH_3)_4 Cl_2]^+$ (b) $[Co(NH_3)_3 Cl_3]$

(c) $[Co(en)_2 Cl_2]^+$ (d) $[Co(en)_3]^{3+}$

55/ Match the List-I with List-II

List-I (Name reaction)		List-II (Product obtainable)	
A.	Swarts reaction	I.	Ethyl benzene
B.	Sandmeyer's reaction	II.	Ethyl iodide
C.	Wurtz Fittig reaction	III.	Cyanobenzene
D.	Finkelstein reaction	IV.	Ethyl fluoride

Choose the correct answer from the options given below:

(a) A-IV, B-I, C-III, D-II (b) A-IV, B-III, C-I, D-II

(c) A-II, B-III, C-I, D-IV (d) A-II, B-I, C-III, D-IV

56/ Ice at – 5°C is heated to become vapour with temperature of 110°C at atmospheric pressure. The entropy change associated with this process can be obtained from

(a) $\int_{268K}^{383K} C_p \, dT + \frac{\Delta H_{\text{melting}}}{273} + \frac{\Delta H_{\text{boiling}}}{373}$

(b) $\int_{268K}^{383K} C_p dT + \frac{q_{\text{rev}}}{T}$

(c) $\int_{268K}^{273K} C_{p,m} \, dT + \frac{\Delta H_m, \text{fusion}}{T_f} + \frac{\Delta H_m, \text{vaporsation}}{T_b} + \int_{273K}^{373K} C_{p,m} dT + \int_{373K}^{383K} C_{p,m} \, dT$

(d) $\int_{268K}^{273K} \frac{C_{p,m}}{T} dT + \frac{\Delta H_m, \text{fusion}}{T_f} + \frac{\Delta H_m, \text{vaporisation}}{T_b} + \int_{273K}^{373K} \frac{C_{p,m} dT}{T} + \int_{373K}^{383K} \frac{C_{p,m} \, dT}{T}$

57/ The correct set of ions (aqueous solution) with same colour from the following is

(a) V^{2+}, Cr^{3+}, Mn^{3+} (b) Zn^{2+}, V^{3+}, Fe^{3+}

(c) $Sc^{3+}, Ti^{3+}, Cr^{2+}$ (d) Ti^{4+}, V^{4+}, Mn^{2+}

58/ Given below are two statements.

Statement I In Lassaigne's test, the covalent organic molecules are transformed into ionic compounds.

Statement II The sodium fusion extract of an organic compound having N and S gives prussian blue colour with $FeSO_4$ and $Na_4[Fe(CN)_6]$.

In the light of the above statements, choose the correct answer from the options given below

(a) Statement I is false but Statement II is true.

(b) Both Statement I and Statement II is true.

(c) Both Statement I and Statement II are false.

(d) Statement I is true but Statement II is false.

59/ 2.8×10^{-3} mol of CO_2 is left after removing 10^{21} molecules from its '*x*' mg sample. The mass of CO_2 taken initially is

(Given, $N_A = 6.02 \times 10^{23} \text{ mol}^{-1}$)

(a) 150.4 mg (b) 196.2 mg

(c) 48.2 mg (d) 98.3 mg

60/ The correct stability order of the following species/molecules is

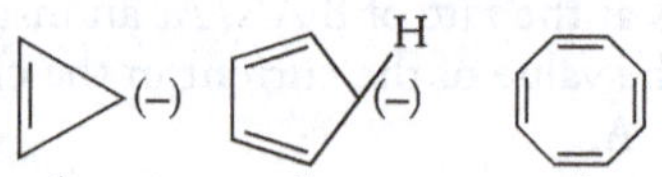

p *q* *r*

(a) $q > p > r$ (b) $p > q > r$

(c) $q > r > p$ (d) $r > q > p$

61/ What amount of bromine will be required to convert 2 g of phenol into 2, 4, 6-tribromophenol?

(Given, molar mass in g mol^{-1} of C, H, O, Br are 12, 1, 16, 80 respectively.)

(a) 6.0 g (b) 10.22 g

(c) 4.0 g (d) 20.44 g

62/ Propane molecule on chlorination under photochemical condition gives two di-chloro products. "*x*" and "*y*". Amongst "*x*" and "*y*", "*x*" is an optically active molecule.

How many tri-chloro products (consider only structural isomers) will be obtained from "*x*" when it is further treated with chlorine under the photochemical condition?

(a) 4 (b) 2

(c) 3 (d) 5

63/ The element that does not belong to the same period of the remaining elements (modern periodic table) is

(a) platinum (b) iridium

(c) palladium (d) osmium

64/ The major product of the following reaction is

$CH_3CH_2CH{=}O \xrightarrow[\text{Reflux}]{\text{Excess HCHO alkali}} ?$

(a) $CH_3-C(CH_2-OH)_2-CH_2-OH$

(b) $CH_3-CH(CH_2-OH)-CH{=}O$

(c) $CH_3-CH_2-CH_2-OH$

(d) $CH_3-C(=CH_2)-CH{=}O$

65/ $CrCl_3 \cdot xNH_3$ can exist as a complex. 0.1 molal aqueous solution of this complex shows a depression in freezing point of 0.558°C. Assuming 100% ionisation of this complex and coordination number of Cr is 6, the complex will be

(Given, $K_f = 1.86$ K kg mol^{-1})

(a) $[Cr(NH_3)_3Cl_3]$
(b) $[Cr(NH_3)_4Cl_2]Cl$
(c) $[Cr(NH_3)_6]Cl_3$
(d) $[Cr(NH_3)_5Cl]Cl_2$

Imp.

66/ Which among the following react with Hinsberg's reagent?

(A) 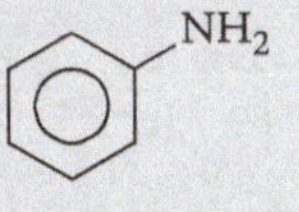

(B) 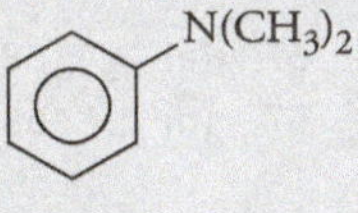

(C) CH_3-NH_2 (D) $N(CH_3)_3$

(E) 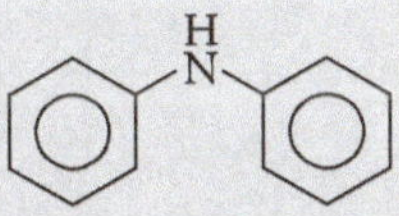

Choose the correct answer from the options given below

(a) C and D only (b) B and D only
(c) A, C and E only (d) A, B and E only

Test of Amines

67/ Which of the following happens when NH_4OH is added gradually to the solution containing 1M A^{2+} and 1 M B^{3+} ions?

(Given, $K_{sp}[A(OH)_2] = 9 \times 10^{-10}$ and $K_{sp}[B(OH)_3] = 27 \times 10^{-18}$ at 298 K.)

(a) $A(OH)_2$ will precipitate before $B(OH)_3$
(b) $B(OH)_3$ will precipitate before $A(OH)_2$
(c) $A(OH)_2$ and $B(OH)_3$ will precipitate together
(d) Both $A(OH)_2$ and $B(OH)_3$ do not show precipitation with NH_4OH

68/ $FeO_4^{2-} \xrightarrow{+2.0\text{ V}} Fe^{3+} \xrightarrow{0.8\text{ V}} Fe^{2+} \xrightarrow{-0.5\text{ V}} Fe^0$

In the above diagram, the standard electrode potentials are given in volts (over the arrow).

The value of $E^\circ_{FeO_4^{2-}/Fe^2}$ is

(a) 2.1 V (b) 1.2 V (c) 1.7 V (d) 1.4 V

69/ The incorrect statement among the following is

(a) PH_3 shows lower proton affinity than NH_3.
(b) NO_2 can dimerise easily.
(c) SO_2 can act as an oxidising agent, but not as a reducing agent.
(d) PF_5 exists but NF_5 does not.

70/ Heat treatment of muscular pain involves radiation of wavelength of about 900 nm. Which spectral line of H atom is suitable for this?

(Given, Rydberg constant $R_H = 10^5$ cm^{-1}, $h = 6.6 \times 10^{-34}$ J s, $c = 3 \times 10^8$ m/s)

(a) Balmer series, $\infty \rightarrow 2$
(b) Paschen series, $5 \rightarrow 3$
(c) Lyman series, $\infty \rightarrow 1$
(d) Paschen series, $\infty \rightarrow 3$

Section B

Numerical Type Questions

71/ The standard enthalpy and standard entropy of decomposition of N_2O_4 to NO_2 are 55.0 kJ mol^{-1} and 175.0 J/K/mol respectively. The standard free energy change for this reaction at 25°C in J mol^{-1} is ...... (Nearest integer)

72/ For the thermal decomposition of N_2O_5 (g) at constant volume, the following table can be formed. For the reaction mentioned below.

$$2N_2O_5(g) \rightarrow 2N_2O_4(g) + O_2(g)$$

S.No.	Time	Total pressure (atm)
1.	0	0.6
2.	100	'x'

$x =$ .... $\times 10^{-3}$ atm [Nearest integer]

(Given, rate constant for the reaction is $4.606 \times 10^{-2} s^{-1}$.)

73/ Consider the following sequence of reactions to produce major product (*A*)

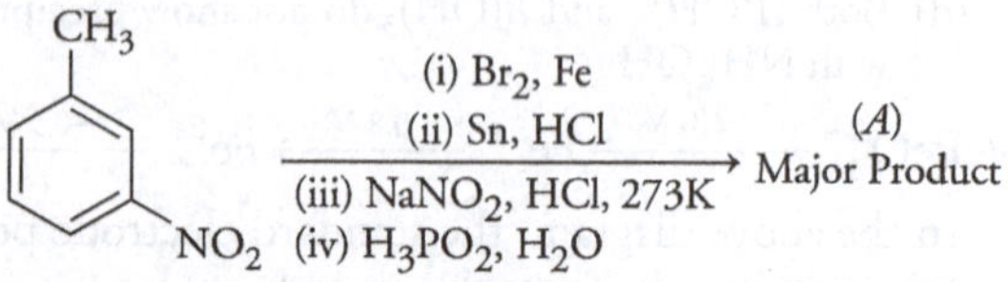

Molar mass of product (*A*) is .... g mol^{-1}.

(Given, molar mass in g mol^{-1} of C : 12, H : 1, O : 16, Br : 80, N : 14, P : 31)

74/ If 1 mM solution of ethylamine produces pH = 9, then the ionisation constant (K_b) of ethylamine is 10^{-x}. The value of x is .... (Nearest integer). [The degree of ionisation of ethylamine can be neglected with respect to unity.]

Imp.

75/ During "S" estimation, 160 mg of an organic compound gives 466 mg of barium sulphate. The percentage of sulphur in the given compound is ....%.

(Given, molar mass in g mol^{-1} of Ba : 137, S : 32, O : 16)

Carius Method for Estimation of Sulphur

Answer Key

Mathematics

1 (d)	2 (b)	3 (b)	4 (a)	5 (a)	6 (a)	7 (b)	8 (c)	9 (d)	10 (c)
11 (d)	12 (b)	13 (c)	14 (c)	15 (c)	16 (a)	17 (d)	18 (d)	19 (c)	20 (d)
21 (19)	22 (30)	23 (117)	24 (77)	25 (612)					

Physics

26 (d)	27 (a)	28 (a)	29 (d)	30 (b)	31 (b)	32 (c)	33 (b)	34 (c)	35 (b)
36 (a)	37 (a)	38 (d)	39 (d)	40 (c)	41 (b)	42 (*)	43 (c)	44 (b)	45 (c)
46 (3)	47 (3)	48 (273)	49 (152)	50 (*)					

Chemistry

51 (c)	52 (a)	53 (d)	54 (b)	55 (b)	56 (d)	57 (a)	58 (d)	59 (b)	60 (c)
61 (b)	62 (c)	63 (c)	64 (a)	65 (d)	66 (c)	67 (b)	68 (c)	69 (c)	70 (d)
71 (2850)	72 (897)	73 (171)	74 (7)	75 (40)					

***Note** (*) None of the option is correct.*

Detailed Explanations

Mathematics

1. *(d)* $A = \{1, 2, 3, 4\}$

$R = \{(1, 2), (2, 3), (3, 3)\}$

Minimum elements will be added in R

$(1, 1), (2, 2), (4, 4)\ (2, 1), (3, 2), (3, 1), (1, 3)$, then R become an equivalence relation

$\therefore$ Minimum number of elements = 7

2. *(b)*

> **Key idea**
> Sum of 'n' term in AP is
> $$S_n = \frac{n}{2}[2a + (n-1)d]$$

Given that $a = 3$

and $S_4 = \frac{1}{5}(S_8 - S_4)$

$$6S_4 = S_8$$

$$\Rightarrow \quad 6 \times \frac{4}{2}[2 \times 3 + 3d] = \frac{8}{2}[2 \times 3 + 7d]$$

$$\Rightarrow \quad 3[6 + 3d] = 6 + 7d$$

$$\Rightarrow \quad 2d = -12$$

$$\Rightarrow \quad d = -6$$

$$S_{20} = 10[6 + 19 \times (-6)]$$

$$= 10[6 - 114] = -1080$$

3. *(b)*

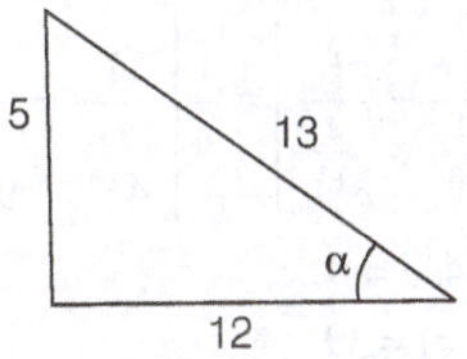

Let $\cos\alpha = \frac{12}{13} \Rightarrow \sin\alpha = \frac{5}{13}$ and

$\tan\alpha = \frac{5}{12}$

$$\because \cos^{-1}\left[\frac{12}{13}\cos x + \frac{5}{13}\sin x\right]$$

$$= \cos^{-1}[\cos x \cos\alpha + \sin x \sin\alpha]$$

$$= \cos^{-1}(\cos(x - \alpha)) = x - \alpha \quad \left[\because x - \alpha \in \left(\frac{-\pi}{2}, \frac{\pi}{2}\right)\right]$$

$$= x - \tan^{-1}\frac{5}{12}$$

$$\left[\because \cos\alpha = \frac{12}{13} \Rightarrow \tan\alpha = \frac{5}{12}\right]$$

4. *(a)* Given that

$$2(3 + y)e^{2x}dx - (7 + e^{2x})dy = 0$$

$$\Rightarrow \quad \frac{dy}{dx} = \frac{2(3 + y)e^{2x}}{7 + e^{2x}}$$

$$\frac{dy}{dx} - \frac{2ye^{2x}}{7 + e^{2x}} = \frac{6 \cdot e^{2x}}{7 + e^{2x}} \quad \ldots \text{(i)}$$

This is linear differential equation and it's form is

$$\frac{dy}{dx} + py = Q \qquad [\because \text{IF} = e^{\int pdx}]$$

$$\text{IF} = e^{\int \frac{-2e^{2x}}{7 + e^{2x}}dx}$$

Put $7 + e^{2x} = t$

$$\Rightarrow \quad 2e^{2x}dx = dt = e^{-\log t} = \frac{1}{t} = \frac{1}{7 + e^{2x}}$$

Solution of Eq. (i), we get

$$\therefore \quad y \cdot \frac{1}{7 + e^{2x}} = \int \frac{6e^{2x}}{(7 + e^{2x})^2}dx$$

$$\Rightarrow \quad y \cdot \frac{1}{7 + e^{2x}} = \frac{-3}{7 + e^{2x}} + c$$

This curve passes through (0, 5), then

$$\frac{5}{8} = \frac{-3}{8} + c \Rightarrow c = 1$$

$$\therefore \quad y = -3 + 7 + e^{2x}$$

$$\Rightarrow \quad y = e^{2x} + 4$$

Now, put $x = \log_e 2$, $y = k$, we get

$$k = e^{2\log_e 2} + 4 \Rightarrow k = 8$$

> **Concept Enhancer**
> Solution linear differential equation $\frac{dy}{dx} + py = Q$ is given by
> $$y(\text{IF}) = \int Q \cdot (\text{IF})dx + C,$$
> where $\text{IF} = e^{\int Pdx}$

5. *(a)* Given, $f(x) = \log_e x$

$$g(x) = \frac{x^4 - 2x^3 + 3x^2 - 2x + 2}{2x^2 - 2x + 1}$$

Domain of $g(x) = x \in R$

$[2x^2 - 2x + 1 > 0, \forall x \in R]$

domain of $f(x) = x \in (0, \infty)$

For domain of $fog \Rightarrow g(x) > 0$

$$\therefore \frac{x^4 - 2x^3 + 3x^2 - 2x + 2}{2x^2 - 2x + 1} > 0$$

$$[2x^2 + 2x + 1 > 0, \forall x \in R, D = 4 - 16 < 0]$$

$$\Rightarrow x^4 - 2x^3 + 3x^2 - 2x + 2 > 0$$

$$\Rightarrow (x^2 + 1)(x^2 - 2x + 2) > 0, \neq \alpha \in R$$

$\therefore$ Domain of $fog = x \in R$

6. *(a)* $\mathbf{OC} = \alpha\mathbf{OA} + \beta\mathbf{OB}$

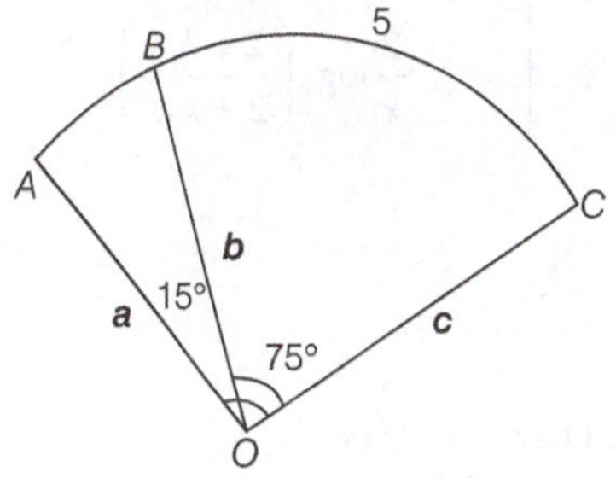

$$\mathbf{c} = \alpha\mathbf{a} + \beta\mathbf{b} \quad \ldots \text{(i)}$$

Taking dot product of **a** both side, we get

$$\Rightarrow \quad \mathbf{a} \cdot \mathbf{c} = \mathbf{a} \cdot \mathbf{a}\,\alpha + (\mathbf{a} \cdot \mathbf{b})\beta$$

$$\Rightarrow \quad \mathbf{a} \cdot \mathbf{c} = \alpha + \beta\cos 15^\circ \quad [\because |\mathbf{a}| = 1]$$

$$\alpha + \beta\cos 15^\circ = 0 \quad \ldots \text{(ii)}$$

Taking dot product of **b** from Eq (i), we get

$$\therefore \quad \mathbf{b} \cdot \mathbf{c} = \alpha\,(\mathbf{a} \cdot \mathbf{b}) + \beta$$

$$\Rightarrow \quad \cos 75^\circ = \alpha\cos 15^\circ + \beta$$

$$\Rightarrow \quad \cos 75^\circ = -\beta\cos^2 15^\circ + \beta$$

$$= \beta(1 - \cos^2 15^\circ)$$

[$\because$ from Eq. (ii)]

$$\Rightarrow \quad \beta = \frac{\cos 75^\circ}{\sin^2 15^\circ}$$

$$\Rightarrow \quad \beta = \frac{1}{\sin 15^\circ} = \frac{2\sqrt{2}}{\sqrt{3}-1}$$

From Eq. (ii), we get

$$\alpha = \frac{-\cos 15^\circ}{\sin 15^\circ} = \frac{-(\sqrt{3}+1)}{(\sqrt{3}-1)}$$

$$\therefore \mathbf{c} = \frac{-(\sqrt{3}+1)}{(\sqrt{3}-1)}\mathbf{a} + \left(\frac{2\sqrt{2}}{\sqrt{3}-1}\right)\mathbf{b}$$

$$\text{Now, } \alpha + \sqrt{2}(\sqrt{3}-1)\beta = \frac{-(\sqrt{3}+1)}{(\sqrt{3}-1)} + \sqrt{2}\,\frac{(\sqrt{3}-1)2\sqrt{2}}{\sqrt{3}-1}$$

$$= 4 - \frac{(\sqrt{3}+1)^2}{2} = 2 - \sqrt{3}$$

7. *(b)*

Key idea

A function $f(x)$ is continuous at $x = a$, then $\lim_{x \to a^+} f(x) = \lim_{x \to a^-} f(x) = f(a)$

$$f(x) = \begin{cases} \frac{2}{x}\{\sin(k_1+1)x + \sin(k_2-1)x\}, & x < 0 \\ 4, & x = 0 \\ \frac{2}{x}\log_e\left(\frac{2+k_1x}{2+k_2x}\right) & x > 0 \end{cases}$$

$$\text{LHL} = \lim_{x \to 0^-} f(x)$$

$$= \lim_{x \to 0^-} \frac{2}{x}\{\sin(k_1+1)x + \sin(k_2-1)x\}$$

$$\lim_{x \to 0^-} 2\left\{\frac{\sin(k_1+1)x}{(k_1+1)x}\cdot(k_1+1) + \frac{\sin(k_2-1)x}{(k_2-1)x}\times(k_2-1)\right\}$$

$$= 2[k_1+1+k_2-1]$$

$$= 2[k_1+k_2]$$

$$\text{RHL} = \lim_{x \to 0^+} f(x) = \lim_{x \to 0^+} \frac{2}{x}\log\left(\frac{2+k_1x}{2+k_2x}\right)$$

$$= \lim_{x \to 0^+} \frac{2}{x}\log\left[1 + \frac{(k_1-k_2)x}{2+k_2x}\right]$$

$k_1 - k_2$ and since, we know that if $f(x)$ is continuous at $x = 0$

$$\Rightarrow \quad \text{LHL} = \text{RHL} = f(0)$$

$$\Rightarrow 2(k_1+k_2) = 4 \Rightarrow k_1 + k_2 = 2 \quad \ldots\text{(i)}$$

$$\Rightarrow \quad k_1 - k_2 = 4 \quad \ldots\text{(ii)}$$

From Eqs (i) and (ii), we get

$$k_1 = 3, k_2 = -1$$

Now $k_1^2 + k_2^2 = 9 + 1 = 10$

8. *(c)* Given word is DAUGHTER

Total number of words = 8!

Total number of words all the vowels come together

AUE, DGHTR

$$1 + 5 = 6 \Rightarrow 6! \times 3!$$

∴ Required number of words when all the vowels never come together

$$= 8! - 6! \times 3!$$
$$= 6!(8 \times 7 - 6)$$
$$= 6! \times 50 = 720 \times 50$$
$$= 36000$$

9. *(d)* Given,

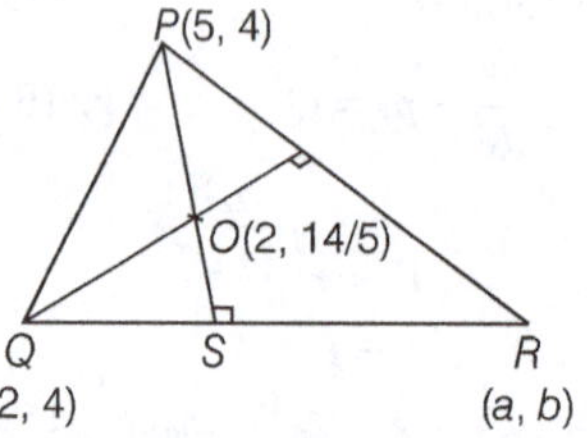

$$\Rightarrow \text{Slope of } PS = \frac{\frac{14}{5} - 4}{2-5} = \frac{-6}{5 \times -3} = \frac{2}{5}$$

$$\Rightarrow \text{Slope of } QR = \frac{-5}{2}$$

$$\Rightarrow \text{Equation of } QR \text{ is } y - 4 = \frac{-5}{2}(x+2)$$

$$5x + 2y + 2 = 0 \quad \ldots\text{(i)}$$

Similarly, equation of line PR

$$10x - 3y - 38 = 0 \quad \ldots\text{(ii)}$$

From, (i) and (ii), we get

$$10x + 4y + 4 = 0$$
$$10x - 3y - 38 = 0$$
$$\underline{- \quad + \quad +}$$
$$7y + 42 \Rightarrow y = -6$$
$$y = -6$$

$$\Rightarrow \quad x = 2$$

Point $R \equiv (2, -6) = (a, b)$ [given]

$\Rightarrow a = 2, b = -6$

Centroid

$$= \left(\frac{5-2+2}{3}, \frac{4+4-6}{3}\right) \equiv \left(\frac{5}{3}, \frac{2}{3}\right)$$

$= (c, d)$ [given]

$$\Rightarrow c = \frac{5}{2}, d = \frac{2}{3}$$

$$\therefore \quad c + 2d = \frac{5}{3} + \frac{4}{3} = 3$$

10. *(c)* $I(x) = \int \frac{dx}{(x-11)^{\frac{11}{13}}(x+15)^{\frac{15}{13}}}$

$$\text{Put } \frac{x-11}{x+15} = t; \quad \frac{x-11+15-15}{x+15} = t$$

$$\Rightarrow dt = \frac{26}{(x+15)^2}dx$$

$$I(x) = \frac{1}{26}\int \frac{(x+15)^2}{\left(\frac{x-11}{x+15}\right)^{\frac{11}{13}}(x+15)^2}dt$$

$$= \frac{1}{26}\int \frac{dt}{(t)^{\frac{11}{13}}} = \frac{1}{26}\frac{t^{\frac{2}{13}}}{\frac{2}{13}} + C$$

$$I(x) = \frac{1}{4}\left(\frac{x-11}{x+15}\right)^{\frac{2}{13}} + C$$

∴

$$I(37) - I(24) = \frac{1}{4}\left[\left(\frac{26}{52}\right)^{\frac{2}{13}} - \left(\frac{13}{39}\right)^{\frac{2}{13}}\right]$$

$$= \frac{1}{4}\left(\frac{1}{2^{\frac{2}{13}}} - \frac{1}{3^{\frac{2}{13}}}\right)$$

$$= \frac{1}{4}\left[\frac{1}{b^{\frac{1}{13}}} - \frac{1}{c^{\frac{1}{13}}}\right] = \frac{1}{4}\left[\frac{1}{4^{\frac{1}{13}}} - \frac{1}{9^{\frac{1}{13}}}\right]$$

$\Rightarrow b = 4, c = 9$

$\therefore 3(b + c) = 39$

11. *(d)* $\sin 70^\circ(\cot 10^\circ \cot 70^\circ - 1)$

$$= \sin 70^\circ \left[\frac{\cos 10^\circ \cos 70^\circ - \sin 10^\circ \sin 70^\circ}{\sin 10^\circ \sin 70^\circ}\right]$$

$$= \frac{\cos 80^\circ}{\sin 10^\circ} = 1$$

12. *(b)* We know that

$$\text{Median} = l + \left(\frac{\frac{N}{2} - cf}{f}\right) \times h$$

$$\Rightarrow \quad 14 = 12 + \left(\frac{\frac{N}{2} - 18}{12}\right) \times 6$$

$$\Rightarrow \quad 2 \times 2 = \frac{N}{2} - 18$$

$$\Rightarrow \quad N = 44$$

13. *(c)* $(\text{Die})_1 = \{1, 1, 2, 2, 3, 4\}$

$(\text{Die})_2 = \{1, 2, 2, 3, 3, 4\}$

Probability (getting sum 4 or 5)

$= P(\text{Sum } 4) + P(\text{Sum } 5)$

$$= \frac{2}{6} \times \frac{1}{6} + \frac{2}{6} \times \frac{2}{6} \times \frac{1}{6} + \frac{1}{6} \times \frac{1}{6} + \frac{2}{6} \times \frac{2}{6}$$

$$+ \frac{2}{6} \times \frac{1}{6} + \frac{1}{6} \times \frac{2}{6} + \frac{1}{6} \times \frac{1}{6} = \frac{16}{36}$$

$$\therefore \text{Probability} = \frac{4}{9}$$

14. *(c)*

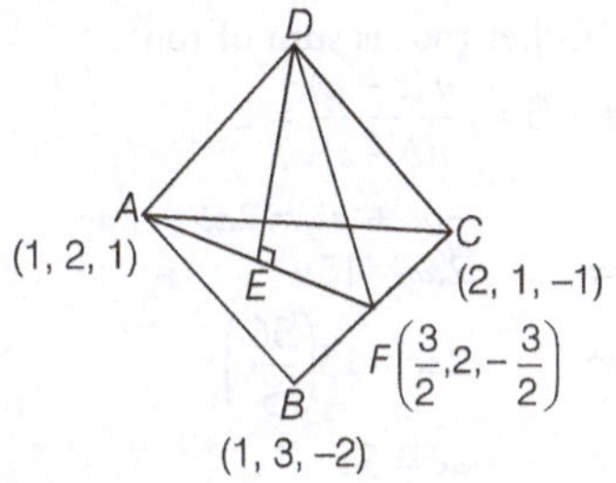

$$\text{Area of } \Delta ABC = \frac{1}{2} |\mathbf{AB} \times \mathbf{AC}|$$

$$= \frac{1}{2} |5\hat{i} + 3\hat{j} + \hat{k}| = \frac{1}{2}\sqrt{35}$$

$\therefore$ Volume of tetrahedron

$$= \frac{1}{3} \text{ Base area} \times \text{Height}$$

$$\Rightarrow \frac{\sqrt{805}}{6\sqrt{2}} = \frac{1}{3} \times \frac{1}{2}\sqrt{35} \times h$$

$$\therefore \quad h = \sqrt{\frac{23}{2}}$$

Again, $AE^2 = AD^2 - DE^2$

$$\Rightarrow AE^2 = \frac{110}{9} - \frac{23}{2} = \frac{13}{18} \Rightarrow AE = \sqrt{\frac{13}{18}}$$

$$\mathbf{AE} = |AE| \cdot \left|\frac{\hat{i} - 5\hat{k}}{\sqrt{26}}\right|$$

$$= \sqrt{\frac{13}{18}}\left(\frac{\hat{i} - 5\hat{k}}{\sqrt{26}}\right) = \frac{\hat{i} - 5\hat{k}}{6}$$

$\therefore$ Position vector of

$$E = \frac{\hat{i} - 5\hat{k}}{6} + \hat{i} + 2\hat{j} + \hat{k}$$

$$= \frac{1}{6}(7\hat{i} + 12\hat{j} + \hat{k})$$

15. *(c)* Given that $\left|\dfrac{\bar{z} - i}{2\bar{z} + i}\right| = \dfrac{1}{3}$

$$\left|\frac{\bar{z} - i}{\bar{z} + 1/2}\right| = \frac{2}{3}$$

Let $z = x + iy$, $x, y \in R$

$\Rightarrow \bar{z} = x - iy$

$$3|x - iy - i| = 2\left|x - iy + \frac{1}{2}\right|$$

$$\Rightarrow 9[x^2 + (y + 1)^2] = 4\left[x^2 + \left(y - \frac{1}{3}\right)^2\right]$$

$$\Rightarrow 5x^2 + 5y^2 + 22y + 8 = 0$$

$$\text{Centre} \equiv \left(0, \frac{-11}{5}\right)$$

$$\text{Area of } \Delta = \left|\frac{1}{2}\begin{vmatrix} 0 & 0 & 1 \\ 0 & -\frac{11}{5} & 1 \\ \alpha & 0 & 1 \end{vmatrix}\right| = 11$$

$$\left(-\frac{11}{5}\alpha\right)^2 = (11 \times 2)^2$$

$$\Rightarrow \quad \alpha^2 = 100$$

16. *(a)* $[A(\text{adj}(A^{-1}) + \text{adj}(B^{-1}))^{-1} B]^{-1}$

$B^{-1}(\text{adj}(A^{-1}) + \text{adj}(B^{-1})A^{-1}$

$B^{-1}\text{adj}(A^{-1})A^{-1} + B^{-1}(\text{adj}(B^{-1}))A^{-1}$

$B^{-1}|A^{-1}|I + |B^{-1}|IA^{-1}$

$$\frac{B^{-1}}{|A|} + \frac{A^{-1}}{|B|}$$

$$= \frac{\text{adj}B}{|B||A|} + \frac{\text{adj}A}{|A||B|}$$

$$= \frac{1}{|A||B|}(\text{adj}B + \text{adj}A)$$

Concept Enhancer

(i) $(ABC)^{-1} = C^{-1}B^{-1}A^{-1}$

(ii) $(A^{-1})^{-1} = A$

(iii) $A \text{ adj } A = |A|I$

17. *(d)*

Key idea

King property

$$\int_a^b f(x)dx = \int_a^b f(a + b - x)dx$$

Let $I =$

$$\int_{e^2}^{e^4} \frac{1}{x}\left(\frac{e^{(\log_e x)^2 + 1)^{-1}}}{e^{((\log_e x)^2 + 1)^{-1}} + e^{((6 - \log_e x)^2 + 1)^{-1}}}\right) dx$$

Put $\log_e x = t \Rightarrow \dfrac{1}{x} dx = dt$

When $x = e^2, t = 2$

and $x = e^4, t = 4$

$$\therefore \; I = \int_2^4 \frac{e^{\frac{1}{1+t^2}}}{e^{\frac{1}{1+t^2}} + e^{\frac{1}{1+(6-t)^2}}} dt \quad \ldots\text{(i)}$$

$$I = \int_2^4 \frac{e^{\frac{1}{1+(6-t)^2}}}{e^{\frac{1}{1+(6-t)^2}} + e^{\frac{1}{1+t^2}}} dt \quad \ldots\text{(ii)}$$

Use property

$$\int_a^b f(x)dx = \int_a^b f(a + b - x)dx$$

From Eqs. (i) and (ii), we get

$$2I = \int_2^4 dt \Rightarrow 2I = 2$$

$$\Rightarrow \quad I = 1$$

18. *(d)* Given, line

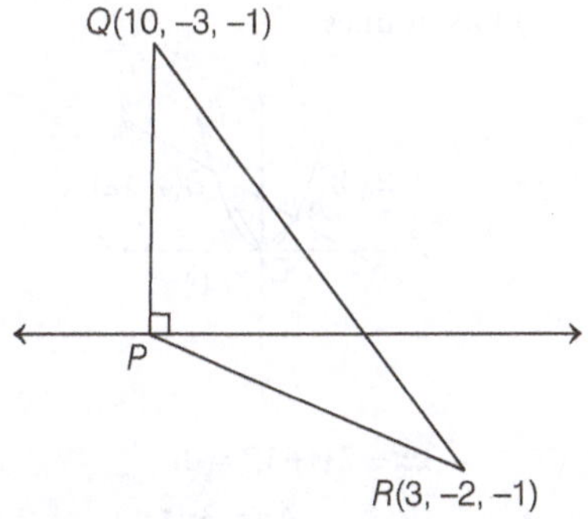

$$L: \frac{x - 3}{7} = \frac{y - 2}{-1} = \frac{z + 1}{-2} = \lambda \text{ (say)}$$

Any point on line be (P)

$(7\lambda + 3, -\lambda + 2, -2\lambda - 1)$

Direction ratio of

$$QP \equiv 7\lambda - 7, -\lambda + 5, -2\lambda$$

Now, $PQ \perp$ line (i)

$$7(7\lambda - 7) - 1(-\lambda + 5) - 2(-2\lambda) = 0$$

$$\Rightarrow \quad \lambda = 1$$

$$P \equiv (10, 1, -3)$$

$$PQ = -4\hat{j} + 2\hat{k}$$

$$PR = -7\hat{i} - 3\hat{j} + 4\hat{k}$$

$$\therefore \text{Required area} = \left| \frac{1}{2} \begin{vmatrix} \hat{i} & \hat{j} & \hat{k} \\ 0 & -4 & 2 \\ -7 & -3 & 4 \end{vmatrix} \right|$$

$$= 3\sqrt{30}$$

19. *(c)*

Key idea
A system of linear equations in 3 variables has infinite solution, then
$\Delta = \Delta_x = \Delta_y = \Delta_z = 0$

For infinitely many solution

$$D \equiv \begin{vmatrix} \lambda - 1 & \lambda - 4 & \lambda \\ \lambda & \lambda - 1 & \lambda - 4 \\ \lambda + 1 & \lambda + 2 & -(\lambda + 2) \end{vmatrix} = 0$$

$$\Rightarrow \quad (\lambda - 3)(2\lambda + 1) = 0 \quad \ldots(i)$$

$$D_x = \begin{vmatrix} 5 & \lambda - 4 & \lambda \\ 7 & \lambda - 1 & \lambda - 4 \\ 9 & \lambda + 2 & -(\lambda + 2) \end{vmatrix} = 0$$

$$\Rightarrow \quad 2(3 - \lambda)(23 - 2\lambda) = 0$$

$$\Rightarrow \quad \lambda = 3 \quad \ldots(ii)$$

[common solution of Eqs. (i) and (ii)]

$$\therefore \lambda^2 + \lambda = 9 + 3 = 12$$

20. *(d)* Given lines

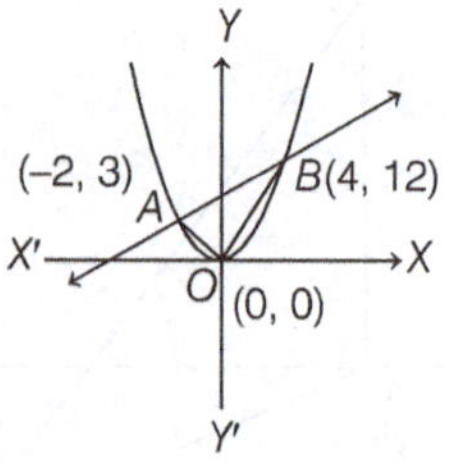

$$3x - 2y + 12 = 0 \quad \ldots(i)$$

$$4y = 3x^2 \quad \ldots(ii)$$

Form Eqs. (i) and (ii), we get

$$2(3x + 12) = 3x^2$$

$$x^2 - 2x - 8 = 0$$

$$x = -2, 4$$

$$\Rightarrow \quad y = 3, 12$$

$$\text{Slope of } OA = \frac{-3}{2}$$

Slope of $OB = 3$

Let AB subtend angle θ at vertex

$$\tan\theta = \left| \frac{-\frac{3}{2} - 3}{1 - \frac{9}{2}} \right|$$

$$\Rightarrow \quad \tan\theta = \frac{9}{7} \quad \Rightarrow \quad \theta = \tan^{-1}\left(\frac{9}{7}\right)$$

21. *(19)* Given that

$$x - y + 1 = 0 \quad \ldots(i)$$

$$\text{CM} = \left| \frac{-3\alpha - 1}{\sqrt{13}} \right|$$

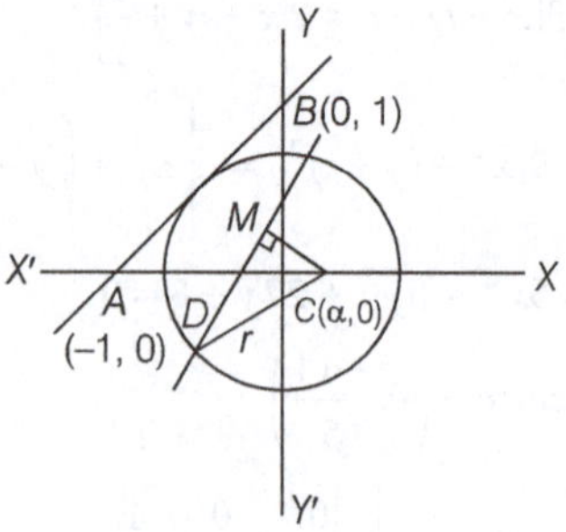

$$r = \left| \frac{\alpha + 1}{\sqrt{2}} \right|$$

$$\Rightarrow \quad (\alpha + 1)^2 = 2r^2 \quad \ldots(i)$$

Now, in ΔCMD

$$CD^2 = DM^2 + CM^2$$

$$r^2 = \frac{(3\alpha + 1)^2}{13} + \frac{4}{13}$$

$$(3\alpha + 1)^2 + 4 = 13r^2 \quad \ldots(ii)$$

From Eqs. (i) and (ii), we get

$$(3\alpha + 1)^2 + 4 = 13\frac{(\alpha + 1)^2}{2}$$

$$\Rightarrow 18\alpha^2 + 12\alpha + 10 = 13\alpha^2 + 26\alpha + 13$$

$$\Rightarrow 5\alpha^2 - 14\alpha - 3 = 0$$

$$\Rightarrow 5\alpha^2 - 15\alpha + \alpha - 3 = 0$$

$$\alpha = \frac{-1}{5}, 3 \quad [\because \alpha > 0 \Rightarrow \alpha = 3]$$

From Eq. (i), we get $r = 2\sqrt{2}$

Now, $\alpha e = 3$, $2\alpha = 4\sqrt{2} \Rightarrow \alpha = 2\sqrt{2}$

$$\alpha^2 e^2 = 9$$

$$\Rightarrow \alpha^2\left(1 + \frac{\beta^2}{\alpha^2}\right) = 9$$

$$\Rightarrow \alpha^2 + \beta^2 = 9 \Rightarrow \beta^2 = 1$$

$$\therefore 2\alpha^2 + 3\beta^2 = 2(8) + 3 = 19$$

22. *(30)* We have, $5x^3 - 15x - a = 0$

Let $f(x) = 5x^3 - 15x$

$$f'(x) = 15x^2 - 15 = 15(x - 1)(x + 1)$$

$$\therefore \quad f(1)f(-1) < 0$$

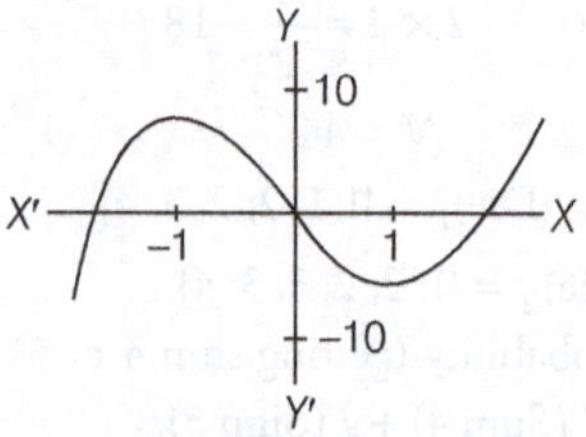

$$\Rightarrow (-10 - a)(10 - a) < 0 \Rightarrow a \in (-10, 10)$$

$$\therefore \quad \alpha = -10, \beta = 10$$

$$\therefore \quad \beta - 2\alpha = 30$$

23. *(117)* We have,

$$a(b - c)x^2 + b((c - a)x + c(a - b) = 0 \ldots(i)$$

$\therefore \quad x = 1$ is a root of Eq. (i), we get

$\therefore$ Other root is sum of root

$$\alpha + \beta = \frac{-b(c - a)}{a(b - c)} = 2$$

$$\Rightarrow \quad -bc + ab = 2ab - 2ac$$

$$\Rightarrow \quad 2ac = 15b \quad \ldots(i)$$

$$\Rightarrow \quad 2ac = 15\left(\frac{36}{5}\right)$$

$$\Rightarrow \quad ac = 54 \quad \ldots(ii)$$

$$a + c = 15$$

$$\Rightarrow a^2 + c^2 + 2ac = 225$$

$$\Rightarrow \quad a^2 + c^2 = 225 - 108 = 117$$

24. *(77)*

Key idea
$$\int \sqrt{a^2 - x^2}\,dx = \frac{1}{2}\left[x\sqrt{a^2 - x^2} + a^2 \sin^{-1}\left(\frac{x}{a}\right)\right] + C$$

We have, $x^2 + y^2 = 25$

$$y = \begin{cases} x - 1, & x \geq 1 \\ -x + 1 & x < 1 \end{cases}$$

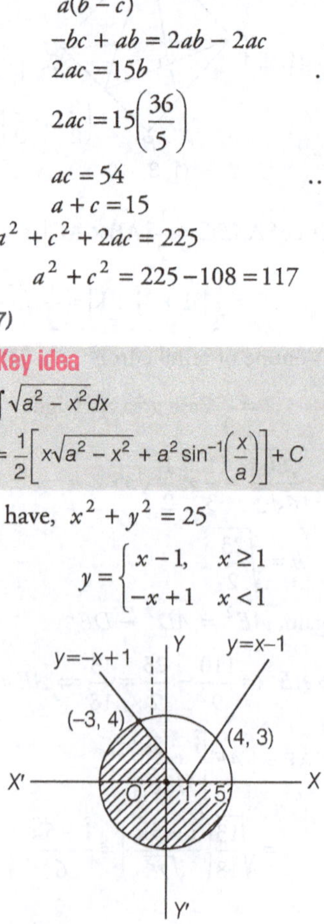

$$\Rightarrow \quad x^2 + (x-1)^2 = 25$$

$$\Rightarrow \quad x = 4$$

$$x^2 + (-x+1)^2 = 25$$

$$\Rightarrow \quad x = -3$$

Area of shaded region

$$= 25\pi - \int_{-3}^{4} \sqrt{25 - x^2}\, dx + \frac{1}{2} \times 4 \times 4 + \frac{1}{2} \times 3 \times 3$$

$$= 25\pi - \left[\frac{x}{2}\sqrt{25 - x^2} + \frac{25}{2}\sin^{-1}\frac{x}{5}\right]_{-3}^{4} + \frac{25}{2}$$

$$A = 25\pi + \frac{25}{2} - \left[(6+3) + \frac{25}{2}\frac{\pi}{2}\right]$$

$$= \frac{75\pi}{4} + \frac{1}{2}$$

$$\Rightarrow \frac{75\pi}{4} + \frac{1}{2} = \frac{1}{4}(b\pi + c)$$

$$\Rightarrow \quad b = 75, c = 2$$

$$\therefore \quad b + c = 75 + 2 = 77$$

25. *(612)* We have, $(1 + 2^{\frac{1}{3}} + 3^{\frac{1}{2}})^6$

$$\Rightarrow \frac{6!}{r_1!r_2!r_3!}(1)^{r_1}(2)^{\frac{r_2}{3}}3^{\frac{r_3}{2}}$$

r_1	r_2	r_3
6	0	0
4	0	2
2	0	4
0	0	6
3	3	0
1	3	2
0	6	0

∴ Required sum

$$= \frac{6!}{6!0!0!} + \frac{6!(3)}{4!0!2!} + \frac{6!}{2!0!4!}(3)^2 + \frac{6!}{0!0!6!}(3)^3 + \frac{6!}{3!3!0!}(2) + \frac{6!}{1!3!2!}(2^1)(3^1) + \frac{6!}{0!6!0!}2^2$$

$$= 1 + 45 + 135 + 27 + 40 + 360 + 4$$

$$= 612$$

Physics

26. *(d)* $\phi = \alpha\sigma + \beta\lambda$

$$[\phi] = [\alpha\sigma] = [\beta\lambda]$$

$$[\alpha] = \frac{[\phi]}{[\sigma]}, [\beta] = \frac{[\phi]}{[\lambda]}$$

$$\frac{[\alpha]}{[\beta]} = \frac{\frac{[\phi]}{[\sigma]}}{\frac{[\phi]}{[\lambda]}} = \frac{[\lambda]}{[\sigma]} = \frac{\left[\frac{Q}{L}\right]}{\left[\frac{Q}{A}\right]} = [L]$$

$$\frac{[\alpha]}{[\beta]} = [L] = \text{Displacement}$$

27. *(a)* **Statement I** Hot water is less viscous than cold water, hence hot water flows faster than cold water, so statement I is true.

Statement II Soap water has a lower surface tension as compared to fresh water because soap molecules disrupt the hydrogen bond between water molecules at the surface, effectively weaking the cohesive forces that create surface tension, so statement II is false.

28. *(a)* $m = 10\text{ g} = 10 \times 10^{-3}$ kg

$A = h \times h = 10 \times 10\text{ cm}^2 = (0.1)^2\text{ m}^2$

When the body is floating

$$mg = F_b$$

$$mg = \rho_f gAx \qquad \ldots\text{(i)}$$

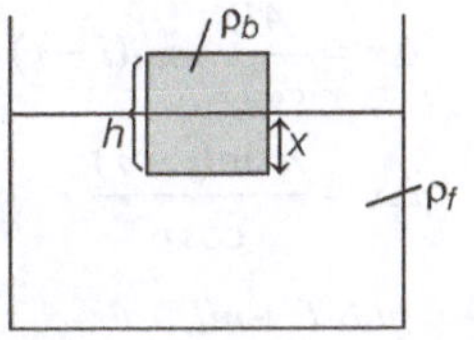

Let it is slightly displaced inside the filuid. Let at any instant, it is further y inside the fluid.

$$F = ma$$

$$F_b' - mg = ma$$

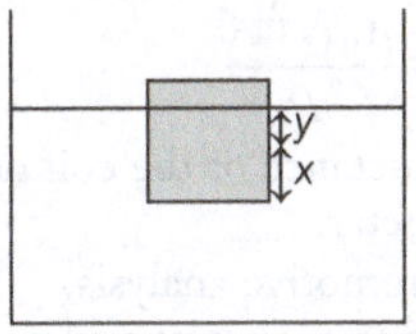

$$\rho_f A(x+y)g - mg = m\left(-\frac{d^2y}{dt^2}\right)$$

$$\rho_f Axg + \rho_f Ayg - mg = -\frac{md^2y}{dt^2}$$

Using Eq. (i)

$$\rho_f Ayg = -m\frac{d^2y}{dt^2}$$

$$\frac{d^2y}{dt^2} + \frac{\rho_f Agy}{m} = 0$$

$$\omega = \sqrt{\frac{\rho_f Ag}{m}}$$

$$T = 2\pi\sqrt{\frac{m}{\rho_f Ag}} = 2\pi\sqrt{\frac{10^{-2}}{10^3 \times (0.1)^2 \times 10}}$$

$$y\pi \times 10^{-2} = 2\pi\sqrt{10^{-4}}$$

$$y\pi \times 10^{-2} = 2\pi \times 10^{-2}$$

$$\Rightarrow \quad y = 2$$

For a cube/cuboidal block floating in a liquid, the time period is given by

> **Concept Enhancer**
>
> 1. $T = 2\pi\sqrt{\frac{m}{\rho_f Ag}}$
>
> 2. $T = 2\pi\sqrt{\frac{x}{g}}$
>
> 3. $T = 2\pi\sqrt{\frac{h\rho_b}{\rho_f}}$

29. *(d)* Initially capacitor behave as a short circuit.

So, current will be maximum. Charge on capacitor will be zero. Potential difference across capacitor will be zero. (B, C, D only)

30. *(b)* Nucleus n_1 Nucleus n_2

No   No

λ 3λ

$(t_{1/2})_1 = \dfrac{\ln 2}{\lambda}$ $(t_{1/2})_2 = \dfrac{\ln 2}{3\lambda}$

$\Rightarrow (t_{1/2})_2 = \dfrac{(t_{1/2})_1}{3}$

$\Rightarrow$ Nucleus n_2 covers 3 half life in the same time nucleus n_1 covers 1 half life.

After 1 half-life of n_1, number of nuclei left

$$= \frac{N_0}{2^n} = \frac{N_0}{2^1} = \frac{N_0}{2}$$

After 3 half-life of n_2, number of nuclei left

$$= \frac{N_0}{2^n} = \frac{N_0}{2^3} = \frac{N_0}{8}$$

According to question,

$$\frac{\text{Number of nuclei of } n_2}{\text{Number of nuclei of } n_1} = \frac{N_0}{8 \times \frac{N_0}{2}} = \frac{1}{4}$$

31. *(b)*

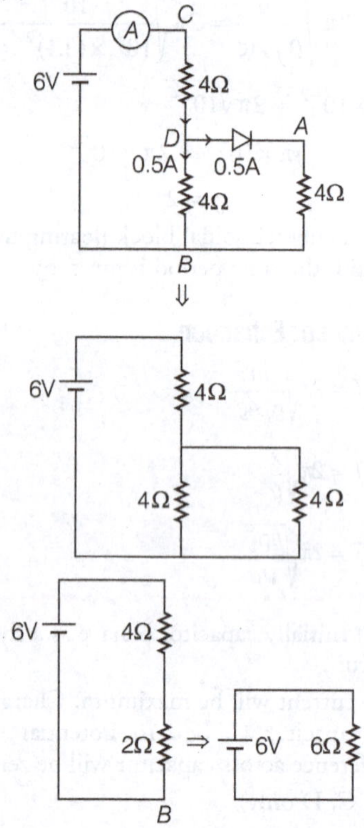

Current through ammeter = 1 A

$R_{\text{net}} = 6\Omega$

$V_{AB} = 0.5 \times 4 = 2\text{ V}$

$V_{CD} = 1 \times 4 = 4\text{ V}$

A, B and D are correct.

32. *(c)*

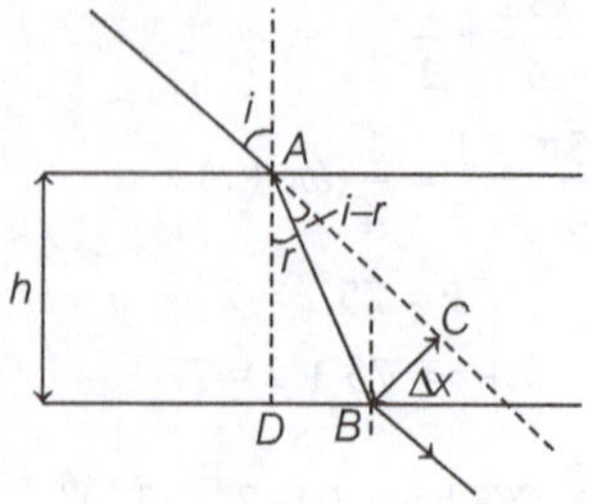

In triangle ABC

$$\sin(i - r) = \frac{\Delta x}{AB}$$

$$\Delta x = AB \sin(i - r) \quad \text{...(i)}$$

In ΔABD

$$\cos r = \frac{AD}{AB}$$

$$AB = \frac{AD}{\cos r} \quad \text{...(ii)}$$

Put the value in Eq. (i)

$$\Delta x = \frac{AD}{\cos r} \cdot \sin(i - r) \; [AD = h]$$

$$\Delta x = \frac{h \sin(i - r)}{\cos r}$$

33. *(b)* $Q = ms\Delta T + mL$

$$625 = m[125 \times 300 + 2.5 \times 10^4]$$

$$625 = m[37500 + 25000]$$

$$m = \frac{625}{62500} = \frac{1}{100}\text{ kg}$$

$$m = 10\text{ g}$$

34. *(c)* $L = \dfrac{\mu_0 N^2 A}{l}$

Self-inductance of the coil depends on its geometry.

In electromotive analysis, "self-inductance is considered the electrical equivalent of mass" in mechanics.

Work needs to be done against the self induced emf because this opposing force tried to resist the change in current flow.

A, C, D, E only correct.

35. *(b)* $E = 57 \cos[7.5 \times 10^6 t - 5 \times 10^{-3} (3x + 4y)] (4\hat{i} - 3\hat{j})$ N/C

$\mathbf{k} = 3\hat{i} + 4\hat{j}$

$$\hat{k} = \frac{3\hat{i} + 4\hat{j}}{\sqrt{(3)^2 + (4)^2}} = \frac{3\hat{i} + 4\hat{j}}{5}$$

$$\mathbf{E} = 4\hat{i} - 3\hat{j}$$

$$\hat{E} = \frac{4\hat{i} - 3\hat{j}}{\sqrt{(4)^2 + (3)^2}} = \frac{4\hat{i} - 3\hat{j}}{5}$$

$$\hat{B} = \hat{k} \times \hat{E}$$

$$\hat{B} = -\hat{z}$$

$$B_0 = \frac{E_0}{c} = \frac{57}{3 \times 10^8}$$

$$\mathbf{B} = \frac{-57}{3 \times 10^8} [\cos(7.5 \times 10^6 t - 5 \times 10^{-3} (3x + 4y)) 5\hat{k}$$

36. *(a)*

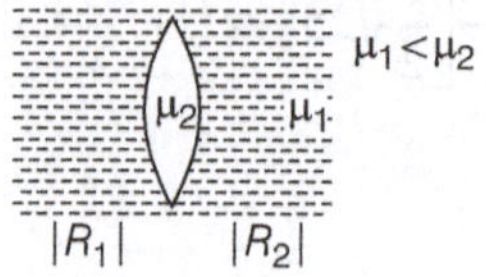

$$\frac{1}{f_{eq}} = \frac{2}{f_L} - \frac{1}{f_m}$$

$$f_m = -\frac{|R|}{2}$$

$$\frac{1}{f_L} = \left(\frac{\mu_2}{\mu_1} - 1\right)\left(\frac{1}{R_1} + \frac{1}{R_2}\right)$$

$$\frac{1}{f_{eq}} = 2\left(\frac{\mu_2 - \mu_1}{\mu_1}\right)\left(\frac{R_1 + R_2}{R_1 R_2}\right) + \frac{2}{R_2}$$

$$= \frac{2}{R_2}\left[\frac{(\mu_2 - \mu_1)(R_1 + R_2) + \mu_1 R_1}{\mu_1 R_1}\right]$$

$$= \frac{2}{R_2}\left[\frac{\mu_2 R_1 + \mu_2 R_2 - \mu_1 R_1 - \mu_1 R_2 + \mu_1 R_1}{\mu_1 R_1}\right]$$

$$\frac{1}{f_{eq}} = \frac{2[\mu_2 R_1 + \mu_2 R_2 - \mu_1 R_2]}{\mu_1 R_1 R_2}$$

For same size of image, $u = 2f$

$$u = \frac{\mu_1 R_1 R_2}{\mu_2 R_1 + \mu_2 R_2 - \mu_1 R_2}$$

$$= \frac{\mu_1 |R_1||R_2|}{\mu_2 (|R_1| + |R_2|) - \mu_1 |R_2|}$$

Option (a) is correct.

37. *(a)*

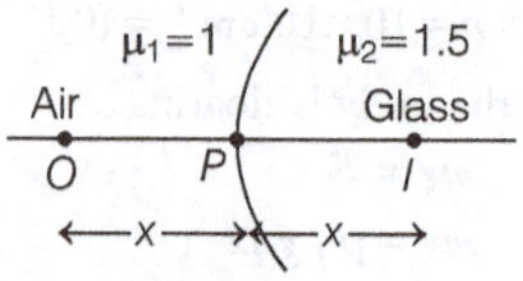

$$PO = u = -x$$

$$PI = v = x$$

$$PO = PI$$
$$\frac{\mu_2}{v} - \frac{\mu_1}{u} = \frac{\mu_2 - \mu_1}{R}$$
$$\frac{1.5}{x} + \frac{1}{x} = \frac{1}{2R}$$
$$\frac{5}{2x} = \frac{1}{2R}$$
$$x = 5R$$

38. *(d)*

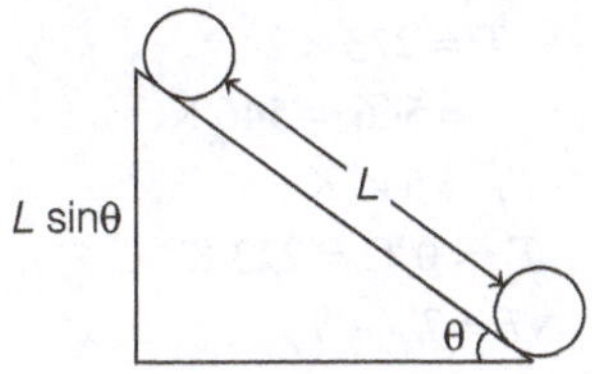

Using work energy theorem

$$W_g = K_f - K_i$$
$$mg\,L\sin\theta = K_f - K_i$$

KE in pure rolling $= \frac{1}{2}mv_{cm}^2 + \frac{1}{2}I_{cm}\omega^2$

$$= \frac{1}{2}mv^2 + \frac{1}{2}\times\frac{2}{5}mR^2\frac{v^2}{R^2} = \frac{7}{10}mv^2$$
$$mgL\sin\theta = \frac{7}{10}mv_f^2 - 0$$
$$v_f^2 \propto \sin\theta$$
$$\left(\frac{v_1}{v_2}\right)^2 = \frac{\sin\theta_1}{\sin\theta_2} = \frac{\sin 30°}{\sin 45°} = \frac{1}{\sqrt{2}}$$
$$\frac{v_1^2}{v_2^2} = \frac{1}{\sqrt{2}}$$

Alternate Solution

Speed of the body at the bottom is

$$v = \sqrt{\frac{2gL\sin\theta}{1+\left(\frac{K}{r}\right)^2}}$$

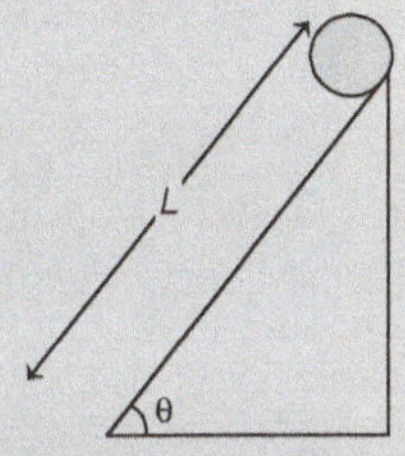

$\Rightarrow \quad v^2 \propto \sin\theta$

$\Rightarrow \quad \left(\frac{v_1}{v_2}\right)^2 = \frac{\sin\theta_1}{\sin\theta_2} = \frac{\sin 30°}{\sin 45°} = \frac{1}{\sqrt{2}}$

39. *(d)* Area under v-t graph represents distance covered.

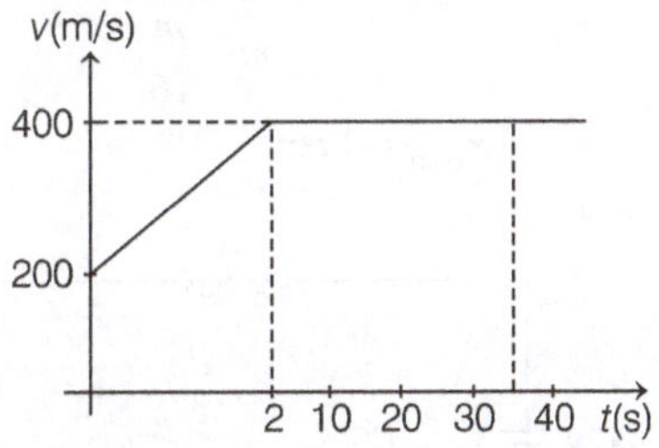

Total area under curve

$$= \frac{1}{2}\times(400+200)\times 2 + 28.5\times 400$$
$$= 600 + 11400$$
$$= 12000 \text{ m}$$
$$= 12 \text{ km}$$

40. *(c)* $x(t) = A\sin t + B\cos^2 t + Ct^2 + D$

$[x(t)] = [\text{L}]$

$[A] = [\text{L}]$

($\sin t$ and $\cos^2 t$ both are dimensionless)

$[B] = [\text{L}]$

$x = Ct^2$

$C = \frac{x}{t^2} = [\text{LT}^{-2}]$

$[C] = [\text{LT}^{-2}]$

$[D] = [\text{L}]$

$$\left[\frac{ABC}{D}\right] = \left[\frac{\text{L}\times\text{L}\times\text{LT}^{-2}}{\text{L}}\right] = [\text{L}^2\text{T}^{-2}]$$

41. (b) A. $p \propto \frac{1}{V}$

$pV = \text{constant}$

$pV = nRT$

$T = \frac{pV}{nR} = \text{constant}$

$T = \text{constant} \rightarrow$ Isothermal process

A $\rightarrow$ III

B. $W \neq 0, \Delta U \neq 0, \Delta Q \neq 0$

Only in isobaric process

B $\rightarrow$ IV

C. $\Delta Q = 0 \rightarrow$ Adiabatic process

C $\rightarrow$ I

D. $W = 0 \rightarrow$ Isochoric process

D $\rightarrow$ II

III, IV, I, II

42. *(*)*

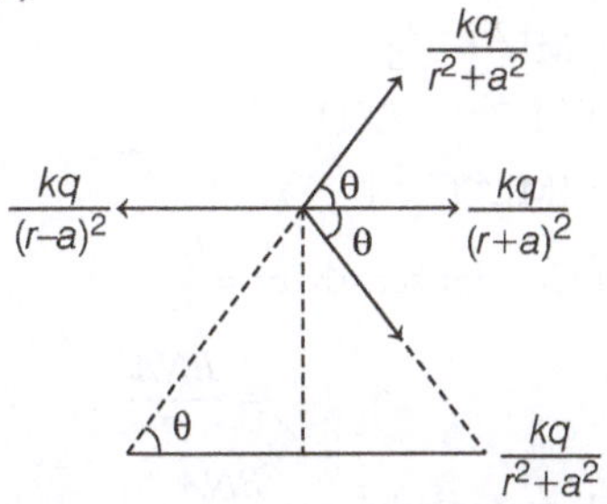

$$\frac{kq}{(r-a)^2} = \frac{kq}{(r+a)^2} + \frac{2kq}{(r^2+a^2)}\cos\theta$$
$$\frac{1}{(r-a)^2} = \frac{1}{(r+a)^2} + \frac{2a}{(r^2+a^2)^{\frac{3}{2}}}$$
$$\frac{1}{(r-a)^2} - \frac{1}{(r+a)^2} = \frac{2a}{(r^2+a^2)^{\frac{3}{2}}}$$
$$\frac{4ra}{(r^2-a^2)^2} = \frac{2a}{(r^2+a^2)^{\frac{3}{2}}}$$
$$\frac{2r}{(r^2-a^2)^2} = \frac{1}{(r^2+a^2)^{3/2}}$$
$$\frac{4r^2}{(r^2-a^2)^4} = \frac{1}{(r^2+a^2)^3}$$
$$4r^2(r^2+a^2)^3 = (r^2-a^2)^4$$
$$4r^8\left(1+\frac{a^2}{r^2}\right)^3 = r^8\left(1-\frac{a^2}{r^2}\right)^4$$
$$4\left(1+\frac{a^2}{r^2}\right)^3 = \left(1-\frac{a^2}{r^2}\right)^4$$

No option is matching.

Alternate Solution

Field on axis of a dipole ($2a << r$) is

$$E_{\text{axis}} = \frac{2kp}{r^3}$$

Field on equatorial axis of a dipole ($2a << r$) is $E_2 = E_{\text{equator}} = \frac{kp}{r^3}$

Now, when charge Q at point P is placed, it is acted upon by two forces;

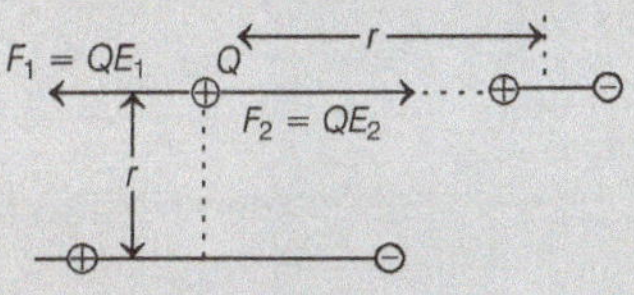

Given net force on charge Q is zero.

$\Rightarrow \quad QE_1 = QE_2$

But this is not possible for any value of a and r.

43. *(c)* (A) $\tau = C\theta$

$\tau = [ML^2T^{-2}]$

$[ML^2T^{-2}] = C[1]$

$C = [ML^2T^{-2}]$

(B) Current senstivity $= \dfrac{\theta}{I}$

$= \dfrac{BNA}{C}$

Voltage sensitivity $= \dfrac{BNA}{RC}$

R also depends on N.

(C) Voltage sensitivity does not change on increasing N or $\dfrac{N}{R}$ = constant.

(D) MCG cannot be converted into an ammeter by introducing a shunt resistance of large value in parallel with galvanometer

(E) $CS = \dfrac{NAB}{C}$

$CS \propto N$

Ans. A, B only correct.

44. *(b)* $\lambda = \dfrac{h}{p} = \dfrac{6.63 \times 10^{-34}}{10^{-30} \times 2.21 \times 10^{6}}$

$= 3 \times 10^{-10}$ m

Hence, particle will behave as X-ray.

45. *(c)*

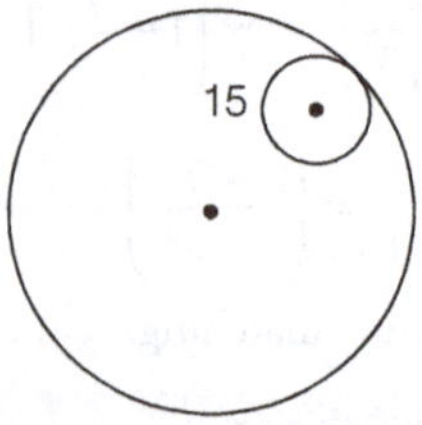

Mass of disc $= m$

Mass of cut part $= \dfrac{m}{16}$

$$x_{com} = \frac{m \times 0 - \frac{m}{16} \times 15}{m - \frac{m}{16}}$$

$x_{com} = 1$ cm

46. *(3)*

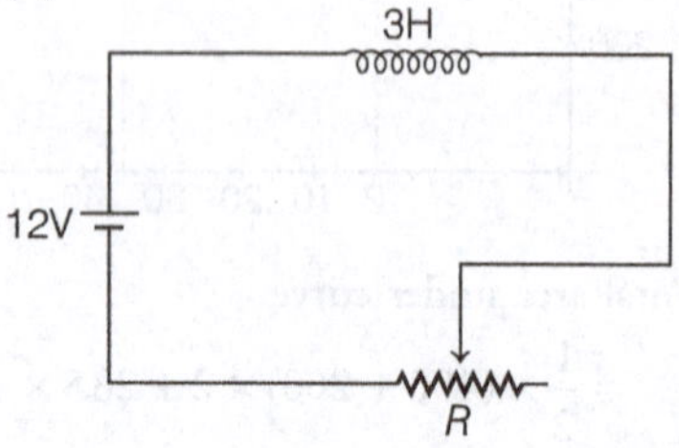

$$\varepsilon - L\frac{dI}{dt} - IR = 0$$

$12 - 3(-8) - I \times 12 = 0$

$I = 3$

47. *(3)* $\mathbf{A} \cdot \mathbf{B} = 0$

$(2\hat{i} + 3n\hat{j} + 2\hat{k}) \cdot (2\hat{i} - 2\hat{j} + 4p\hat{k}) = 0$

$4 - 6n + 8p = 0$ ...(i)

$|\mathbf{A}| = |\mathbf{B}|$

$4 + 9n^2 + 4 = 4 + 4 + 16p^2$

$9n^2 = 16p^2$

So, $p = +\dfrac{3}{4}n,\ p = -\dfrac{3n}{4}$

$4 - 6n + 8\left(\dfrac{3}{4}n\right) = 0$

$4 - 6n + 8\left(-\dfrac{3}{4}n\right) = 0$

$4 - 6n + 6n = 0$

$4 - 6n - 6n = 0$

$4 = 12n,\ n = \dfrac{1}{3} = 3^{-1}$

$\Rightarrow\ n^{-1} = 3$

48. *(273)* $\gamma = \dfrac{3}{2}$

$TV^{\gamma - 1} = C$

$273V_0^{1.5-1} = T\left(\dfrac{V_0}{4}\right)^{1.5-1}$

$273V_0^{0.5} = T\left(\dfrac{V_0}{4}\right)^{0.5}$

$273V_0^{\frac{1}{2}} = T\dfrac{V_0^{\frac{1}{2}}}{4^{\frac{1}{2}}}$

$T = 273 \times 2$

$= 546 = 546$ K

$T_f = 546$ K

$T_1 = 0°C = 273$ K

$\Delta T = T_f - T_i$

$= 546 - 273$

$\Delta T = 273$ K

49. *(152)* $F = x^2y\hat{i} + y^2\hat{j}$

$x + y = 10,\ y = 10 - x$

$\int dW = \int F_x dx + \int F_y dy$

$W = \int_0^4 x^2(10 - x)dx + \int_0^2 y^2 dy$

$= \left[\dfrac{10x^2}{3} - \dfrac{x^4}{4}\right] + \left[\dfrac{y^3}{3}\right]_0^2$

$= \dfrac{640}{3} - 64 + \dfrac{8}{3} = 152$

50. *(*)* $P \times 10^{-13} = \dfrac{F_e}{F_G} = \dfrac{\frac{kq_1q_2}{r^2}}{\frac{Gm_1m_2}{r^2}}$

$$= \frac{9 \times 10^9 \times 6.67 \times 10^{-19} \times 9.6 \times 10^{-10}}{6.67 \times 10^{-11} \times 19.2 \times 10^{-27} \times 9 \times 10^{-27}}$$

$= \dfrac{1}{2} \times 10^{45}$

Chemistry

51. *(c)* The correct match is A-IV, B-II, C-I, D-III.

52. *(a)* Both Statement I and Statement II are true.

53. *(d)* The electronic configuration of Co^{2+} is given by $[Ar]3d^7 4s^0$. Also, since it is given that magnetic moment of the octahedral complex of Co(II) is 3.95. So, it has 3 unpaired electrons. Thus, its configuration is $t_{2g}^5 e_g^2$.

54. *(b)* Complexes of the type Ma_3b_3 shows facial-meridional isomerism. So, among the given options, compound given in option (b) i.e $[Co(NH_3)_3Cl_3]$ is of type Ma_3b_3. Thus, it will show facial-meridional isomerism.

Time Saver Tech

Coordination compound of the type $[Ma_3b_3]$ shows facial-meridional isomerism.
So, among the given options only option (b) is of type $[Ma_3b_3]$

55. *(b)* The correct match is A-IV, B-III, C-I, D-II.

- Swarts reaction →
  $C_2H_5—Br + AgF \rightarrow C_2H_5—F + AgBr$
  (Ethyl fluoride)
- Sandmeyer's reaction → $PhN_2^+Cl^- \xrightarrow{CuCN/KCN} PhCN + N_2$
  (Cyanobenzene)
- Wurtz-Fittig reaction →
  $Ph—Cl + EtCl \xrightarrow[Ether]{Na} Ph-Et$
  (Ethyl benzene)
- Finkelstein reaction →
  $Et-Cl \xrightarrow[Acetone]{NaI} Et-I + NaCl$
  (Ethyl iodide)

56. *(d)*

Key idea

To solve this problem, first of all write all the phase changes involved in the reaction. Then calculate individual entropy change for each phase. After that take sum of all the individual entropy change.

The phase change involved as per the question is given by

Ice → Ice ⇌ Water → Water ⇌ Water vapour → Water vapour

268 K —(A)— 273 K —(B)— 273 K —(C)— 373 K —(D)— 373 K —(E)— 383 K

So, the overall entropy change is given by

$$\Delta S_{overall} = \Delta S_A + \Delta S_B + \Delta S_C + S_D + S_E$$

$$\Delta S_A = \int_{268\,K}^{273\,K} C_{p,n} \cdot dT$$

$$\Delta S_B = \frac{\Delta H_{m\,fusion}}{T_f}, \quad \Delta S_C = \int_{273}^{373} \frac{C_{p,m}\,dT}{T}$$

$$\Delta S_D = \frac{\Delta H_{m\,vaporisation}}{T}$$

$$\Delta S_E = \int_{373}^{383} \frac{C_{p,m}\,dT}{T}$$

So, option (d) is the correct answer.

57. *(a)* The colour of the given species are as follows

(a) V^{2+} (Violet), Cr^{3+} (Violet), Mn^{3+} (Violet)

(b) Zn^{2+} (Colourless), V^{3+} (Green), Fe^{3+} (Yellow)

(c) Ti^{4+} (Colourless), V^{4+} (Blue), Mn^{2+} (Pink)

(d) Sc^{3+} (Colourless), Ti^{3+} (Purple), Cr^{2+} (Blue)

So, option (a) contains all the species with same colour i.e violet.

58. *(d)* Statement I is true but Statement II is false. The correct form of statement II is :

The sodium fusion extract of an organic compound having N and S gives blood red colour with $FeSO_4$ and $Na_4[Fe(CN)_6]$.

59. *(b)* Let the mass of CO_2 taken initially $= x$

No. of moles of CO_2 left $= 2.8 \times 10^{-3}$ mol

No. of molecules removed $= 10^{21}$ molecules

Now, no. of initial moles $= \frac{x \times 10^{-3}}{44}$

No. of moles removed $= \frac{10^{21}}{6.02 \times 10^{23}}$

No. of moles left = initial moles − moles removed

$$2.8 \times 10^{-3} = \frac{x \times 10^{-3}}{44} - \frac{10^{21}}{6.02 \times 10^{23}}$$

$$x = 196.2 \text{ mg}$$

60. *(c)* 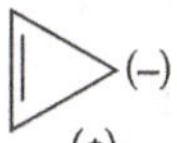
(p) is an anti-aromatic compound, 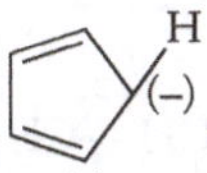
(q) is an aromatic compound, while 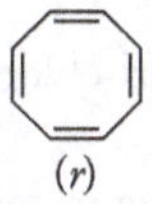
(r) is non-aromatic compound. As we know more is the aromatic character higher is the stability. Thus, the correct order of stability is $q > r > p$.

61. *(b)* The reaction involved in conversion of phenol to 2, 4, 6-tribromophenol is as follows

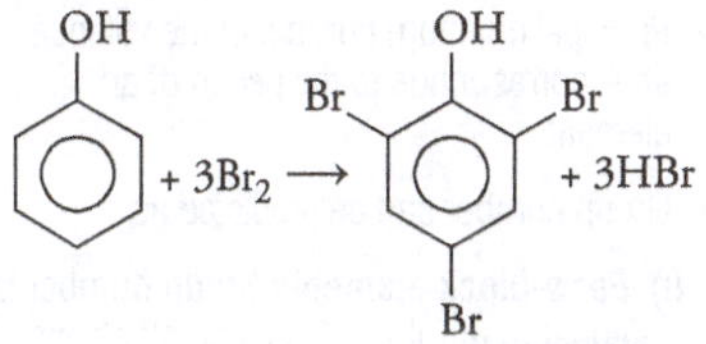

Molar mass of phenol = 94 g/mol

Molar mass of Br_2 = 80 g/mol

∴ Mass of $3Br_2 = 80 \times 3 \times 2 = 480$ g

So, 94 g phenol required 480 g bromine

∴ 2 g phenol required

$$= \frac{480}{94} \times 2 \text{ g of bromine.}$$

$$= 10.22 \text{ g of bromine}$$

62. *(c)* According to question, the product 'x' formed is

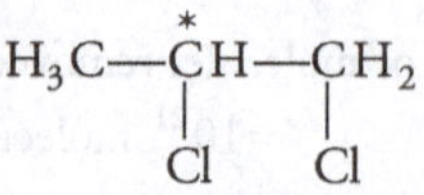

So, reaction of 'x' with chlorine under the photochemical condition is as follows

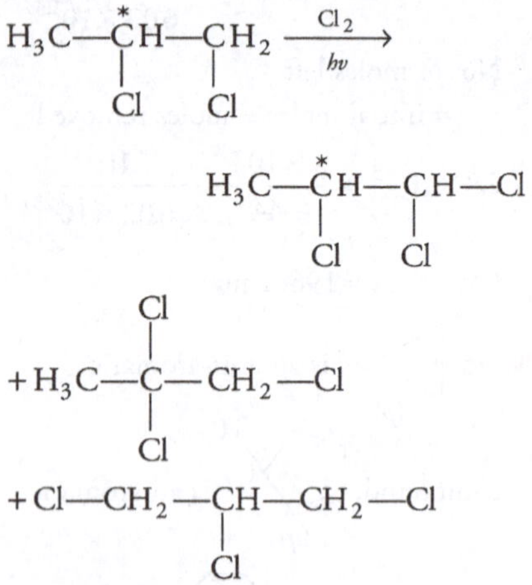

So, total '3' tri-chloro products are obtained.

63. *(c)* According to modern periodic table, elements platinum, iridium and osmium are placed in 6th period. White palladium is placed in 5th period.
So, option (c) is the correct answer.

Concept Enhancer

- Principal quantum number or the valence shell corresponds to the period of an element.
- Group number can be predicted as
  - (i) **For *s*-block elements** Group number is equal to number of valence electrons.
  - (ii) **For *p*-block elements** Group number is equal to 10+ number of electron in the valence shell.
  - (iii) **For *d*-block elements** Group number is equal to the number of electrons in $(n-1)d$ subshell + number of electrons in valence shell.

64. *(a)* Complete reaction sequence is as follows

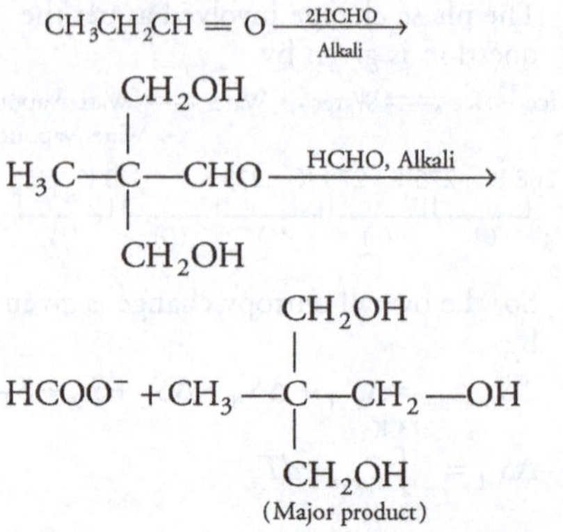

So, the major product obtained is given in option (a).

65. *(d)* Given, $K_f = 1.86$ K kg mol^{-1}

$\Delta T_f = 0.558\,°C$

Molality = 0.1 molal

Using formula,

$\Delta T_f = iK_f m$

$0.558 = i \times 1.86 \times 0.1$

On solving, we get

$i = 3$

So, among the given options, compound given in option (d) have the value of $i = 3$. So, the complex is $[Cr(NH_3)_5Cl]Cl_2$.

66. *(c)* Hinsberg reagent reacts with primary and secondary amines to form sulphonamides. But Hinsberg reagent do not react with tertiary amines.
So, among the given options compound given in A, C and E will react Hinsberg's reagent while compound given in B and D will not react.

67. *(b)* Given, $K_{sp}[A(OH)_2] = 9 \times 10^{-10}$

$K_{sp}[B(OH)_3] = 27 \times 10^{-18}$

$[A^{2+}] = 1\text{ M}$

$[B^{3+}] = 1\text{ M}$

For precipitation $Q_{ip} > K_{sp}$

So, $[A^{2+}][OH^-]^2 > 9 \times 10^{-10}$

$\Rightarrow [OH^-] > 3 \times 10^{-5}$ M

For $[B(OH)_3]$

$[B^{3+}][OH^-]^3 > 27 \times 10^{-18}$

$[OH^-] > 3 \times 10^{-6}$ M

So, $B(OH)_3$ will precipitate before $A(OH)_2$.

68. *(c)*

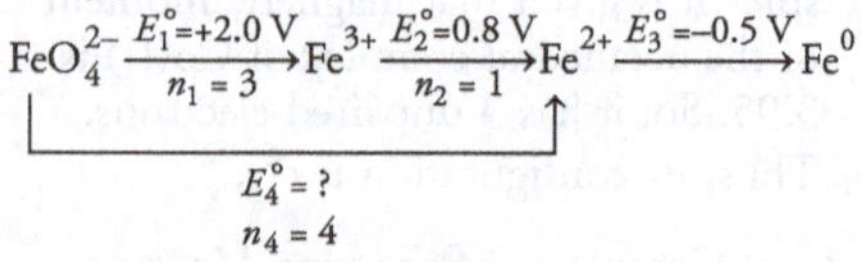

$-n_4 FE^\circ_4 = -n_1 FE^\circ_1 - n_2 FE^\circ_2$

On substituting the values, we get

$+4E^\circ_4 = 3 \times 2 + (1 \times 0.8)$

$E^\circ_4 = \dfrac{6.8}{4} = 1.7$ V

So, the value of $E^\circ_{FeO_4^{2-}/Fe^{2+}} = 1.7$ V

69. *(c)* Incorrect statement is given in option (c). Its correct form is SO_2 can acts as both, oxidising and reducing agent.

70. *(d)* Given, $\lambda = 900\text{ nm} = 9 \times 10^{-5}$ cm

$R_H = 10^5\text{ cm}^{-1}$

$h = 6.6 \times 10^{-34}$ Js

$c = 3 \times 10^8$ m/s

Using equation,

$$\frac{1}{\lambda} = R_H Z^2 \times \left[\frac{1}{n_1^2} - \frac{1}{n_2^2}\right]$$

On substituting the given value of we get

$$\frac{1}{9 \times 10^{-5}} = 10^5 (1)^2 \left[\frac{1}{n_1^2} - \frac{1}{n_2^2}\right]$$

($\because Z = 1$ for hydrogen atom)

$$\Rightarrow \frac{1}{n_1^2} - \frac{1}{n_2^2} = \frac{1}{9}$$

So, the given condition is possible only when $n_1 = 3$, $n_2 = \infty$.

So, possible series is Paschen series with $\infty \rightarrow 3$

71. *(2850)* Given, $\Delta H^\circ = 55$ kJ/mol

$$\Delta S^\circ = 175 \text{ J/K/mol}$$
$$T = 25\ ^\circ\text{C}$$
$$= 298 \text{ K}, \Delta G^\circ = ?$$

Using equation,

$$\Delta G^\circ = \Delta H^\circ - T\Delta S^\circ$$

On substituting the value in above equation, we get

$$\Delta G^\circ = 55000 - 298 \times 175$$
$$= 55000 - 52150$$
$$= 2850 \text{ J/mol}$$

72. *(897)*

$$2N_2O_5(g) \longrightarrow 2N_2O_4(g) + O_2(g)$$

p_i	0.6	0	0
p_t	$0.6 - p$	p	$p/2$

Using $k = \frac{2.303}{t} \log \frac{p_i}{p_t}$

$$4.606 \times 10^{-2} = \frac{2.303}{100} \log \frac{0.6}{0.6 - p}$$
$$10^2 = \frac{0.6}{0.6 - p}$$
$$10^4 = \frac{0.6}{0.6 - p}$$

On solving, we get $p = 0.594$

$$p_{\text{total}} = 0.6 + \frac{p}{2}$$
$$= 0.6 + 0.297$$
$$p_{\text{Total}} = 897 \times 10^{-3}$$

So, the value of approx is 897.

73. *(171)* The complete reaction sequence involved is as follows

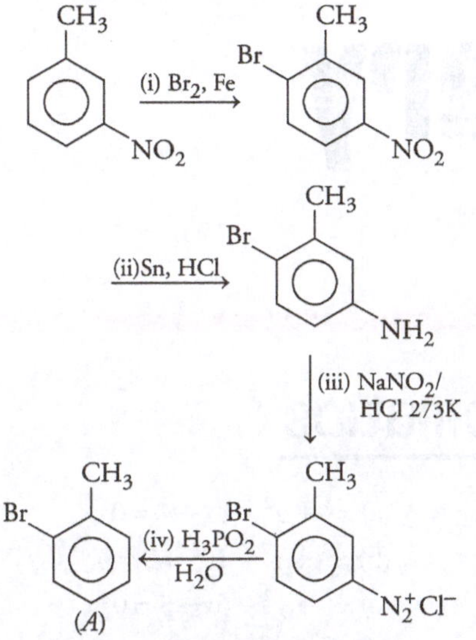

So, the compound A is C_7H_7Br

The molar mass of C_7H_7Br (A)

$$= 7 \times 12 + 7 \times 1 + 80$$
$$= 171 \text{ g/mol}$$

So, the molar mass of compound A is 171 g/mol.

74. *(7)* Given, pH = 9

$$C_{\text{initial}} = 1 \text{ mM}$$
$$K_b = 10^{-x}$$
$$x = ?$$

Since pH $= 9 \Rightarrow$ pOH $= 5$

So, $[\bar{O}H] = 10^{-5}$ M

Chemical reaction involved in ionisation of ethyl amine is as follows

$C_2H_5NH_2(aq) + H_2O \rightleftharpoons$	$C_2H_5NH_3^+$	$+\bar{O}H$
$C = 10^{-3}$ M	—	—
$C(1-\alpha)$	$C\alpha$	$C\alpha$
$\Rightarrow C = 10^{-3}$	$= 10^{-5}$	$= 10^{-5}$

According to question, $1 - \alpha \approx 1$

Now we know that,

$$K_b = \frac{[C_2H_5NH_3^+][\bar{O}H]}{[C_2H_5NH_2]}$$
$$K_b = \frac{10^{-5} \times 10^{-5}}{10^{-3}}$$
$$= 10^{-7}$$

So, the value of x is 7.

75. *(40)* Given, mass of $BaSO_4$ = 466 mg

mass of organic compound = 160 mg

% of sulphur in given compound = ?

Using the formula for calculation of % of sulphur

$$= \frac{\text{Milli. moles of } BaSO_4 \times 32}{\text{Mass of organic compound}} \times 100$$
$$= \frac{\frac{466}{233} \times 32}{160} \times 100$$
$$= 40\%$$

So, the % of sulphur in given compound is 40%.

Time 3:00-6:00 | **MM** : 300 Marks

23 January
Shift-II

Online Question Paper — JEE Main 2025

Mathematics

Section A
Objective Type Questions

01/ Let the point A divide the line segment joining the points $P(-1,-1,2)$ and $Q(5,5,10)$ internally in the ratio $r:1(r>0)$, If O is the origin and $(\text{OQ}\cdot\text{OA})-\frac{1}{5}|\text{OP}\times\text{OA}|^2=10$, then the value of r is

(a) 7 (b) 3 (c) 14 (d) $\sqrt{7}$

02/ Let $\int x^3 \sin x\, dx = g(x)+C$, where C is the constant of integration. If $8\left(g\left(\frac{\pi}{2}\right)+g'\left(\frac{\pi}{2}\right)\right)$ $=\alpha\pi^3+\beta\pi^2+\gamma, \alpha,\beta,\gamma,\in Z$, then $\alpha+\beta-\gamma$ equals

(a) 55 (b) 48 (c) 62 (d) 47

03/ The system of equations $x+y+z=6, x+2y+5z=9$ $x+5y+\lambda z=\mu$, has no solution, if

(a) $\lambda=15,\mu\neq17$ (b) $\lambda\neq17,\mu\neq18$
(c) $\lambda=17,\mu=18$ (d) $\lambda=17,\mu\neq18$

04/ Let the shortest distance from $(a,0)$ $a>0$, to the parabola $y^2=4x$ be 4. Then, the equation of the circle passing through the point $(a,0)$ and the focus of the parabola and having its centre on the axis of the parabola is

(a) $x^2+y^2-8x+7=0$
(b) $x^2+y^2-10x+9=0$
(c) $x^2+y^2-6x+5=0$
(d) $x^2+y^2-4x+3=0$

05/ The distance of the line $\frac{x-2}{2}=\frac{y-6}{3}=\frac{z-3}{4}$ from the point (1, 4, 0) along the line $\frac{x}{1}=\frac{y-2}{2}=\frac{z+3}{3}$ is

(a) $\sqrt{13}$ (b) $\sqrt{15}$ (c) $\sqrt{17}$ (d) $\sqrt{14}$

Imp.

06/ $\lim_{x\to\infty}\frac{(2x^2-3x+5)(3x-1)^{\frac{x}{2}}}{(3x^2+5x+4)\sqrt{(3x+2)^x}}$ is equal to

(a) $\frac{2}{3\sqrt{e}}$ (b) $\frac{2e}{\sqrt{3}}$ (c) $\frac{2e}{3}$ (d) $\frac{2}{\sqrt{3e}}$

Limits of the form 1^∞

07/ The number of complex numbers z satisfying $|z|=1$ and $\left|\frac{z}{\bar{z}}+\frac{\bar{z}}{z}\right|=1$, is

(a) 10 (b) 6 (c) 8 (d) 4

08/ If in the expansion of $(1+x)^p(1-x)^q$, then the coefficients of x and x^2 are 1 and -2, respectively, then p^2+q^2 is equal to

(a) 8 (b) 20 (c) 13 (d) 18

09/ A board has 16 squares as shown in the figure

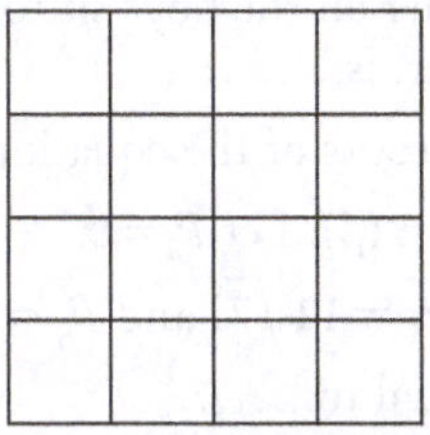

Out of these 16 squares, two squares are chosen at random. The probability that they have no side in common is

(a) $\frac{7}{10}$ (b) $\frac{23}{30}$

(c) $\frac{3}{5}$ (d) $\frac{4}{5}$

10/ Let $A=[a_{ij}]$ be a 3×3 matrix such that $A\begin{bmatrix}0\\1\\0\end{bmatrix}=\begin{bmatrix}0\\0\\1\end{bmatrix}$, $A\begin{bmatrix}4\\1\\3\end{bmatrix}=\begin{bmatrix}0\\1\\0\end{bmatrix}$ and $A\begin{bmatrix}2\\1\\2\end{bmatrix}=\begin{bmatrix}1\\0\\0\end{bmatrix}$, then a_{23} equals

(a) 2 (b) −1

(c) 1 (d) 0

11/ Let the range of the function

$$f(x)=6+16\cos x\cdot\cos\left(\frac{\pi}{3}-x\right)\cdot\cos\left(\frac{\pi}{3}+x\right)\cdot\sin 3x\cdot\cos 6x, x\in R$$

be $[\alpha,\beta]$. Then, the distance of the point (α,β) from the line $3x+4y+12=0$ is

(a) 9 (b) 10

(c) 8 (d) 11

12/ If the area of the region $\{(x,y):-1\le x\le 1, 0\le y\le a+e^{|x|}-e^{-x}, a>0\}$ is $\frac{e^2+8e+1}{e}$, then the value of a is

(a) 6 (b) 8

(c) 5 (d) 7

13/ Let $x=x(y)$ be the solution of the differential equation $y=\left(x-y\frac{dx}{dy}\right)\sin\left(\frac{x}{y}\right)$, $y>0$ and $x(1)=\frac{\pi}{2}$

Then, $\cos(x(2))$ is equal to

(a) $2(\log_e 2)^2-1$

(b) $2(\log_e 2)-1$

(c) $1-2(\log_e 2)$

(d) $1-2(\log_e 2)^2$

14/ A spherical chocolate ball has a layer of ice-cream of uniform thickness around it. When the thickness of the ice-cream layer is 1 cm, the ice-cream melts at the rate of 81 cm^3/min and the thickness of the ice-cream layer decreases at the rate of $\frac{1}{4\pi}$ cm/min. The surface area (in cm^2) of the chocolate ball (without the ice-cream layer) is

(a) 128π (b) 196π

(c) 256π (d) 225π

15/ A rod of length eight units moves such that its ends A and B always lie on the lines $x-y+2=0$ and $y+2=0$, respectively. If the locus of the point P that divides the rod AB internally in the ratio 2 : 1 is $9(x^2+\alpha y^2+\beta xy+\gamma x+28y)-76=0$, then $\alpha-\beta-\gamma$ is equal to

(a) 22 (b) 24

(c) 23 (d) 21

16/ If $I=\int_0^{\frac{\pi}{2}}\frac{\sin^{\frac{3}{2}}x}{\sin^{\frac{3}{2}}x+\cos^{\frac{3}{2}}x}dx$, then $\int_0^{2I}\frac{x\sin x\cos x}{\sin^4 x+\cos^4 x}dx$ equals

(a) $\frac{\pi^2}{4}$ (b) $\frac{\pi^2}{16}$ (c) $\frac{\pi^2}{8}$ (d) $\frac{\pi^2}{12}$

17/ Let $X=R\times R$, Define a relation R on X as $(a_1,b_1)R(a_2,b_2)\Leftrightarrow b_1=b_2$

Statement I R is an equivalence relation

Statement II For some $(a,b)\in X$, then set $S=\{(x,y)\in X:(x,y)R(a,b)\}$ represents a line parallel to $y=x$.

In the light of the above statements, choose the correct answer from the options given below

(a) Both statement I and Statement II are true.

(b) Statement I is true but Statement II is false.

(c) Both statement I and Statement II are false.

(d) Statement I is false but Statement II is true.

18/ If the square of the shortest distance between the lines $\frac{x-2}{1}=\frac{y-1}{2}=\frac{z+3}{-3}$ and $\frac{x+1}{2}=\frac{y+3}{4}=\frac{z+5}{-5}$ is $\frac{m}{n}$, where m,n are coprime number, then $m+n$ is equal to

(a) 9 (b) 14

(c) 6 (d) 21

19/ Let $A = \{(x, y) \in R \times R : |x + y| \ge 3\}$ and $B = \{(x, y) \in R \times R : |x| + |y| \le 3\}$. If $C = \{(x, y) \in A \cap B : x = 0 \text{ or } y = 0\}$, then $\sum_{(x, y) \in C} |x + y|$ is

(a) 12 (b) 24 (c) 18 (d) 15

20/ The length of the chord of the ellipse $\frac{x^2}{4} + \frac{y^2}{2} = 1$, whose mid-point is $\left(1, \frac{1}{2}\right)$ is

(a) $\sqrt{15}$ (b) $\frac{5}{3}\sqrt{15}$ (c) $\frac{2}{3}\sqrt{15}$ (d) $\frac{1}{3}\sqrt{15}$

Section B
Numerical Type Questions

Imp.

21/ The variance of the number 8, 21, 34, 47, ..., 320 is............

Variance and Standard Deviation

22/ The number of ways, 5 boys and 4 girls can sit in a row so that either all the boys sit together or no two boys sit together, is.........

23/ Let α, β be the roots of the equation $x^2 - ax - b = 0$ with Im (α) < Im (β). Let $P_n = \alpha^n - \beta^n$. If $P_3 = -5\sqrt{7}\,i$, $P_4 = -3\sqrt{7}\,i$, $P_5 = 11\sqrt{7}i$ and $P_6 = 45\sqrt{7}i$, then $|\alpha^4 + \beta^4|$ is equal to.......

24/ The focus of the parabola $y^2 = 4x + 16$ is the centre of the circle C of radius 5. If the value of λ, for which C passes through the point of intersection of the lines $3x - y = 0$ and $x + \lambda y = 4$ are λ_1 and λ_2, $\lambda_1 < \lambda_2$, then $12\lambda_1 + 29\lambda_2$ is equal to........

25/ The roots of the quadratic equation $3x^2 - px + q = 0$ are 10th and 11th terms of an arithmetic progression with common difference $\frac{3}{2}$. If the sum of the first 11 terms of this arithmetic progression is 88, then $q - 2p$ is equal to.....

Physics

Section A
Objective Type Questions

26/ A massless spring gets elongated by amount x_1 under a tension of 5 N. Its elongation is x_2 under the tension of 7 N. For the elongation of $(5x_1 - 2x_2)$, then tension in the spring will be

(a) 11 N (b) 39 N
(c) 20 N (d) 15 N

27/ Given below are two statements. One is labelled as Assertion (A) and the other is labelled as Reason (R).

Assertion (A) The binding energy per nucleon is found to be practically independent of the atomic number A, for nuclei with mass numbers between 30 and 170.

Reason (R) Nuclear force is long range.

In the light of the above statements, choose the correct answer from the options given below.

(a) Both A and R are true but R is not the correct explanation of A.
(b) A is false but R is true.
(c) A is true but R is false.
(d) Both A and R are true and R is the correct explanation of A.

28/ Match List-I with List-II.

	List-I		List-II
(A)	Permeability of free space	(I)	$[ML^2T^{-2}]$
(B)	Magnetic field	(II)	$[MT^{-2}A^{-1}]$
(C)	Torsional moment	(III)	$[MLT^{-2}A^{-2}]$
(D)	Torsional constant	(IV)	$[L^2A]$

Choose the correct answer from the options given below.

(a) (A)-(III), (B)-(II), (C)-(IV), (D)-(I)
(b) (A)-(I), (B)-(IV), (C)-(II), (D)-(III)
(c) (A)-(II), (B)-(I), (C)-(III), (D)-(IV)
(d) (A)-(IV), (B)-(III), (C)-(I), (D)-(II)

29/ The width of one of the two slits in Young's double slit experiment is d while that of the other slit is xd. If the ratio of the maximum of the minimum intensity in the interference pattern on the screen is 9 : 4, then what is the value of x?

(Assume that the field strength varies according to the slit width)

(a) 2 (b) 5 (c) 4 (d) 3

Imp.

30/ The refractive index of the material of a glass prism is $\sqrt{3}$. The angle of minimum deviation is equal to the angle of the prism. What is the angle of the prism?
(a) 58° (b) 48° (c) 60° (d) 50°

Refraction through Prism

31/ Two charges 7 μC and −4 μC are placed at (−7 cm, 0, 0) and (7 cm, 0, 0) respectively. Given, $\varepsilon_0 = 8.85 \times 10^{-12} C^2N^{-1}m^{-2}$, the electrostatic potential energy of the charge configuration is
(a) −1.5 J (b) −1.2 J (c) −2.0 J (d) −1.8 J

32/ If a satellite orbiting the Earth is 9 times closer to the Earth than the Moon, what is the time period of rotation of the satellite? Given rotational time period of Moon = 27 days and gravitational attraction between the satellite and the Moon is neglected.
(a) 27 days (b) 3 days (c) 81 days (d) 1 day

33/ A galvanometer having a coil of resistance 30 Ω need 20 mA of current for full-scale deflection. If a maximum current of 3 A is to be measured using this galvanometer, the resistance of the shunt to be added to the galvanometer should be $\frac{30}{X}$ Ω, where X is
(a) 149 (b) 298 (c) 596 (d) 447

34/ Two point charges −4 μC and 4 μC constituting an electric dipole, are placed at (−9,0,0) cm and (9, 0, 0) cm in a uniform electric field of strength 10^4 NC^{-1}. The work done on the dipole in rotating it from the equilibrium through 180° is
(a) 12.4 mJ (b) 16.4 mJ (c) 14.4 mJ (d) 18.4 mJ

35/ Water flows in a horizontal pipe whose one end is closed with a valve. The reading of the pressure gauge attached to the pipe is p_1. The reading of the pressure gauge falls to p_2 when the valve is opened. The speed of water flowing in the pipe is proportional to
(a) $(p_1 - p_2)^2$ (b) $p_1 - p_2$ (c) $(p_1 - p_2)^4$ (d) $\sqrt{p_1 - p_2}$

36/ Water of mass m gram is slowly heated to increase the temperature from T_1 to T_2. The change in entropy of the water, given specific heat of water is $1 Jkg^{-1}K^{-1}$, is
(a) zero (b) $m \ln\left(\frac{T_1}{T_2}\right)$ (c) $m(T_2 - T_1)$ (d) $m \ln\left(\frac{T_2}{T_1}\right)$

37/ A ball having kinetic energy KE, is projected at an angle of 60° from the horizontal. What will be the kinetic energy of ball at the highest point of its flight?
(a) $\frac{(KE)}{4}$ (b) $\frac{(KE)}{16}$ (c) $\frac{(KE)}{2}$ (d) $\frac{(KE)}{8}$

38/ A circular disk of radius R metre and mass M kg is rotating around the axis perpendicular to the disk. An external torque is applied to the disk such that $\theta(t) = 5t^2 - 8t$, where $\theta(t)$ is the angular position of the rotating disc as a function of time t.
How much power is delivered by the applied torque, when $t = 2$ s?
(a) $108MR^2$ (b) $8MR^2$ (c) $60MR^2$ (d) $72MR^2$

39/ The energy of a system is given as $E(t) = \alpha^3 e^{-\beta t}$, where t is the time and $\beta = 0.3 s^{-1}$. The errors in the measurement of α and t are 1.2% and 1.6%, respectively. At $t = 5$ s, maximum percentage error in the energy is
(a) 8.4% (b) 6% (c) 11.6% (d) 4%

40/ In photoelectric effect an EM wave is incident on a metal surface and electrons are ejected from the surface. If the work function of the metal is 2.14 eV and stopping potential is 2 V, what is the wavelength of the EM wave?
(Given, hc = 1242 eV-nm where h is the Planck's constant and c is the speed of light in vacuum)
(a) 400 nm (b) 300 nm (c) 600 nm (d) 200 nm

41/ A concave mirror of focal length f in air is dipped in a liquid of refractive index μ. Its focal length in the liquid will be
(a) $f/(\mu - 1)$ (b) f/μ (c) f (d) μf

42/ A plane electromagnetic wave of frequency 20 MHz travels in free space along the +x-direction. At a particular point in space and time, the electric field vector of the wave is $E_y = 9.3 \text{ Vm}^{-1}$. Then, the magnetic field vector of the wave at that point is
(a) $B_z = 1.55 \times 10^{-8}$ T (b) $B_z = 3.1 \times 10^{-8}$ T
(c) $B_z = 9.3 \times 10^{-8}$ T (d) $B_z = 6.2 \times 10^{-8}$ T

43/ The equation of a transverse wave travelling along a string is $y(x, t) = 4.0 \sin[20 \times 10^{-3} x + 600t]$ mm, where x is in mm and t is in second. The velocity of the wave is
(a) −60 m/s (b) +30 m/s (c) +60 m/s (d) −30 m/s

44/

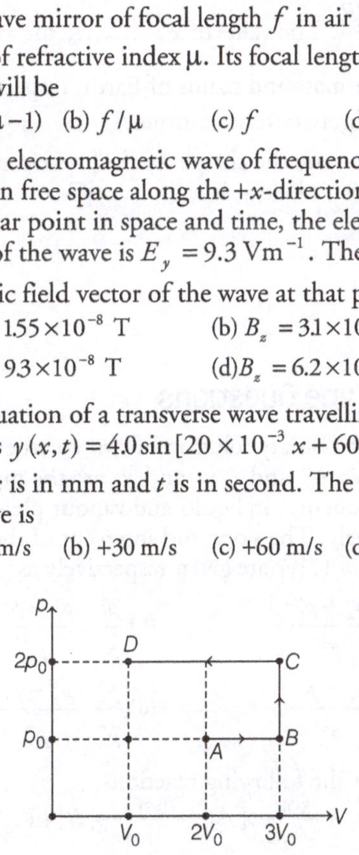

Using the given p-V diagram, the work done by an ideal gas along the path $ABCD$ is

(a) $-4p_0 V_0$ (b) $3p_0 V_0$
(c) $-3p_0 V_0$ (d) $4p_0 V_0$

45/ What is the current through the battery in the circuit shown below?

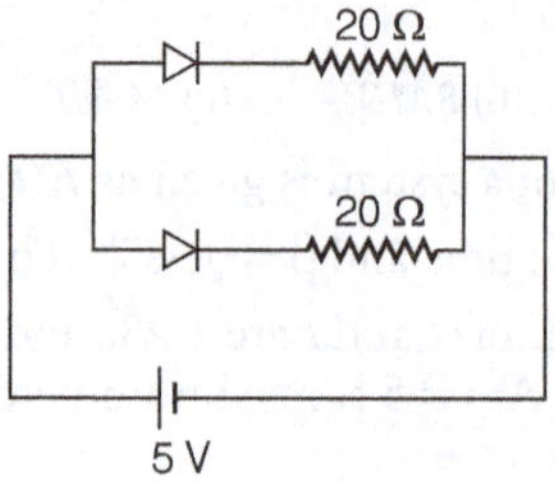

(a) 1.5 A (b) 1.0 A
(c) 0.5 A (d) 0.25 A

Section B
Numerical Type Questions

46/ A satellite of mass $M/2$ is revolving around Earth in a circular orbit at a height of $R/3$ from Earth surface. The angular momentum of the satellite is $M\sqrt{\frac{GMR}{x}}$. The value of x is.........m, where M and R are the mass and radius of Earth, respectively. (G is the gravitational constant)

47/ At steady state the charge on the capacitor, as shown in the circuit below, is .... μC.

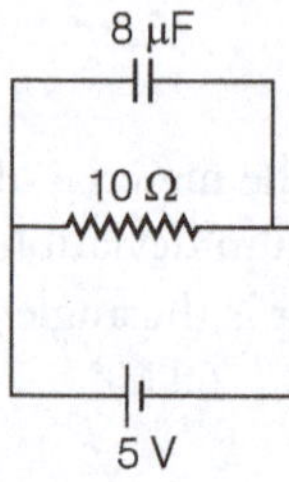

48/ An air bubble of radius 1.0 mm is observed at a depth of 20 cm below the free surface of a liquid having surface tension 0.095 J/m^2 and density 10^3 kg/m^3. The difference between pressure inside the bubble and atmospheric pressure is ....N/m^2. (Take, $g = 10 \text{ m/s}^2$)

49/ A time varying potential difference is applied between the plates of a parallel plate capacitor of capacitance 25 μF. The dielectric constant of the medium between the capacitor plates is 1. It produces an instantaneous displacement current of 0.25 mA in the intervening space between the capacitor plates, the magnitude of the rate of change of the potential difference will be... Vs^{-1}.

Imp.

50/ In a series L-C-R circuit, a resistor of 300 Ω, a capacitor of 25 nF and an inductor of 100 mH are used. For maximum current in the circuit, the angular frequency of the AC source is ....$\times 10^4$ rad s^{-1}.

L-C-R Series Resonance Circuit

Chemistry

Section A
Objective Type Questions

51/ Consider a binary solution of two volatile liquid components 1 and 2. x_1 and y_1 are the mole fractions of component 1 in liquid and vapour phase, respectively. The slope and intercept of the linear plot of $1/x_1$ vs $1/y_1$ are given respectively as

(a) $\frac{p_2^\circ}{p_1^\circ}, \frac{p_2^\circ - p_1^\circ}{p_2^\circ}$ (b) $\frac{p_1^\circ}{p_2^\circ}, \frac{p_1^\circ - p_2^\circ}{p_2^\circ}$

(c) $\frac{p_2^\circ}{p_1^\circ}, \frac{p_1^\circ - p_2^\circ}{p_2^\circ}$ (d) $\frac{p_1^\circ}{p_2^\circ}, \frac{p_2^\circ - p_1^\circ}{p_2^\circ}$

52/ Consider the following reactions,

$$K_2Cr_2O_7 \xrightarrow[-H_2O]{KOH} [A] \xrightarrow[-H_2O]{H_2SO_4} [B] + K_2SO_4$$

The products $[A]$ and $[B]$, respectively are

(a) K_2CrO_4 and $K_2Cr_2O_7$
(b) $K_2Cr(OH)_6$ and Cr_2O_3
(c) K_2CrO_4 and CrO
(d) K_2CrO_4 and Cr_2O_3

53/ The α- helix and β - pleated sheet structures of protein are associated with its

(a) tertiary structure
(b) primary structure
(c) quaternary structure
(d) secondary structure

54/ Given below are two statements.

Statement I For a given shell, the total number of allowed orbitals is given by n^2.

Statement II For any subshell, the spatial orientation of the orbitals is given by $-l$ to $+l$ values including zero.

In the light of the above statements, choose the correct answer from the options given below
(a) Statement I is false but Statement II is true.
(b) Statement I is true but Statement II is false.
(c) Both statement I and Statement II are true.
(d) Both statement I and Statement II are false.

55/ The ascending order of relative rate of solvolysis of following compounds is

A. 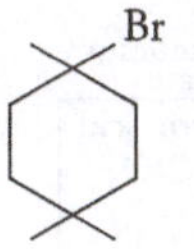

B. 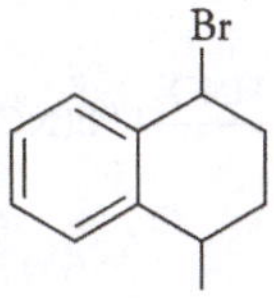

C. 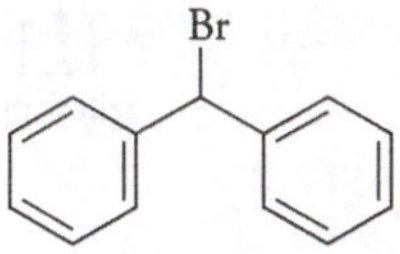

D. 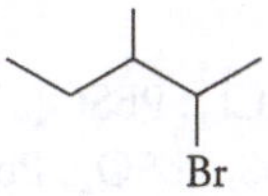

(a) (C) < (B) < (A) < (D)
(b) (D) < (B) < (A) < (C)
(c) (C) < (D) < (B) < (A)
(d) (D) < (A) < (B) < (C)

56/ Identify the products [*A*] and [*B*], respectively in the following reaction

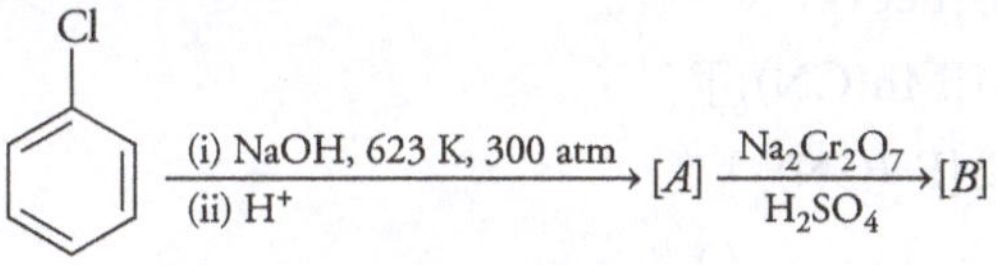

(a) (A) 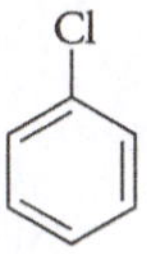

, (B) 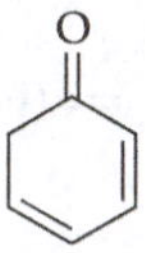

(b) (A) 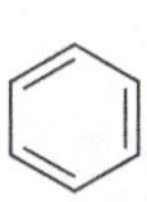, (B) 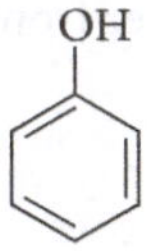

(c) (A) 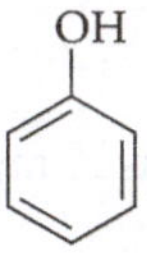

, (B) 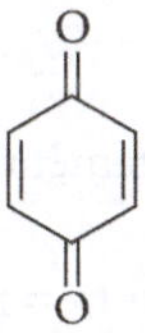

(d) (A) 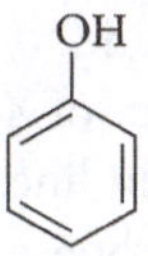

, (B) 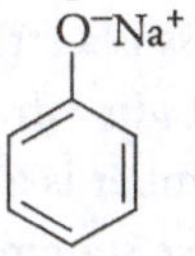

57/ Given below are two statements
Consider the following reaction,

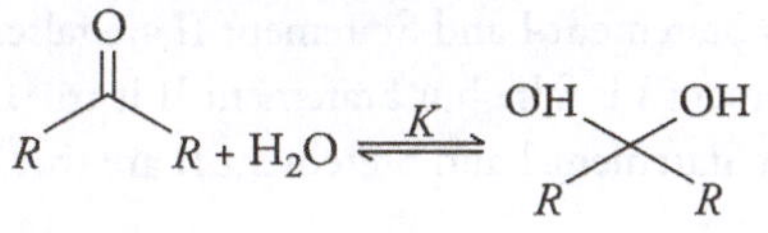

Statement I In the case of formaldehyde 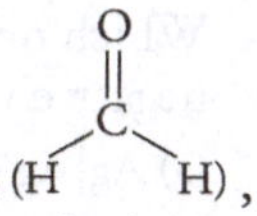

, *K* is about 2280, due to small substituents hydration is faster.

Statement II In the case of trichloro acetaldehyde 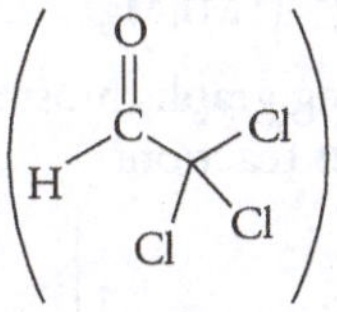

, *K* is about 2000 due to $-I$ effect of $-$Cl.

In the light of the above statements, choose the correct answer from the options given below
(a) Statement I is true but Statement II is false.
(b) Both Statement I and Statement II are false.
(c) Both Statement I and Statement II are true.
(d) Statement I is false but Statement II is true.

58/ Given below are two statements.
Statement I The boiling points of alcohols and phenols increase with increase in the number of C-atoms.
Statement II The boiling points of alcohols and phenols are higher in comparison to other class of compounds such as ethers, haloalkanes.
In the light of the above statements, choose the **correct** answer from the options given below
(a) Statement I is false but Statement II is true.
(b) Both Statement I and Statement II are true.
(c) Statement I is true but Statement II is false.
(d) Both Statement I and Statement II are false.

59/ Given below are the atomic numbers of some group 14 elements. The atomic number of the element with lowest melting point is
(a) 82
(b) 6
(c) 50
(d) 14

60/ pH of water is 7 at 25°C. If water is heated to 80°C, it's pH will
(a) increase
(b) decrease
(c) H^+ concentration increases, OH^- concentration decreases
(d) remains the same

61/ Standard electrode potentials for a few half cells are mentioned below.

$$E^\circ_{Cu^{2+}/Cu} = 0.34\text{ V}, E^\circ_{Zn^{2+}/Zn} = -0.76\text{ V}$$

$$E^\circ_{Ag^+/Ag} = 0.80\text{ V}, E^\circ_{Mg^{2+}/Mg} = -2.37\text{ V}$$

Which one of the following cells given the most negative value of ΔG° ?

(a) $Ag|Ag^+(1M)||Mg^{2+}(1M)|Mg$
(b) $Cu|Cu^{2+}(1M)||Ag^+(1M)|Ag$
(c) $Zn|Zn^{2+}(1M)||Ag^+(1M)|Ag$
(d) $Zn|Zn^{2+}(1M)||Mg^{2+}(1M)|Mg$

62/ Which of the following graphs most appropriately represents a zero order reaction?

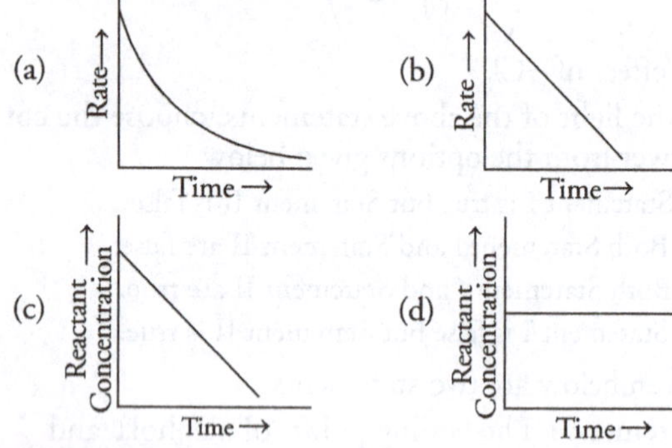

Imp.

63/ Match List -I with List -II.

	List-I		List-II
(A)	Bronze	(I)	Cu, Ni
(B)	Brass	(II)	Fe, Cr, Ni, C
(C)	UK silver coin	(III)	Cu, Zn
(D)	Stainless steel	(IV)	Cu, Sn

Choose the correct answer from the options given below.

(a) (A)-(IV), (B)-(III), (C)-(I), (D)-(II)
(b) (A)-(III), (B)-(I), (C)-(IV), (D)-(II)
(c) (A)-(III), (B)-(IV), (C)-(II), (D)-(I)
(d) (A)-(IV), (B)-(II), (C)-(III), (D)-(I)

Composition of Alloys

64/ When a non-volatile solute is added to the solvent, the vapour pressure of the solvent decreases by 10 mm of Hg. The mole fraction of the solute in the solution is 0.2. What would be the mole fraction of the solvent if decrease in vapour pressure is 20 mm of Hg ?

(a) 0.4 (b) 0.2 (c) 0.8 (d) 0.6

65/ Consider the reaction,

$$X_2Y(g) \rightleftharpoons X_2(g) + \frac{1}{2}Y_2(g)$$

The equation representing correct relationship between the degree of dissociation (x) of $X_2Y(g)$ with its equilibrium constant Kp is ....

Assume x to be very small.

(a) $x = \sqrt[3]{\frac{2Kp^2}{p}}$ (b) $x = \sqrt[3]{\frac{Kp}{p}}$

(c) $x = \sqrt[3]{\frac{2Kp}{p}}$ (d) $x = \sqrt[3]{\frac{Kp}{2p}}$

66/ Identify A, B and C in the given below reaction sequence

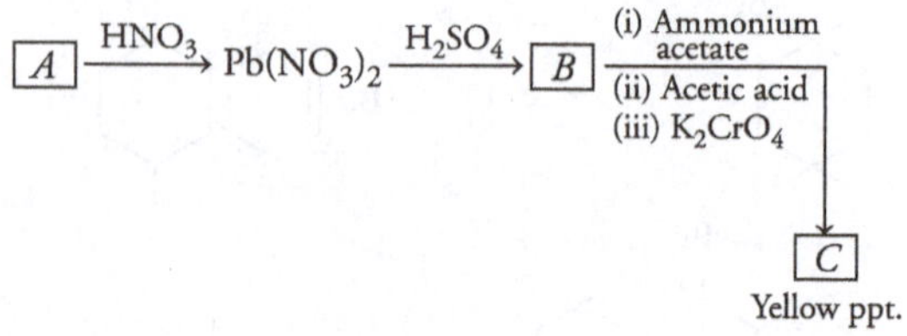

(a) $PbCl_2, PbSO_4, PbCrO_4$
(b) $PbS, PbSO_4, PbCrO_4$
(c) $PbS, PbSO_4, Pb(CH_3COO)_2$
(d) $PbCl_2, Pb(SO_4)_2, PbCrO_4$

67/ Identify the coordination complexes in which the central metal ion has d^4 configuration.

(A) $[FeO_4]^{2-}$
(B) $[Mn(CN)_6]^{3-}$
(C) $[Fe(CN)_6]^{3-}$
(D) $Cr_2(O-\overset{\overset{O}{\|}}{C}-Me)_4 = (H_2O)_2$
(E) $[NiF_6]^{2-}$

Choose the correct answer from the option given below.

(a) (A), (B) and (E) only
(b) (C) and (E) only
(c) (B) and (D) only
(d) (B), (C) and (D) only

68/ Given below are two statement abouts X-ray spectra of elements :

Statement I A plot of $\sqrt{v}$ (v = frequency of X-rays emitted) *vs* atomic mass is a straight line.

Statement II A plot of v(v = frequency of X-rays emitted) *vs* atomic number is a straight line.

In the light of the above statements, choose the correct answer from the options given below

(a) Statement I is true but Statement II is false.
(b) Both Statement I and Statement II are false.
(c) Statement I is false but Statement II is true.
(d) Both Statement I and Statement II are true.

69/ The effect of temperature on spontaneity of reactions are represented as

	ΔH	ΔS	Temperature	Spontaneity
(A)	+	−	any T	Non-spontaneous
(B)	+	+	low T	Spontaneous
(C)	−	−	low T	Non-spontaneous
(D)	−	+	any T	Spontaneous

The incorrect combinations are
(a) (B) and (D) only
(b) (A) and (C) only
(c) (B) and (C) only
(d) (A) and (D) only

70/ Match List-I with List-II.

	List-I (Isomers of $C_{10}H_{14}$)		List-II (Ozonolysis product)
(a)	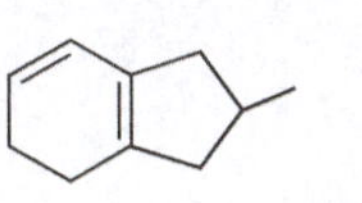	I.	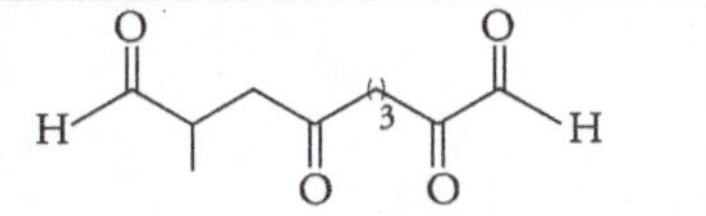
(b)	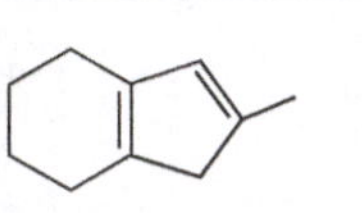	II.	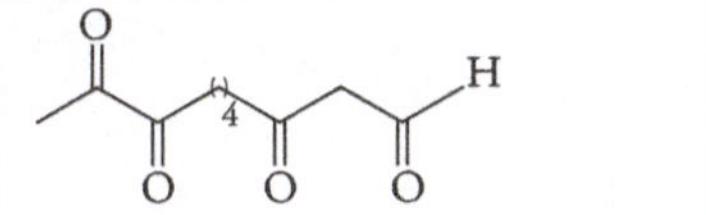
(c)	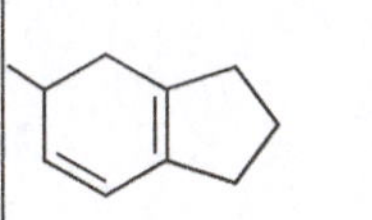	III.	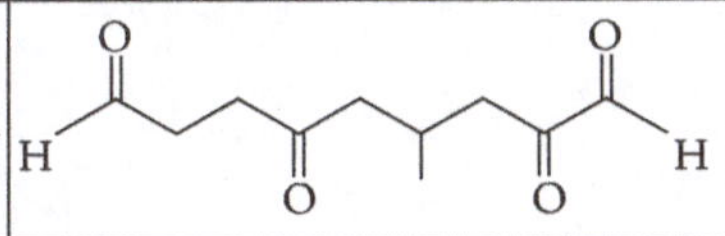
(d)	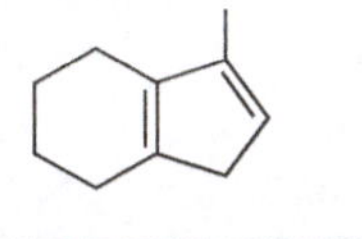	IV.	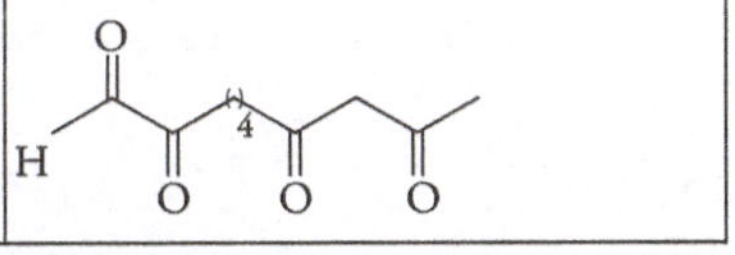

Choose the **correct** answer from the options given below
(a) (A)-(II), (B)-(III), (C)-(I), (D)-(IV)
(b) (A)-(I), (B)-(IV), (C)-(III), (D)-(II)
(c) (A)-(III), (B)-(II), (C)-(I), (D)-(IV)
(d) (A)-(III), (B)-(IV), (C)-(I), (D)-(II)

Section B

Numerical Type Questions

71/ 0.01 mole of an organic compound (X) containing 10% hydrogen, on complete combustion produced 0.9 g H_2O. Molar mass of (X) is ....g mol^{-1}.

Imp.

72/ The bond dissociation enthalpy of X_2, ΔH°_{bond} calculated from the given data is......... kJ mol^{-1}. (Nearest integer)

$$M^+X^-(s) \longrightarrow M^+(g) + X^-(g); \quad \Delta H^\circ_{lattice} = 800 \text{ kJ mol}^{-1}$$

$$M(s) \longrightarrow M(g); \Delta H^\circ_{sub} = 100 \text{ kJmol}^{-1}$$

$$M(g) \longrightarrow M^+(g) + e^-; \Delta H^\circ_i = 500 \text{ kJ mol}^{-1}$$

$$X(g) + e^-(g) \longrightarrow X^-(g); \Delta H^\circ_{e_g} = -300 \text{ kJ mol}^{-1}$$

$$M(s) + \frac{1}{2}X_2(g) \longrightarrow M^+X^-(s); \quad \backslash\Delta H^\circ_f = -400 \text{ kJ mol}^{-1}$$

[Given, M^+X^- is a pure ionic compound and X forms a diatomic molecule X_2 in gaseous state].

Bond Dissociation Enthalpy Calculation

73/ When 81.0 g of aluminium is allowed to react with 128.0 g of oxygen gas, the mass of aluminium oxide produced in grams is......(Nearest integer)
[Given; Molar mass of Al is 27.0 g mol^{-1};
Molar mass of O is 16.0 g mol^{-1}]

74/ A compound 'X' absorbs 2 moles of hydrogen and 'X' upon oxidation with $KMnO_4|H^+$ gives

$$CH_3-\underset{\underset{O}{\|}}{C}-CH_3,\ CH_3-\underset{\underset{O}{\|}}{C}-OH$$

$$\text{and } CH_3-\underset{\underset{O}{\|}}{C}-CH_2CH_2-\underset{\underset{O}{\|}}{C}-OH$$

The total number of σ-bonds present in the compound 'X' is ....

75/ Consider the following sequence of reactions.

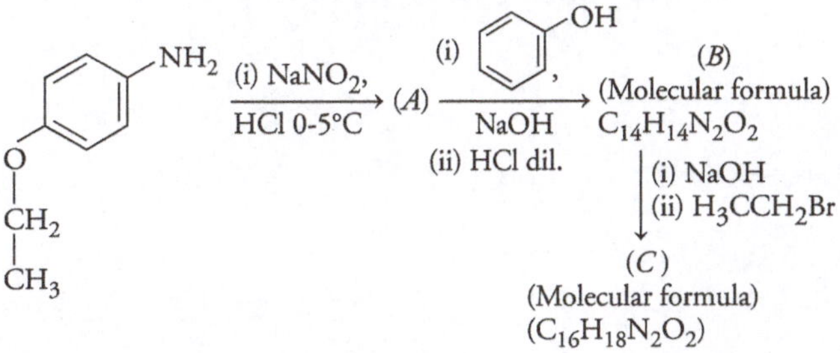

Total number of sp^3 hybridised carbon atoms in the major product C formed is ....

Answer Key

Mathematics

1 (a)	2 (a)	3 (d)	4 (c)	5 (d)	6 (a)	7 (c)	8 (c)	9 (d)	10 (b)
11 (d)	12 (c)	13 (a)	14 (c)	15 (c)	16 (b)	17 (b)	18 (a)	19 (a)	20 (c)
21 (8788)	22 (17280)	23 (31)	24 (15)	25 (474)					

Physics

26 (a)	27 (c)	28 (a)	29 (b)	30 (c)	31 (d)	32 (d)	33 (a)	34 (c)	35 (d)
36 (d)	37 (a)	38 (c)	39 (b)	40 (b)	41 (c)	42 (b)	43 (d)	44 (c)	45 (c)
46 (0.75)	47 (16)	48 (2190)	49 (100)	50 (2)					

Chemistry

51 (d)	52 (a)	53 (d)	54 (c)	55 (d)	56 (c)	57 (c)	58 (b)	59 (c)	60 (b)
61 (c)	62 (c)	63 (a)	64 (d)	65 (a)	66 (b)	67 (c)	68 (b)	69 (c)	70 (d)
71 (100)	72 (200)	73 (153)	74 (27)	75 (4)					

***Note** (*) None of the option is correct.*

Detailed Explanations

Mathematics

1. *(a)*

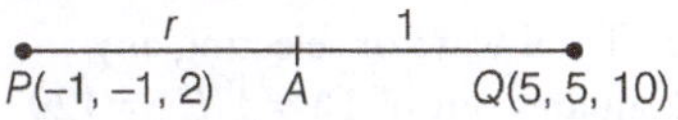

$\therefore$ Coordinates of A

$$= \left(\frac{5r-1}{r+1}, \frac{5r-1}{r+1}, \frac{10r+2}{r+1}\right)$$

$$\because \mathbf{OQ} \cdot \mathbf{OA} - \frac{1}{5}|\mathbf{OP} \times \mathbf{OA}|^2 = 10 \quad \ldots\text{(i)}$$

Now, $\mathbf{OP} \times \mathbf{OA}$

$$= \begin{vmatrix} \hat{\mathbf{i}} & \hat{\mathbf{j}} & \hat{\mathbf{k}} \\ -1 & -1 & 2 \\ \frac{5r-1}{r+1} & \frac{5r-1}{r+1} & \frac{10r+2}{r+1} \end{vmatrix}$$

$$= \left(\frac{-10r-2-10r+2}{r+1}\right)\hat{\mathbf{i}} - \left(\frac{-10r-2-10r+2}{r+1}\right)\hat{\mathbf{j}} + (0)\,\hat{\mathbf{k}}$$

$$= \left(\frac{-20r}{r+1}\right)\hat{\mathbf{i}} + \left(\frac{20r}{r+1}\right)\hat{\mathbf{j}}$$

$$= \left(\frac{20r}{r+1}\right)(-\hat{\mathbf{i}} + \hat{\mathbf{j}})$$

$$\Rightarrow |\mathbf{OP} \times \mathbf{OA}|^2 = \frac{400r^2}{(r+1)^2} \cdot 2 = \frac{800r^2}{(r+1)^2} \quad \ldots\text{(ii)}$$

and $\mathbf{OQ} \cdot \mathbf{OA} = (5\hat{\mathbf{i}} + 5\hat{\mathbf{j}} + 10\hat{\mathbf{k}}) \cdot \left(\frac{5r-1}{r+1}\hat{\mathbf{i}} + \frac{5r-1}{r+1}\hat{\mathbf{j}} + \frac{10r+2}{r+1}\hat{\mathbf{k}}\right)$

$$= \frac{5}{r+1}(5r - 1 + 5r - 1 + 20r + 4)$$

$$= \frac{5}{r+1}(30r + 2) \quad \ldots\text{(iii)}$$

From Eqs. (i) and (ii), we get

$$\frac{10}{r+1}(15r+1) - \frac{1}{5}\frac{r^2(800)}{(r+1)^2} = 10$$

$$\Rightarrow (15r+1)(r+1) - 16r^2 = (r+1)^2$$

$$\Rightarrow 15r^2 + 15r + r + 1 - 16r^2 = r^2 + 2r + 1$$

$$\Rightarrow 2r^2 - 14r = 0$$

$$\Rightarrow r = 0, 7$$

$$\because r > 0 \Rightarrow r = 7$$

2. *(a)*

Key idea
Successive integration by part
$\int U_{\rm I} \cdot V_{\rm II} = UV_1 - U'V_2 + U''V_3 - U'''V_4 \ldots\ldots$
Where U' = derivative of U
$V_n = \int V_{n-1}$

$$I = \int x^3 \sin x \, dx$$

Using successive integration

$$= x^3(-\cos x) - 3x^2(-\sin x) + 6x(\cos x) - 6(\sin x) + C$$

$$= -x^3 \cos x + 3x^2 \sin x + 6x \cos x - 6 \sin x + C$$

$$\therefore g(x) = -x^3 \cos x + 3x^2 \sin x + 6x \cos x - 6 \sin x$$

$$g\left(\frac{\pi}{2}\right) = 0 + 3\left(\frac{\pi}{2}\right)^2(1) + 0 - 6(1)$$

$$= \frac{3\pi^2}{4} - 6$$

$$\because g'(x) = x^3 \sin x$$

$$\Rightarrow g'\left(\frac{\pi}{2}\right) = \frac{\pi^3}{8}$$

$$\therefore 8\left(g\left(\frac{\pi}{2}\right) + g'\left(\frac{\pi}{2}\right)\right) = 6\pi^2 - 48 + \pi^3$$

$$\therefore \alpha = 1, \beta = 6 \text{ and } \gamma = -48$$

$$\therefore \alpha + \beta - \gamma = 55$$

3. *(d)*

$$x + y + z = 6$$
$$x + 2y + 5z = 9$$
$$x + 5y + \lambda z = \mu$$

For no solution,
$\Delta = 0$ and atleast one in Δ_x, Δ_y or Δ_z is non-zero.

Now, $$\Delta = \begin{vmatrix} 1 & 1 & 1 \\ 1 & 2 & 5 \\ 1 & 5 & \lambda \end{vmatrix} = 0$$

$$\Rightarrow 1(2\lambda - 25) - 1(\lambda - 5) + 1(5 - 2) = 0$$

$$\Rightarrow 2\lambda - 25 - \lambda + 5 + 3 = 0$$

$$\Rightarrow \lambda = 17$$

$$\Delta_z \neq 0$$

$$\Rightarrow \begin{vmatrix} 1 & 1 & 6 \\ 1 & 2 & 9 \\ 1 & 5 & \mu \end{vmatrix} \neq 0$$

$$\Rightarrow 1(2\mu - 45) - 1(\mu - 9) + 6(3) \neq 0$$

$$\Rightarrow 2\mu - 45 - \mu + 9 + 18 \neq 0 \Rightarrow \mu \neq 18$$

Concept Enhancer
If a system of linear equations in 3 variables has no solution than $\Delta = 0$ and in $\Delta_x, \Delta_y, \Delta_z$ atleast one is non-zero.

4. *(c)* Parabola $y^2 = 4x$

Normal at $P(t)$ to the parabola

$$y + tx = 2t + t^3$$

$\because$ It passes through $R(a, 0)$.

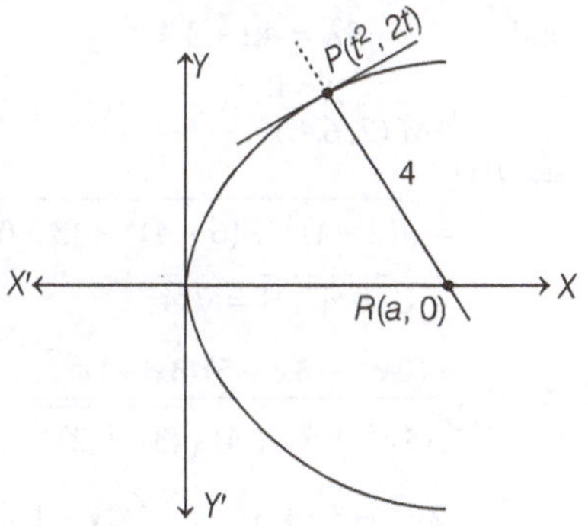

$$\Rightarrow at = 2t + t^3$$

$$\Rightarrow a = 2 + t^2 \Rightarrow a - t^2 = 2$$

$$\because PR = 4$$

$$PR^2 = 16$$

$$\Rightarrow (a - t^2)^2 + (2t)^2 = 16$$

$$\Rightarrow (2)^2 + 4t^2 = 16 \Rightarrow t^2 = 3$$

$$\therefore t = \pm\sqrt{3}$$

$$a = 2 + 3 = 5 \Rightarrow R(5, 0)$$

$\because$ Focus of parabola = (1, 0)

$\because$ (1, 0) and (5, 0) are end points of diameter.

$\therefore$ Equation of circle

$$(x - 1)(x - 5) + (y - 0)(y - 0) = 0$$

$$x^2 + y^2 - 6x + 5 = 0$$

5. *(d)* $L_1 : \frac{x-2}{2} = \frac{y-6}{3} = \frac{z-3}{4} = \mu$ (say)

$L_2 : \frac{x}{1} = \frac{y-2}{2} = \frac{z+3}{3}$

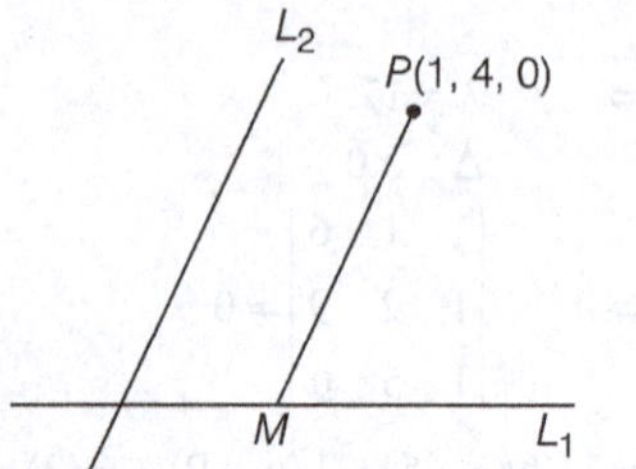

Equation of line passing through (1, 4, 0) and parallel to line L_2 is

$$\frac{x-1}{1} = \frac{y-4}{2} = \frac{z-0}{3} = \lambda \text{ (say)}$$

Let $M(\lambda + 1, 2\lambda + 4, 3\lambda)$

or $M(2\mu + 2, 3\mu + 6, 4\mu + 3)$

$\lambda + 1 = 2\mu + 2$

$\Rightarrow \quad \lambda = 2\mu + 1,$

and $\quad 2\lambda + 4 = 3\mu + 6$

$\Rightarrow \quad 2\lambda = 3\mu + 2$

and $\quad 3\lambda = 4\mu + 3$

$\therefore \quad \mu = 0$

$\therefore \quad M(2, 6, 3)$

$\Rightarrow PM$

$$= \sqrt{(2-1)^2 + (6-4)^2 + (3-0)^2}$$

$$= \sqrt{1+4+9} = \sqrt{14}$$

6. *(a)* $\lim\limits_{x \to \infty} \frac{(2x^2 - 3x + 5)(3x-1)^{x/2}}{(3x^2 + 5x + 4)\sqrt{(3x+2)^x}}$

$$= \lim_{x \to \infty} \frac{2x^2 - 3x + 5}{3x^2 + 5x + 4} \cdot \lim_{x \to \infty} \left(\frac{3x-1}{3x+2}\right)^{x/2}$$

$$= \lim_{x \to \infty} \frac{x^2\left(2 - \frac{3}{x} + \frac{5}{x^2}\right)}{x^2\left(3 + \frac{5}{x} + \frac{4}{x^2}\right)}$$

$$e^{\lim\limits_{x \to \infty}\left(\frac{3x-1}{3x+2} - 1\right) \cdot \frac{x}{2}}$$

$$= \frac{2}{3} e^{\lim\limits_{x \to \infty}\left(\frac{-3}{3x+2}\right)\left(\frac{x}{2}\right)}$$

$$= \frac{2}{3} e^{\lim\limits_{x \to \infty} \frac{-3}{\left(3 + \frac{2}{x}\right)}\left(\frac{1}{2}\right)}$$

$$= \frac{2}{3} e^{\frac{-3}{3 \times 2}} = \frac{2}{3} \cdot e^{-\frac{1}{2}} = \frac{2}{3\sqrt{e}}$$

7. *(c)* $|z| = 1$

Let $\quad z = e^{i\theta}$

$\bar{z} = e^{-i\theta}$

$$\frac{z}{\bar{z}} = \frac{e^{i\theta}}{e^{-i\theta}} = e^{2i\theta}$$

$$\frac{\bar{z}}{z} = \frac{e^{-i\theta}}{e^{i\theta}} = \frac{1}{e^{2i\theta}} = e^{-2i\theta}$$

$$\frac{z}{\bar{z}} + \frac{\bar{z}}{z} = (\cos 2\theta + i \sin 2\theta)$$
$$+ (\cos 2\theta - i \sin 2\theta)$$
$$= 2 \cos 2\theta$$

$\because \quad \left|\frac{z}{\bar{z}} + \frac{\bar{z}}{z}\right| = 1$

$\Rightarrow \quad |\cos 2\theta| = \frac{1}{2}$

$\therefore \quad \cos 2\theta = \pm \frac{1}{2}$

$\therefore$ Number of solutions $= 4 + 4 = 8$

> **Concept Enhancer**
> A complex number z is written in the following form $z = a + ib = re^{i\theta} = r(\cos\theta + i\sin\theta)$

8. *(c)* General term of $(1+x)^p (1-x)^q$ is

$$= {}^pC_{r_1} x^{r_1} \cdot {}^qC_{r_2} (-1)^{r_2} x^{r_2}$$

$$= {}^pC_{r_1} {}^qC_{r_2} (-1)^{r_2} x^{r_1 + r_2}$$

For coefficient of x,

$r_1 + r_2 = 1$

$\therefore \quad r_1 = 1, r_2 = 0$ and $r_1 = 0, r_2 = 1$

$\therefore$ Coefficient

$\Rightarrow {}^pC_1 {}^qC_0 (-1)^0 + {}^pC_0 {}^qC_1 (-1)^1 = 7$

$\Rightarrow \quad p - q = 1 \quad \ldots(\text{i})$

For coefficient of x^2, $r_1 + r_2 = 2$

r_1	r_2
0	2
1	1
2	0

$\therefore$ Coefficient

$$= {}^pC_0 {}^qC_2 (-1)^2 + {}^pC_1 {}^qC_1 (-1)^1$$
$$+ {}^pC_2 {}^qC_0 (-1)^0$$

$$\Rightarrow -2 = \frac{q(q-1)}{2} - pq + \frac{p(p-1)}{2}$$

$\Rightarrow -4 = q^2 - q - 2pq + p^2 - p$

$\Rightarrow -4 = (p-q)^2 - p - q$

$\Rightarrow -4 = 1 - p - q$

$\Rightarrow p + q = 5 \quad \ldots(\text{ii})$

On solving Eqs. (i) and (ii), we get

$p = 3, q = 2$

$\Rightarrow p^2 + q^2 = 9 + 4 = 13$

9. *(d)* Total ways of selecting any 2 squares out of $16 = {}^{16}C_2 = 120$

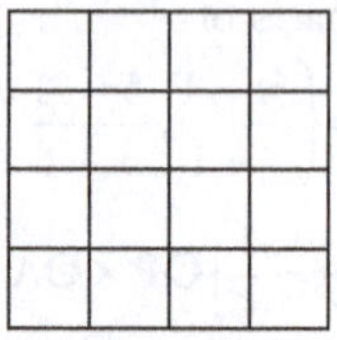

Total ways for selecting common side squares

$= 3 \times 4 + 3 \times 4$

$\downarrow \quad \downarrow$

Horizontal Vertical

$= 24$

$\therefore$ Required probability

$$= 1 - \frac{24}{120} = 1 - \frac{1}{5} = \frac{4}{5}$$

10. *(b)* Let $A = \begin{bmatrix} a_{11} & a_{12} & a_{13} \\ a_{21} & a_{22} & a_{23} \\ a_{31} & a_{32} & a_{33} \end{bmatrix}$

$$\because \quad A\begin{bmatrix} 0 \\ 1 \\ 0 \end{bmatrix} = \begin{bmatrix} 0 \\ 0 \\ 1 \end{bmatrix}$$

$$\Rightarrow \begin{bmatrix} a_{12} \\ a_{22} \\ a_{32} \end{bmatrix} = \begin{bmatrix} 0 \\ 0 \\ 1 \end{bmatrix}$$

$\Rightarrow a_{12} = 0, a_{22} = 0, a_{32} = 1$

and $\quad A\begin{bmatrix} 4 \\ 1 \\ 3 \end{bmatrix} = \begin{bmatrix} 0 \\ 1 \\ 0 \end{bmatrix}$

$4a_{11} + a_{12} + 3a_{13} = 0$

$\Rightarrow \quad 4a_{11} + 3a_{13} = 0 \quad \ldots(\text{i})$

$4a_{21} + a_{22} + 3a_{23} = 1$

$\Rightarrow \quad 4a_{21} + 3a_{23} = 1 \quad \ldots(\text{ii})$

$4a_{31} + a_{32} + 3a_{33} = 0$

$\Rightarrow \quad 4a_{31} + 3a_{33} = 0 \quad \ldots(\text{iii})$

and $\quad A\begin{bmatrix} 2 \\ 1 \\ 2 \end{bmatrix} = \begin{bmatrix} 1 \\ 0 \\ 0 \end{bmatrix}$

$\Rightarrow \quad 2a_{11} + a_{12} + 2a_{13} = 1$

$$\Rightarrow \quad 2a_{11} + 2a_{13} = 1 \quad \ldots(iv)$$
$$2a_{21} + a_{22} + 2a_{23} = 0$$
$$\Rightarrow \quad a_{21} + a_{23} = 0 \quad \ldots(v)$$
$$2a_{31} + a_{32} + 2a_{33} = 0$$
$$\Rightarrow \quad a_{31} + a_{33} = 0 \quad \ldots(vi)$$

Solving Eqs. (ii) and (v), we get

$$-a_{23} = 1 \Rightarrow a_{23} = -1$$

11. *(d)* $f(x) = 6 + 16 \cos x \cos\left(\frac{\pi}{3} - x\right)$

$$\cos\left(\frac{\pi}{3} + x\right) \times \sin 3x \cos 6x$$

$\because \cos A \cos(60° - A) \cos(60° + A)$

$$= \frac{1}{4} \cos 3A$$

$$\therefore f(x) = 6 + 16\left(\frac{1}{4}\cos 3x\right)\sin 3x \cos 6x$$
$$= 6 + 4\cos 3x \sin 3x \cos 6x$$
$$= 6 + 2\sin 6x \cos 6x$$
$$= 6 + \sin 12x$$

$\therefore$ Range of $f(x) = [6 - 1, 6 + 1]$

$$= [5, 7]$$

$\therefore (\alpha, \beta)$ is $(5, 7)$.

$\therefore$ Distance between $(5, 7)$ and line $3x + 4y + 12 = 0$ is

$$= \left|\frac{3(5) + 4(7) + 12}{\sqrt{3^2 + 4^2}}\right|$$
$$= \frac{15 + 28 + 12}{5} = 11$$

12. *(c)* $R = \{(x, y) : -1 \le x \le 1,$

$$0 \le y \le a + e^{|x|} - e^{-x}, a > 0\}$$

$$y = a + e^{|x|} - e^{-x}$$

$$y = \begin{bmatrix} a + e^{-x} - e^{-x} = a, & -1 < x < 0 \\ a + e^{x} - e^{-x}, & 0 \le x < 1 \end{bmatrix}$$

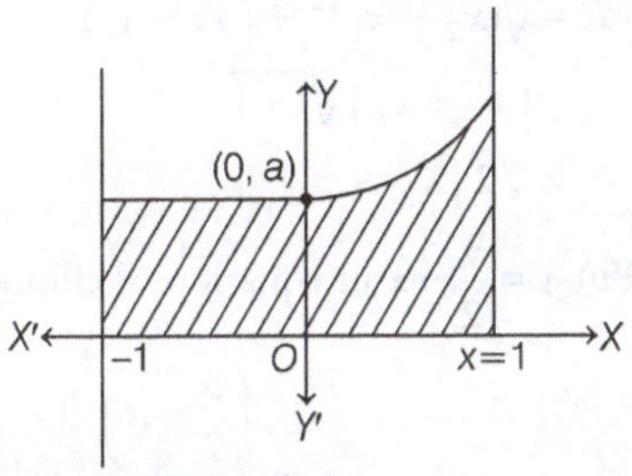

Required area

$$= a + \int_0^1 (a + e^x - e^{-x})\, dx$$
$$= a + a + (e^1 - 1) + (e^{-1} - 1)$$
$$= 2a + e + \frac{1}{e} - 2$$

Given, $2a + e + \frac{1}{e} - 2 = e + 8 + \frac{1}{e}$

$$\Rightarrow \quad 2a - 2 = 8$$
$$\therefore \quad a = 5$$

13. *(a)* $y = \left(x - y\frac{dx}{dy}\right)\sin\left(\frac{x}{y}\right)$

$$\Rightarrow \quad 1 = \left(\frac{x}{y} - \frac{dx}{dy}\right)\sin\left(\frac{x}{y}\right)$$

$$\Rightarrow \quad \frac{dx}{dy} = \frac{x}{y} - \operatorname{cosec}\left(\frac{x}{y}\right)$$

Put $\quad x = vy$

$$\Rightarrow \quad \frac{dx}{dy} = v + y\frac{dv}{dy}$$

$\therefore$ Then, equation

$$v + y\frac{dv}{dy} = v - \operatorname{cosec} v$$

$$\Rightarrow \quad \sin v\, dv = -\frac{dy}{y}$$

Integrating both sides, we get

$$-\cos v = -\log_e y - C$$
$$\Rightarrow \quad \cos v = \log_e y + C$$
$$\Rightarrow \quad \cos\left(\frac{x}{y}\right) = \log_e y + C$$
$$\therefore \quad y = 1 \text{ and } x = \frac{\pi}{2}$$
$$\cos\frac{\pi}{2} = \log_e 1 + C \Rightarrow C = 0$$
$$\therefore \quad \log_e y = \cos\left(\frac{x}{y}\right)$$
$$\frac{x}{y} = \cos^{-1}(\log_e y)$$
$$x = y(\cos^{-1} \log_e y)$$
$$x(2) = 2\cos^{-1} \log_e 2$$
$$\cos(x(2)) = \cos(2\cos^{-1} \log_e 2)$$
$$= 2(\log_e 2)^2 - 1$$
$$= 2(\log_e 2)^2 - 1$$

14. *(c)* Given, $\frac{dr}{dt} = \frac{1}{4\pi}$ cm/min

Let volume of sphere be V at any time t.

$$V = \frac{4}{3}\pi r^3$$

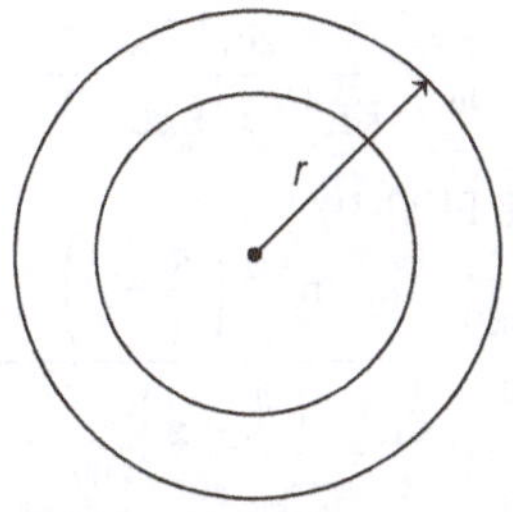

$$\frac{dV}{dt} = 4\pi r^2 \frac{dr}{dt}$$
$$\Rightarrow \quad 81 = 4\pi r^2\left(\frac{1}{4\pi}\right)$$
$$\Rightarrow \quad r^2 = 81$$
$$\Rightarrow \quad r = 9$$

$\because$ Surface area of chocolate

$$S = 4\pi(r - 1)^2 = 4\pi(8)^2$$
$$= 256\pi$$

15. *(c)* $h = \frac{2\beta + \alpha}{3}, k = \frac{-4 + \alpha + 2}{3}$

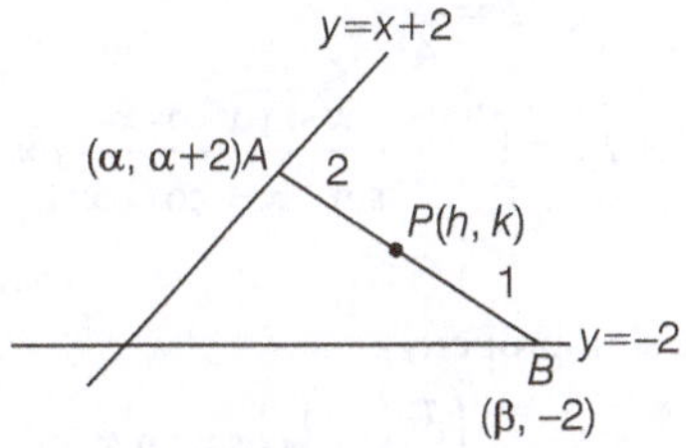

$$\Rightarrow \quad 2\beta = 3h - \alpha, \alpha = 3k + 2$$
$$\Rightarrow \quad 2\beta = 3h - 3k - 2$$

$\because AB = 8$

$$(\alpha - \beta)^2 + (\alpha + 4)^2 = (8)^2$$

$$\left(3k + 2 - \left(\frac{3h - 3k - 2}{2}\right)\right)^2 + (3k + 2 + 4)^2 = 64$$

$$\Rightarrow \frac{(9k - 3h + 6)^2}{4} + (3k + 6)^2 = 64$$

$$\Rightarrow 9[(3k - h + 2)^2 + 4(k + 2)^2] = 64 \times 4$$

Put $h = x$ and $k = y$

$$9(x^2 + 13y^2 - 6xy - 4x + 28y) - 76$$
$$\therefore \quad \alpha = 13, \beta = -6$$
$$\text{and} \quad \gamma = -4$$
$$\Rightarrow \alpha - \beta - \gamma = 13 + 6 + 4 = 23$$

16. *(b)*

Key idea
Use property $\int_0^a f(x)dx = \int_0^a f(a - x)dx$

$$I = \int_0^{\pi/2} \frac{\sin^{3/2} x}{\sin^{3/2} x + \cos^{3/2} x}\, dx \quad \ldots(i)$$

Using property,

$$I = \int_0^{\pi/2} \frac{\sin^{3/2}\left(\frac{\pi}{2} - x\right)}{\sin^{3/2}\left(\frac{\pi}{2} - x\right) + \cos^{3/2}\left(\frac{\pi}{2} - x\right)}\, dx$$

$$I = \int_0^{\pi/2} \frac{\cos^{3/2}\left(\frac{\pi}{2} - x\right)}{\cos^{3/2}\left(\frac{\pi}{2} - x\right) + \sin^{3/2}\left(\frac{\pi}{2} - x\right)} \quad \ldots(ii)$$

On adding Eqs. (i) and (ii), we get

$$2I = \int_0^{\pi/2} 1\, dx$$

$$I = \frac{\pi}{4}$$

Let $I_2 = \int_0^{2(\pi/4)} \frac{x \sin x \cos x}{\sin^4 x + \cos^4 x}\, dx$ $\quad \ldots(iii)$

Using property,

$$I_2 = \int_0^{\pi/2} \frac{\left(\frac{\pi}{2} - x\right) \cos x \sin x}{\sin^4 x + \cos^4 x} \quad \ldots(iv)$$

On adding Eqs. (iii) and (iv), we get

$$2I_2 = \int_0^{\pi/2} \frac{\frac{\pi}{2} \sin x \cos x}{\sin^4 x + \cos^4 x}\, dx$$

$$\Rightarrow \quad 2I_2 = \frac{\pi}{2} \int_0^{\pi/2} \frac{\sin x \cos x\, dx}{\sin^4 x + \cos^4 x}$$

$$\Rightarrow \quad 2I_2 = \frac{\pi}{2} \int_0^{\pi/2} \frac{\tan x \sec^2 x\, dx}{\tan^4 x + 1}$$

let $\tan^2 x = t$

$$\Rightarrow \quad 2 \tan x \sec^2 x\, dx = dt$$

$$\Rightarrow \quad 2I_2 = \frac{\pi}{4} \int_0^{\infty} \frac{dt}{1 + t^2}$$

$$\Rightarrow \quad I_2 = \frac{\pi}{8} [\tan^{-1} t]_0^{\infty} = \frac{\pi^2}{16}$$

17. *(b)* **Statement I**

Reflexive $(a_1, b_1) R (a_1, b_1) \Rightarrow b_1 = b_1$

It is always true.

$\therefore R$ is reflexive relation.

Symmetric

$(a_1, b_1) R (a_2, b_2) \Rightarrow b_1 = b_2$

$\Rightarrow \quad b_2 = b_1$

$\Rightarrow \quad (a_2, b_2) R (a_1, b_1)$

$\therefore R$ is symmetric.

Transitive

$(a_1, b_1) R (a_2, b_2) \Rightarrow b_1 = b_2$

and $\quad (a_2, b_2) R (a_3, b_3) \Rightarrow b_2 = b_3$

$\therefore \quad b_1 = b_3$

$\Rightarrow \quad (a_1, b_1) R (a_3, b_3)$

$\therefore R$ is transitive.

$\therefore R$ is equivalence relation.

$\therefore$ Statement I is true.

Statement II

$$S = \{(x, y) \in X : (x, y) R (a, b)\}$$

$\because \quad (a, b) \in X$

$\because \quad (x, y) R (a, b)$

$\Rightarrow \quad y = b$

It does not represent a line parallel to $y = x$

$\therefore$ Statement II is not true.

Equivalence relation Any relation on a set A which is reflexive, symmetric and transitive is called an equivalence relation.

18. *(a)* d = Shortest distance between line

$$= \left| \frac{(\mathbf{a}_2 - \mathbf{a}_1) \cdot (\mathbf{b}_1 \times \mathbf{b}_2)}{\mathbf{b}_1 \times \mathbf{b}_2} \right|$$

$$\mathbf{b}_1 \times \mathbf{b}_2 = \begin{vmatrix} \hat{\mathbf{i}} & \hat{\mathbf{j}} & \hat{\mathbf{k}} \\ 1 & 2 & -3 \\ 2 & 4 & -5 \end{vmatrix}$$

$$= \hat{\mathbf{i}}(2) - \hat{\mathbf{j}}(1) + \hat{\mathbf{k}}(0) = 2\hat{\mathbf{i}} - \hat{\mathbf{j}}$$

$$\Rightarrow |\mathbf{b}_1 \times \mathbf{b}_2| = \sqrt{5}$$

$$\therefore (\mathbf{a}_2 - \mathbf{a}_1) - (\mathbf{b}_1 \times \mathbf{b}_2) = \begin{vmatrix} -3 & -4 & -2 \\ 1 & 2 & -3 \\ 2 & 4 & -5 \end{vmatrix}$$

$$= (-3(2) + 4(1) - 20)$$

$$= -6 + 4 = -2$$

$$\therefore \quad |d| = \left| \frac{-2}{\sqrt{5}} \right| = \frac{2}{\sqrt{5}}$$

$$\Rightarrow \quad d^2 = \frac{4}{5} = \frac{m}{n} \quad \text{[given]}$$

$$\therefore \quad m + n = 4 + 5 = 9$$

19. *(a)* $A = \{(x, y) \in R \times R : |x + y| \geq 3\}$

$\Rightarrow x + y \geq 3 \quad$ or $\quad x + y \leq -3 \quad \ldots(i)$

$B = \{(x, y) \in R \times R : |x| + |y| \leq 3\}$

I quadrant : $x + y \leq 3$

II quadrant : $-x + y \leq 3 \ldots(ii)$

III quadrant : $-x - y \leq 3$

IV quadrant : $x - y \leq 3$

$C : \{(x, y)\ A \cap B : x = 0$ or $y = 0\}$

Put $x = 0$ in conditions (i) and (ii), we get common solution $y = -3, 3$.

Similarly, $y = 0$ we get $x = -3, 3$

$\therefore \ C = \{(0, 3), (0, -3), (3, 0), (-3, 0)\}$

$$\therefore \quad \sum_{(x, y) \in C} |x + y| = 3 + 3 + 3 + 3 = 12$$

20. *(c)* Ellipse : $\frac{x^2}{4} + \frac{y^2}{2} = 1$

Equation of chord whose mid-point is $\left(1, \frac{1}{2}\right)$ is

$$T = S_1$$

$$\Rightarrow \quad \frac{x(1)}{4} + \frac{y(1/2)}{2} - 1 = \frac{1}{4} + \frac{1}{8} - 1$$

$$\Rightarrow \quad \frac{x}{4} + \frac{y}{4} = \frac{3}{8}$$

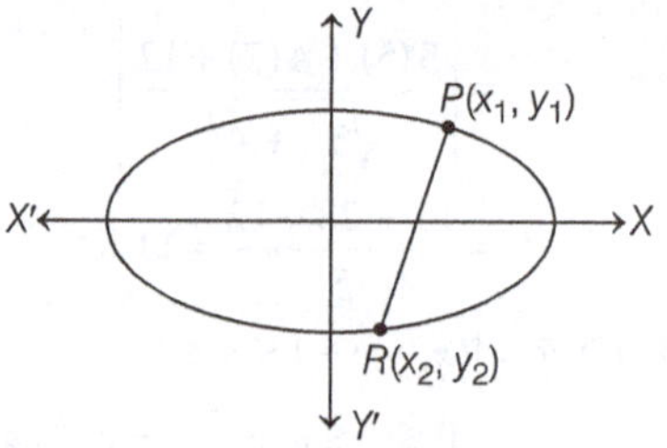

$$\Rightarrow \quad x + y = \frac{3}{2} \quad \ldots(i)$$

Eq. (i) represents equation of chord PR.

$$PR = \sqrt{(x_2 - x_1)^2 + (y_2 - y_1)^2}$$

$$= |x_2 - x_1| \sqrt{1 + 1}$$

$$= \sqrt{2}\, |x_2 - x_1|$$

Put $y = \frac{3}{2} - x$ in equation of ellipse

$$\Rightarrow \quad \frac{x^2}{4} + \frac{\left(\frac{3}{2} - x\right)^2}{2} = 1$$

$$\Rightarrow \quad x^2 + 2\left(\frac{9}{4} + x^2 - 3x\right) = 4$$

$\Rightarrow \quad x^2 + \frac{9}{2} + 2x^2 - 6x = 4$

$\Rightarrow \quad 3x^2 - 6x + \frac{1}{2} = 0$

$\Rightarrow \quad 6x^2 - 12x + 1 = 0$

$\Rightarrow \quad |x_2 - x_1| = \frac{\sqrt{D}}{a}$

$= \frac{\sqrt{144-24}}{6} = \frac{\sqrt{120}}{6}$

$\therefore \; PR = \sqrt{2}\,\frac{\sqrt{120}}{6} = \frac{2}{6}\sqrt{60} = \frac{2}{3}\sqrt{15}$

21. *(8788)* 8, 21, 34, 47, ..., 320

This series is in AP.

For n, $n = \frac{320-8}{13} + 1 = \frac{312}{13} + 1$

$= 24 + 1 = 25$

$\Rightarrow \quad \bar{X} = \frac{\frac{n}{2}(8+320)}{n} = 4 + 160 = 164$

$\Rightarrow \quad a_n = 8 + (n-1)13 = 13n - 5$

$a_n^2 = 169n^2 + 25 - 130n$

$\sum a_i^2 = 169\left(\frac{n}{6}\right)(n+1)(2n+1) + 25n$

$-130\,\frac{n}{2}(n+1)$

$= n\left(\frac{169}{6}(n+1)(2n+1) + 25 - 65(n+1)\right)$

$n = 25$

$\sum x_i^2 = 25\left(\frac{169}{6}(26)(51) + 25 - 65\times 26\right)$

$\frac{\sum x_i^2}{25} = 35684$

$\therefore \text{Var}(X)$

$= \frac{\sum x_i^2}{25} - (\bar{X})^2 = 35684 - (164)^2$

$= 8788$

Time Saver Tech

Variance of $a, a+d, a+2d, \ldots\ldots,$

$a + (n-1)d$ is $d^2\left(\frac{n^2-1}{12}\right)$

$\therefore$ Variance of 8, 21, 34, ..., 320

$\therefore$ Here, $d = 13, n = 25$

$\therefore$ Variance

$= 13^2\left(\frac{25^2-1}{12}\right) = 169 \times \left(\frac{625-1}{12}\right) = 8788$

22. *(17280)* 5 boys and 4 girls

E_1 = All boys sit together.

E_2 = No 2 boys sit together

$E_1 \cap E_2 = \phi$

$n(E_1) = 5!\,5!$

$n(E_2) = 4!\;5!$

$n(E_1 \cap E_2) = 0$

$\therefore \quad n(E_1 \cup E_2) = n(E_1) + n(E_2)$

$= 5!\,5! + 5!\,4!$

$= 6!\,4! = 17280$

23. *(31)* $x^2 - ax - b = 0 \begin{smallmatrix}\nearrow \alpha \\ \searrow \beta\end{smallmatrix}$

$\alpha + \beta = a$ and $\alpha\beta = -b$

$\because \quad P_n = \alpha^n - \beta^n$

$\therefore \quad P_n - aP_{n-1} - bP_{n-2} = 0$

$\Rightarrow \quad P_n = aP_{n-1} + bP_{n-2}$

$P_6 = aP_5 + bP_4$

$45\sqrt{7}i = a(11\sqrt{7}i) + b(-3\sqrt{7}i)$

$45 = 11a - 3b \quad \ldots$(i)

and $\quad P_5 = aP_4 + bP_3$

$(11\sqrt{7}i) = a(-3\sqrt{7}i) + b(-5\sqrt{7}i)$

$11 = -3a - 5b \quad \ldots$(ii)

Solving Eqs. (i) and (ii), we get

$a = 3$ and $b = -4$

$\therefore \quad \alpha + \beta = 3$ and $\alpha\beta = 4$

$\alpha^2 + \beta^2 = (\alpha+\beta)^2 - 2\alpha\beta = 1$

$\alpha^4 + \beta^4 = (\alpha^2 + \beta^2)^2 - 2\alpha^2\beta^2$

$= 1 - 32 = -31$

$\therefore \; |\alpha^4 + \beta^4| = 31$

24. *(15)* Equation of parabola

$y^2 = 4(x+4)$

$\therefore \quad 4a = 4 \Rightarrow a = 1$

Focus $(-4+1, 0) = (-3, 0)$

$\therefore$ Equation of circle C,

$(x+3)^2 + y^2 = 5^2$

Point of intersection P of $3x - y = 0$ and $x + \lambda y = 4$

$= x(1 + 3\lambda) = 4$

$\Rightarrow \quad x = \frac{4}{3\lambda+1}$

and $\quad y = \frac{12}{3\lambda+1}$

$\because P$ lies on circle C.

$\Rightarrow \left(\frac{4}{3\lambda+1} + 3\right)^2 + \left(\frac{12}{3\lambda+1}\right)^2 = 25$

$\Rightarrow (9\lambda+7)^2 + 144 = 25(9\lambda^2 + 6\lambda + 1)$

$\Rightarrow 81\lambda^2 + 49 + 126\lambda + 144$

$= 225\lambda^2 + 150\lambda + 25$

$\Rightarrow \quad 144\lambda^2 + 24\lambda - 168 = 0$

$\Rightarrow \quad 36\lambda^2 + 6\lambda - 42 = 0$

$\Rightarrow 6\lambda^2 + \lambda - 7 = 0$

$\lambda = -\frac{7}{6}, 1$

$\therefore \quad \lambda_1 = -\frac{7}{6}, \lambda_2 = 1$

$[\because \lambda_2 > \lambda_1]$

$\Rightarrow \quad 12\lambda_1 + 29\lambda_2 = -14 + 29 = 15$

25. *(474)*

$3x^2 - px + q = 0 \begin{smallmatrix}\nearrow a_{10} \\ \searrow a_{11}\end{smallmatrix}$

a_{10} = 10th term of AP

a_{11} = 11th term of AP

$d = \frac{3}{2}$

$\Rightarrow \quad S_{11} = 88$

$\Rightarrow \quad \frac{11}{2}(2a + 10d) = 88$

$\Rightarrow \quad a + 5d = 8$

$\Rightarrow \quad a = 8 - 5\left(\frac{3}{2}\right) = \frac{1}{2}$

$\therefore \quad a_{10} = a + 9d = \frac{1}{2} + \frac{27}{2} = 14$

$\therefore \quad a_{11} = a + 10d = \frac{1}{2} + \frac{30}{2} = \frac{31}{2}$

$\because \; a_{10} + a_{11} = \frac{p}{3}$

$\Rightarrow \; 14 + \frac{31}{2} = \frac{p}{3}$

$\Rightarrow \quad \frac{59}{2} = \frac{p}{3} \Rightarrow 2p = 177$

and $\quad a_{10}a_{11} = \frac{q}{3}$

$\Rightarrow \quad 14 \times \frac{31}{2} = \frac{q}{3}$

$\Rightarrow \quad q = 21 \times 31 = 651$

$\therefore \; q - 2p = 651 - 177 = 474$

Physics

26. *(a)*

$$F = kx$$

$$5 = kx_1 \quad \text{...(i)}$$

$$7 = kx_2 \quad \text{...(ii)}$$

Multiply by 5 in Eq. (i)

$$5 \times 5 = k5x_1 \quad \text{...(iii)}$$

Multiply by 2 in Eq. (ii)

$$7 \times 2 = k \times 2x_2 \quad \text{...(iv)}$$

From Eq. (iii) and Eq. (iv)

$$k \cdot 5x_1 - k \cdot 2x_2 = 25 - 14$$

$$k(5x_1 - 2x_2) = 25 - 14$$

$$k(5x_1 - 2x_2) = 11$$

$\Rightarrow$ Tension in the string,

$$T = k(5x_1 - 2x_2) = 11\text{ N}$$

27. *(c)* Binding energy per nucleon is found to be practically independent of the atomic number A, for nuclei with mass number between 30 and 170.

Nuclear force is short range force.

A is true but R is false.

28. *(a)* Magnetic moment, $M = NIA$

$$[M] = [L^2A]$$

$$\therefore \quad \tau = C\theta$$

$$\Rightarrow \quad C = \frac{\tau}{\theta} = [ML^2T^{-2}]$$

$$\therefore \quad F = qvB$$

$$B = \frac{F}{qv} = \frac{[MLT^{-2}]}{[AT][LT^{-1}]}$$

$$= \frac{[MLT^{-2}]}{[AL]}$$

$$= [MT^{-2}A^{-1}]$$

$$\therefore \quad \frac{B^2}{2\mu_0} = \text{Energy density}$$

$$\mu_0 = \frac{B^2}{2(\text{Energy density})}$$

$$= \frac{[M^2T^{-4}A^{-2}]}{\frac{[ML^2T^{-2}]}{[L^3]}} = [MT^{-2}LA^{-2}]$$

$$= [MLT^{-2}A^{-2}]$$

29. *(b)* $\dfrac{I_{max}}{I_{min}} = \dfrac{9}{4}$ (given)

$$I_{max} = (\sqrt{I_1} + \sqrt{I_2})^2$$

$$I_{min} = (\sqrt{I_1} - \sqrt{I_2})^2$$

Put the value

$$\frac{(\sqrt{I_1} + \sqrt{I_2})^2}{(\sqrt{I_1} - \sqrt{I_2})^2} = \frac{9}{4}$$

$$\frac{\sqrt{I_1} + \sqrt{I_2}}{\sqrt{I_1} - \sqrt{I_2}} = \frac{3}{2} \quad \text{...(i)}$$

Given intensity is directly proportional to the square of the slit width

$$I \propto d^2$$

Thus, according to question $\sqrt{I_1} = xd$, $\sqrt{I_2} = d$ put the value in Eq. (i)

$$\frac{xd + d}{xd - d} = \frac{3}{2}$$

$$\frac{d(x+1)}{d(x-1)} = \frac{3}{2}$$

$$2(x+1) = 3(x-1)$$

$$2x + 2 = 3x - 3$$

$$5 = 3x - 2x$$

$$x = 5$$

30. *(c)* $\mu = \dfrac{\sin\left(\dfrac{A + \delta_{min}}{2}\right)}{\sin\left(\dfrac{A}{2}\right)}$

$$\delta_{min} = A$$

$$\sqrt{3} = \frac{\sin\left(\frac{A+A}{2}\right)}{\sin\left(\frac{A}{2}\right)} = \frac{\sin\left(\frac{2A}{2}\right)}{\sin\left(\frac{A}{2}\right)}$$

$$= \frac{2\sin\left(\frac{A}{2}\right) \cdot \cos\left(\frac{A}{2}\right)}{\sin\left(\frac{A}{2}\right)}$$

$$\sqrt{3} = 2\cos\left(\frac{A}{2}\right)$$

$$\frac{\sqrt{3}}{2} = \cos\left(\frac{A}{2}\right)$$

$$\frac{A}{2} = 30°$$

$$\Rightarrow A = 60°$$

31. *(d)* $U = \dfrac{kq_1q_2}{r}$

$$= \frac{1}{4\pi\varepsilon_0}\frac{q_1q_2}{r}$$

$$= \frac{9 \times 10^9 \times (7 \times 10^{-6}) \times (-4 \times 10^{-6})}{(14) \times 10^{-2}}$$

$$= \frac{-9 \times 10^9 \times 28 \times 10^{-12}}{14 \times 10^{-2}}$$

$$= \frac{-252 \times 10^{-3}}{14 \times 10^{-2}}$$

$$= \frac{-18 \times 10^{-3}}{10^{-2}} = -1.8\text{ J}$$

32. *(d)* Given in question, $9R_s = R_m$,

$T_m = 27$ days

According to Kepler's law

$$T^2 \propto R^3$$

So, $$\frac{T_m^2}{T_s^2} = \frac{R_m^3}{(R_s)^3} \quad \text{...(i)}$$

Given, $9R_s = R_m$

$$R_s = \frac{R_m}{9}$$

Put the value of R_s in Eq. (i)

$$\frac{T_m^2}{T_s^2} = \frac{R_m^3}{\left(\frac{R_m}{9}\right)^3}$$

$$\frac{T_m^2}{T_s^2} = \frac{R_m^2}{\frac{R_m^3}{(9)^3}} = (9)^3$$

$$\left(\frac{T_m}{T_s}\right)^2 = 9^3 \Rightarrow \frac{T_m}{T_s} = 9^{\frac{3}{2}}$$

$$T_s = \frac{T_m}{9^{\frac{3}{2}}} = \frac{27}{27} = 1\text{ day}$$

33. *(a)* $I_gG = (I - I_g)S$ or $I_gG \approx IS$

$$S = \frac{I_gG}{I} = \frac{(20 \times 10^{-3})(30\ \Omega)}{3\text{A}}$$

$$= \frac{600 \times 10^{-3}}{3} = \frac{6 \times 10^{-1}}{3}$$

$$= \frac{6}{30} = \frac{6 \times 5}{30 \times 5} = \frac{30}{150}$$

$$X \approx 150,\ X = 149$$

34. *(c)* $U = -pE\cos\theta$

$$W_{ext} = \Delta U = U_f - U_i$$
$$= -pE\cos 180° + pE\cos 0°$$
$$= 2pE = 2qlE$$
$$= 2 \times 4 \times 10^{-6} \times 18 \times 10^{-2} \times 10^4$$
$$= 144 \times 10^{-4} = 14.4 \text{ mJ}$$

35. *(d)* By Bernoulli equation

$$p_1 + \frac{1}{2}\rho \times 0^2 = p_2 + \frac{1}{2}\rho v^2$$
$$v = \sqrt{\frac{2(p_1 - p_2)}{\rho}}$$
$$v \propto \sqrt{(p_1 - p_2)}$$

36. *(d)*

Key idea

Entropy (S) is a measure of randomness (disorder) in a system. It's SI unit is JK^{-1}. For a reversible process change in entropy dS due to a transfer of heat dQ at finite temperature T is,

$$dS = \frac{dQ}{T}$$

$$d(\text{Entropy})_{sys} = \frac{dQ}{T}$$
$$\int d(\Delta E)_{sys} = \int_{T_1}^{T_2} \frac{dQ}{T}$$
$$(\Delta E)_{sys} = \int_{T_1}^{T_2} \frac{msdT}{T}$$
$$= ms\int_{T_1}^{T_2} \frac{dT}{T} = ms\,[\ln T]_{T_1}^{T_2}$$
$$= ms\,[\ln T_2 - \ln T_1]$$
$$= ms\left[\ln\left(\frac{T_2}{T_1}\right)\right]$$
$$= ms\ln\frac{T_2}{T_1} = m(1)\ln\frac{T_2}{T_1}$$
$$= m\ln\frac{T_2}{T_1}$$

37. *(a)*

Key idea

At highest point of it's hight a projectile has only horizontal velocity component ($v = u\cos\theta$). It's vertical velocity component vanishes at highest point.

$$KE = \frac{1}{2}mv^2$$
$$KE_H = \frac{1}{2}m(v\cos 60°)^2$$

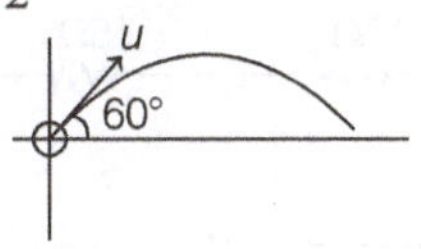

$$KE_H = \frac{1}{2}mv^2 \times \left(\frac{1}{2}\right)^2$$
$$KE_H = \frac{mv^2}{8} = \frac{1}{2}\frac{mv^2}{4} = \frac{1}{4}KE$$

38. *(c)* $\theta = 5t^2 - 8t$

$$\omega = \frac{d\theta}{dt} = 10t - 8$$
$$\alpha = \frac{d\omega}{dt} = 10$$
$$\therefore P = \tau\omega = I\alpha\omega \quad [\tau = I\alpha]$$
$$= \left(\frac{MR^2}{2}\right)\alpha\omega \quad \left[I = \frac{MR^2}{2}\right]$$
$$= \frac{MR^2}{2} \times (10)(10t - 8)$$

At $t = 2$s,

$$P = 60\,MR^2$$

39. *(b)* $E = \alpha^3 e^{-\beta t}$

$$\ln E = \ln(\alpha^3 e^{-\beta t}) = \ln\alpha^3 + \ln e^{-\beta t}$$
$$= 3\ln\alpha - \beta t\ln e$$
$$\ln E = 3\ln\alpha - \beta t \quad \text{...(i)} \; [\ln e = 1]$$

Differentiate Eq. (i)

$$\frac{dE}{E} \times 100 = 3\frac{d\alpha}{\alpha} \times 100 - \beta(100)dt \quad \text{...(ii)}$$
$$\frac{d\alpha}{\alpha} \times 100 = 1.2\%, \; \beta = 0.3\text{ s}^{-1}$$
$$\frac{dt}{t} \times 100 = 1.6\%$$

Put the value in Eq. (ii)

$$\frac{dE}{E} \times 100 = 3(1.2) + 0.3 \times 1.6 \times 5$$
$$= 3.6 + 1.5 \times 1.6$$
$$= 3.6 + 2.4 = 6\%$$

40. *(b)* $E = 2 + 2.14 = 4.14$ eV

$$E = \frac{hc}{\lambda}$$
$$\lambda = \frac{hc}{E} = \frac{1242}{4.14} = 300 \text{ nm}$$

41. *(c)* Focal length of spherical mirror does not depends on medium in which it placed.

42. *(b)* $E_y = cB_z$

$$B_z = \frac{E_y}{c} = \frac{9.3}{3 \times 10^8}$$
$$B_z = 3.1 \times 10^{-8}\text{ T}$$

43. *(d)* $y(x, t) = 4.0\sin(20 \times 10^{-3}x + 600t)$ mm

Compare $y(x, t) = A\sin(kx + \omega t)$

$A = 4.0$, $k = 20 \times 10^{-3}$, $\omega = 600$

$$\text{Velocity of wave} = \frac{\text{Coefficient of } t}{\text{Coefficient of } x}$$
$$= \frac{600}{20 \times 10^{-3}}$$
$$= 3 \times 10^4 \text{ mm/s}$$
$$v = 30 \text{ m/s}$$

Also wave is travelling in negative x-direction.

44. *(c)* Work done = Area under curve

$$= -4p_0V_0 + p_0V_0$$
$$= -3p_0V_0$$

Alternate Solution

Work done in an isobaric process work A to B

$= p\Delta V$

$= p(V_f - V_i)$

$= p_0(3V_0 - 2V_0) = p_0V_0$

Work $BC = 0$ (process is isochoric)

Work $CD = 2p_0(V_0 - 3V_0)$

$2p_0 = (-2V_0) = -4p_0V_0$

Total work $A \to B \to C \to D$

$= p_0V_0 + 0 + (-4p_0V_0) = -3p_0V_0$

45. *(c)*

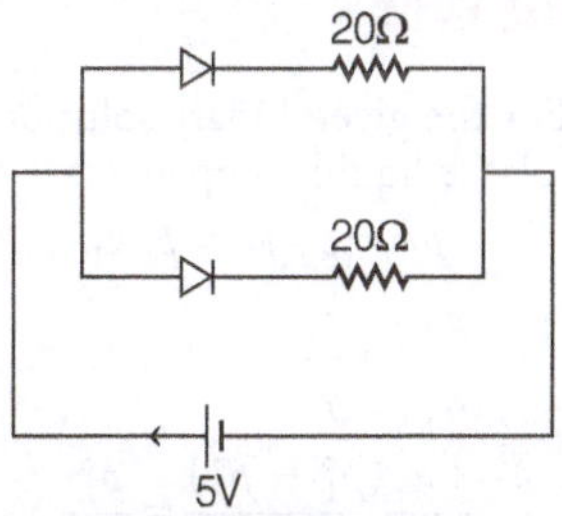

Both diode are forward bias,
In forward bias diode is act as conducting path
So, resistor are in parallel

$$\frac{1}{R_{eq}} = \frac{1}{20} + \frac{1}{20} \Rightarrow R_{eq} = 10\,\Omega$$
$$I = \frac{V}{R_{eq}} = \frac{5}{10} = 0.5\text{ A}$$

46. *(0.75)*

Given, $L = M\sqrt{\frac{GMR}{x}}$

$h = R + \frac{R}{3} = \frac{4R}{3}$

$L = Mvr$

$$M\sqrt{\frac{GMR}{x}} = M\sqrt{\frac{GM}{\left(R+\frac{R}{3}\right)}} \cdot \left(R + \frac{R}{3}\right)$$

$$M\sqrt{\frac{GMR}{x}} = M\sqrt{\frac{3GM}{4R}} \cdot \frac{4R}{3}$$

$$\sqrt{\frac{GMR}{x}} = \sqrt{\frac{3GM}{4R}} \cdot \frac{4R}{3}$$

$$\frac{GMR}{x} = \frac{3GM}{4R} \cdot \frac{16R^2}{9}$$

$$x = \frac{3}{4}$$

$$= 0.75$$

47. *(16)*

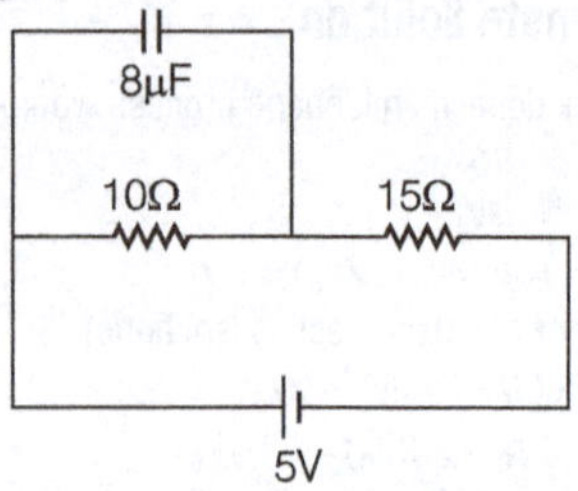

In steady state capacitor acts as open circuit

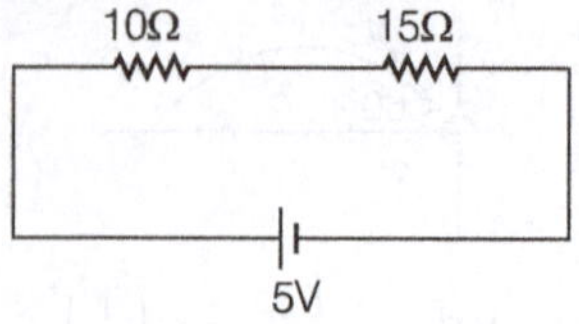

$$R_{eq} = 10 + 15 = 25\ \Omega$$

$$I = \frac{V}{R_{eq}} = \frac{5}{25}$$

$$I = \frac{1}{5}\text{A},$$

Potential across 10Ω,

$$V_{10} = IR = \frac{1}{5} \times 10 = 2\text{V}$$

$$Q = CV = 8 \times 10^{-6} \times 2$$

$$= 16 \times 10^{-6}$$

$$= 16\ \mu\text{C}$$

48. *(2190)* Given,

$$S = 0.095\ \text{J/m}^2$$

$$\rho = 10^3\ \text{kg/m}^3$$

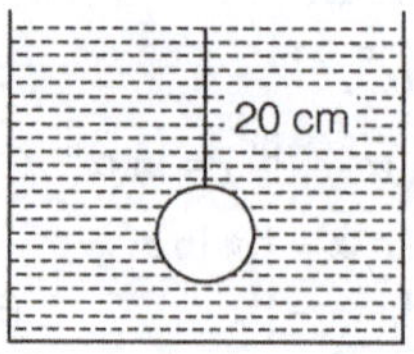

Pressure just outside the bubble,

$$p_A = p_0 + \rho g h$$

Pressure of air inside the bubble,

$$p_{in} = p_A + \frac{2S}{R}$$

$$p_{in} = p_0 + \rho g h + \frac{2S}{R}$$

$$p_{in} - p_0 = \rho g h + \frac{2S}{R}$$

$$= 10^3 \times 10 \times \frac{20}{100} + \frac{2 \times 0.095}{1 \times 10^{-3}}$$

$$p_{in} - p_0 = 2000 + 190 = 2190\ \text{N/m}^2$$

49. *(100)* $V = \frac{Q}{C}$

$$\frac{dV}{dt} = \frac{1}{C} \cdot \frac{dQ}{dt}$$

$$\frac{dV}{dt} = \frac{1}{C}(I) = \frac{0.25 \times 10^{-3}}{2.5 \times 10^{-6}}$$

$$= \frac{1000}{10} = 100\ \text{Vs}^{-1}$$

50. *(2)* The current is maximum at resonance,

$$\omega = \frac{1}{\sqrt{LC}} = \frac{1}{\sqrt{(100 \times 10^{-3})(25 \times 10^{-9})}}$$

$$= \sqrt{\frac{1}{10^{-10} \times 25}} = \sqrt{\frac{10^{10}}{25}}$$

$$= \frac{10^5}{5} = 2 \times 10^4\ \text{rad/s}$$

Chemistry

51. *(d)* For the given binary solution of two volatile liquid components 1 and 2.

$$p_1 = p_T y_1 = p_1^\circ x_1$$

$$\frac{p_T}{x_1} = \frac{p_1^\circ}{y_1}$$

So, $$\frac{p_2^\circ + x_1(p_1^\circ - p_2^\circ)}{x_1} = \frac{p_1^\circ}{y_1}$$

$$\therefore \quad \frac{p_2^\circ}{x_1} + (p_1^\circ - p_2^\circ) = \frac{p_1^\circ}{y_1}$$

$$\therefore \quad \frac{1}{x_1} = \left(\frac{p_1^\circ}{p_2^\circ}\right)\left(\frac{1}{y_1}\right) + \left(\frac{p_2^\circ - p_1^\circ}{p_2^\circ}\right)$$

$$\therefore \quad \text{Slope} = \left(\frac{p_1^\circ}{p_2^\circ}\right)$$

So, Intercept $= \left(\frac{p_2^\circ - p_1^\circ}{p_2^\circ}\right)$

Thus, option (d) is the correct answer.

52. *(a)* The complete reaction sequence is as follows

$$K_2Cr_2O_7 \xrightarrow[-H_2O]{KOH} \underset{[A]}{K_2CrO_4} \xrightarrow[-H_2O]{H_2SO_4} \underset{[B]}{K_2Cr_2O_7} + K_2SO_4$$

Thus, $[A]$ and $[B]$ respectively are K_2CrO_4 and $K_2Cr_2O_7$.

53. *(d)* The α-helix and β-pleated sheet structures of protein are associated with its secondary structure, that contains hydrogen bonds.

> **Concept Enhancer**
>
> **α-helix**
>
> - It is right-handed, as it coils clockwise.
> - The structure is stabilised by hydrogen bonds between the CO and NH groups of the main chain.
>
> **β-pleated**
>
> - These are made up of beta strands.
> - All peptide chains are stretched out to nearly maximum extension and then laid side by side which are held together by intermolecular hydrogen bonds.

54. *(c)* Both statement I and statement II are true.

55. *(d)* Rate of solvolysis depends on stability of carbocation. As the stability of carbocation increases the rate of solvolysis also increases. The order of stability of the carbocations of the given compounds is

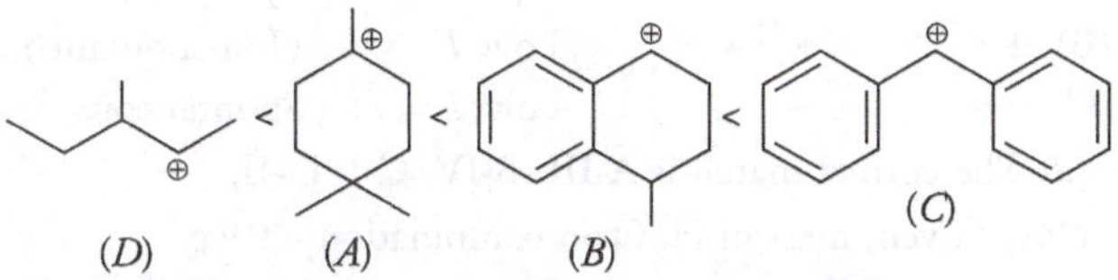

So, the order of solvolysis is also $D < A < B < C$.

56. *(c)* The complete reaction sequence is as follows.

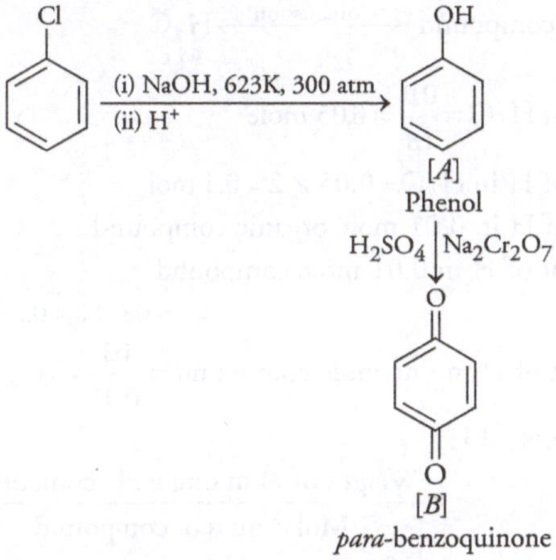

So, option (c) is the correct answer.

57. *(c)* Both statement I and Statement II are true.

58. *(b)* Both statement I and statement II are true.

59. *(c)* The order of melting point of group 14 elements is C > Si > Ge > Pb > Sn.

So, Sn has lowest melting point. Its atomic number is 50.

60. *(b)* When there is an increases in temperature, the k_w of water increases. Thus, the degree of dissociation of water also increases. Hence, its pH will decrease.

61. *(c)*

> **Key idea**
> To solve this problem first of all calculate the value of $E°$ of corresponding option. After that use this value of $E°$ into the equation, $\Delta G° = -nFE°$ to obtain the value of $\Delta G°$.

Using the formula,

$$\Delta G° = -nFE°$$

For option (a),

$$E° = -2.37 - 0.8 = -3.17\text{ V}$$

$$\Delta G° = -2 \times F \times (-3.17) = +6.34\ FV$$

For option (b),

$$E° = 0.8 - 0.34 = 0.46\text{ V}$$

$$\Delta G° = -2 \times F \times 0.46 = -0.92\ FV$$

For option (c),

$$E° = 0.8 + 0.76 = 1.56\text{ V}$$

Now, $\Delta G° = -2 \times F \times 1.56 = -3.12\ FV$

For option (d),

$$E° = -2.37 + 0.76 = -1.61\text{ V}$$

Now, $\Delta G° = -2 \times F \times (-1.61) = +3.22\ FV$

Thus, the most negative value of $\Delta G°$ is obtained for option (c).

62. *(c)* For zero order reaction,

$$A \longrightarrow P,\ \text{rate} = k$$

$$[A]_t = [A]_0 - kt$$

So, the most suitable graph is given in option (c).

i.e.,

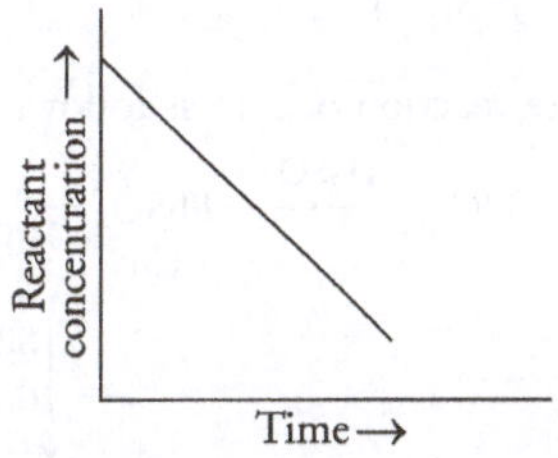

63. *(a)* The correct match is A-IV, B-III, C-I, D-II.

64. *(d)* Given, $\Delta p_1 = 10$ mm Hg

Initial mole fraction of solute $\chi_{B_1} = 0.2$

$\Delta p_2 = 20$ mm Hg

Final mole fraction of solute, $\chi_{B_2} = ?$

Mole fraction of solvent when vapour pressure is 20 mm Hg

$\chi_{A_2} = ?$

$$\frac{\Delta p_1}{p°} = \chi_{B_1} \Rightarrow \frac{\Delta p_2}{p°} = \chi_{B_2}$$

$$\frac{\Delta p_1}{\Delta p_2} = \frac{\chi_{B_1}}{\chi_{B_2}} \Rightarrow \frac{10}{20} = \frac{0.2}{\chi_{B_2}}$$

$$\chi_{B_2} = 0.2 \times 2 = 0.4$$

$$\chi_{A_2} = 1 - \chi_{B_2} = 1 - 0.4 = 0.6$$

65. *(a)* $X_2Y(g) \rightleftharpoons X_2(g) + \frac{1}{2}Y_2(g)$

$1-x$ mol $\quad$ x mol $\quad$ $x/2$ mol

$$\therefore \quad p_{X_2Y} = \frac{1-x}{1+x/2} \times p$$

$$p_{X_2} = \frac{x}{1+x/2} \times p$$

$$p_{Y_2} = \frac{x/2}{1+x/2} \times p$$

Therefore, for the given reaction,

$$K_p = \frac{\left(\frac{x}{1+x/2}p\right)\left(\frac{x}{2(1+x/2)}p\right)^{1/2}}{\left(\frac{1-x}{1+x/2}\right)p}$$

$$\therefore \quad K_p = \left(\frac{x}{1-x}\right)\left(\frac{x}{2(1+x/2)}\right)^{1/2} \times p^{1/2}$$

Also it is given that, the value of x is very small.

So, $K_p = \frac{x^{3/2}}{(2)^{1/2}} \times p^{1/2}$

or $x^{3/2} = \frac{K_p \times 2^{1/2}}{p^{1/2}}$

or $x = \left(\frac{K_p^2 \times 2}{p}\right)^{1/3}$

66. *(b)* The complete reaction occurs as follows.

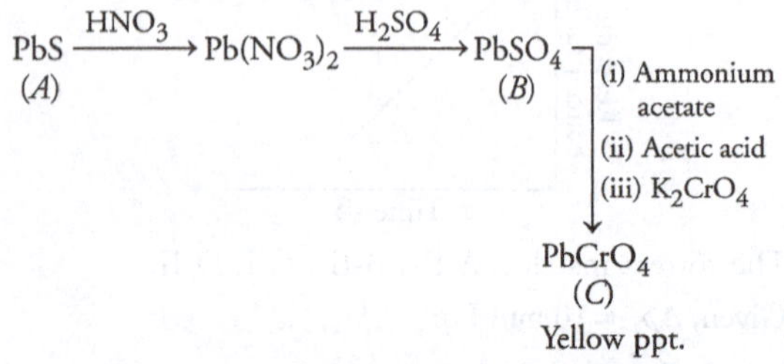

Thus, A, B, C respectively are PbS, $PbSO_4$, $PbCrO_4$.

67. *(c)* The electronic configuration of the central metal ion in the given complex with their respective oxidation state are as follows

$[A]\,Fe^{6+} = [Ar]\,3d^2$

$[B]\,Mn^{3+} = [Ar]\,3d^4$

$[C]\,Fe^{3+} = [Ar]\,3d^5$

$[D]\,Cr^{2+} = [Ar]\,3d^4$

$[E]\,Ni^{4+} = [Ar]\,3d^6$

So, among the given complexes $[Mn(CN)_6]^{3-}$ and $Cr_2(O—\overset{\overset{O}{||}}{C}—Me)_4(H_2O)_2$ has their central metal in d^4 configuration.

68. *(b)* Both statement I and statement II are false.

The correct form of the statement are as follows.

Statement I. A plot of $\sqrt{\nu}$ *vs* atomic mass is a *zig-zag* plot.

Statement II. A plot of $\sqrt{\nu}$ *vs* atomic number gives a straight line.

69. *(c)* The incorrect combination is given in B and C i.e option (c).

The correct match is given as

	ΔH	ΔS	Temperature	Spontanity
(B)	+	+	Low T	Non-spontaneous
(C)	−	−	Low T	Spontaneous

70. *(d)* The correct match is A-III, B-IV, C-I, D-II.

71. *(100)* Given, mass of H_2O on combination = 0.9 g

Number of moles of organic compound (α) = 0.01 mol

Molar mass of x = ?

Organic compound $\xrightarrow{\text{Combustion}} \underset{0.9\text{ g}}{H_2O}$

$\therefore$ Mole of $H_2O = \frac{0.9}{18} = 0.05$ mole

$\therefore$ Mole of H in $H_2O = 0.05 \times 2 = 0.1$ mol

= mole of H in 0.01 mole organic compound

$\therefore$ Weight of H in 0.01 mole compound

$= 0.1 \times 1 = 0.1$ g

$\therefore$ Weight of H in one mole compound $= \frac{0.1}{0.01} = 10$ g

$\because$ Weight % of H

$$= \frac{\text{Weight of H in one mole compound}}{\text{Molar mass of compound}} \times 100$$

$$10 = \frac{10}{M} \times 100$$

$\therefore \quad M = 100$ g/mol

So, the molar mass of x is 100 g/mol.

72. *(200)* Bond dissociaton enthalpy of X_2 = ?

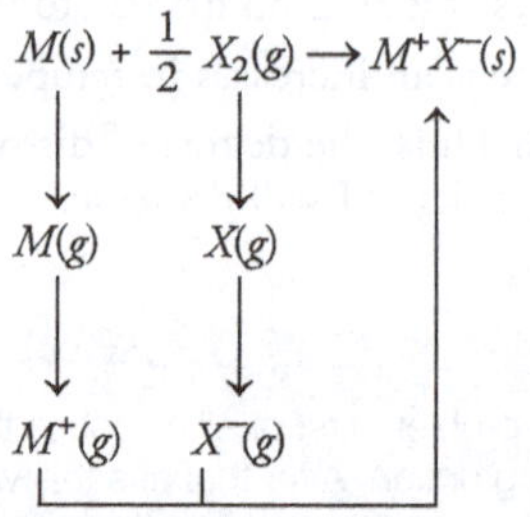

So, the value of $\Delta H_f(MX)$ can be calculated by using

$$\Delta H_f(MX) = \Delta H_{\text{sub}}(M) + \text{IE}(M) + \frac{1}{2}[\text{BE}(X-X)] + \text{EG}(X) + \text{LE}(MX)$$

Substitute the values from given data

$$-400 = 100 + 500 + \frac{1}{2}(\text{BE}) + (-300) + (-800)$$

$\therefore \quad$ BE = 200 kJ/mole

73. *(153)* Given,

Mass of Al = 81 g

Mass of oxygen = 128 g

Mass of Al_2O_3 = ?

The reaction involved is

$$\underset{\frac{81}{27}=3\text{ mol}}{4Al} + \underset{\frac{128}{32}=4\text{ mol}}{3O_2} \longrightarrow 2Al_2O_3$$

Limiting reagent,

$\therefore$ Mole of Al_2O_3 formed $= \frac{1}{2} \times 3$ mole

$\therefore$ Weight of Al_2O_3 formed $= \frac{3}{2} \times 102$

$= 153g$

74. *(27)*

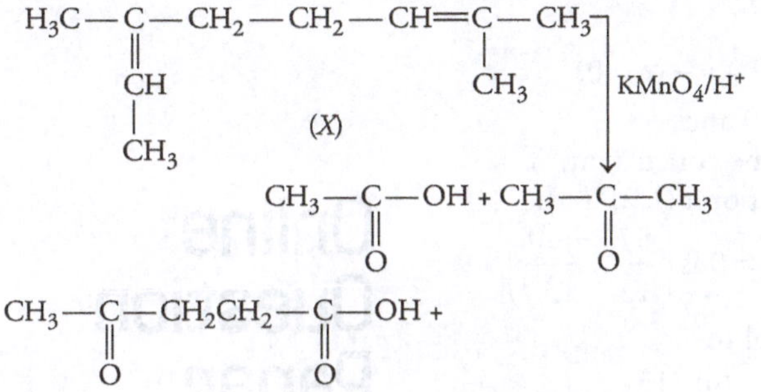

Thus, in the compound (*X*) total 27 σ-bonds are present.

75. *(4)* The complete reaction involved is as follows

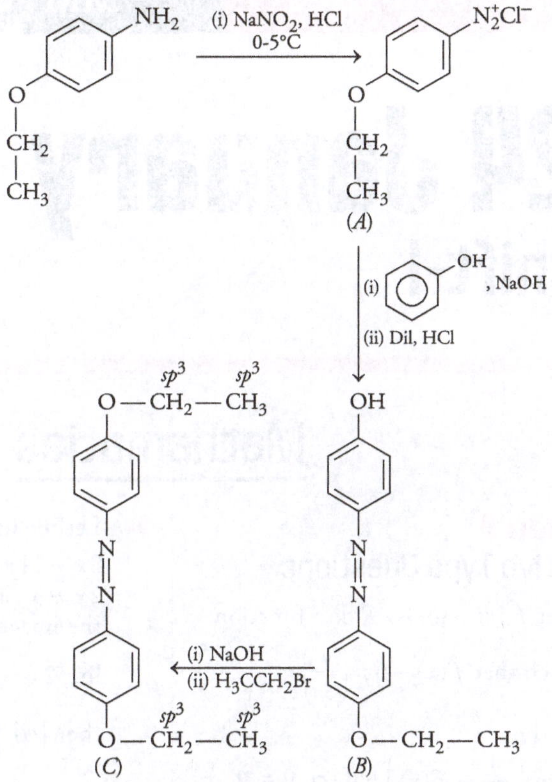

So, total four sp^3 hybridised carbon are present in product *C*.

Time 9:00-12:00 | MM : 300 Marks

24 January
Shift-I

Online Question Paper

—

JEE Main 2025

Mathematics

Section A
Objective Type Questions

01/ Let $f : R - \{0\} \to R$ be a function such that $f(x) - 6f\left(\frac{1}{x}\right) = \frac{35}{3x} - \frac{5}{2}$. If the $\lim_{x \to 0}\left(\frac{1}{\alpha x} + f(x)\right) = \beta$; $\alpha, \beta \in R$, then $\alpha + 2\beta$ is equal to

(a) 6 (b) 3
(c) 4 (d) 5

02/ Let the product of the focal distances of the point $\left(\sqrt{3}, \frac{1}{2}\right)$ on the ellipse $\frac{x^2}{a^2} + \frac{y^2}{b^2} = 1$, $(a > b)$ be $\frac{7}{4}$. Then, the absolute difference of the eccentricities of two such ellipses is

(a) $\frac{3 - 2\sqrt{2}}{2\sqrt{3}}$ (b) $\frac{1 - 2\sqrt{2}}{\sqrt{3}}$
(c) $\frac{1 - \sqrt{3}}{\sqrt{2}}$ (d) $\frac{3 - 2\sqrt{2}}{3\sqrt{2}}$

03/ If $I(m, n) = \int_0^1 x^{m-1}(1-x)^{n-1}\,dx$, $m, n > 0$, then $I(9, 14) + I(10, 13)$ is

(a) $I(9, 1)$ (b) $I(1, 13)$
(c) $I(19, 27)$ (d) $I(9, 13)$

04/ Let the lines $3x - 4y - \alpha = 0$, $8x - 11y - 33 = 0$ and $2x - 3y + \lambda = 0$ be concurrent. If the image of the point (1, 2) in the line $2x - 3y + \lambda = 0$ is $\left(\frac{57}{13}, \frac{-40}{13}\right)$, then $|\alpha\lambda|$ is equal to

(a) 84 (b) 113
(c) 101 (d) 91

05/ Let $y = y(x)$ be the solution of the differential equation $(xy - 5x^2\sqrt{1 + x^2})dx + (1 + x^2)dy = 0$, $y(0) = 0$. Then, $y(\sqrt{3})$ is equal to

(a) $\frac{5\sqrt{3}}{2}$ (b) $2\sqrt{2}$
(c) $\sqrt{\frac{15}{2}}$ (d) $\sqrt{\frac{14}{3}}$

06/ $\lim_{x \to 0} \operatorname{cosec} x(\sqrt{2\cos^2 x + 3\cos x} - \sqrt{\cos^2 x + \sin x + 4})$ is

(a) $\frac{1}{\sqrt{15}}$ (b) $-\frac{1}{2\sqrt{5}}$
(c) 0 (d) $\frac{1}{2\sqrt{5}}$

07/ The product of all the rational roots of the equation $(x^2 - 9x + 11)^2 - (x - 4)(x - 5) = 3$ is equal to

(a) 28 (b) 14 (c) 7 (d) 21

08/ For some $n \neq 10$, let the coefficients of the 5th, 6th and 7th terms in the binomial expansion of $(1+x)^{n+4}$ be in AP. Then, the largest coefficient in the expansion of $(1+x)^{n+4}$ is

(a) 70 (b) 35 (c) 20 (d) 10

09/ Let in a ΔABC, the length of the side AC be 6, the vertex B be (1, 2, 3) and the vertices A, C lie on the line $\frac{x-6}{3} = \frac{y-7}{2} = \frac{z-7}{-2}$. Then, the area (in sq. units) of ΔABC is

(a) 21 (b) 56 (c) 42 (d) 17

10/ For a statistical data $x_1, x_2, \ldots, x_{10}$ statistical of 10 values a student obtained the mean as 5.5 and $\sum_{i=1}^{10} x_i^2 = 371$. He later found that he had noted two values in the data incorrectly as 4 and 5, instead of the correct values 6 and 8, respectively. The variance of the corrected data is

(a) 5 (b) 9 (c) 4 (d) 7

11/ The area of the region $\{(x, y) : x^2 + 4x + 2 \leq y \leq |x+2|\}$ is equal to

(a) 7 (b) 5 (c) $\frac{20}{3}$ (d) $\frac{24}{5}$

Imp.

12/ Let $f(x) = \frac{2^{x+2} + 16}{2^{2x+1} + 2^{x+4} + 32}$. Then, the value of $8\left(f\left(\frac{1}{15}\right) + f\left(\frac{2}{15}\right) + \ldots + f\left(\frac{59}{15}\right)\right)$ is equal to

(a) 118 (b) 102 (c) 108 (d) 92

Functional Equation

13/ Let $\mathbf{a} = \hat{i} + 2\hat{j} + 3\hat{k}$, $\mathbf{b} = 3\hat{i} + \hat{j} - \hat{k}$ and $\mathbf{c}$ be three vectors such that $\mathbf{c}$ is coplanar with $\mathbf{a}$ and $\mathbf{b}$. If the vector $\mathbf{c}$ is perpendicular to $\mathbf{b}$ and $\mathbf{a} \cdot \mathbf{c} = 5$, then $|\mathbf{c}|$ is equal to

(a) 18 (b) 16 (c) $\sqrt{\frac{11}{6}}$ (d) $\frac{1}{3\sqrt{2}}$

14/ A and B alternately throw a pair of dice. A wins if he throws a sum of 5 before B throws a sum of 8 and B wins if he throws a sum of 8 before A throws a sum of 5. The probability that A wins if A makes the first throw, is

(a) $\frac{8}{17}$ (b) $\frac{8}{19}$ (c) $\frac{9}{19}$ (d) $\frac{9}{17}$

15/ If α and β are the roots of the equation $2z^2 - 3z - 2i = 0$, where $i = \sqrt{-1}$, then

$$16 \cdot \text{Re}\left(\frac{\alpha^{19} + \beta^{19} + \alpha^{11} + \beta^{11}}{\alpha^{15} + \beta^{15}}\right) \cdot \text{Im}\left(\frac{\alpha^{19} + \beta^{19} + \alpha^{11} + \beta^{11}}{\alpha^{15} + \beta^{15}}\right)$$ is equal to

(a) 409 (b) 441
(c) 312 (d) 398

16/ Let $S_n = \frac{1}{2} + \frac{1}{6} + \frac{1}{12} + \frac{1}{20} + \ldots +$ upto n terms. If the sum of the first six terms of an AP with first term $-p$ and common difference p is $\sqrt{2026 S_{2025}}$, then the absolute difference between 20th and 15th terms of the AP is

(a) 45 (b) 20
(c) 25 (d) 90

17/ Consider the region

$$R = \{(x, y) : x \leq y \leq 9 - \frac{11}{3}x^2, x \geq 0\}.$$

The area of the largest rectangle of sides parallel to coordinate axes and inscribed in R is

(a) $\frac{625}{111}$ (b) $\frac{567}{121}$
(c) $\frac{821}{123}$ (d) $\frac{730}{119}$

18/ Let circle C be the image of $x^2 + y^2 - 2x + 4y - 4 = 0$ in the line $2x - 3y + 5 = 0$ and A be the point on C such that OA is parallel to X-axis and A lies on the right hand side of the centre O of C. If $B(\alpha, \beta)$ with $\beta < 4$, lies on C such that the length of the arc AB is $\left(\frac{1}{6}\right)$th of the perimeter of C, then $\beta - \sqrt{3}\alpha$ is equal to

(a) $3 + \sqrt{3}$ (b) $4 - \sqrt{3}$
(c) 3 (d) 4

19/ If the system of equations

$2x - y + z = 4$
$5x + \lambda y + 3z = 12$
$100x - 47y + \mu z = 212.$

has infinitely many solutions, then $\mu - 2\lambda$ is equal to

(a) 59 (b) 57
(c) 55 (d) 56

20/ Let the line passing through the points $(-1, 2, 1)$ and parallel to the line $\frac{x-1}{2}=\frac{y+1}{3}=\frac{z}{4}$ intersect the line $\frac{x+2}{3}=\frac{y-3}{2}=\frac{z-4}{1}$ at the point P. Then, the distance of P from the point $Q(4, -5, 1)$ is

(a) 10 (b) $5\sqrt{5}$

(c) 5 (d) $5\sqrt{6}$

Section B

Numerical Type Questions

21/ If for some $\alpha, \beta : \alpha \leq \beta, \alpha + \beta = 8$ and $\sec^2(\tan^{-1}\alpha) + \text{cosec}^2(\cot^{-1}\beta) = 36$, then $\alpha^2 + \beta$ is …… .

22/ Let A be a 3×3 matrix such that $X^T AX = O$ for all non-zero 3×1 matrices $X = \begin{bmatrix} x \\ y \\ z \end{bmatrix}$.

If $A\begin{bmatrix} 1 \\ 1 \\ 1 \end{bmatrix} = \begin{bmatrix} 1 \\ 4 \\ -5 \end{bmatrix}, A\begin{bmatrix} 1 \\ 2 \\ 1 \end{bmatrix} = \begin{bmatrix} 0 \\ 4 \\ -8 \end{bmatrix}$ and det (adj $(2(A+I))$) $= 2^\alpha 3^\beta 5^\gamma, \alpha, \beta, \gamma \in N$, then $\alpha^2 + \beta^2 + \gamma^2$ is

Imp.

23/ Let $S = [p_1, p_2, ..., p_{10}]$ be the set of first ten prime number. Let $A = S \cup P$, where P is the set of all possible products of distinct elements of S. Then, the number of all ordered pair $(x, y), x \in S, y \in A$ such that x divides y is ………..

Cartesian Product of Sets

24/ Let f be a differentiable function such that $2(x+2)^2 f(x) - 3(x+2)^2 = 10\int_0^x (t+2) f(t)\,dt, x \geq 0$, then $f(2)$ is equal to

25/ The number of 3-digit numbers that are divisible by 2 and 3 but not divisible by 4 and 9, is

Physics

Section A

Objective Type Questions

26/ A force $F = \alpha + \beta x^2$ acts on an object in the x-direction. The work done by the force is 5 J, when the object is displaced by 1 m. If the constant α = 1 N, then β will be

(a) 12 N/m^2 (b) 10 N/m^2

(c) 15 N/m^2 (d) 8 N/m^2

27/ An alternating current is given by $I = I_A \sin \omega t + I_B \cos \omega t$. The rms current will be

(a) $\frac{\sqrt{I_A^2 + I_B^2}}{2}$ (b) $\frac{|I_A + I_B|}{\sqrt{2}}$

(c) $\sqrt{I_A^2 + I_B^2}$ (d) $\sqrt{\frac{I_A^2 + I_B^2}{2}}$

28/ A satellite is launched into a circular orbit of radius R around the earth. A second satellite is launched into an orbit of radius 1.03 R. The time period of revolution of the second satellite is larger than the first one approximately by

(a) 2.5% (b) 9%

(c) 4.5% (d) 3%

Imp.

29/ A car of mass m moves on a banked road having radius r and banking angle θ. To avoid slipping from banked road, the maximum permissible speed of the car is v_0. The coefficient of friction μ between the wheels of the car and the banked road is

(a) $\mu = \frac{v_0^2 - rg \tan\theta}{rg - v_0^2 \tan\theta}$ (b) $\mu = \frac{v_0^2 - rg \tan\theta}{rg + v_0^2 \tan\theta}$

(c) $\mu = \frac{v_0^2 + rg \tan\theta}{rg + v_0^2 \tan\theta}$ (d) $\mu = \frac{v_0^2 + rg \tan\theta}{rg - v_0^2 \tan\theta}$

Banking of Raod

30/ The amount of work done to break a big water drop of radius R into 27 small drops of equal radius is 10 J. The work done required to break the same big drop into 64 small drops of equal radius will be

(a) 20 J (b) 15 J (c) 5 J (d) 10 J

31/ A particle is executing simple harmonic motion with time period 2 s and amplitude 1 cm. If D and d are the total distance and displacement covered by the particle in 12.5 s, then $\frac{D}{d}$ is

(a) 15/4 (b) 16/5 (c) 10 (d) 25

32/ A uniform solid cylinder of mass m and radius r rolls along an inclined rough plane of inclination 45°. If it starts to roll from rest from the top of the plane then, the linear acceleration of the cylinder's axis will be

(a) $\frac{1}{3\sqrt{2}}g$ (b) $\frac{\sqrt{2}g}{3}$

(c) $\frac{1}{\sqrt{2}}g$ (d) $\sqrt{2}g$

33/ A thin plano-convex lens made of glass of refractive index 1.5 is immersed in a liquid of refractive index 1.2. When the plane side of the lens is silver coated for complete reflection, the lens immersed in the liquid behaves like a concave mirror of focal length 0.2 m. The radius of curvature of the curved surface of the lens is

(a) 0.20 m (b) 0.10 m

(c) 0.25 m (d) 0.15 m

34/ Consider the following statements.

A. The junction area of solar cell is made very narrow compared to a photo diode.

B. Solar cells are not connected with any external bias.

C. LED is made of lightly doped p-n junction.

D. Increase of forward current results in continuous increase of LED light intensity.

E. LEDs have to be connected in forward bias for emission of light.

Choose the correct answer from the options given below.

(a) B, D, E Only (b) A, C Only

(c) B, E Only (d) A, C, E Only

35/ The Young's double slit interference experiment is performed using light consisting of 480 nm and 600 mm wavelengths to form interference patterns. The least number of the bright fringes of 480 nm light that are required for the first coincidence with the bright fringes formed by 600 nm light is

(a) 5 (b) 4 (c) 6 (d) 8

36/ For an experimental expression $y = \frac{32.3 \times 1125}{27.4}$, where all the digits are significant. Then, to report the value of y we should write

(a) $y = 1330$ (b) $y = 1326.19$

(c) $y = 1326.186$ (d) $y = 1326.2$

37/ What is the relative decrease in focal length of a lens for an increase in optical power by 0.1 D from 2.5 D? ['D' stands for dioptre]

(a) 0.1 (b) 0.01

(c) 0.04 (d) 0.40

38/ An object of mass m is projected from origin in a vertical XY-plane at an angle 45° with the X-axis with an initial velocity v_0. The magnitude and direction of the angular momentum of the object with respect to origin, when maximum height, will be
(g is acceleration due to gravity)

(a) $\frac{mv_0^3}{2\sqrt{2}g}$ along positive Z-axis

(b) $\frac{mv_0^3}{4\sqrt{2}g}$ along positive Z-axis

(c) $\frac{mv_0^3}{2\sqrt{2}g}$ along negative Z-axis

(d) $\frac{mv_0^3}{4\sqrt{2}g}$ along negative Z-axis

39/ An electron of mass m with an initial velocity $\mathbf{v} = v_0\hat{\mathbf{i}} (v_0 > 0)$ enters an electric field $\mathbf{E} = -E_0\hat{\mathbf{k}}$. If the initial de-Broglie wavelength is λ_0, the value after time t would be

(a) λ_0 (b) $\frac{\lambda_0}{\sqrt{1 - \frac{e^2E_0^2t^2}{m^2v_0^2}}}$

(c) $\frac{\lambda_0}{\sqrt{1 + \frac{e^2E_0^2t^2}{m^2v_0^2}}}$ (d) $\lambda_0\sqrt{1 + \frac{e^2E_0^2t^2}{m^2v_0^2}}$

40/ An air bubble of radius 0.1 cm lies at a depth of 20 cm below the free surface of a liquid of density 1000 kg/m^3. If the pressure inside the bubble is 2100 N/m^2 greater than the atmospheric pressure, then the surface tension of the liquid in SI unit is (use $g = 10 \text{m/s}^2$)

(a) 0.02 (b) 0.25 (c) 0.05 (d) 0.1

41/ An ideal gas goes from an initial state to final state. During the process, the pressure of gas increases linearly with temperature.

A. The work done by gas during the process is zero.

B. The heat added to gas is different from change in its internal energy.

C. The volume of the gas is increased.

D. The internal energy of the gas is increased

E. The process is isochoric (constant volume process)

Choose the correct answer from the options given below.

(a) E only (b) A, C only

(c) A, B, C, D only (d) A, D, E only

42/ A parallel plate capacitor was made with space two rectangular plates, each with a length of $l = 3$ cm and breath of $b = 1$ cm. The distance between the plates is 3 μm. Out of the following, which are the ways to increase the capacitance by a factor of 10 ?

A. $l = 30\,\text{cm}, b = 1\,\text{cm}, d = 1\,\mu\text{m}$
B. $l = 3\,\text{cm}, b = 1\,\text{cm}, d = 30\,\mu\text{m}$
C. $l = 6\,\text{cm}, b = 5\,\text{cm}, d = 3\,\mu\text{m}$
D. $l = 1\,\text{cm}, b = 1\,\text{cm}, d = 10\,\mu\text{m}$
E. $l = 5\,\text{cm}, b = 2\,\text{cm}, d = 1\,\mu\text{m}$

Choose the correct answer from the options given below.

(a) C only (b) A only
(c) B and D only (d) C and E only

Imp.

43/ A plano-convex lens having radius of curvature of first surface 2 cm exhibits focal length of f_1 in air. Another plano-convex lens with first surface radius of curvature 3 cm has focal length of f_2 when it is immersed in a liquid of refractive index 1.2. If both the lenses are made of same glass of refractive index 1.5, the ratio of f_1 and f_2 will be

(a) 1 : 3 (b) 2 : 3 (c) 1 : 2 (d) 3 : 5

Lens Maker's Formula

44/ Consider a parallel plate capacitor of area A (of each plate) and separation d between the plates. If E is the electric field and ε_0 is the permittivity of free space between the plates, then potential energy stored in the capacitor is

(a) $\frac{1}{2}\varepsilon_0 E^2 Ad$ (b) $\varepsilon_0 E^2 Ad$
(c) $\frac{1}{4}\varepsilon_0 E^2 Ad$ (d) $\frac{3}{4}\varepsilon_0 E^2 Ad$

45/ During the transition of electron from state A to state C of a Bohr atom, the wavelength of emitted radiation is 2000 Å and it becomes 6000 Å when the electron jumps from state B to state C. Then, the wavelength of the radiation emitted during the transition of electrons from state A to state B is

(a) 3000 Å (b) 6000 Å
(c) 2000 Å (d) 4000 Å

Section B

Numerical Type Questions

46/ The least count of a screw guage is 0.01 mm. If the pitch is increased by 75% and number of divisions on the circular scale is reduced by 50%, the new least count will be ....$\times 10^{-3}$ mm.

47/ A square loop of sides $a = 1$m is held normally in front of a point charge $q = 1\,\text{C}$. The flux of the electric field through the shaded region is $\frac{5}{p} \times \frac{1}{\varepsilon_0} \frac{\text{Nm}^2}{\text{C}}$, where the value of p is ....

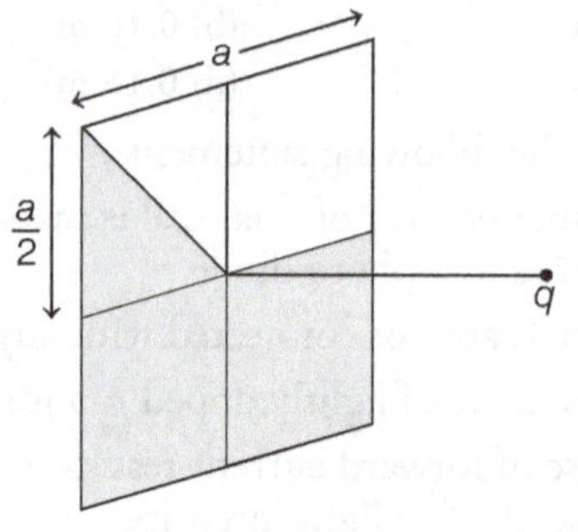

48/ A wire of resistance 9 Ω is bent to form an equilateral triangle. Then, the equivalent resistance across any two vertices will be...... ohm.

49/ A current of 5 A exists in a square loop of side $\frac{1}{\sqrt{2}}$ m. Then, the magnitude of the magnetic field B at the centre of the square loop will be $p \times 10^{-6}$T, where value of p is ....
[Take, $\mu_0 = 4\pi \times 10^{-7}\,\text{mA}^{-1}$]

50/ The temperature of 1 mole of an ideal monoatomic gas is increased by 50°C at constant pressure. The total heat added and change in internal energy are E_1 and E_2, respectively. If $\frac{E_1}{E_2} = \frac{x}{9}$, then the value of x is............

Chemistry

Section A

Objective Type Questions

51/ The product (A) formed in the following reaction sequence is

$$CH_3-C\equiv CH \xrightarrow[\text{(iii) } H_2/Ni]{\text{(i) } Hg^{2+}, H_2SO_4 \text{ (ii) } HCN} \underset{\text{Product}}{(A)}$$

(a) $CH_3-CH_2-CH(NH_2)-CH_2-OH$

(b) $CH_3-C(NH_2)(CH_3)-CH_2-OH$

(c) $CH_3-C(OH)(CH_3)-CH_2-NH_2$

(d) $CH_3-CH_2-CH(OH)-CH_2-NH_2$

52/ The large difference between the melting and boiling points of oxygen and sulphur may be explained on the basis of

(a) electronegativity

(b) atomic size

(c) electron gain enthalpy

(d) atomicity

53/ Which one of the carbocations from the following is most stable?

(a) 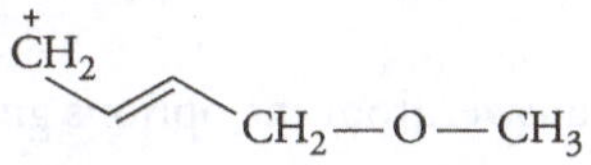

(b) 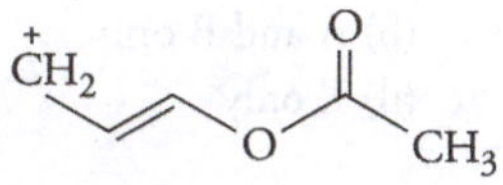

(c) 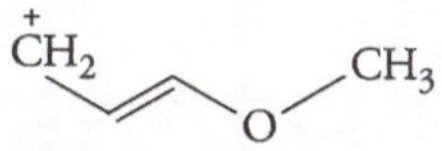

(d) 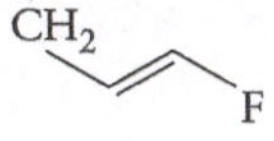

54/ Following are the four molecules "P", "Q", "R" and "S".

Which one among the four molecules will react with $H-Br(aq)$ at the fastest rate?

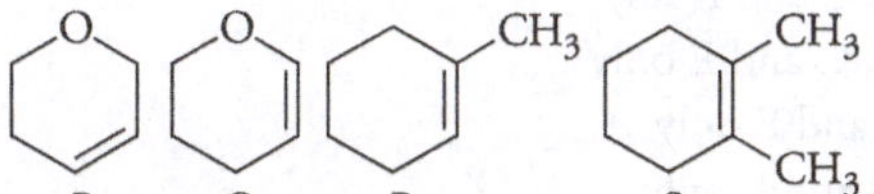

P Q R S

(a) R (b) S

(c) P (d) Q

55/ Let us consider an endothermic reaction which is non-spontaneous at the freezing point of water. However, the reaction is spontaneous at boiling point of water. Choose the correct option.

(a) Both ΔH and ΔS are (−ve)

(b) Both ΔH and ΔS are (+ve)

(c) ΔH is (+ve) but ΔS is (−ve)

(d) ΔH is (−ve) but ΔS is (+ve)

56/ Aman has been asked to synthesise the molecule

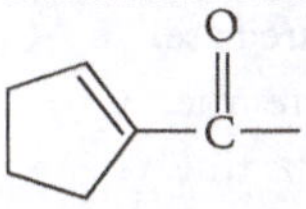 CH_3(x). He thought of preparing the molecule using an aldol condensation reaction. He found a few cyclic alkenes in his laboratory. He thought of performing ozonolysis reaction on alkene to produce a dicarbonyl compound followed by aldol reaction to prepare "x". Predict the suitable alkene that can lead to the formation of "x".

(a) 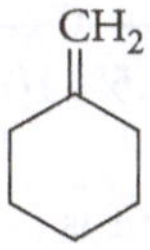

(b) 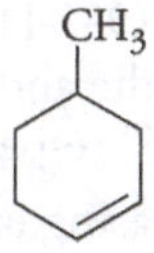

(c) 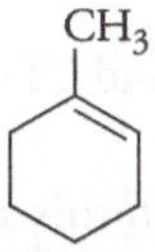

(d) 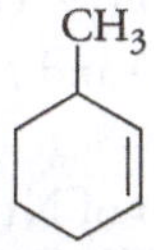

57/ Which of the following statements are not true about the periodic table?

A. The properties of elements are function of atomic weights.

B. The properties of elements are function of atomic numbers.

C. Elements having similar outer electronic configurations are arranged in same period.

D. An element's location reflects the quantum numbers of the last filled orbital.

E. The number of elements in a period is same as the number of atomic orbitals available in energy level that is being filled.

Choose the correct answer from the options given below

(a) B, C and E only
(b) A, C and E only
(c) A and E only
(d) D and E only

Imp.

58/ Given below are two statements I and II.

Statement I Dumas method is used for estimation of "Nitrogen" in an organic compound.

Statement II Dumas method involves the formation of ammonium sulphate by heating the organic compound with conc. H_2SO_4.

In the light of the above statements, choose the correct answer from the options given below.

(a) Statement I is true but Statement II is false.
(b) Both Statement I and Statement II are false.
(c) Both Statement I and Statement II are true.
(d) Statement I is false but Statement II is true.

Duma's Method

59/ Which of the following statements is true with respect to H_2O, NH_3 and CH_4?

A. The central atoms of all the molecules are sp^3 hybridised.

B. The H—O—H, H—N—H and H—C—H angles in the above molecules are 104.5°, 107.5° and 109.5° respectively.

C. The increasing order of dipole moment is $CH_4 < NH_3 < H_2O$.

D. Both H_2O and NH_3 are Lewis acids and CH_4 is a Lewis base.

E. A solution of NH_3 in H_2O is basic. In this solution NH_3 and H_2O act as Lowry-Bronsted acid and base respectively.

Choose the correct answer from the options given below

(a) A, D and E only
(b) C, D and E only
(c) A, B, C and E only
(d) A, B and C only

60/ Preparation of potassium permanganate from MnO_2, involves two step process in which the 1st step is a reaction with KOH and KNO_3 to produce

(a) K_3MnO_4
(b) $K_4[Mn(OH)_6]$
(c) K_2MnO_4
(d) $KMnO_4$

61/ The carbohydrate "Ribose" present in DNA, is

A. a pentose sugar
B. present in pyranose form
C. in "D" configuration
D. a reducing sugar, when free
E. in α-anomeric form

Choose the correct answer from the options given below:

(a) A, C and D only
(b) A, B and E only
(c) B, D and E only
(d) A, D and E only

62/ Which of the following arrangements with respect to their reactivity in nucleophilic addition reaction is correct?

(a) acetophenone < *p*-tolualdehyde < benzaldehyde < *p*-nitrobenzaldehyde
(b) *p*-nitrobenzaldehyde < benzaldehyde < *p*-tolualdehyde < acetophenone
(c) benzaldehyde < acetophenone < *p*- nitrobenzaldehyde < *p*-tolualdehyde
(d) acetophenone < benzaldehyde < *p*-tolualdehyde < *p*-nitrobenzaldehyde

63/ Which of the following linear combination of atomic orbitals will lead to formation of molecular orbitals in homonuclear diatomic molecules [internuclear axis in *z*- direction] ?

(A) $2p_z$ and $2p_x$
(B) $2s$ and $2p_x$
(C) $3d_{xy}$ and $3d_{x^2-y^2}$
(D) $2s$ and $2p_z$
(E) $2p_z$ and $3d_{x^2-y^2}$

Choose the correct answer from the options given below

(a) C and D only
(b) A and B only
(c) D only
(d) E only

Imp.

64/ Consider the given plots of vapour pressure (VP) *vs* temperature (T / K). Which amongst the following options is correct graphical representation showing ΔT_f depression in the freezing point of a solvent in a solution?

(a)
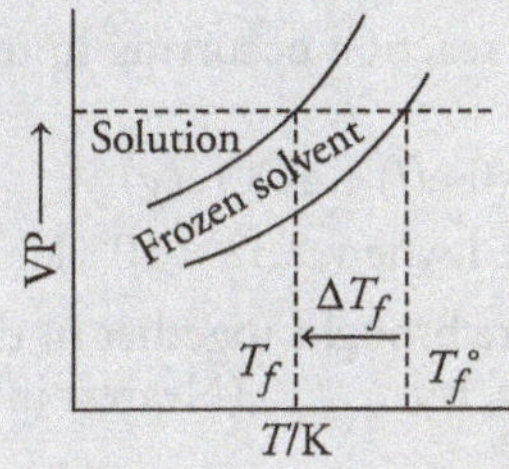

(b)
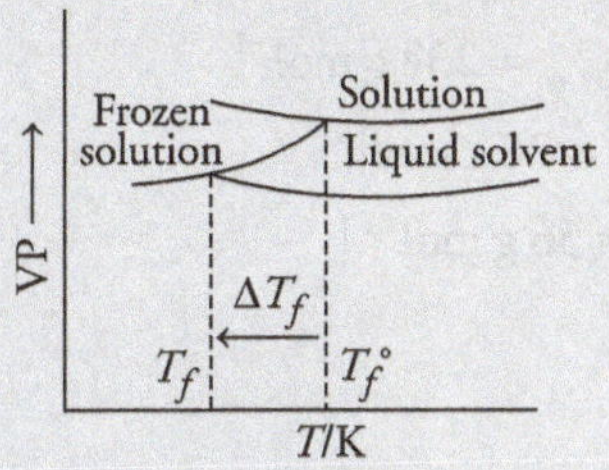

(c)
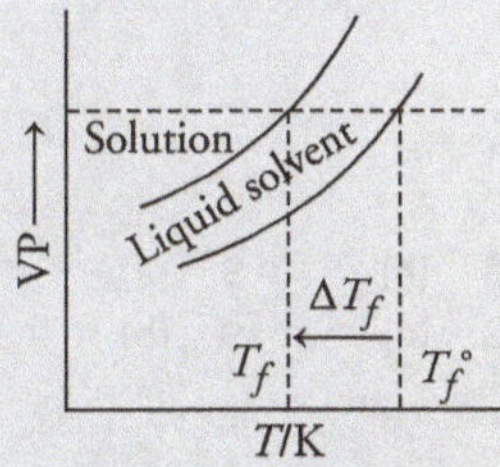

(d)
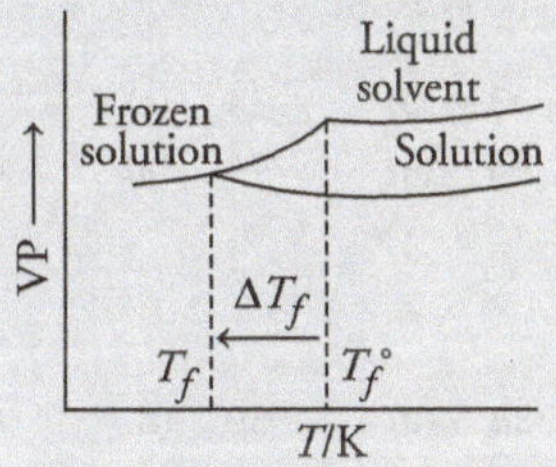

Graph of ΔT_f (Freezing Poing Depression)

65/ Which of the following ions is the strongest oxidising agent? (Atomic Number of $Ce = 58, Eu = 63, Tb = 65, Lu = 71$)

(a) Lu^{3+} (b) Ce^{3+}
(c) Tb^{4+} (d) Eu^{2+}

66/ Given below are two statements:

Statement I The conversion proceeds well in the less polar medium.

$$CH_3-CH_2-CH_2-CH_2-Cl \xrightarrow{HO^-} CH_3-CH_2-CH_2-CH_2-OH + Cl^{\ominus}$$

Statement II The conversion proceeds well in the more polar medium.

$$CH_3-CH_2-CH_2-CH_2-Cl \xrightarrow{R_3\ddot{N}} CH_3-CH_2-CH_2-CH_2-\overset{(+)}{N}R_3\ Cl^{\ominus}$$

In the light of the above statements, choose the correct answer from the options given below

(a) Both Statement I and Statement II are true.
(b) Statement I is false but Statement II is true.
(c) Statement I is true but Statement II is false.
(d) Both Statement I and Statement II are false.

67/ K_{sp} for $Cr(OH)_3$ is 1.6×10^{-30}. What is the molar solubility of this salt in water?

(a) $\sqrt[4]{\dfrac{1.6\times10^{-30}}{27}}$ (b) $\dfrac{1.8\times10^{-30}}{27}$
(c) $\sqrt[5]{1.8\times10^{-30}}$ (d) $\sqrt[2]{1.6\times10^{-30}}$

68/ For a reaction,

$N_2O_5(g) \longrightarrow 2NO_2(g) + \frac{1}{2}O_2(g)$ in a constant volume container, no products were present initially. The final pressure of the system when 50% of reaction gets completed is

(a) 7/2 times of initial pressure
(b) 7/4 times of initial pressure
(c) 5/2 times of initial pressure
(d) 5 times of initial pressure

69/ One mole of the octahedral complex compound $Co(NH_3)_5Cl_3$ gives 3 moles of ions on dissolution in water. One mole of the same complex reacts with excess of $AgNO_3$ solution to yield two moles of $AgCl(s)$. The structure of the complex is

(a) $[Co(NH_3)_4Cl_2]Cl \cdot NH_3$
(b) $[Co(NH_3)_3Cl_3] \cdot 2NH_3$
(c) $[Co(NH_3)_5Cl]Cl_2$
(d) $[Co(NH_3)_4Cl] \cdot Cl_2 \cdot NH_3$

70/ For the given cell

$$Fe^{2+}(aq) + Ag^+(aq) \rightarrow Fe^{3+}(aq) + Ag(s)$$

The standard cell potential of the above reaction is (Given :

$Ag^+ + e^- \rightarrow Ag$ $E^\circ = xV$
$Fe^{2+} + 2e^- \rightarrow Fe$ $E^\circ = yV$
$Fe^{3+} + 3e^- \rightarrow Fe$ $E^\circ = zV$)

(a) $y - 2x$ (b) $x + 2y - 3z$
(c) $x + y - z$ (d) $x + 2y$

Section B

Numerical Type Questions

71/ Standard entropies of X_2, Y_2 and XY_5 are 70, 50 and 110 $JK^{-1}mol^{-1}$ respectively. The temperature in Kelvin at which the reaction

$$\frac{1}{2}X_2 + \frac{5}{2}Y_2 \rightleftharpoons XY_5; \Delta H^\ominus = -35 \text{ kJ mol}^{-1}$$

will be at equilibrium is............(Nearest integer)

72/ Xg of benzoic acid on reaction with aq $NaHCO_3$ released CO_2 that occupied 11.2L volume at STP. X is ....g.

73/ 37.8 g N_2O_5 was taken in a 1L reaction vessel and allowed to undergo the following reaction at 500K

$$2N_2O_5(g) \rightleftharpoons 2N_2O_4(g) + O_2(g)$$

The total pressure at equilibrium was found to be 18.65 bar.

Then, K_p = ...$\times 10^{-2}$ [nearest integer]

Assume N_2O_5 to behave ideally under these conditions.

(Given, $R = 0.082$ bar L $mol^{-1}K^{-1}$)

74/ Among the following cations, the number of cations which will give characteristic precipitate in their identification tests with $K_4[Fe(CN)_6]$ is.....

$Cu^{2+}, Fe^{3+}, Ba^{2+}, Ca^{2+}, NH_4^+, Mg^{2+}, Zn^{2+}$

75/ Consider the following reaction occurring in the blast furnace

$$Fe_3O_4(s) + 4CO(g) \rightarrow 3Fe(l) + 4CO_2(g)$$

'x' kg of iron is produced when 2.32×10^3 kg Fe_3O_4 and 2.8×10^2 kg CO are brought together in the furnace. The value of 'x' is ....(Nearest integer)

[Given,

molar mass of Fe_3O_4 = 232 g mol^{-1}

molar mass of CO = 28 g mol^{-1}

molar mass of Fe = 56 g mol^{-1}]

Answer Key

Mathematics

1 (c)	2 (a)	3 (d)	4 (d)	5 (a)	6 (b)	7 (b)	8 (b)	9 (a)	10 (d)
11 (c)	12 (a)	13 (c)	14 (c)	15 (b)	16 (c)	17 (b)	18 (d)	19 (b)	20 (b)
21 (14)	22 (44)	23 (5120)	24 (19)	25 (125)					

Physics

26 (a)	27 (d)	28 (c)	29 (b)	30 (b)	31 (d)	32 (b)	33 (b)	34 (c)	35 (a)
36 (a)	37 (c)	38 (d)	39 (c)	40 (c)	41 (d)	42 (d)	43 (a)	44 (a)	45 (a)
46 (35)	47 (48)	48 (2)	49 (8)	50 (15)					

Chemistry

51 (c)	52 (d)	53 (c)	54 (d)	55 (b)	56 (c)	57 (b)	58 (a)	59 (d)	60 (c)
61 (a)	62 (a)	63 (c)	64 (d)	65 (c)	66 (a)	67 (a)	68 (b)	69 (c)	70 (b)
71 (700)	72 (61)	73 (962)	74 (4)	75 (420)					

***Note** (*) None of the option is correct.*

Detailed Explanations

Mathematics

1. (c) Given $f(x)-6f(1/x)=\dfrac{35}{3x}-\dfrac{5}{2}$ ...(i)

Replace x by $1/x$ in Eq. (i), we get

$$f(1/x)-6f(x)=\frac{35x}{3}-5/2 \quad ...(ii)$$

On solving Eqs. (i) and (ii) we get

$$f(x)=-2x-\frac{1}{3x}+\frac{1}{2}$$

Now, $\beta=\lim\limits_{x\to 0}\left(\dfrac{1}{\alpha x}+f(x)\right)$

$$\Rightarrow \beta=\lim_{x\to 0}\left(\frac{1}{\alpha x}-2x-\frac{1}{3x}+\frac{1}{2}\right)$$

limits exists finitely iff $\alpha=3 \Rightarrow \beta=1/2$

Hence, $\alpha+2\beta=3+2\times\dfrac{1}{2}=4$

2. (a) As, we know that focal distance of an ellipse $=a\pm ex$

$\therefore$ Its product

$$=(a+ex)(a-ex)=a^2-e^2x^2$$
$$=a^2-e^2(\sqrt{3})^2$$
$$=a^2-3e^2=\frac{7}{4} \quad \text{[given]}$$
$$\Rightarrow \quad 4a^2=7+12e^2 \quad ...(i)$$

Also, $(\sqrt{3}, 1/2)$ lies on ellipse

$$\frac{x^2}{a^2}+\frac{y^2}{b^2}=1$$

So, $\dfrac{3}{a^2}+\dfrac{1}{4b^2}=1$

$$\Rightarrow \quad \frac{3}{a^2}+\frac{1}{4a^2(1-e^2)}=1$$

$[\because b^2=a^2(1-e^2)]$

$$\Rightarrow 12(1-e^2)+1=4a^2(1-e^2)$$
$$\Rightarrow 13-12e^2=(7+12e^2)(1-e^2)$$

(using Eq. (i))

$$\Rightarrow 13-12e^2=7-7e^2+12e^2-12e^4$$
$$\Rightarrow 12e^4-17e^2+6=0$$
$$\Rightarrow e^2=\frac{17\pm\sqrt{289-288}}{24}=\frac{17\pm 1}{24}$$
$$\Rightarrow e^2=\frac{3}{4},\frac{2}{3}$$
$$\Rightarrow e=\frac{\sqrt{3}}{2} \text{ and } \sqrt{2/3}$$

$[\because 0<e<1$ always for an ellipse$]$

$\therefore$Absolute difference

$$=\left|\frac{\sqrt{3}}{2}-\sqrt{2/3}\right|=\left|\sqrt{2/3}-\frac{\sqrt{3}}{2}\right|$$
$$=\frac{3-2\sqrt{2}}{2\sqrt{3}}$$

3. (d) As, we know that beta function

$$\beta(p,q)=\int_0^1 x^{p-1}(1-x)^{q-1}\,dx$$
$$=\frac{(p-1)!(q-1)!}{(p+q-1)!}, \text{ where } p, q\in N$$

So, $I(9,14)=\dfrac{(8!)(13!)}{(22!)}$

and $I(10,13)=\dfrac{(9!)(12!)}{(22!)}$

Now, $I(9,14)+I(10,13)$

$$=\frac{1}{22!}(8!13!+9!12!)$$
$$=\frac{1}{22!}(8!)(12!)(13+9)=\frac{(8!)(12!)}{(21!)}$$
$$=\frac{(9-1)!(13-1)!}{(9+13-1)!}=I(9,13)$$

4. (d) $A\equiv(1,2)$, $B=\left(\dfrac{57}{13},-\dfrac{40}{13}\right)$

According to the question R is the mid-point of AB

$$\therefore R\equiv\left(\frac{\frac{57}{13}+1}{2},\frac{\frac{-40}{13}+2}{2}\right)$$

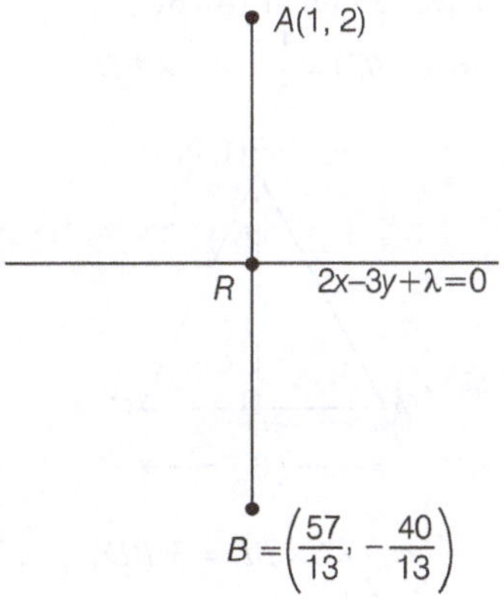

$$=\left(\frac{35}{13},\frac{-7}{13}\right)$$

$\because R$ lies on the line $2x-3y+\lambda=0$

So, $2\left(\dfrac{35}{13}\right)-3\left(\dfrac{-7}{13}\right)+\lambda=0$

$$\Rightarrow \lambda=\frac{-91}{13}=-7$$

Also, $3x-4y-\alpha=0$, $8x-11y-33=0$ and $2x-3y+\lambda=0$ are concurrent.

So, $\begin{vmatrix}3 & -4 & -\alpha\\ 8 & -11 & -33\\ 2 & -3 & \lambda\end{vmatrix}=0$

$$\Rightarrow 3(-11\lambda-99)+4(8\lambda+66)-\alpha(-24+22)=0$$
$$\Rightarrow -33\lambda-297+32\lambda+264+24\alpha-22\alpha=0$$
$$\Rightarrow -\lambda+2\alpha-33=0$$
$$\Rightarrow -(-7)+2\alpha-33=0 \quad (\because \lambda=-7)$$
$$\Rightarrow 2\alpha=26$$
$$\Rightarrow \alpha=13$$
$$\therefore |\alpha\lambda|=|13\times(-7)|=91$$

> **Concept Enhancer**
>
> Lines $a_1x+b_1y+c_1=0$, $a_2x+b_2y+c_2=0$ and $a_3x+b_3y+c_3=0$ are concurrent, if
>
> $\begin{vmatrix}a_1 & b_1 & c_1\\ a_2 & b_2 & c_2\\ a_3 & b_3 & c_3\end{vmatrix}=0$

5. (a) Given, differential equation is

$$(xy-5x^2\sqrt{1+x^2})dx+(1+x^2)dy=0$$
$$\Rightarrow \left(\frac{xy}{1+x^2}-\frac{5x^2\sqrt{1+x^2}}{1+x^2}\right)dx=-dy$$
$$\Rightarrow \frac{dy}{dx}=\frac{5x^2}{\sqrt{1+x^2}}-\frac{x}{1+x^2}y$$

$\Rightarrow \dfrac{dy}{dx}+\dfrac{x}{1+x^2}y=\dfrac{5x^2}{\sqrt{1+x^2}}$ which is linear differential equation

$$\text{IF}=e^{\int\frac{x}{1+x^2}dx}=e^{\frac{1}{2}\int\frac{2x}{1+x^2}dx}$$
$$=e^{1/2\ln(1+x^2)}=e^{\ln(1+x^2)^{1/2}}=\sqrt{1+x^2}$$

So, its solution is

$$y\times(\text{IF})=\int Q\times(\text{IF})dx$$

$$\Rightarrow y\times\sqrt{1+x^2}=\int\frac{5x^2}{\sqrt{1+x^2}}\times\sqrt{1+x^2}\,dx$$
$$=5\int x^2dx$$
$$=5\times\frac{x^3}{3}+C \quad \ldots(\text{i})$$

But given $y(0)=0$

From (i), we get

$$0\times\sqrt{1+0^2}=5\times\frac{(0)^3}{3}+C$$
$$\Rightarrow \quad C=0$$
$$\therefore \quad y=\frac{5x^3}{3\sqrt{1+x^2}}$$
$$\therefore \quad y(\sqrt{3})=\frac{5(\sqrt{3})^3}{3\sqrt{1+3}}=\frac{5\sqrt{3}}{2}$$

6. *(b)* $\lim\limits_{x\to 0}\operatorname{cosec}x(\sqrt{2\cos^2x+3\cos x}-\sqrt{\cos^2x+\sin x+4})$

$$=\lim_{x\to 0}\left\{\frac{\operatorname{cosec}x(\cos^2x+3\cos x-\sin x-4)}{\sqrt{2\cos^2x+3\cos x}+\sqrt{\cos^2x+\sin x+4}}\right\}$$

(rationalising)

$$=\frac{1}{2\sqrt{5}}\lim_{x\to 0}\frac{\cos^2x+3\cos x-\sin x-4)}{\sin x}$$
$$=\frac{1}{2\sqrt{5}}\lim_{x\to 0}\left[\frac{-2\cos x\cdot\sin x-3\sin x-\cos x}{\cos x}\right]$$

[Using L'Hospital Rule]

$$=\frac{-1}{2\sqrt{5}}$$

7. *(b)* Given equation is

$(x^2-9x+11)^2-(x-4)(x-5)=3 \ldots(\text{i})$

Let $x^2-9x=y$, then

Eq. (i) becomes, $(y+11)^2-(y+20)=3$

$(\because (x-4)(x-5)=x^2-9x+20)$

$$\Rightarrow y^2+121+22y-y-20-3=0$$
$$\Rightarrow y^2+21y+98=0$$
$$\Rightarrow (y+14)(y+7)=0 \Rightarrow y=-14,-7$$
$$\Rightarrow x^2-9x=-14 \text{ or } x^2-9x=-7$$
$$\Rightarrow x^2-9x+14=0 \text{ or } x^2-9x+7=0$$
$$\Rightarrow x=\frac{9\pm\sqrt{81-4\times 14}}{2}$$
$$\text{or } x=\frac{9\pm\sqrt{81-4\times 7}}{2}$$
$$\Rightarrow x=\frac{9\pm 5}{2} \text{ or } x=\frac{9\pm\sqrt{53}}{2}$$
$$\Rightarrow \quad x=7,2 \quad \text{or } x=\frac{9+\sqrt{53}}{2},\frac{9-\sqrt{53}}{2}$$

(Rational roots) (Irrational roots)

$\therefore$ Product of all rational roots

$=7\times 2=14$

8. *(b)* Given coefficient of 5th, 6th and 7th terms in the binomial expansion of $(1+x)^{n+4}$ are in AP

i.e. ${}^{n+4}C_4,\ {}^{n+4}C_5,\ {}^{n+4}C_6$ are in AP

$$\Rightarrow 2\times{}^{n+4}C_5={}^{n+4}C_4+{}^{n+4}C_6$$
$$\Rightarrow 4\times{}^{n+4}C_5=({}^{n+4}C_4+{}^{n+4}C_5)+({}^{n+4}C_5+{}^{n+4}C_6)$$
$$\Rightarrow 4\times{}^{n+4}C_5={}^{n+5}C_5+{}^{n+5}C_6$$
$$\Rightarrow 4\times\frac{(n+4)!}{5!(n-1)!}=\frac{(n+6)!}{6!n!}$$
$$\Rightarrow 4=\frac{(n+6)(n+5)}{6n}$$
$$\Rightarrow n^2+11n+30=24n$$
$$\Rightarrow n^2-13n+30=0$$

$\Rightarrow n=3, 10$ (rejected) $(\because n\neq 10$ given)

$\therefore n+4=3+4=7$

$\therefore$ Largest binomial coefficient in expansion of $(1+x)^7$ is the coefficient of middle term i.e. 5th term.

$={}^7C_4={}^7C_3=35$

9. *(a)* Given, $B\equiv(1,2,3)$

$AC=6$

Equation of line

$$\frac{x-6}{3}=\frac{y-7}{2}=\frac{z-7}{-2} \quad \ldots(\text{i})$$

We have to find ar (ΔABC).

So, $\text{ar}(\Delta ABC)=\frac{1}{2}\times AC\times BD$

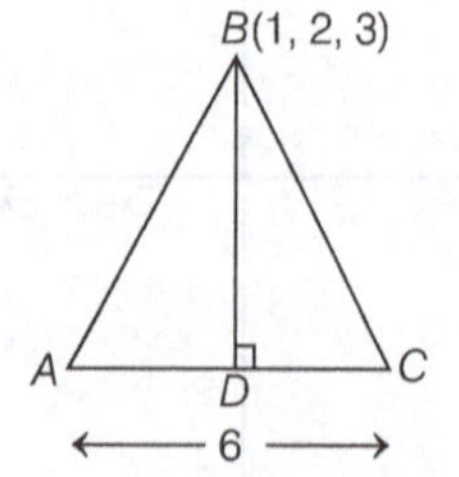

$$=\frac{1}{2}\times 6\times BD=3(BD) \quad \ldots(\text{ii})$$

Now, any point taken on line (i) is $D(3\lambda+6, 2\lambda+7, -2\lambda+7)$

So, $\mathbf{BD}=(3\lambda+5)\hat{\mathbf{i}}+(2\lambda+5)\hat{\mathbf{j}}+(-2\lambda+4)\hat{\mathbf{k}}$

$\because \mathbf{AC}\perp\mathbf{BD}$

$\therefore \mathbf{AC}\cdot\mathbf{BD}=0$

($\because$ **AC** is parallel to $3\hat{\mathbf{i}}+2\hat{\mathbf{j}}-2\hat{\mathbf{k}}$)

$$\Rightarrow 3(3\lambda+5)+2(2\lambda+5)-2(-2\lambda+4)=0$$
$$\Rightarrow 9\lambda+4\lambda+4\lambda+15+10-8=0$$
$$\Rightarrow 17\lambda=-17 \Rightarrow \lambda=-1$$
$$\therefore \mathbf{BD}=2\hat{\mathbf{i}}+3\hat{\mathbf{j}}+6\hat{\mathbf{k}}$$
$$\Rightarrow |\mathbf{BD}|=\sqrt{(2)^2+(3)^2+(6)^2}$$
$$=\sqrt{4+9+36}=\sqrt{49}=7$$

Hence, $\text{ar}(\Delta ABC)=3(BD)$

[from Eq. (ii)]

$=3\times 7=21$

10. *(d)* Given, mean $=\bar{x}=5.5$

$$\Rightarrow \sum_{i=1}^{10}x_i=5.5\times 10=55 \text{ and } \sum_{i=1}^{10}x_i^2=371$$
$$(\Sigma x_i)_{\text{new}}=55-(4+5)+(6+8)=60$$
$$(\Sigma x_i^2)_{\text{new}}=371-(4^2+5^2)+(6^2+8^2)$$
$$=430$$
$$\text{Variance }(\sigma^2)=\frac{\Sigma x_i^2}{10}-\left(\frac{\Sigma x_i}{10}\right)^2$$
$$\Rightarrow \sigma^2=\frac{430}{10}-\left(\frac{60}{10}\right)^2=43-36=7$$

11. *(c)* Given, $x^2+4x+2\leq y\leq|x+2|$

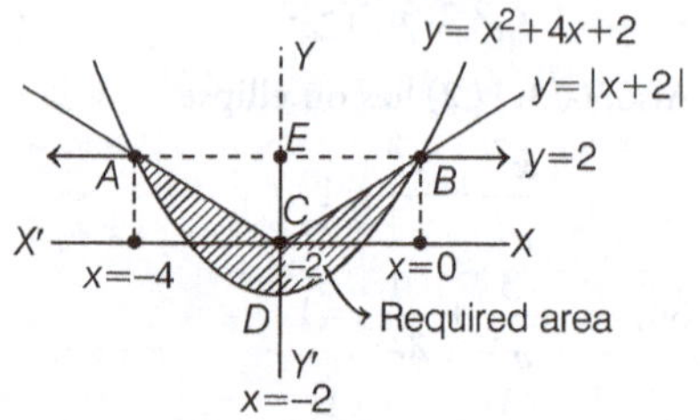

So, required area $=\text{ar}(ABDA)-\text{ar}(\Delta ACB)$

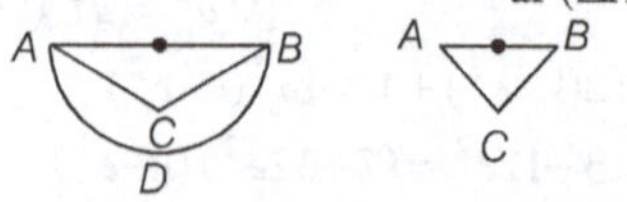

$$=\int_{-4}^{0}[2-(x^2+4x+2)]dx-\frac{1}{2}\times AB\times CE$$
$$=\left(\frac{-x^3}{3}-2x^2\right)\Bigg|_{-4}^{0}-\frac{1}{2}\times 4\times 2$$
$$=0-\left(\frac{64}{3}-32\right)-4=32-\frac{64}{3}-4$$
$$=\frac{20}{3}\text{ sq units}$$

12. (a) Given, $f(x)=\dfrac{2^{x+2}+16}{2^{2x+1}+2^{x+4}+32}$

$$\Rightarrow f(x)=\frac{2^x\cdot 2^2+2^4}{(2^x)^2\cdot 2+2^x\cdot 2^4+2^5}=\frac{2(2\cdot 2^x+2^3)}{2((2^x)^2+8\cdot 2^x+16)}$$

$$\Rightarrow f(x)=\frac{2(2^x+4)}{(2^x+4)^2}=\frac{2}{2^x+4} \quad \ldots(i)$$

Also, $f(4-x)=\dfrac{2}{2^{4-x}+4}=\dfrac{2}{\dfrac{2^4+4\cdot 2^x}{2^x}}$

$$\Rightarrow f(4-x)=\frac{2\cdot 2^x}{4(2^x+4)}=\frac{2^x}{2(2^x+4)} \quad \ldots(ii)$$

On adding Eqs. (i) and (ii), we get

$$f(x)+f(4-x)=\frac{2}{2^x+4}+\frac{2^x}{2(2^x+4)}=\frac{(4+2^x)}{2(2^x+4)}$$

$$\Rightarrow f(x)+f(4-x)=\frac{1}{2} \quad \ldots(iii)$$

Now, $f(1/15)+f\left(4-\dfrac{1}{15}\right)$

$$=f(1/15)+f\left(\frac{59}{15}\right)=\frac{1}{2}$$

Similarly, $f(2/15)+f(58/15)=\dfrac{1}{2}$

$$f(3/15)+f\left(\frac{57}{15}\right)=\frac{1}{2}$$

and soon.

$f(29/15)+f(31/15)=1/2$ and

$$f(30/15)=f(2)=\frac{2}{2^2+4}=\frac{2}{8}=\frac{1}{4}$$

Hence,

$8[f(1/15)+f(2/15)+\ldots+f(59/15)]$

$$=8\left[\underbrace{1/2+1/2+1/2+\ldots+1/2}_{29\text{ terms}}+f(30/15)\right]$$

$$=8[29\times 1/2+1/4)=118$$

13. (c)

Key idea

Three vectors **a**, **b** and **c** are coplanar iff **a** · (**b** × **c**) = 0

Given, $\mathbf{a}=\hat{\mathbf{i}}+2\hat{\mathbf{j}}+3\hat{\mathbf{k}}$

$\mathbf{b}=3\hat{\mathbf{i}}+\hat{\mathbf{j}}-\hat{\mathbf{k}}$

and **c** is coplanar with **a** and **b**

$\therefore [\mathbf{c}\ \mathbf{a}\ \mathbf{b}]=0$

$$\Rightarrow \begin{vmatrix} c_1 & c_2 & c_3 \\ 1 & 2 & 3 \\ 3 & 1 & -1 \end{vmatrix}=0,$$

where $\mathbf{c}=c_1\hat{\mathbf{i}}+c_2\hat{\mathbf{j}}+c_3\hat{\mathbf{k}}$ (say)

$\Rightarrow c_1-2c_2+c_3=0$ ...(i)

Also, $\mathbf{a}\cdot\mathbf{c}=5$ (given)

$\Rightarrow c_1+2c_2+3c_3=5$ ...(ii)

and $\mathbf{b}\cdot\mathbf{c}=0$ (given)

$\Rightarrow 3c_1+c_2-c_3=0$ ...(iii)

On solving Eqs. (i), (ii) and (iii), we get

$$c_1=\frac{1}{6}, c_2=\frac{4}{6}, c_3=\frac{7}{6}$$

$$\therefore \quad |\mathbf{c}|=\sqrt{\frac{1}{36}+\frac{16}{36}+\frac{49}{36}}$$

$$(\because |\mathbf{c}|=\sqrt{c_1^2+c_2^2+c_3^2})$$

$$=\sqrt{\frac{11}{6}}$$

14. (c) For sum $=5$, {(1, 4), (2, 3), (3, 2), (4, 1)}

$$\therefore \quad P(A)=\frac{4}{36}=\frac{1}{9}$$

For sum $=8$

$\Rightarrow$ {(2, 6), (3, 5), (4, 4), (5, 3), (6, 2)}

$\therefore P(B)=5/36$

Now, $P(\bar{A})=1-P(A)=1-\dfrac{4}{36}=\dfrac{32}{36}$

and $P(\bar{B})=1-P(B)=1-\dfrac{5}{36}=\dfrac{31}{36}$

$P(A \text{ win } S)=P(A)+P(\bar{A})\cdot P(\bar{B})\cdot P(A)$

$+\overbrace{P(\bar{A})\,P(\bar{B})}^{(P(\bar{A})\cdot P(\bar{B}))^2}P(\bar{A})P(\bar{B})\cdot P(A)\ldots$ (takes the form sum of infinite GP)

$$\left(S_\infty=\frac{a}{1-r}\right)$$

$$=\frac{P(A)}{1-P(\bar{A})P(\bar{B})}=\frac{\frac{4}{36}}{1-\frac{32}{36}\times\frac{31}{36}}=\frac{9}{19}$$

15. (b) Given quadratic equation is

$$2z^2-3z-2i=0 \quad \ldots(i)$$

$$\Rightarrow 2z^2-2i=3z$$

$$\Rightarrow 2z\left(z-\frac{i}{z}\right)=3z$$

$$\Rightarrow \quad 2\left(z-\frac{i}{z}\right)=3$$

$$\Rightarrow \quad z-\frac{i}{z}=\frac{3}{2}$$

$\because \alpha, \beta$ are the roots of Eq. (i)

So, $\left(\alpha-\dfrac{i}{\alpha}\right)=\dfrac{3}{2}$

On squaring both sides, we get

$$\alpha^2-\frac{1}{\alpha^2}-2i=9/4$$

$$\Rightarrow \quad \alpha^2-\frac{1}{\alpha^2}=\frac{9}{4}+2i$$

Again, squaring both sides, we get

$$\alpha^4+\frac{1}{\alpha^4}-2=\frac{81}{16}-4+9i$$

$$\Rightarrow \quad \alpha^4+\frac{1}{\alpha^4}=\frac{49}{16}+9i \quad \ldots(ii)$$

Similarly, we can find

$$\beta^4+\frac{1}{\beta^4}=\frac{49}{16}+9i \quad \ldots(iii)$$

$(\because \beta$ is also a root of Eq. (i))

Now, $\dfrac{\alpha^{19}+\beta^{19}+\alpha^{11}+\beta^{11}}{\alpha^{15}+\beta^{15}}$

$$=\frac{\alpha^{15}\left(\alpha^4+\frac{1}{\alpha^4}\right)+\beta^{15}\left(\beta^4+\frac{1}{\beta^4}\right)}{\alpha^{15}+\beta^{15}}$$

$$=\frac{(\alpha^{15}+\beta^{15})\left(\frac{49}{16}+9i\right)}{(\alpha^{15}+\beta^{15})}$$

(using Eqs. (ii) and (iii))

$$=\frac{49}{16}+9i$$

$\therefore$ Real part $=\dfrac{49}{16}$ and imaginary part $=9$

Hence, 16[(real part) × (imaginary part]

$$=\left[\frac{49}{16}\times 9\right]\times 16=441$$

16. (c) Given, $S_n=\dfrac{1}{2}+\dfrac{1}{6}+\dfrac{1}{12}+\dfrac{1}{20}+\ldots+n$ terms

$$S_{2025}=\sum_{n=1}^{2025}\frac{1}{n(n+1)}=\sum_{n=1}^{2025}\left(\frac{1}{n}-\frac{1}{n+1}\right)$$

$$=\left(\frac{1}{1}-\frac{1}{2}\right)+\left(\frac{1}{2}-\frac{1}{3}\right)+\left(\frac{1}{3}-\frac{1}{4}\right)+\ldots\ldots+\left(\frac{1}{2025}-\frac{1}{2026}\right)$$

$$=1-\frac{1}{2026}=\frac{2025}{2026}$$

Now, $\sqrt{2026\cdot S_{2025}}=\sqrt{2026\times\dfrac{2025}{2026}}$

$$=\sqrt{2025}=45$$

According to the question,

$$\frac{6}{2}[-2p+(6-1)p]=45$$

$$\Rightarrow \quad qp=45 \Rightarrow p=5$$

Hence, $|A_{20}-A_{15}|$

$$|[-5+19\times 5]-[-5+14\times 5]|$$
$$=|90-65|=25$$

17. *(b)* Let $Q(t, t)$ be any point on $y=x$.

and P be any point on the curve

$$y=9-\frac{11}{3}x^2$$

So, coordinates of P is $\left(t, 9-\frac{11}{3}t^2\right)$

Let us consider $PQRS$ be the required rectangle

So, $ar(\Delta PQRS)=A=QR\times PQ$

$$=QR\times(SO-RO)$$
$$=t\left(9-\frac{11t^2}{3}-t\right)$$
$$=9t-t^2-\frac{11t^3}{3}$$

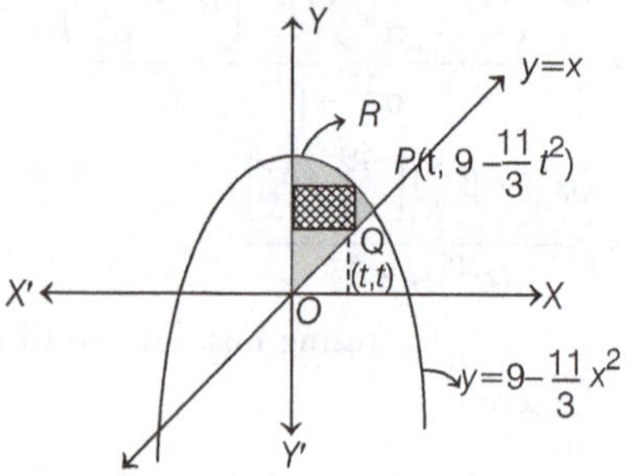

Now, $\frac{dA}{dt}=9-2t-11t^2$

For maxima on minima put $\frac{dA}{dt}=0$

$$\Rightarrow 11t^2+2t-9=0$$
$$\Rightarrow 11t^2+11t-9t-9=0$$
$$\Rightarrow 11t(t+1)-9(t+1)=0$$
$$\Rightarrow (t+1)(11t-9)=0$$
$$\Rightarrow t=-1, 9/11$$

Hence, at $t=9/11$ gives us maximum value.

So, put $t=9/11$ in A.

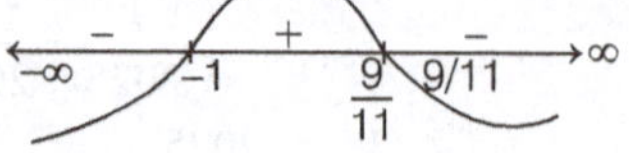

$\therefore$ Largest area

$$=\frac{9}{11}\left(9-\frac{11}{3}\times\frac{81}{121}-\frac{9}{11}\right)$$

$$=\frac{9}{11}\times\frac{63}{11}=\frac{567}{121}\text{ sq units}$$

18. *(d)* Centre and radius of given circle are $(1, -2)$ and 3, respectively.

Now, image of $(1, -2)$ in $2x-3y+5=0$ is

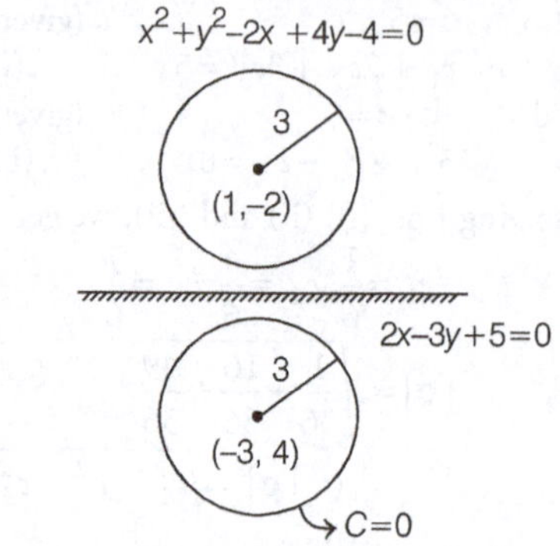

$$\frac{x-1}{2}=\frac{y+2}{-3}=\frac{-2(2+6+5)}{13}=-2$$

$\Rightarrow x=-3,\ y=4$

$\therefore$ Centre of $C=0$ is $(-3, 4)$ and radius of $C=0$ will be 3 (as given circle)

So, equation of C_2 is

$$(x+3)^2+(y-4)^2=(3)^2$$

According to the question,

$$l(\text{arc } AB)=\frac{1}{6}\times 2\pi r$$

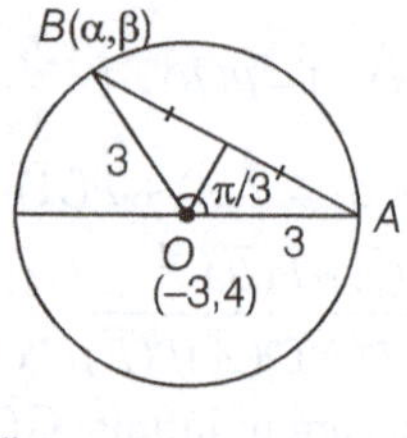

$$\Rightarrow r\theta=\frac{\pi r}{3} \qquad (\because l=r\theta)$$

$$\Rightarrow \quad \theta=\pi/3$$

$\because (\alpha, \beta)$ lies on $C=0$,

$\therefore (\alpha+3)^2+(\beta-4)^2=9$ ...(i)

$(\alpha+6)^2+(\beta-4)^2=27$ ...(ii)

(ii) − (i) $\Rightarrow (\alpha+6)^2-(\alpha+3)^2=18$

$$\Rightarrow 6\alpha=-9 \Rightarrow \alpha=\frac{-3}{2}$$

$$\therefore \quad \beta=4-\frac{3\sqrt{3}}{2}$$

Hence, $\beta-\sqrt{3}\alpha=4-\frac{3\sqrt{3}}{2}-\sqrt{3}\left(\frac{-3}{2}\right)$

$$=4-\frac{3\sqrt{3}}{2}+\frac{3\sqrt{3}}{2}=4$$

19. *(b)* According to the question,

$$\Delta=0\Rightarrow\begin{vmatrix}2 & -1 & 1\\ 5 & \lambda & 3\\ 100 & -47 & \mu\end{vmatrix}=0$$

$$\Rightarrow 2(\lambda\mu+141)+1(5\mu-300)$$
$$+1(-235-100\lambda)=0 \quad \text{... (i)}$$

Now, $\Delta_3=0\Rightarrow\begin{vmatrix}2 & -1 & 4\\ 5 & \lambda & 12\\ 100 & -47 & 212\end{vmatrix}=0$

$$\Rightarrow 2(212\lambda+564)+1(212\times 5-1200)$$
$$+4(-47\times 5-100-\lambda)=0$$
$$\Rightarrow 6\lambda=-12$$
$$\Rightarrow \lambda=-2$$

On putting in Eq. (i), we get

$$2(-2\mu+141)+5\mu-300-235+200=0$$
$$\Rightarrow \mu=53$$

Hence, $\mu-2\lambda=53-2\times(-2)=57$

> **Concept Enhancer**
>
> The system of equations $a_1x+b_1y+c_1=d_1$, $a_2x+b_2y+c_2=d_2$ and $a_3x+b_3y+c_3=d_3$ is consistent with many infinite solutions if $\Delta=0=\Delta_1=\Delta_2=\Delta_3$

20. *(b)* Equation of any line passing through $(-1, 2, 1)$ and parallel to the line $\frac{x-1}{2}=\frac{y+1}{3}=\frac{z}{4}$ ...(i) is given by

$$\frac{x+1}{2}=\frac{y-2}{3}=\frac{z-1}{4}=\lambda\text{ (say)} \quad \text{...(ii)}$$

Any point on line (ii), we get

$$x=2\lambda-1,\ y=3\lambda+2,\ z=4\lambda+1$$

$$\frac{x+2}{3}=\frac{y-3}{2}=\frac{z-4}{1}=\mu\text{ (say)} \quad \text{...(iii)}$$

Any point on line (iii)

$$x=3\mu-2,\ y=2\mu+3, z=\mu+4$$

Since, lines (ii) and (iii) intersect at point P.

So,
$$x=2\lambda-1=3\mu-2 \quad \text{... (iv)}$$
$$y=3\lambda+2=2\mu+3 \quad \text{...(v)}$$
$$z=4\lambda+1=\mu+4 \quad \text{...(vi)}$$

On solving Eqs. (iv), (v) and (vi), we get

$$\lambda=1, \mu=1$$

$\therefore$ Coordinates of point $P\equiv(1, 5, 5)$

Hence, distance of point P from $Q(4, -5, 1)$ is

$$PQ=\sqrt{(4-1)^2+(-5-5)^2+(1-5)^2}$$
$$=\sqrt{9+100+16}=\sqrt{125}=5\sqrt{5}$$

21. *(14)* Given,

$$\sec^2(\tan^{-1}\alpha)+\text{cosec}^2(\cot^{-1}\beta)=36 \quad \text{...(i)}$$

let $\tan^{-1}\alpha = x$ and $\cot^{-1}\beta = y$

So, that $\alpha = \tan x$ and $\beta = \cot y$

From Eq. (i), we get

$\sec^2 x + \text{cosec}^2 y = 36$

$\Rightarrow (1+\tan^2 x) + (1+\cot^2 y) = 36$

$\Rightarrow 2 + \tan^2 x + \cot^2 y = 36$

$\Rightarrow 2 + \alpha^2 + \beta^2 = 36$

$\Rightarrow \alpha^2 + \beta^2 = 34$ ... (ii)

Also, given $\alpha + \beta = 8$ ...(iii)

On solving Eqs. (ii) and (iii), we get

$\because (\alpha+\beta)^2 = \alpha^2 + \beta^2 + 2\alpha\beta$

$\Rightarrow (8)^2 = 34 + 2\alpha\beta \Rightarrow \dfrac{64-34}{2} = \alpha\beta$

$\Rightarrow \alpha\beta = 15$ ...(iv)

So, we observe that α, β are the roots of the equation

$x^2 - 8x + 15 = 0$

$[\because \alpha + \beta = 8 \text{ and } \alpha\beta = 15]$

$\Rightarrow (x-3)(x-5) = 0$

$\Rightarrow x = 3, 5$

$\therefore \alpha = 3, \beta = 5$ $[\because \alpha \le \beta \text{ given}]$

$\alpha = 3, \beta = 5$ because $\alpha \le \beta$ [given]

$\therefore \quad \alpha^2 + \beta = (3)^2 + 5 = 14$

22. *(44)* Given,

$X^T AX = 0$

$$[x\ y\ z]\begin{bmatrix} a_1 & a_2 & a_3 \\ b_1 & b_2 & b_3 \\ c_1 & c_2 & c_3 \end{bmatrix}\begin{bmatrix} x \\ y \\ z \end{bmatrix} = 0$$

$$\Rightarrow [x\ y\ z]\begin{bmatrix} a_1x + a_2y + a_3z \\ b_1x + b_2y + b_3z \\ c_1x + c_2y + c_3z \end{bmatrix} = 0$$

$\Rightarrow x(a_1x + a_2y + a_3z) + y(b_1x + b_2y + b_3z) + z(c_1x + c_2y + c_3z) = 0$

$\Rightarrow a_1 = 0, b_2 = 0, c_3 = 0, a_2 + b_1 = 0,$
$a_3 + c_1 = 0, b_3 = c_2 = 0$

So, A will be skew-symmetric matrix.

$$A = \begin{bmatrix} 0 & x & y \\ -x & 0 & z \\ -y & -z & 0 \end{bmatrix}$$

Given $A\begin{bmatrix} 1 \\ 1 \\ 1 \end{bmatrix} = \begin{bmatrix} 1 \\ 4 \\ -5 \end{bmatrix}$

$$\Rightarrow \begin{bmatrix} 0 & x & y \\ -x & 0 & z \\ -y & -z & 0 \end{bmatrix}\begin{bmatrix} 1 \\ 1 \\ 1 \end{bmatrix} = \begin{bmatrix} 1 \\ 4 \\ -5 \end{bmatrix}$$

$\Rightarrow x + y = 1, -x + z = 4, y + z = 5$

Also, given $A\begin{bmatrix} 1 \\ 2 \\ 1 \end{bmatrix} = \begin{bmatrix} 0 \\ 4 \\ -8 \end{bmatrix}$

$$\Rightarrow \begin{bmatrix} 0 & x & y \\ -x & 0 & z \\ -y & -z & 0 \end{bmatrix}\begin{bmatrix} 1 \\ 2 \\ 1 \end{bmatrix} = \begin{bmatrix} 0 \\ 4 \\ -8 \end{bmatrix}$$

$$\Rightarrow \left.\begin{matrix} 2x + y = 0 \\ -x + z = 4 \\ -y - 2z = -8 \end{matrix}\right\} \text{on solving these we}$$

get $\quad x = -1, y = 2, z = 3$

$$\therefore A = \begin{bmatrix} 0 & -1 & 2 \\ 1 & 0 & 3 \\ -2 & -3 & 0 \end{bmatrix}$$

$$\text{Now, } 2(A+I) = \begin{bmatrix} 2 & -2 & 4 \\ 2 & 2 & 6 \\ -4 & -6 & 2 \end{bmatrix}$$

$\Rightarrow |2(A+I)| = 120$

$\because |\text{adj}\, 2(A+I)| = |2(A+I)|^{(n-1)}$

$(\because |\text{adj}\, A| = |A|^{n-1})$

$= |2(A+I)|^2 = (120)^2$

$= 2^6 \cdot 3^2 \cdot 5^2$

By comparing with $2^\alpha \cdot 3^\beta \cdot 5^\gamma$, we get

$\alpha = 6, \beta = 2, \gamma = 2$

$\therefore \alpha^2 + \beta^2 + \gamma^2 = 6^2 + 2^2 + 2^2$

$= 36 + 4 + 4 = 44$

23. *(5120)* Given,

$S = \{p_1, p_2, p_3, \ldots, p_{10}\}$

and $A = \{$product of distinct element of $S\}$

$P = \{(x, y) : x \in S, y \in A$ and y is divided by$x\}$

Since, $x \in S$

Let $x = P_i \in S$; selected from S is ${}^{10}C_1$ ways

Now, $x \mid y$ iff y contains P_i in product. i.e., P_i is there and at rest one other element as a product

$\therefore \quad n(P) = {}^{10}C_1[(1) \cdot (2^9)]$

$= 10 \times 512 = 5120$

24. *(19)* Given,

$2(x+2)^2 \cdot f(x) - 3(x+2)^2$

$= 10\int_0^x (t+2) f(t)\, dt$ ... (i)

On differentiating Eq. (i) both side, we get

$4(x+2)f(x) + 2(x+2)^2 \cdot f'(x) - 6(x+2)$
$= 10(x+2)f(x)$

$\Rightarrow 2(x+2)^2 \cdot f'(x) - 6(x+2)f(x)$
$= 6(x+2)$

$\Rightarrow (x+2)\dfrac{dy}{dx} - 3y = 3$, where $y = f(x)$

and $\dfrac{dy}{dx} = f'(x)$

$\Rightarrow (x+2)\dfrac{dy}{dx} = 3(y+1)$

$\Rightarrow \displaystyle\int \frac{1}{y+1}\, dy = \int \frac{3}{x+2}\, dx$

$\Rightarrow \ln(y+1) = 3\ln(x+2) + \ln c$

$= \ln(c(x+2)^3)$

$\Rightarrow (y+1) = c(x+2)^3$

For initial condition put $x = 0$ in Eq. (i) we get

$2 \times (0+2)^2 \cdot f(0) - 3(0+2)^2$

$= 10\int_0^0 (t+2) f(t)\, dt = 0$

$\Rightarrow f(0) = \dfrac{3}{2}$

$\therefore \quad c = \dfrac{5}{16}$

So, $y + 1 = \dfrac{5}{16}(x+2)^3$ ...(ii)

$\therefore f(2) = \dfrac{5}{16} \times 64 - 1$

$= 20 - 1 = 19$

25. *(125)* number of 3-digit numbers which are divisible by 6 is 102, 108, 114, ... , 996

$\therefore$Total number $= 150$

And number divisible by 36 is 108, 144, ... 972

$\therefore$Total number $= 25$

Hence, required number = (divisible by 6) – (divisible by 36)

$= 150 - 25 = 125$

Method II

$\because$ Total number of 3-digit numbers

$= 999 - 99 = 900$

Now, total number of 3-digit numbers which are divisible by 2 and 3 i.e., 6

$= \dfrac{900}{6} = 150$

and total number of 2-digit numbers which are divisible by 4 and 9 i.e., by

$36 = \dfrac{900}{36} = 25$

Hence, total number of 3-digit number which are divisible by 2 and 3 but not by 4 and 9 $= 150 - 25 = 125$

Physics

26. *(a)* Given force, $F = \alpha + \beta x^2$

Work done is given by, $W = \int F \cdot dx$

$$5 = \int (\alpha + \beta x^2) dx$$

$$5 = \alpha x + \frac{\beta x^3}{3} \Big|_0^1$$

$$5 = \alpha + \frac{\beta}{3} \qquad [\alpha = 1]$$

$$4 = \frac{\beta}{3}, \beta = 12 \text{ N/m}^2$$

27. *(d)* Given current value is,

$$I = I_A \sin \omega t + I_B \cos \omega t$$

I_{rms} is given by

$$I_{\text{rms}} = \sqrt{\frac{\int_0^T I^2 dt}{\int_0^T dt}}$$

$$= \sqrt{\frac{\int I_A^2 \sin^2 \omega t dt + \int I_B^2 \cos^2 \omega t dt + 2\int I_A I_B \sin \omega t \cos \omega t dt}{\int dt}}$$

$$= \sqrt{\frac{I_A^2 \times \frac{T}{2} + I_B^2 \times \frac{T}{2} + I_A I_B \times 0}{T}}$$

$$= \sqrt{\frac{I_A^2 + I_B^2}{2}}$$

28. *(c)* According to Kepler's third law,

$$T^2 = KR^3$$

$$\frac{2\Delta T}{T} = \frac{3\Delta R}{R}$$

$$\frac{2\Delta T}{T} = 3 \times \frac{0.03R}{R}$$

$$\frac{\Delta T}{T} = \frac{3 \times 0.03}{2} \times 100 = 4.5\%$$

29. *(b)* Drawing the free body diagram,

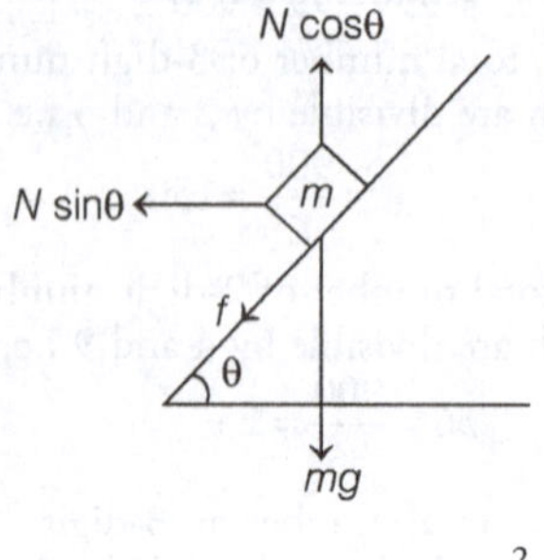

$$N \sin \theta + f \cos \theta = \frac{mv_0^2}{r}$$

$$N \cos \theta - f \sin \theta = mg$$

$$\frac{\sin \theta + \mu \cos \theta}{\cos \theta - \mu \sin \theta} = \frac{v_0^2}{rg}$$

$$rg \tan \theta + \mu rg = v_0^2 - v_0^2 \mu \tan \theta$$

$$\mu = \frac{v_0^2 - rg \tan \theta}{rg + v_0^2 \tan \theta}$$

30. *(b)* Work done, W = Surface energy

$$W = \Delta U = S\Delta A$$

where, S is surface tension

One drop change to n drop

$$\frac{4}{3}\pi R^3 = n\frac{4}{3}\pi r^3$$

$$r = \frac{R}{n^{1/3}}$$

(Small r is radius of small drop)

$$W = S[n 4\pi r^2 - 4\pi R^2]$$

$$= S 4\pi R^2 [n^{\frac{1}{3}} - 1]$$

For one drop to 27 drops

$$W = S4\pi R^2 [27^{1/3} - 1] = 10 \quad \ldots \text{(i)}$$

For one drop to 64 drops

$$W' = S4\pi R^2 (64^{1/3} - 1) \quad \ldots \text{(ii)}$$

Dividing Eq. (ii) by Eq. (i)

$$\frac{W'}{W} = \frac{4-1}{3-1} = \frac{3}{2}$$

$$W' = \frac{3}{2} W = 15 \text{ J}$$

Alternate Solution

Work required to break a drop into n smaller drops is given by

$$W \propto (n^{1/3} - 1)$$

So, as per question

For $n_1 = 27$, $W_1 = 10$ J

For $n_2 = 64$, $W_2 = ?$

Applying the formula

$$\frac{W_2}{W_1} = \frac{(n_2^{1/3} - 1)}{(n_1^{1/3} - 1)} = \frac{64^{1/3} - 1}{27^{1/3} - 1}$$

$$\frac{W_2}{10} = \frac{3}{2}$$

$$W_2 = 15 \text{ J}$$

31. *(d)* Particle perform SHM

Time period = 2s

Amplitude = 1 cm

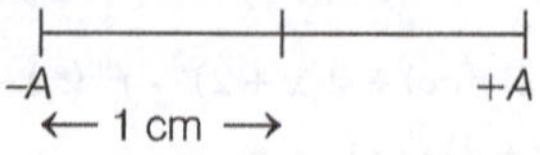

$$n = \frac{12.5}{2} = 6.25 \text{ cycle}$$

$$\therefore \quad D = 4 \times 6 + 1 = 25$$

$$d = 1$$

$$\frac{D}{d} = 25$$

32. *(b)* Linear acceleration is,

$$a = \frac{g \sin \theta}{1 + \frac{I}{mR^2}}$$

$$a = \frac{\frac{g}{\sqrt{2}}}{1 + \frac{1}{2}} = \frac{2 \times \frac{g}{\sqrt{2}}}{3} = \frac{\sqrt{2} g}{3}$$

33. *(b)*

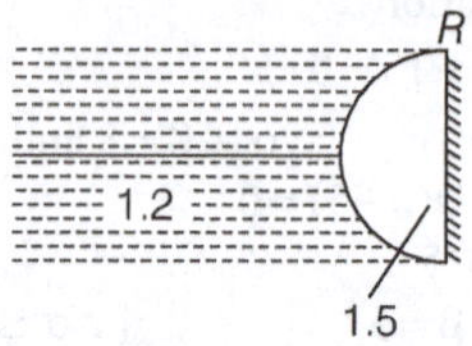

$\mu_2 = 1.5$, $\mu_1 = 1.2$, $R_1 = R$, $R_2 \to \infty$

Using thin lens formula

$$\frac{1}{f_{\text{lens}}} = \left(\frac{\mu_2}{\mu_1} - 1\right)\left(\frac{1}{R_2} - \frac{1}{R_2}\right)$$

$$\Rightarrow \quad \frac{1}{f_{\text{lens}}} = \left(\frac{1.5}{1.2} - 1\right)\left(\frac{1}{R} - \frac{1}{\infty}\right)$$

$$\Rightarrow \quad \frac{1}{f_{\text{lens}}} = \frac{1}{4R} \quad \ldots \text{(i)}$$

Now, lens is silvered, thus its focal length is

$$\frac{1}{f_{\text{lens}}} = \frac{2}{R} = \frac{2}{\infty} = 0$$

By combination of less

$$\frac{1}{f_{\text{eq}}} = \frac{1}{f_{\text{lens}}} - \frac{1}{f'} + \frac{1}{f_{\text{lens}}}$$

$$\frac{1}{f_{\text{eq}}} = \frac{2}{f_{\text{lens}}} = \frac{2}{4R}$$

$$f_{\text{eq}} = 2R \Rightarrow R = 0.1$$

34. *(c)* Statements given in B and E are correct, while A, C and D are incorrect.

Junction area of solar cell is made broad compared to photodiode.

LED is made of heavily doped p - n junction.

Increase of forward current does not always result in increase in light intensity.

35. *(a)* The fringe width is given by

$$\frac{n_1\lambda_1 D}{d}=\frac{n_2\lambda_2 D}{d}$$

$$\Rightarrow \quad n_1\times 480=n_2\times 600$$

$$(n_1)_{\min}=5$$

36. *(a)* Given,

$$y=\frac{32\cdot 3\times 1125}{27\cdot 4}=1326\cdot 186$$

Least significant digits are 3 in operands, so results should be rounded off to 3 digits.

$$\therefore \quad y=1330$$

37. *(c)* Focal length is given by,

$$f=\frac{1}{P}$$

For small change in P

$$\frac{\Delta f}{f}=-\frac{\Delta P}{P}$$

As given, $\Delta P=0.1$ D and $P=2.5$ D

$$\Rightarrow \frac{\Delta f}{f}=-\frac{0.1}{2.5}=-0.04$$

So, relative decrease in focal length is 0.04.

38. *(d)* Angle of projection, $\theta=45°$, initial velocity v_0.

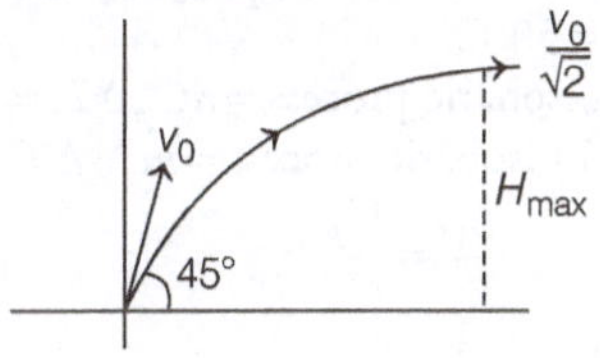

At maximum height

$$H=\frac{\left(\frac{v_0}{\sqrt{2}}\right)^2}{2g}=\frac{v_0^2}{4g}$$

Angular momentum is given by

$$L=mvh \qquad (h=H)$$

$$=\frac{mv_0v_0^2}{\sqrt{2}4g}=\frac{mv_0^3}{4\sqrt{2}g}$$

along negative Z-axis

39. *(c)* Initial velocity of electron $=v_0\hat{\mathbf{i}}$

Electric field $\mathbf{E}=-E_0\hat{\mathbf{k}}$

Velocity when it enters the field

$$\mathbf{v}=\mathbf{v}_0\hat{\mathbf{i}}+\frac{E_0e}{m}t\,\hat{\mathbf{k}}$$

$$|\mathbf{v}|=\sqrt{v_0^2+\frac{E_0^2e^2t^2}{m^2}}$$

Now, according to de-Broglie

$$\lambda_0=\frac{h}{mv_0}$$

and

$$\lambda'=\frac{h}{mv_0\sqrt{1+\frac{E_0^2e^2t^2}{v_0^2m^2}}}$$

$$\lambda'=\frac{\lambda_0}{\sqrt{1+\frac{E_0^2e^2t^2}{v_0^2m^2}}}$$

40. *(c)* Let T is the surface tension.

Excess pressure p_e in the air bubble

$$=\rho gh+\frac{2T}{R}$$

$$p_{\text{in}}-p_0=\rho gh+\frac{2T}{R}=2100$$

$$\frac{2T}{R}=2100-\rho gh$$

$$T=\frac{R}{2}(2100-10^3\times 10\times 0.2)$$

$$=0.05 \qquad [\text{where } R=0.1 \text{ cm}]$$

41. *(d)* Among the given statements, A, D and E are correct, while B and C are incorrect.

The correct form of B and C are

$$p=kT$$

$$\frac{p}{T}=\text{constant}$$

$\therefore$ Volume is constant or isochoric process.

$$\therefore \quad W_\infty=0$$

$$\therefore \quad Q=\Delta U$$

Hence, heat added is equal to change in internal energy.

42. *(d)* The capacitance is given by the formula

$$C=\frac{A\varepsilon_0}{d}$$

Initial capacitance $=\frac{(l\times b)\varepsilon_0}{d}$

$$=\frac{3\times 1}{3}\varepsilon_0=\varepsilon_0$$

For option C, $l=6$ cm, $b=5$ cm, $d=3$ cm.

$$C=\frac{6\times 5}{3}\times\varepsilon_0=10\varepsilon_0 \text{ units}$$

For option $E=l=5$ cm $b=2$ cm, $d=1$ cm

$$C=\frac{5\times 2}{1}\varepsilon_0=10\varepsilon_0 \text{ units}$$

43. *(a)* Using lens Maker formula

$$\frac{1}{f}=({}_1\mu_2-1)\left[\frac{1}{R_1}-\frac{1}{R_2}\right]$$

$$\frac{1}{f_1}=(1.5-1)\left[\frac{1}{2}-0\right]$$

$$f_1=4 \text{ cm}$$

For f_2, $\frac{1}{f_2}=\left(\frac{1.5}{1.2}-1\right)\left(\frac{1}{3}-0\right)$

$$f_2=12 \text{ cm}$$

$$f_1:f_2$$

$$\Rightarrow \quad 4:12=1:3$$

44. *(a)* Potential energy per unit volume is given as $\frac{U}{V}=\frac{1}{2}\varepsilon_0E^2$,

where E is electric field.

$$U=\frac{1}{2}\varepsilon_0E_0^2V=\left(\frac{1}{2}\varepsilon_0E_0^2A\cdot d\right)$$

Hence, option (a) is correct.

45. *(a)* The energy transition from A to C is given by

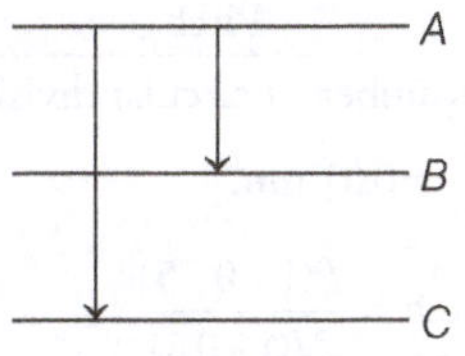

$$E_A-E_C=\frac{hc}{\lambda}$$

where, $\lambda=2000$ Å ... (i)

Similarly, when transition take place from A to B

$$E_A-E_B=[E_A-E_C)-(E_B-E_C)$$

$$\frac{hc}{\lambda_{AB}}=\frac{hc}{2000\text{Å}}-\frac{hc}{6000\text{ Å}}$$

$$\therefore \quad \lambda_{AB}=3000 \text{ Å}$$

Alternate Solution

The relationship between the wavelengths for step-by-step jump of electron and direct jump of electron is given by

$$\frac{1}{\lambda}=\frac{1}{\lambda_1}+\frac{1}{\lambda_2}+\frac{1}{\lambda_3}+\ldots\ldots$$

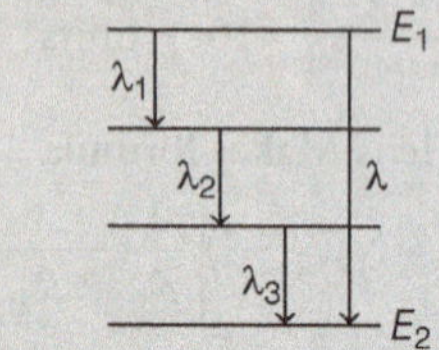

So, according to question

$\lambda_1 = 6000$ Å

$\lambda = 2000$ Å

$\lambda_2 = ?$

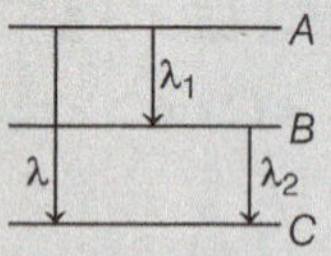

Using the formula

$$\frac{1}{\lambda}=\frac{1}{\lambda_1}+\frac{1}{\lambda_2}$$

$$\frac{1}{2000}=\frac{1}{6000}+\frac{1}{\lambda_2}$$

$\Rightarrow \quad \lambda_2 = 3000$ Å

46. *(35)* Given, least count of screw gauge $=0.01$ mm

Least count is given by

$$\text{LC} = \frac{\text{Pitch}}{\text{Number of circular division}}$$

$$=\frac{P}{N}=0.01 \text{ mm}$$

$$\text{New pitch} = \frac{P(1+0.75)}{N(1-0.5)}$$

$$=\frac{P}{N}\left(\frac{1.75}{0.5}\right)$$

$$=(0.01)3.5$$

$$=35\times10^{-3} \text{ mm}$$

47. *(48)* Considering charge at a distance of $\frac{a}{2}$ from the square.

Total flux through the square

$$=\frac{q}{\varepsilon_0}\left(\frac{1}{6}\right)$$

As flux is same in the plane, therefore flux is divided into equal parts

$\therefore$ Flux through shaded portion is $\frac{5}{8}$

$$=\frac{5}{8}\times\frac{q}{\varepsilon_0}\left(\frac{1}{6}\right)$$

$$=\frac{5}{48}\frac{1}{\varepsilon_0}$$

48. *(2)* The resistor diagram is as follows,

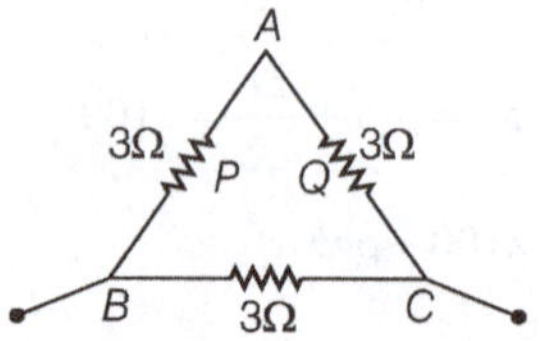

$\Rightarrow$ As 3Ω and 3Ω (P and Q) are in series.

$$R_{eq} = R_1 + R_2 = 3+3 = 6\,\Omega$$

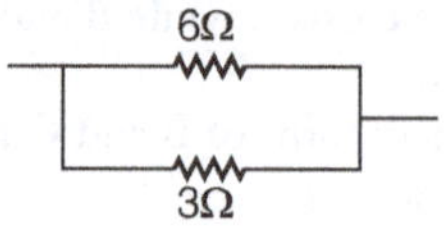

Now 6Ω and 3Ω are in parallel

$$R_{eq} = \frac{R_1 \times R_2}{R_1 + R_2}$$

$$=\frac{6\times3}{6+3}=2\,\Omega$$

Time Saver Tech

For a polygon of n sides, the equivalent resistance across two adjacent vertices is given by

$$R_{eq} = \frac{(n-1)R}{n^2}$$

where, R is the total resestance.

So, according to question $n = 3$ and $R = 9\,\Omega$

$$R_{eq} = \frac{(n-1)R}{n^2} = \frac{(3-1)\times 9}{3^2} = 2\Omega$$

49. *(8)* Magnetic field due to single side is

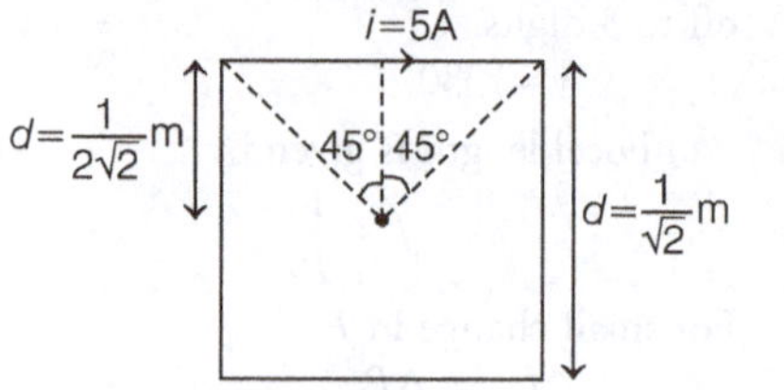

$$B=\frac{\mu_0 i}{4\pi d}[\sin\theta_1+\sin\theta_2]$$

$$=\frac{10^{-7}\times5\times2}{\frac{1}{2\sqrt{2}}}\times\frac{1}{\sqrt{2}}=2\times10^{-6}\text{ T}$$

$\therefore B_{net}$ at centre $O = 4B = 8\times10^{-6}$ T

$$p=8$$

50. *(15)* Given that the process is isobaric

$\Delta T = 50°$C

Q in isobaric process $= nC_p\Delta T = E_1$

ΔU in isobaric process $= nC_V\Delta T = E_2$

$$\therefore \quad \frac{E_1}{E_2}=\frac{C_p}{C_V}=\gamma$$

As the gas is monoatomic

$$\gamma = 1+\frac{2}{f} = 1+\frac{2}{3}=\frac{5}{3}$$

As per question

$$\frac{5}{3}=\frac{x}{9} \Rightarrow x = 15$$

Chemistry

51. *(c)* The complete reaction involved is as follows

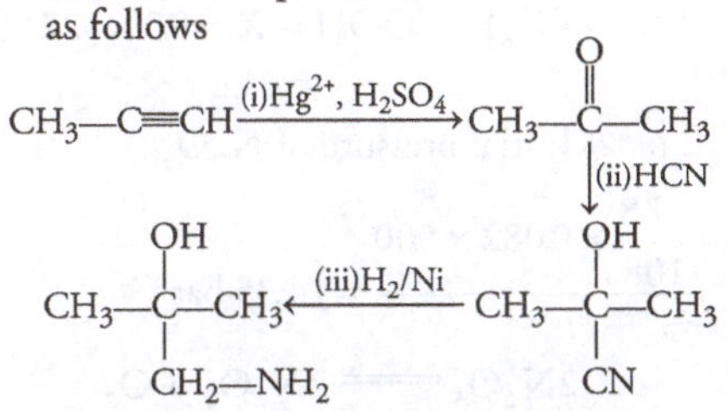

52. *(d)* The large difference between the melting point and boiling point of oxygen and sulphur is explained by atomicity.

Oxygen exists as O_2 = Atomicity = 2

Sulphur exists as S_8 = Atomicity = 8

53. *(c)* Carbocation intermediate is stabilised by +*I*, +*M* and hyperconjugation effect. Since in

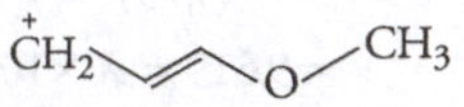

Carbocation is in conjugation with stronger +*M* with group $—OCH_3$, hence it is most stable.

54. *(d)* Addition of H—Br (*aq*) to an alkene follows electrophillic addition mechanism. In the rate determining step a carbocation intermediate is formed. Among *P*, *Q*, *R*, and *S* compound, *Q* will form most stable carbocation intermediate since it is resonance stabilised.

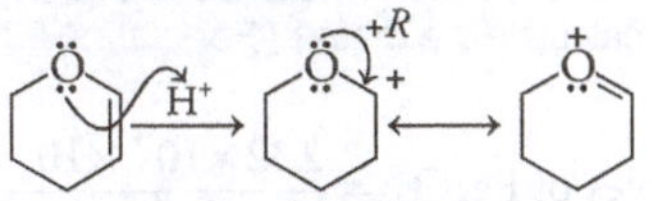

55. *(b)* Reaction is spontaneous at relatively high temperature and non spontaneous at low temperature.

$$\Delta G = \Delta H - T\Delta S$$

$\Delta G < 0$ is only possible when ΔH and ΔS both are positive.

Hence, option (b) is correct.

56. *(c)* The complete reaction is as follows

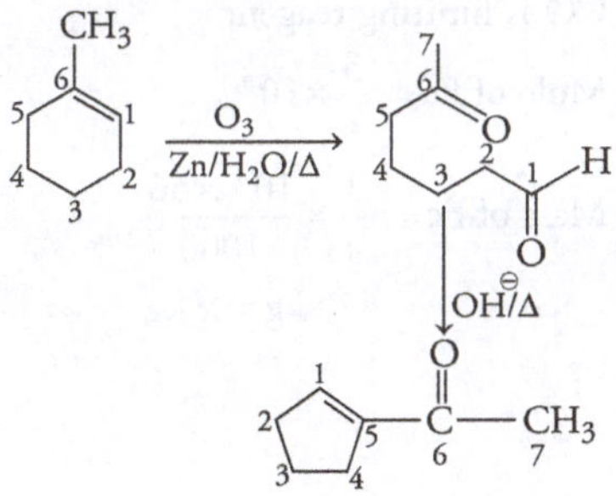

So, the suitable alkene is given in option (c).

57. *(b)* Statements given is *A*, *C* and *E* are incorrect while *B* and *D* are correct. The correct form of incorrect statements are as follows

- Properties of elements are periodic function of their atomic number.
- Element with similar outer electronic configuration are arranged in same group.
- Number of elements in a period is not equal to number of atomic orbitals available in energy level that is being filled.

58. *(a)* Statement I is true but II is false. The correct from of statement II is,

In Dumas method nitrogen present in organic compound is converted into N_2 gas whose volumetric analysis gives the percentage of nitrogen in the organic compounds.

59. *(d)* Statements given is *A*, *B* and *C* are correct. While *D* and *E* are incorrect.

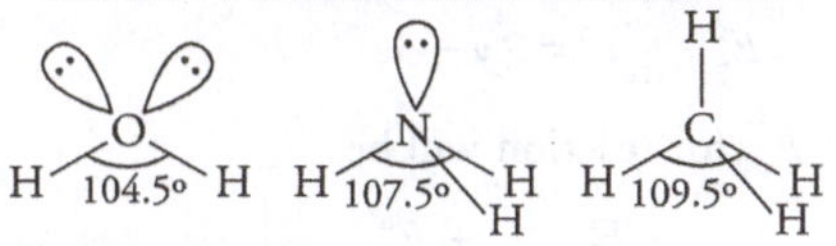

- $O \rightarrow sp^3$ $N \rightarrow sp^3$ $C \rightarrow sp^3$
- Dipole moment order : $CH_4 < NH_3 < H_2O$
- H_2O and NH_3 are Lewis base.
- NH_3 act as Lowry Bronsted base. In solution of NH_3 in H_2O.

> **Concept Enhancer**
> - According to Bronsted-Lowry acids and bases concept, the acid-base pair that differs only by one proton is called conjugate acid-base pair.
> - If Bronsted acid is strong acid then its conjugate base is a weak base and *vice-versa*.

60. *(c)* The first step of the reaction for preparation of potassium permanganate from MnO_2 is

$$MnO_2 \xrightarrow[KNO_3,\ \Delta]{KOH} K_2MnO_4$$

Thus, the product formed is K_2MnO_4.

61. *(a)* Statements given in *A*, *C* and *D* are correct while *B* and *E* are incorrect.

In ribose, carbohydrate present in DNA is β-D-2-deoxyribose whose structure is

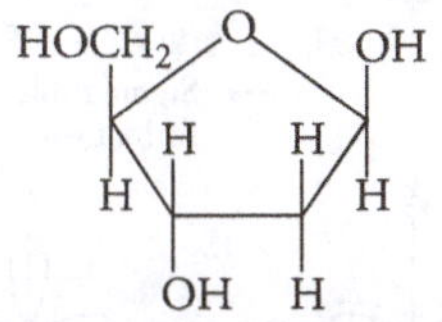

Which is a reducing D-sugar in β anomeric form (at C1) and it is a pentose sugar.

62. *(a)* The rate of nucleophilic addition reaction decreases due to steric crowding around carbonyl carbon and increases by electron withdrawing group if the steric crowding is same and hence the reactivity towards nucleophilic addition will be

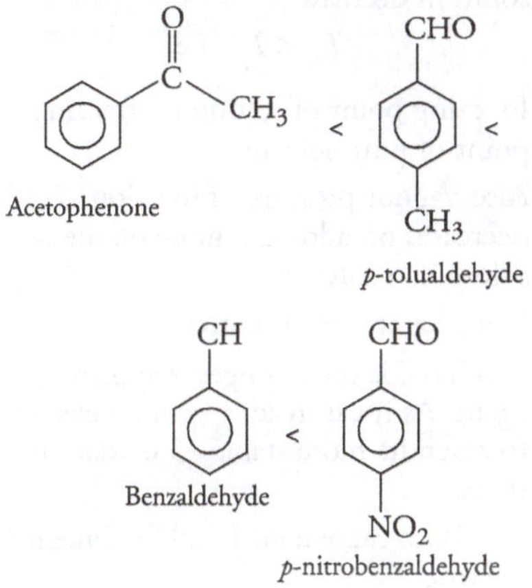

63. *(c)*

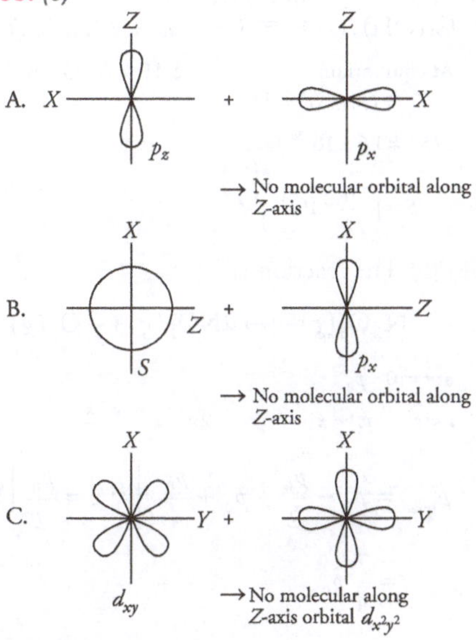

D. 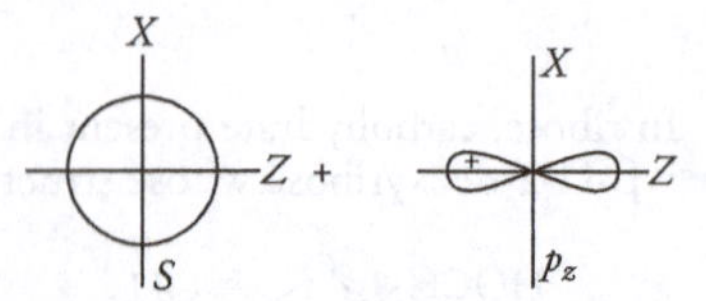

→ Sigma molecular orbit formation

E. 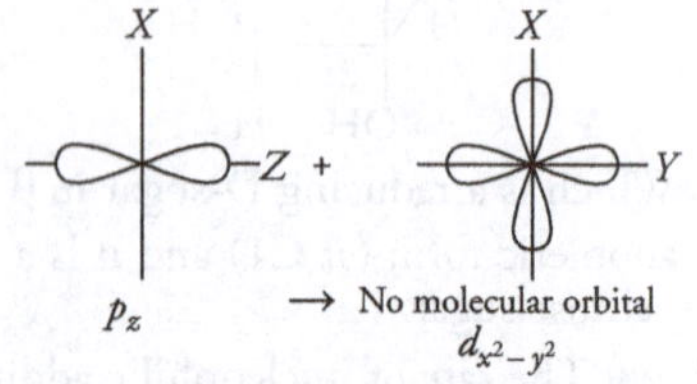

→ No molecular orbital $d_{x^2-y^2}$

So, only $2s$ and $2p_z$ will lead to formation of molecular orbital in a homonuclear diatomic molecule.

64. *(d)* When a non-volatile solute is added to solvent, freezing point of solution decreases.

$$T_f < T_f{}^\circ \text{ i.e.}$$

Freezing point of solution < freezing point of pure solvent.

Also vapour pressure of solution decreases on adding a non-volatile solute in a solvent.

Hence graph (d) is correct.

65. *(c)* Tb^{4+} is the strongest oxidising agent. As it can readily gain an electron to reach its more stable +3 oxidation state.

66. *(a)* Both Statement I and Statement II are true.

67. *(a)* The complete reaction is

$$Cr(OH)_3(s) \rightleftharpoons Cr^{+3}(aq) + 3OH^-(aq)$$

At equilibrium $\quad SM \quad 3SM$

$$K_{sp} = (S)(3S)^3 = 27S^4$$

$$27S^4 = 1.6 \times 10^{-30}$$

$$S = \left(\frac{1.6}{27} \times 10^{-30}\right)^{\frac{1}{4}} M$$

68. *(b)* The reaction is

$$N_2O_5(g) \longrightarrow 2NO_2(g) + \frac{1}{2}O_2(g)$$

at $t = 0$ $\quad p_0$

$t = t$ $\quad p_0 - x \quad 2x \quad \frac{x}{2}$

$$p_{\text{total}} = p_0 - \frac{p_0}{2} + p_0 + \frac{p_0}{4} \quad \left[\because x = \frac{p_0}{2}\right]$$

$$= \frac{7}{4} p_0$$

69. *(c)* The complete reaction is,

$$[Co(NH_3)_5Cl]Cl_2$$
$$\downarrow$$
$$\underbrace{[Co(NH_3)_5Cl]^{2+}(aq) + 2Cl^-(aq)}_{\text{3 ions in water}}$$

$$[Co(NH_3)_5Cl]Cl_2(aq) + 2AgNO_3)(aq) \longrightarrow [Co(NH_3)_5Cl](NO_3)_2(aq) + 2AgCl(s)$$

Thus, the structure of the complex is $[Co(NH_3)_5Cl]Cl_2$.

70. *(b)* For given cell;

$$Fe^{2+}(aq) + Ag^+(aq) \longrightarrow Fe^{3+}(aq) + Ag(s)$$

$$Fe \xrightarrow{\Delta G_1^\circ} Fe^{2+} \xrightarrow{\Delta G_2^\circ} Fe^{3+}$$

$$\Delta G_3^\circ = \Delta G_1^\circ + \Delta G_2^\circ$$

$$-3F(-z) = -2F(-y) + \Delta G_2^\circ$$

$$\Delta G_2^\circ = 3F_z - 2F_y$$

Also $\Delta G_2^\circ = -nFE^\circ_{Fe^{2+}/Fe^{3+}}$

$$3F_z - 2F_y = -lF(E^0_{Fe^{2+}/Fe^{+3}})$$

$$E^\circ_{Fe^{2+}/Fe^{3+}} = 2y - 3z$$

E°_{cell} for reaction will be

$$E^\circ_{Ag^+/Ag} + E^\circ_{Fe^{2+}/Fe^{3+}}$$

$$= x + 2y - 3z$$

71. *(700)* The reaction is

$$\frac{1}{2}X_2 + \frac{5}{2}Y_2 \rightleftharpoons XY_5$$

$$\Delta H^- = -35 \text{ kJ/mol}$$
$$= \Delta_r H^\circ$$

$$\Delta_r S^\circ = 110 - \left[\left(\frac{1}{2} \times 70\right) + \left(\frac{5}{2} \times 50\right)\right]$$

$$= 110 - 160 = -50 \text{ JK}^{-1}\text{mol}^{-1}$$

$\Delta G^\circ = 0$ at equilibrium

At equilibrium

$$T = \frac{|\Delta_r H^\circ|}{|\Delta_r S^\circ|} = \frac{35 \times 1000}{50}$$

$$T = 700 \text{ K}$$

72. *(61)* The complete reaction

$$\underset{X\text{ g}}{C_6H_5COOH} + NaHCO_3 \longrightarrow$$

$$C_6H_5COO^-Na^+ + H_2O + CO_2$$

$$\text{Mole of } CO_2 = \frac{11.2}{22.4} = 0.5$$

$$\text{Mass of } C_6H_5COOH = X = 0.5 \times 122$$
$$= 61 \text{ g}$$

73. *(962)* Initial pressure of N_2O_5

$$\frac{\frac{37.8}{108} \times 0.082 \times 500}{1} = 14.35 \text{ bar}$$

$$2N_2O_5 \rightleftharpoons 2N_2O_4 + O_2$$

at $t = 0$ $\quad 14.35$

$t =$ eq $\quad 14.35-2p \quad 2p \quad p$

p_{total} at eq $= 14.35 + p = 18.65$ bar

$$p = 4.3$$

$$P_{N_2O_5} = 5.75 \text{ bar}$$

$$p_{N_2O_4} = 8.6 \text{ bar}$$

$$p_{O_2} = 4.3 \text{ bar}$$

$$K_p = \frac{(8.6)^2 \times (4.3)}{(5.75)^2}$$

$$= 9.619 = X \times 10^{-2}$$

$$X = 961.9$$

$$\simeq 962$$

74. *(4)* Only Cu^{2+}, Fe^{3+}, Ca^{2+} and Zn^{2+} form precipitate with $K_4[Fe(CN)_6]$

75. *(420)*

Key idea

To solve this question first of calculate the number of moles of Fe_3O_4 and CO. Then with the help of this find out limiting reagent. After that use stoichiometry to calculate the mass of Fe.

$$\text{Moles of } Fe_3O_4 = \frac{232 \times 10^3 \times 10^3}{232}$$

$$= 10000 \text{ mol}$$

$$\text{Moles of CO} = \frac{2.8 \times 10^2 \times 10^3}{28}$$

$$= 10000 \text{ mol}$$

$$\underset{10^4\text{mol}}{Fe_3O_4} + \underset{10^4\text{mol}}{4CO} \longrightarrow 3Fe + 4CO_2$$

CO is limiting reagent

$$\text{Mole of Fe} = \frac{3}{4} \times 10^4$$

$$\text{Mass of Fe} = \frac{3}{4} \times \frac{10^4 \times 56}{1000}$$

$$= 420 \text{ kg} = x \text{ kg}$$

Time 3:00-6:00 **MM** : 300 Marks

24 January
Shift-II

Online Question Paper

—

JEE Main 2025

Mathematics

Section A
Objective Type Questions

01/ Let $A = [a_{ij}]$ be a square matrix of order 2 with entries either 0 or 1. Let E be the event that A is an invertible matrix. Then, the probability $P(E)$ is

(a) $\frac{1}{8}$ (b) $\frac{3}{16}$
(c) $\frac{3}{8}$ (d) $\frac{5}{8}$

02/ The number of real solution(s) of the equation $x^2 + 3x + 2 = \min\{|x-3|, |x+2|\}$ is

(a) 2 (b) 3
(c) 0 (d) 1

03/ Let the point $\left(\frac{11}{2}, \alpha\right)$ lie on or inside the triangle with sides $x + y = 11$, $x + 2y = 16$ and $2x + 3y = 29$.
Then, the product of the smallest and the largest values of α is equal to

(a) 44 (b) 22
(c) 55 (d) 33

04/ If the system of equations
$x + 2y - 3z = 2$
$2x + \lambda y + 5z = 5$
$14x + 3y + \mu z = 33$
has infinitely many solutions, then $\lambda + \mu$ is equal to

(a) 12 (b) 13
(c) 10 (d) 11

05/ Let $f : (0, \infty) \to R$ be a function which is differentiable at all points of its domain and satisfies the condition $x^2 f'(x) = 2xf(x) + 3$ with $f(1) = 4$. Then, $2f(2)$ is equal to

(a) 29 (b) 23
(c) 19 (d) 39

06/ For some a, b let

$$f(x) = \begin{vmatrix} a + \frac{\sin x}{x} & 1 & b \\ a & 1 + \frac{\sin x}{x} & b \\ a & 1 & b + \frac{\sin x}{x} \end{vmatrix},$$

$x \neq 0$, $\lim_{x \to 0} f(x) = \lambda + \mu a + vb$.

Then, $(\lambda + \mu + v)^2$ is equal to

(a) 25 (b) 16
(c) 36 (d) 9

07/ The area of the region enclosed by the curves $y = e^x$, $y = |e^x - 1|$ and Y-axis is

(a) $1 + \log_e 2$ (b) $\log_e 2$
(c) $1 - \log_e 2$ (d) $2\log_e 2 - 1$

08/ Let $[x]$ denote the greatest integer function and let m and n, respectively be the numbers of the points, where the function $f(x) = [x] + |x - 2|$, $-2 < x < 3$ is not continuous and not differentiable.

Then, $m+n$ is equal to

(a) 6 (b) 8 (c) 9 (d) 7

09/ If $S = 5 + \frac{1}{7}(5+\alpha) + \frac{1}{7^2}(5+2\alpha) + \frac{1}{7^3}(5+3\alpha) + \ldots + \infty$, then the value of α is

(a) 1 (b) 1/7 (c) 6 (d) 6/7

10/ Group A consists of 7 boys and 3 girls, while group B consists of 6 boys and 5 girls. The number of ways, 4 boys and 4 girls can be invited for a picnic if 5 of them must be from group A and the remaining 3 from group B, is equal to

(a) 8575 (b) 8750
(c) 8925 (d) 9100

11/ Let the position vectors of three vertices of a triangle be $4\mathbf{p}+\mathbf{q}-3\mathbf{r}$, $-5\mathbf{p}+\mathbf{q}+2\mathbf{r}$ and $2\mathbf{p}-\mathbf{q}+2\mathbf{r}$. If the position vectors of the orthocenter and the circumcenter of the triangle are $\frac{\mathbf{p}+\mathbf{q}+\mathbf{r}}{4}$ and $\alpha\mathbf{p}+\beta\mathbf{q}+\gamma\mathbf{r}$ respectively, then $\alpha+2\beta+5\gamma$ is equal to

(a) 1 (b) 3
(c) 6 (d) 4

12/ Suppose A and B are the coefficients of 30th and 12th terms, respectively in the binomial expansion of $(1+x)^{2n-1}$. If $2A = 5B$, then n is equal to

(a) 21 (b) 20
(c) 19 (d) 22

Imp.

13/ Let $\mathbf{a} = 3\hat{i} - \hat{j} + 2\hat{k}$, $\mathbf{b} = \mathbf{a} \times (\hat{i} - 2\hat{k})$ and $\mathbf{c} = \mathbf{b} \times \hat{k}$. Then, the projection of $\mathbf{c} - 2\hat{j}$ on $\mathbf{a}$ is:

(a) $3\sqrt{7}$ (b) $\sqrt{14}$ (c) $2\sqrt{7}$ (d) $2\sqrt{14}$

Cross Product of Vectors

14/ The function $f:(-\infty,\infty)\to(-\infty,1)$ defined by $f(x) = \frac{2^x - 2^{-x}}{2^x + 2^{-x}}$ is

(a) Neither one-one nor onto (b) Both one-one and onto
(c) One-one but not onto (d) Onto but not one-one

15/ If the equation of the parabola with vertex $V\left(\frac{3}{2}, 3\right)$ and the directrix $x+2y=0$ is $\alpha x^2 + \beta y^2 - \gamma xy - 30x - 60y + 225 = 0$, then $\alpha+\beta+\gamma$ is equal to

(a) 7 (b) 6 (c) 9 (d) 8

16/ Let (2, 3) be the largest open interval in which the function $f(x) = 2\log_e(x-2) - x^2 + ax + 1$ is strictly increasing and (b, c) be the largest open interval, in which the function $g(x) = (x-1)^3(x+2-a)^2$ is strictly decreasing. Then, $100(a+b-c)$ is equal to

(a) 420 (b) 160
(c) 360 (d) 280

17/ In an arithmetic progression, if $S_{40} = 1030$ and $S_{12} = 57$, then $S_{30} - S_{10}$ is equal to

(a) 505 (b) 515
(c) 510 (d) 525

18/ If $\alpha > \beta > \gamma > 0$, then the expression

$$\cot^{-1}\left\{\beta + \frac{(1+\beta^2)}{(\alpha-\beta)}\right\} + \cot^{-1}\left\{\gamma + \frac{(1+\gamma^2)}{(\beta-\gamma)}\right\} + \cot^{-1}\left\{\alpha + \frac{(1+\alpha^2)}{(\gamma-\alpha)}\right\}$$ is equal to

(a) 0 (b) 3π
(c) $\frac{\pi}{2} - (\alpha+\beta+\gamma)$ (d) π

19/ Let

$$A = \left\{x \in (0,\pi) - \left\{\frac{\pi}{2}\right\} : \log_{(2/\pi)}|\sin x| + \log_{(2/\pi)}|\cos x| = 2\right\}$$ and

$B = \{x \geq 0 : \sqrt{x}(\sqrt{x}-4) - 3|\sqrt{x}-2| + 6 = 0\}$. Then $n(A \cup B)$ is equal to

(a) 2 (b) 4
(c) 6 (d) 8

20/ The equation of the chord of ellipse $\frac{x^2}{25} + \frac{y^2}{16} = 1$, whose mid-point is (3, 1) is

(a) $5x + 16y = 31$ (b) $4x + 122y = 134$
(c) $48x + 25y = 169$ (d) $25x + 101y = 176$

Section B

Numerical Type Questions

21/ If $\int \frac{2x^2+5x+9}{\sqrt{x^2+x+1}}\,dx = x\sqrt{x^2+x+1} + \alpha\sqrt{x^2+x+1} + \beta\log_e\left|x + \frac{1}{2} + \sqrt{x^2+x+1}\right| + C$, where C is the constant of integration, then $\alpha + 2\beta$ is equal to ....

Imp.

22/ Let $y = y(x)$ be the solution of the differential equation $2\cos x \dfrac{dy}{dx} = \sin 2x - 4y \sin x$, $x \in \left(0, \dfrac{\pi}{2}\right)$.

If $y\left(\dfrac{\pi}{3}\right) = 0$, then $y'\left(\dfrac{\pi}{4}\right) + y\left(\dfrac{\pi}{4}\right)$ is equal to ....

Linear Differential Equation

23/ Number of functions $f : \{1, 2, ..., 100\} \rightarrow \{0, 1\}$ that assign 1 to exactly one of the positive integers less than or equal to 98 is equal to ....

24/ Let $H_1 : \dfrac{x^2}{a^2} - \dfrac{y^2}{b^2} = 1$ and $H_2 : -\dfrac{x^2}{A^2} + \dfrac{y^2}{B^2} = 1$ be two hyperbolas having length of latus-rectums $15\sqrt{2}$ and $12\sqrt{5}$, respectively. Let their eccentricities be $e_1 = \sqrt{5/2}$ and e_2 respectively. If the product of the lengths of their transverse axes is $100\sqrt{10}$, then $25\, e_2^2$ is equal to ....

25/ Let P be the image of the point $Q(7, -2, 5)$ in the line $L : \dfrac{x-1}{2} = \dfrac{y+1}{3} = \dfrac{z}{4}$ and $R(5, p, q)$ be a point on L. Then, the square of the area of ΔPQR is ....

Physics

Section A

Objective Type Questions

26/ The position vector of a moving body at any instant of time is given as space m. The magnitude and direction of velocity at $t = 2$ s is,

(a) $5\sqrt{17}$ m/s, making an angle of $\tan^{-1}(4)$ with + ve X-axis.

(b) $5\sqrt{15}$ m/s, making an angle of $\tan^{-1}(4)$ with – ve Y-axis.

(c) $5\sqrt{15}$ m/s, making an angle of $\tan^{-1}(4)$ with + ve X-axis.

(d) $5\sqrt{17}$ m/s, making an angle of $\tan^{-1}(4)$ with – ve Y-axis.

27/ In photoelectric effect, the stopping potential (V_0) *versus* frequency (ν) curve is plotted.

(h is the Planck's constant and ϕ_0 is work function of metal)

(A) V_0 *versus* ν is linear.

(B) The slope of V_0 *versus* ν curve $= \dfrac{\phi_0}{h}$.

(C) h constant is related to the slope of V_0 *versus* ν line.

(D) The value of electric charge of electron is not required to determine h using the V_0 *versus* ν curve.

(E) The work function can be estimated without knowing the value of h.

Choose the correct answer from the options given below.

(a) A, C and E only

(b) C and D only

(c) A, B and C only

(d) D and E only

28/ The output of the circuit is low (zero) for

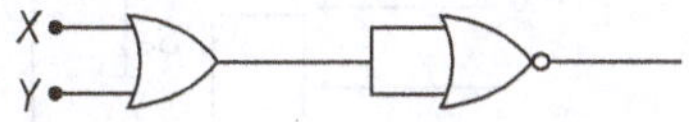

(A) $X = 0, Y = 0$ (B) $X = 0, Y = 1$

(C) $X = 1, Y = 0$ (D) $X = 1, Y = 1$

Choose the correct answer from the options given below.

(a) B, C and D only (b) A, B and D only

(c) A, C and D only (d) A, B and C only

Imp.

29/

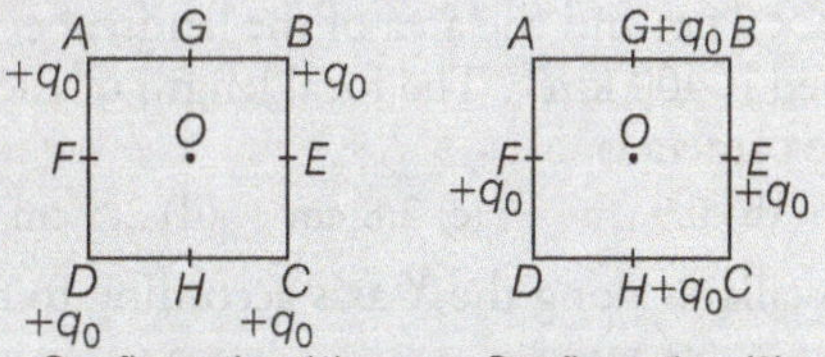

Configuration (1) Configuration (2)

In the first configuration (1) as shown in the figure, four identical charges (q_0) are kept at the corners A, B, C and D of square of side length a. In the second configuration (2), the same charges are shifted to mid-points G, E, H and F, of the square. If $k = \dfrac{1}{4\pi\varepsilon_0}$, the difference between the potential energies of configuration (2) and (1) is given by

(a) $\dfrac{kq_0^2}{a}(3\sqrt{2} - 2)$ (b) $\dfrac{kq_0^2}{a}(4 - 2\sqrt{2})$

(c) $\dfrac{kq_0^2}{a}(4\sqrt{2} - 2)$ (d) $\dfrac{kq_0^2}{a}(3 - \sqrt{2})$

Electrostatic Potential Energy

30/ In a Young's double slit experiment, three polarisers are kept as shown in the figure. The transmission axes of P_1 and P_2 are orthogonal to each other. The polariser P_3 covers both the slits with its transmission axis at 45° to those of P_1 and P_2. An unpolarised light of wavelength λ and intensity I_0 is incident on P_1 and P_2. The intensity at a point after P_3, where the path difference between the light waves from S_1 and S_2 is $\frac{\lambda}{3}$, is

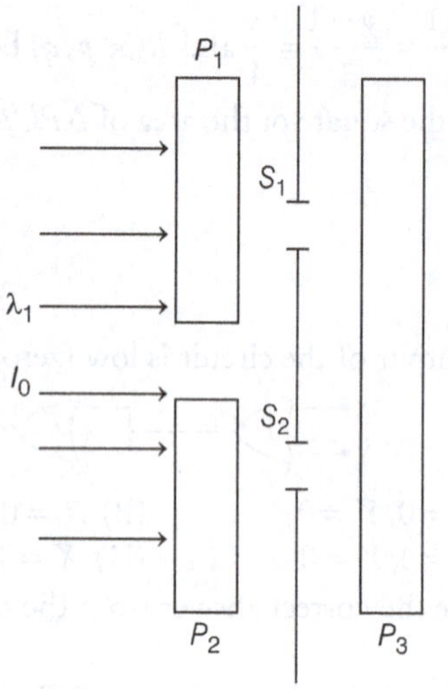

(a) $\frac{I_0}{3}$ (b) I_0 (c) $\frac{I_0}{2}$ (d) $\frac{I_0}{4}$

31/ A photograph of a landscape is captured by a drone camera at a height of 18 km. The size of the camera film is 2 cm × 2 cm and the area of the landscape photographed is 400 km^2. The focal length of the lens in the drone camera is

(a) 1.8 cm (b) 0.9 cm (c) 2.8 cm (d) 2.5 cm

32/ A particle oscillates along the X-axis according to the law, $x(t) = x_0 \sin^2\left(\frac{t}{2}\right)$, where $x_0 = 1$ m. The kinetic energy (K) of the particle as a function of x is correctly represented by the graph

(a)
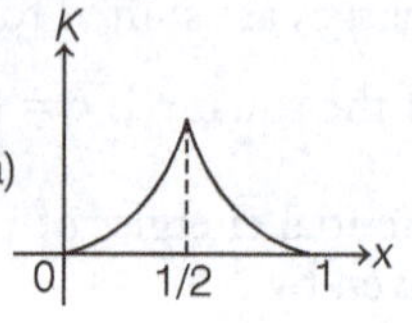

(b)
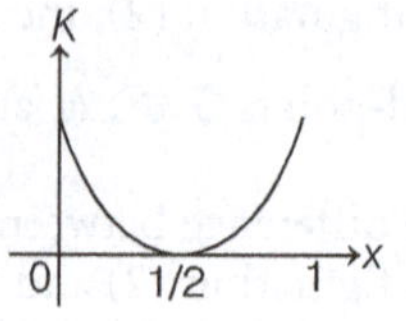

(c)
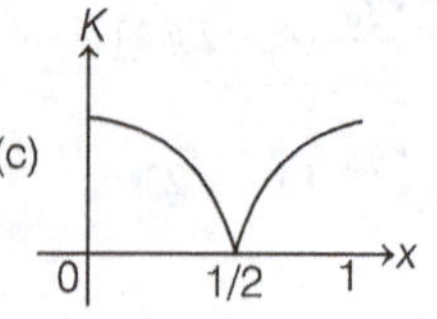

(d)
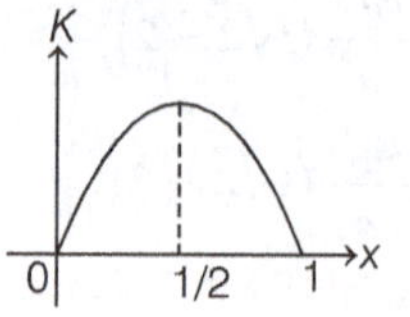

33/ The energy E and momentum p of a moving body of mass m are related by some equation. Given that c represents the speed of light, identify the correct equation

(a) $E^2 = pc^2 + m^2c^4$ (b) $E^2 = p^2c^2 + m^2c^4$
(c) $E^2 = pc^2 + m^2c^2$ (d) $E^2 = p^2c^2 + m^2c^2$

34/ A solid sphere and a hollow sphere of the same mass and of same radius are rolled on an inclined plane. Let the time taken to reach the bottom by the solid sphere and the hollow sphere be t_1 and t_2, respectively, then

(a) $t_1 < t_2$ (b) $t_1 > t_2$ (c) $t_1 = t_2$ (d) $t_1 = 2t_2$

35/ Young's double slit inteference apparatus is immersed in a liquid of refractive index 1.44. It has slit separation of 1.5 mm. The slits are illuminated by a parallel beam of light whose wavelength in air is 690 nm. The fringe width on a screen placed behind the plane of slits at a distance of 0.72 m, will be

(a) 0.23 mm (b) 0.33 mm (c) 0.46 mm (d) 0.63 mm

Imp.

36/ The magnitude of heat exchanged by a system for the given cyclic process $ABCA$ (as shown in figure) is (in SI unit)

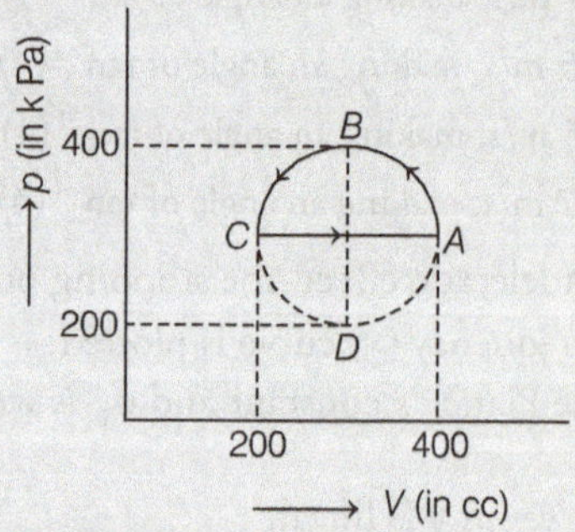

(a) 5π (b) zero (c) 10π (d) 40π

Heat Exchange in Cyclic Process

37/ Given below are two statements. One is labelled as Assertion (A) and the other is labelled as Reason (R).

Assertion (A) In an insulated container, a gas is adiabatically shrunk to half of its initial volume. The temperature of the gas decreases.

Reason (R) Free expansion of an ideal gas is an irreversible and an adiabatic process.

In the light of the above statements, choose the correct answer from the options given below.

(a) A is false but R is true.
(b) Both A and R are true but R is not the correct explanation of A.
(c) Both A and R are true and R is the correct explanation of A.
(d) A is true but R is false.

38/ A small uncharged conducting sphere is placed in contact with an identical sphere but having 4×10^{-8} C charge and then removed to a distance such that the force of repulsion between them is 9×10^{-3} N. The distance between them is

$$\left(\text{Take } \frac{1}{4\pi\varepsilon_0} \text{ as } 9 \times 10^9 \text{ in SI units}\right)$$

(a) 2 cm (b) 3 cm (c) 1 cm (d) 4 cm

39/ A long straight wire of a circular cross-section with radius a carries a steady current I. The current I is uniformly distributed across this cross-section. The plot of magnitude of magnetic field B with distance r from the centre of the wire is given by

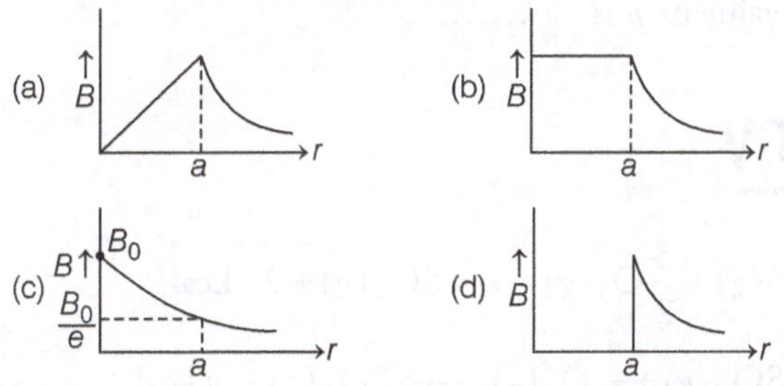

40/ The temperature of a body in air falls from 40°C to 24°C in 4 min. The temperature of the air is 16°C. The temperature of the body in the next 4 min will be

(a) $\frac{28}{3}$ °C (b) $\frac{42}{3}$ °C (c) $\frac{14}{3}$ °C (d) $\frac{56}{3}$ °C

41/

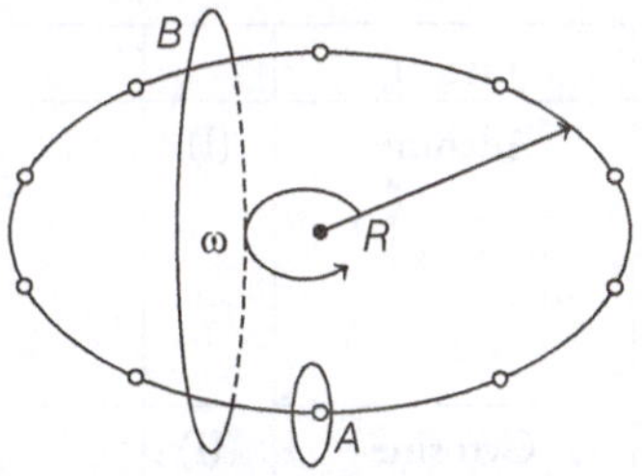

N equally spaced charges each of value q, are placed on a circle of radius R. The circle rotates about its axis with an angular velocity ω as shown in the figure. A bigger amperian loop B encloses the whole circle where as a smaller amperian loop A encloses a small segment. The difference between enclosed currents, $I_A - I_B$, for the given amperian loops is

(a) $\frac{2\pi}{N} q\omega$ (b) $\frac{N}{2\pi} q\omega$ (c) $\frac{N}{\pi} q\omega$ (d) $\frac{N^2}{2\pi} q\omega$

42/ Given below are two statements. One is labelled as Assertion (A) and the other is labelled as Reason (R)

Assertion (A) A electron in a certain region of uniform magnetic field is moving with constant velocity in a straight line path.

Reason (R) The magnetic field in that region is along the direction of velocity of the electron.

In the light of the above statements, choose the correct answer from the options given below.

(a) A is true but R is false.
(b) Both A and R are true and R is the correct explanation of A.
(c) Both A and R are true but R is not the correct explanation of A.
(d) A is false but R is true.

43/ Arrange the following in the ascending order of wavelength (λ)

(A) Microwaves (λ_1) (B) Ultraviolet rays (λ_2)
(C) Infrared rays (λ_3) (D) X-rays (λ_4)

Choose the most appropriate answer from the options given below.

(a) $\lambda_3 < \lambda_4 < \lambda_2 < \lambda_1$ (b) $\lambda_4 < \lambda_2 < \lambda_3 < \lambda_1$
(c) $\lambda_4 < \lambda_3 < \lambda_2 < \lambda_1$ (d) $\lambda_4 < \lambda_3 < \lambda_1 < \lambda_2$

44/ A solid sphere is rolling without slipping on a horizontal plane. The ratio of the linear kinetic energy of the centre of mass of the sphere and rotational kinetic energy is

(a) 2/5 (b) 4/3 (c) 3/4 (d) 5/2

45/ Which of the following figure represents the relation between Celsius and Fahrenheit temperatures?

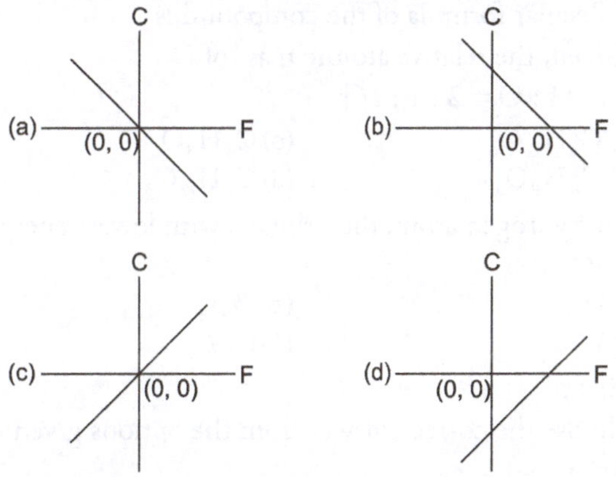

Section B

Numerical Type Questions

46/ A lightly wound long solenoid carries a current of 1.5 A. An electron is executing uniform circular motion inside the solenoid with a time period of 75 ns. The number of turns per metre in the solenoid is ....

[Take mass of electron $m_e = 9 \times 10^{-31}$ kg, charge of electron $|q_e| = 1.6 \times 10^{-19}$ C,

$\mu_0 = 4\pi \times 10^{-7} \frac{\text{N}}{\text{A}^2}$, 1 ns $= 10^{-9}$ s]

47/ Acceleration due to gravity on the surface of earth is g. If the diameter of earth is reduced to one-third of its original value and mass remains unchanged, then the acceleration due to gravity on the surface of the earth is ……….. g.

48/

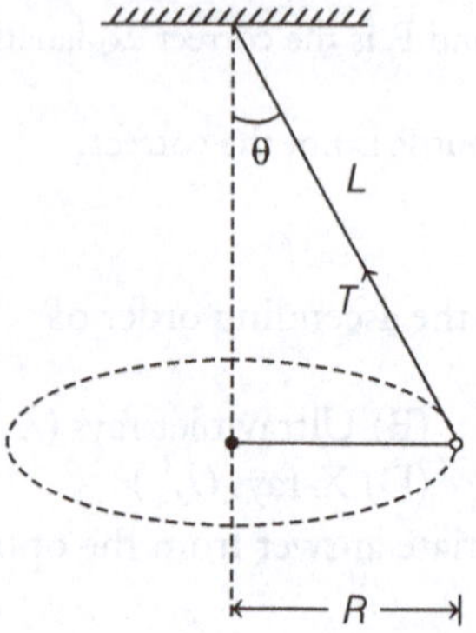

A string of length L is fixed at one end and carries a mass of M at the other end. The mass makes $3/\pi$ rotations per second about the vertical axis passing through end of the string as shown. The tension in the string is …….. ML.

49/ The ratio of the power of a light source S_1 to that the light source S_2 is 2. S_1 is emitting 2×10^{15} photons per second at 600 nm. If the wavelength of the source S_2 is 300 nm, then the number of photons per second emitted by S_2 is ……….. $\times 10^{14}$.

50/ The increase in pressure required to decrease the volume of a water sample by 0.2% is $p \times 10^5$ Nm^{-2}. Bulk modulus of water is 2.15×10^9 Nm^{-2}. The value of p is ……….. .

Chemistry

Section A
Objective Type Questions

51/ The elemental composition of a compound is 54.2% C, 9.2% H and 36.6% O.
If the molar mass of the compound is 132 g mol^{-1}, the molecular formula of the compound is
[Given, the relative atomic mass of C : H : O = 2 : 1 : 16]

(a) $C_6H_{12}O_6$ (b) $C_4H_8O_2$
(c) $C_4H_9O_3$ (d) $C_6H_{12}O_3$

52/ For hydrogen atom, the orbital/s with lowest energy is/are

(A) $4s$ (B) $3p_x$
(C) $3d_{x^2-y^2}$ (D) $3d_{z^2}$
(E) $4p_z$

Choose the correct answer from the options given below

(a) (B), (C) and (D) only
(b) (A) and (E) only
(c) (B) only
(d) (A) only

53/ Based on the data given below

$E^\circ_{Cr_2O_7^{2-}/Cr^{3+}} = 1.33$ V; $E^\circ_{Cl_2/Cl^{(-)}} = 1.36$ V

$E^\circ_{MnO_4^-/Mn^{2+}} = 1.51$ V; $E^\circ_{Cr^{3+}/Cr} = -0.74$ V

the strongest reducing agent is

(a) Cl^- (b) Cr
(c) MnO_4^- (d) Mn^{2+}

54/ $S(g) + \frac{3}{2}O_2(g) \longrightarrow SO_3(g) + 2x$ kcal

$SO_2(g) + \frac{1}{2}O_2(g) \longrightarrow SO_3(g) + y$ kcal

The heat of formation of $SO_2(g)$ is given by

(a) $y - 2x$ kcal (b) $x + y$ kcal (c) $2x + y$ kcal (d) $\frac{2x}{y}$ kcal

55/ Match List - I with List - II.

	List - I		List - II
(A)	Adenine	(I)	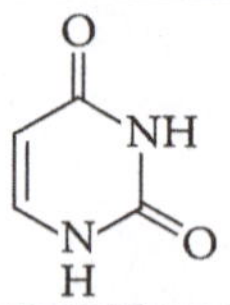
(B)	Cytosine	(II)	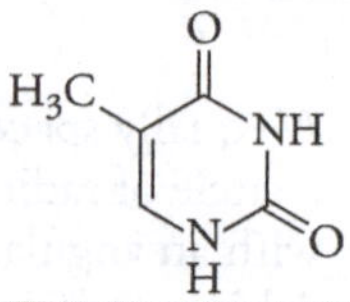
(C)	Thymine	(III)	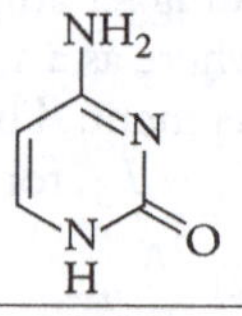
(D)	Uracil	(IV)	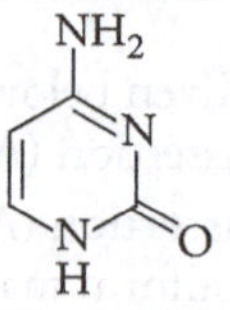

Choose the correct answer from the options given below

(a) (A) - (III), (B) - (I), (C) - (IV), (D) - (II)
(b) (A) - (IV), (B)-(III), (C) - (II), (D) - (I)
(c) (A) - (III), (B) - (IV), (C) - (II), (D) - (I)
(d) (A) - (III), (B)-(IV), (C) - (I), (D) - (II)

56/ Find the compound '*A*' from the following reaction sequences.

$$A \xrightarrow{\textit{Aqua-regia}} B \xrightarrow[\text{(ii) AcOH}]{\text{(i) } KNO_2/[NH_4OH]} \text{yellow ppt.}$$

(a) ZnS (b) MnS
(c) CoS (d) NiS

57/ The successive 5 ionisation energies of an element are 800, 2427, 3658, 25024 and 32824 kJ/mol, respectively. By using the above values predict the group in which the above element is present

(a) Group 2
(b) Group 14
(c) Group 4
(d) Group 13

58. Given below are two statements.

Statement I Experimentally determined oxygen-oxygen bond lengths in the O_3 are found to be same and the bond length is greater than that of a $O{=}O$ (double bond) but less than that of a single (O—O) bond.

Statement II The strong lone pair-lone pair repulsion between oxygen atoms is solely responsible for the fact that the bond length in ozone is smaller than that of a double bond ($O{=}O$) but more than that of a single bond (O—O).

In the light of the above statements, choose the correct answer from the options given below

(a) Both Statement I and Statement II are false.
(b) Statement I is false but Statement II is true.
(c) Statement I is true but Statement II is false.
(d) Both Statement I and Statement II are true.

59/ When ethane-1,2-diamine is added progressively to an aqueous solution of nickel (II) chloride, the sequence of colour change observed will be

(a) Violet $\rightarrow$ Blue $\rightarrow$ Pale blue $\rightarrow$ Green
(b) Green $\rightarrow$ Pale blue $\rightarrow$ Blue $\rightarrow$ Violet
(c) Pale blue $\rightarrow$ Blue $\rightarrow$ Green $\rightarrow$ Violet
(d) Pale blue $\rightarrow$ Blue $\rightarrow$ Violet $\rightarrow$ Green

60/ Match List-I with List-II.

	List-I (Transition metal ion)		List-II (Spin only magnetic moment (BM))
(A)	Ti^{3+}	(I)	3.87
(B)	V^{2+}	(II)	0.00
(C)	Ni^{2+}	(III)	1.73
(D)	Sc^{3+}	(IV)	2.84

Choose the correct answer from the options given below

(a) (A) - (IV), (B) - (II), (C)-(III), (D) - (I)
(b) (A) - (III), (B) - (I), (C) - (IV), (D) - (II)
(c) (A) - (III), (B) - (I), (C) - (II), (D) - (IV)
(d) (A) - (II), (B) - (IV), (C) - (I), (D)-(III)

61/ Given below are two statements.

Statement I The first ionisation energy of Pb is greater than that of Sn.

Statement II The first ionisation energy of Ge is greater than that of Si.

In the light of the above statements, choose the correct answer from the options given below

(a) Both Statement I and Statement II are true.
(b) Statement I is true but Statement II is false.
(c) Both Statement I and Statement II are false.
(d) Statement I is false but Statement II is true.

62/ In the given structure, number of *sp* and sp^2 hybridised carbon atoms present respectively are

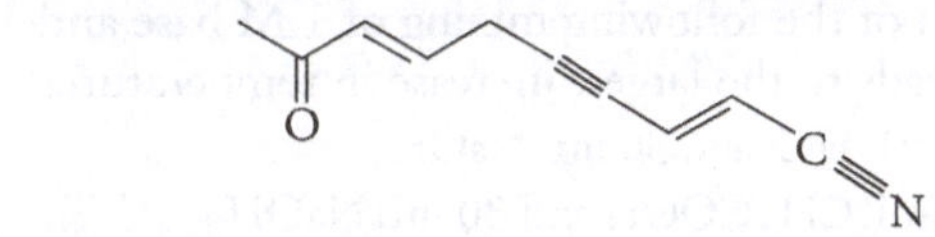

(a) 3 and 5 (b) 3 and 6 (c) 4 and 6 (d) 4 and 5

63/ For reaction,

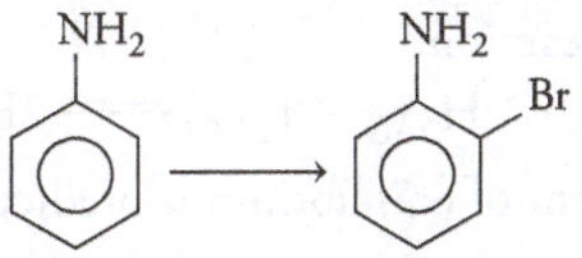

(Major)

The correct order of set of reagents for the above conversion is

(a) H_2SO_4, Ac_2O, Br_2, $H_2O(\Delta)$, NaOH
(b) Ac_2O, H_2SO_4, Br_2, NaOH
(c) Br_2 $|FeBr_3$, $H_2O(\Delta)$, NaOH
(d) Ac_2O, Br_2, $H_2O(\Delta)$, NaOH

Imp.

64/ Given below are two statements.

Statement I

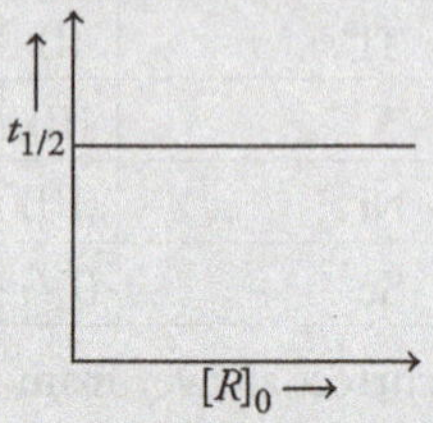

is valid for first order reaction.

Statement II

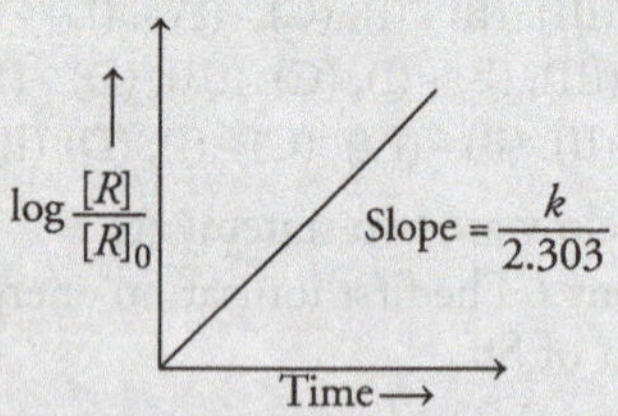

is valid for first order reaction.

In the light of the above statements, choose the correct answer from the options given below

(a) Statement I is true but Statement II is false.
(b) Both Statement I and Statement II are false.
(c) Both Statement I and Statement II are true.
(d) Statement I is false but Statement II is true.

First Order Reactions

65/ Which of the following mixing of 1 M base and 1 M acid leads to the largest increase in temperature?

(a) 30 mL HCl and 30 mL NaOH
(b) 30 mL CH_3COOH and 30 mL NaOH
(c) 50 mL HCl and 20 mL NaOH
(d) 45 mL CH_3COOH and 25 mL NaOH

66/ For the reaction,

$$H_2(g) + I_2(g) \rightleftharpoons 2HI(g)$$

Attainment of equilibrium is predicted correctly by

(a) 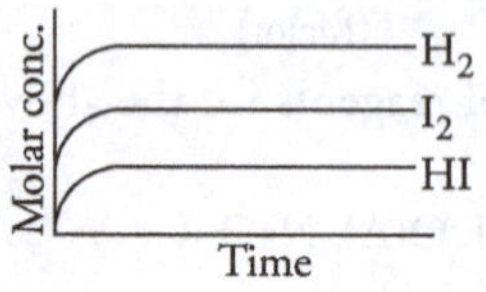

(b) 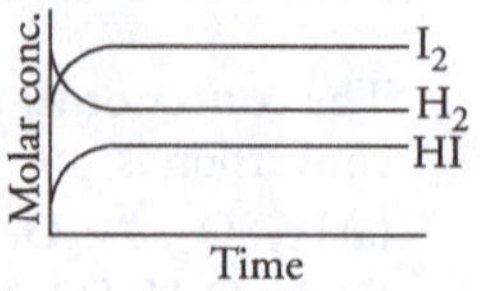

(c) 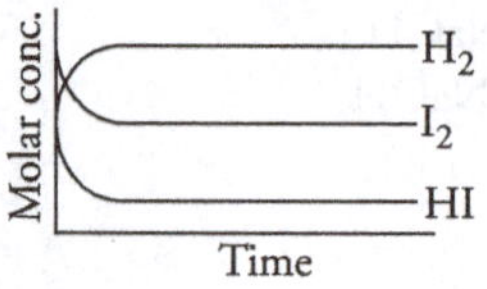

(d) 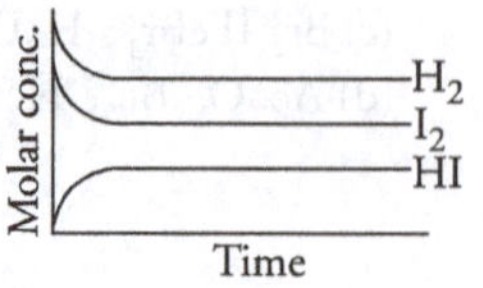

67. Match List-I with List-II.

	List-I		List-II
(A)	$RCN \xrightarrow[\text{(ii) } H_3O^+]{\text{(i) } SnCl_2,\ HCl} RCHO$	(I)	Etard reaction
(B)	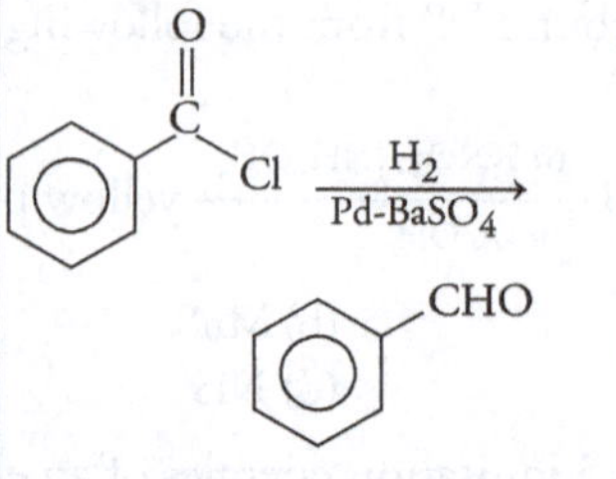	(II)	Gattermann-Koch reaction
(C)	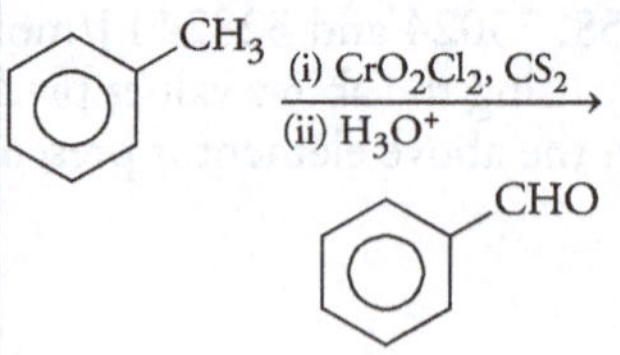	(III)	Rosenmund reduction
(D)	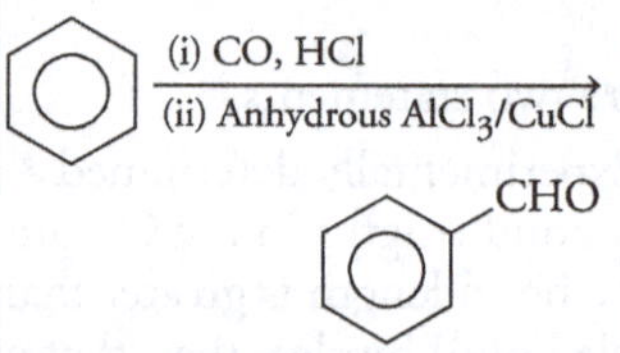	(IV)	Stephen reaction

Choose the correct answer from the options given below.

(a) (A)-(III), (B)-(IV), (C)-(I), (D)-(II)
(b) (A)-(III), (B)-(IV), (C)-(II), (D)-(I)
(c) (A)-(I), (B)-(III), (C)-(II), (D)-(IV)
(d) (A)-(IV), (B)-(III), (C)-(I), (D)-(II)

68/ The conditions and consequence that favours the $t_{2g}^3 e_g^1$ configuration in a metal complex are

(a) weak field ligand, low spin complex
(b) strong field ligand, high spin complex
(c) weak field ligand, high spin complex
(d) strong field ligand, low spin complex

69/ The structure of the major product formed in the following reaction is

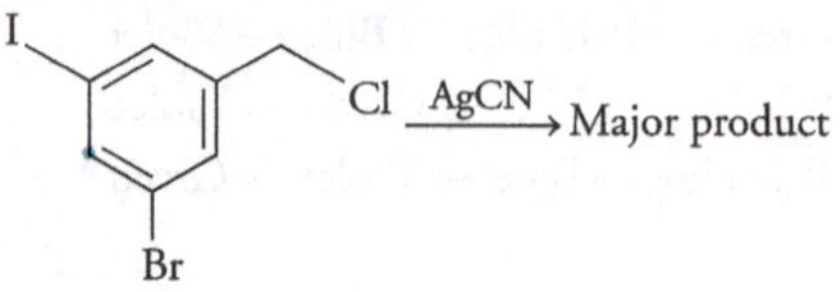

(a) 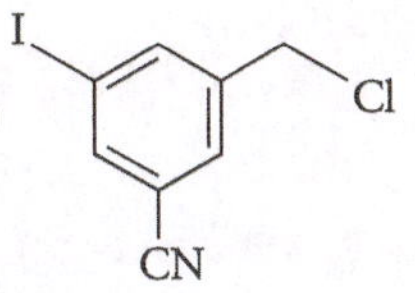 (b) 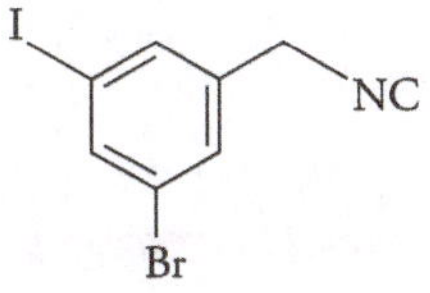

(c) 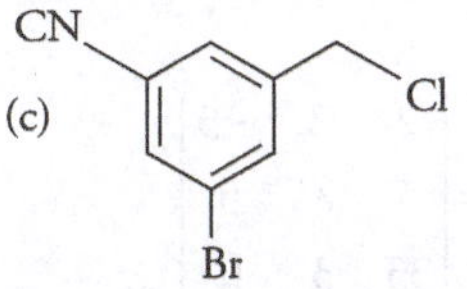 (d) 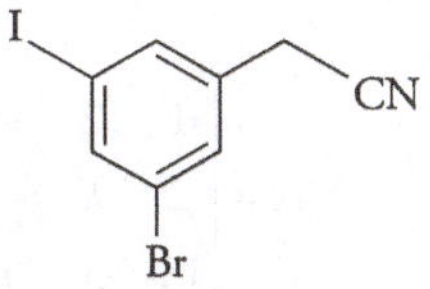

70/ Identify correct statement/s

(A) — OCH_3 and — $NHCOCH_3$ are activating groups.

(B) — CN and — OH are *meta* directing groups.

(C) — CN and — SO_3H are *meta* directing groups.

(D) Activating groups act as *ortho* and *para* directing groups.

(E) Halides are activating groups.

Choose the correct answer from the options given below

(a) (A), (B) and (E) only (b) (A), (C) and (D) only

(c) (A) and (C) only (d) (A) only

Section B

Numerical Type Questions

71/ In Carius method of estimation of halogen, 0.25 g of an organic compound gave 0.15 g of silver bromide (AgBr). The percentage of bromine in the organic compound is .... $\times 10^{-1}$% (Nearest integer)

(Given: Molar mass of Ag is 108 and Br is 80 g mol^{-1})

72/ The observed and normal molar masses of compound MX_2 are 65.6 and 164 respectively. The percent degree of ionisation of MX_2 is ....%. (Nearest integer)

Imp.

73/ The possible number of stereoisomers for 5-phenylpent-4-en-2-ol is ....

Isomerism

74/ The hydrocarbon (*X*) with molar mass 80 g mol^{-1} and 90% carbon has .... degree of unsaturation.

75/ Consider a complex reaction taking place in three steps with rate constants K_1, K_2 and K_3 respectively. The overall rate constant K is given by the expression

$$K = \sqrt{\frac{K_1 K_3}{K_2}}.$$

If the activation energies of the three steps are 60, 30 and 10 kJ mol^{-1} respectively, then the overall energy of activation in kJ mol^{-1} is ....(Nearest integer)

Answer Key

Mathematics

1	(c)	2	(a)	3	(d)	4	(a)	5	(d)	6	(b)	7	(c)	8	(b)	9	(c)	10	(c)
11	(b)	12	(a)	13	(d)	14	(c)	15	(c)	16	(c)	17	(b)	18	(d)	19	(d)	20	(c)
21	(16)	22	(1)	23	(392)	24	(55)	25	(957)										

Physics

26	(d)	27	(a)	28	(a)	29	(a)	30	(d)	31	(a)	32	(d)	33	(b)	34	(a)	35	(a)
36	(a)	37	(a)	38	(a)	39	(a)	40	(d)	41	(b)	42	(b)	43	(b)	44	(d)	45	(d)
46	(250)	47	(9)	48	(36)	49	(5)	50	(43)										

Chemistry

51	(d)	52	(a)	53	(b)	54	(a)	55	(c)	56	(c)	57	(d)	58	(c)	59	(b)	60	(b)
61	(b)	62	(a)	63	(a)	64	(a)	65	(a)	66	(d)	67	(d)	68	(c)	69	(b)	70	(b)
71	(255)	72	(75)	73	(4)	74	(3)	75	(20)										

***Note** (*) None of the option is correct.*

Detailed Explanations

Mathematics

1. *(c)* **Case I** $A = \begin{bmatrix} 1 & 1 \\ 0 & 1 \end{bmatrix} \rightarrow 4$ ways.

Case II $A = \begin{bmatrix} 1 & 0 \\ 0 & 1 \end{bmatrix}$ and $\begin{bmatrix} 0 & 1 \\ 1 & 0 \end{bmatrix} \rightarrow 2$ ways.

$\therefore P(E) = \dfrac{\text{Total number of favourable cases}}{\text{Total cases}} = \dfrac{6}{16} = \dfrac{3}{8}$

2. *(a)* Graphical method is best solution for this question

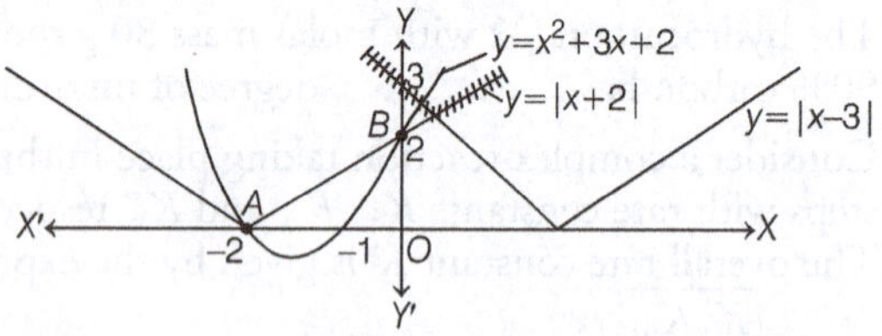

From graph, it is clear that total number of solutions = 2
(Scratched graph rejected)

3. *(d)* Clearly, $x = \dfrac{11}{2}$ intersect

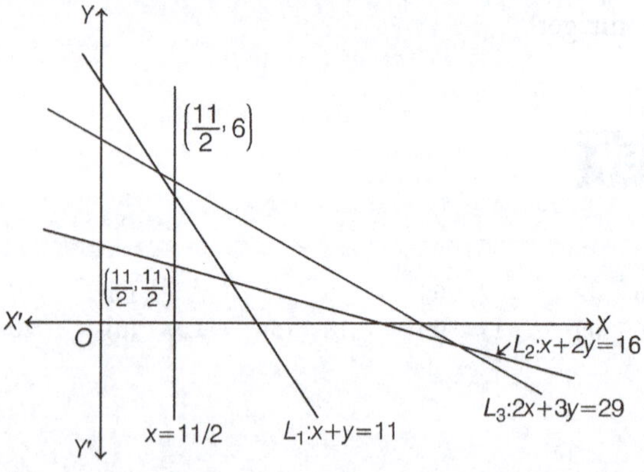

$L_1 : x + y = 11$ at $\left(\dfrac{11}{2}, \dfrac{11}{2}\right)$

and $L_3 : 2x + 3y - 29 = 0$ at $\left(\dfrac{11}{2}, 6\right)$

So, $\alpha \in \left[\dfrac{11}{2}, 6\right]$

i.e., $\alpha_{\min} = \dfrac{11}{2}$ and $\alpha_{\max} = 6$

So, $\alpha_{\min} \times \alpha_{\max} = \dfrac{11}{2} \times 6 = 33$

4. *(a)* From given system of linear equation, we get

$$\Delta = \begin{vmatrix} 1 & 2 & -3 \\ 2 & \lambda & 5 \\ 14 & 3 & \mu \end{vmatrix}, \Delta_1 = \begin{vmatrix} 2 & 2 & -3 \\ 5 & \lambda & 5 \\ 33 & 3 & \mu \end{vmatrix},$$

$$\Delta_2 = \begin{vmatrix} 1 & 2 & -3 \\ 2 & 5 & 5 \\ 14 & 33 & \mu \end{vmatrix},$$

and $$\Delta_3 = \begin{vmatrix} 1 & 2 & 2 \\ 2 & \lambda & 5 \\ 14 & 3 & 33 \end{vmatrix}$$

For infinitely many solutions

Put $\Delta = 0, \Delta_1 = 0, \Delta_2 = 0$ and $\Delta_3 = 0$

For $\Delta = 0 \Rightarrow \lambda\mu + 42\lambda - 4\mu + 107 = 0$ ...(i)

$\Delta_1 = 0 \Rightarrow 2\lambda\mu + 99\lambda - 10\mu + 255 = 0$ ...(ii)

$\Delta_2 = 0 \Rightarrow 13 - \mu = 0 \Rightarrow \mu = 13$

$\Delta_3 = 0 \Rightarrow 5\lambda + 5 = 0 \Rightarrow \lambda = -1$

We observe that $\mu = 13, \lambda = -1$ also satisfies Eqs. (i) and (ii).

Hence, $\lambda + \mu = -1 + 13 = 12$

5. *(d)*

Key idea To solve this problem, use formula

$$\frac{d}{dx}\left(\frac{f(x)}{g(x)}\right) = \frac{f'(x)g(x) - g'(x)f(x)}{(f(x))^2}$$

Given, $x^2 f'(x) - 2xf(x) = 3$

$\Rightarrow \dfrac{x^2 f'(x) - 2xf(x)}{(x^2)^2} = \dfrac{3}{(x^2)^2}$

$\Rightarrow \dfrac{d}{dx}\left(\dfrac{f(x)}{x^2}\right) = \dfrac{3}{x^4}$

Integrating both sides, we get

$\dfrac{f(x)}{x^2} = -\dfrac{1}{x^3} + C$, where C is constant of integration.

$\Rightarrow f(x) = -\dfrac{1}{x} + Cx^2$

Given, $f(1) = 4$ i.e. when $x = 1$, then $f(x) = 4$.

So, $4 = -\dfrac{1}{1} + C \times (1)^2 \Rightarrow C = 5$

$\therefore f(x) = -\dfrac{1}{x} + 5x^2$

Hence, $2f(2) = 2 \times \left[-\dfrac{1}{2} + 5(2)^2\right] = 39$

6. *(b)* Since, we know that $\lim_{x \to 0} \frac{\sin x}{x} = 1$

So,

$$\lim_{x \to 0} f(x) = \lim_{x \to 0} \begin{vmatrix} a + \frac{\sin x}{x} & 1 & b \\ a & 1 + \frac{\sin x}{x} & b \\ a & 1 & b + \frac{\sin x}{x} \end{vmatrix}$$

$$= \begin{vmatrix} a+1 & 1 & b \\ a & 1+1 & b \\ a & 1 & b+1 \end{vmatrix}$$

$$= (a+1)(2(b+1) - b) + 1(ab - a(b+1)) - ba$$

$$= (a+1)(b+2) + (-a) - ab$$

$$= ab + 2a + b + 2 - a - ab$$

$$= a + b + 2 = \lambda + \mu a + vb \quad \text{[given]}$$

$\therefore \quad \lambda = 2, \mu = 1, v = 1$

Hence,

$(\lambda + \mu + \gamma)^2 = (2 + 1 + 1)^2 = 4^2 = 16$

7. *(c)* Given, $y = e^x$ ...(i)

and $y = |e^x - 1|$ ...(ii)

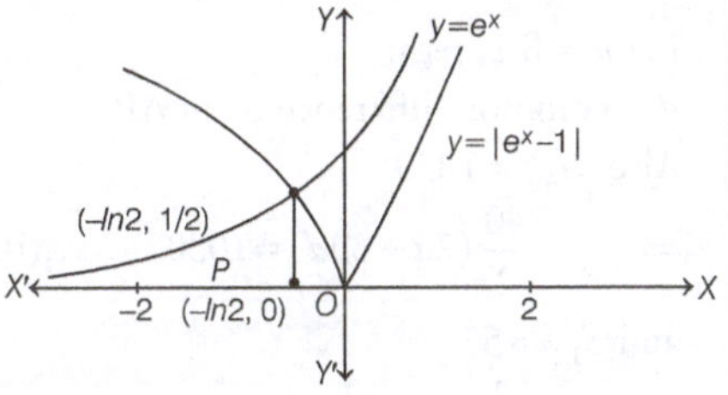

To get coordinates of point P, we have to solve Eqs. (i) and (ii)

So, $e^x = |e^x - 1|$

$\Rightarrow \quad e^x - 1 = \pm e^x \quad [\because e^x > 0]$

$\therefore \quad e^x - 1 = e^x$ or $e^x - 1 = -e^x$

$\Rightarrow \quad e^x - e^x = 1$ (which is false).

So, $2e^x = 1 \Rightarrow x = \ln 1/2 \Rightarrow x = -\ln 2.$

$\therefore \quad y = e^{-\log_e 2} = 1/2$

So, $P \equiv (-\log_e 2, 0)$

Now, area $= \int_{-\log_e 2}^{0} (e^x - |e^x - 1|)\, dx$

$= \int_{-\log_e 2}^{0} (e^x - (1 - e^x))\, dx$

$[\because$ when $-\log_e 2 < x < 0, |e^x - 1| = 1 - e^x]$

$= \int_{-\log_e 2}^{0} (2e^x - 1)\, dx = (2e^x - x)_{-\ln 2}^{0}$

$= 2 - (1 + \log_e 2) = 1 - \log_e 2$

8. *(b)* Given, $f(x) = [x] + |x - 2|, \quad -2 < x < 3$

$$\Rightarrow f(x) = \begin{cases} -x, & -2 < x < -1 \\ -x+1, & -1 \le x < 0 \\ -x+2, & 0 \le x < 1 \\ -x+3, & 1 \le x < 2 \\ x, & 2 \le x < 3 \end{cases}$$

Hints :
$$\begin{pmatrix} [x] = -2, |x-2| = -(x-2) \\ [x] = -1, |x-2| = -(x-2) \\ [x] = 0, |x-2| = -(x-2) \\ [x] = 1, |x-2| = -(x-2) \\ [x] = 2, |x-2| = x-2 \end{pmatrix}$$

We see that at $x = -1$, LHL $\neq$ RHL

at $x = 0$, LHL $\neq$ RHL

at $x = 1, 2$, LHL $\neq$ RHL

So, $f(x)$ is discontinuous at 4 points and hence obviously it is non-differentiable at these points.

So, $\quad m = n = 4$

$\therefore \quad m + n = 4 + 4 = 8$

9. *(c)* Let $S = 5 + \frac{1}{7}(5 + \alpha) + \frac{1}{7^2}(5 + 2\alpha) + \ldots + \infty$ ...(i)

$\frac{1}{7} S = \frac{1}{7}(5) + \frac{1}{7^2}(5 + \alpha) + \ldots + \infty$ ...(ii)

On subtracting Eq. (ii) from Eq. (i), we get

$$S - \frac{S}{7} = 5 + \frac{1}{7}\alpha \left(\frac{1}{1 - 1/7}\right)$$

$$\Rightarrow \quad \frac{6}{7} S = 5 + \frac{\alpha}{7} \times \frac{7}{6}$$

$$\Rightarrow \quad \frac{6}{7} S = 5 + \frac{\alpha}{6} \Rightarrow \frac{6}{7} \times 7 = 5 + \frac{\alpha}{6}$$

$[\because s = 7$ given$]$

$\Rightarrow \quad \alpha = 6$

10. *(c)*

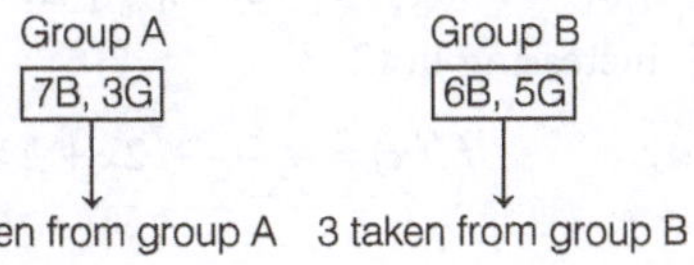

Case I ($3G$ and $2B$) from group A and ($1G$ and $2B$) from group B.

Case II ($2G$ and $3B$) from group A and ($2G$ and $1B$) from group B

Case III ($1G$ and $4B$) from group A and ($3G$ and $0B$) from group B.

So, total = Case I + Case II + Case III

$= {}^7C_2 \times {}^3C_3 \times {}^5C_1 \times {}^6C_2 + {}^7C_3 \times {}^3C_2 \times {}^6C_1 \times {}^5C_2 + {}^7C_4 \times {}^3C_1 \times {}^6C_0 \times {}^5C_3$

$= 8925$

11. *(b)* Given, position of value of vertices of triangle is

$$4\mathbf{p} + \mathbf{q} - 3\mathbf{r}, -5\mathbf{p} + \mathbf{q} + 2\mathbf{r}$$

and $2\mathbf{p} - \mathbf{q} + 2\mathbf{r}$.

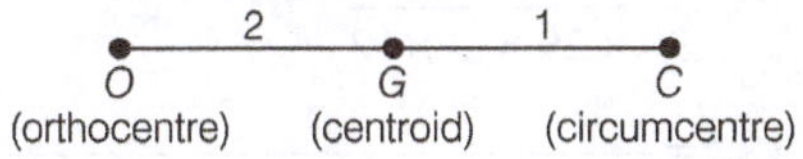

As we know that in a triangle, centroid divides join of orthocentre and circumcentre in 2 : 1

$\therefore \quad O = \frac{\mathbf{p} + \mathbf{q} + \mathbf{r}}{4}$ [given]

$C = \alpha\mathbf{p} + \beta\mathbf{q} + \gamma\mathbf{r}$ [given]

$$\therefore G = \frac{(4\mathbf{p} + \mathbf{q} - 3\mathbf{r}) + (-5\mathbf{p} + \mathbf{q} + 2\mathbf{r}) + (2\mathbf{p} - \mathbf{q} + 2\mathbf{r})}{3}$$

$$= \frac{\mathbf{p} + \mathbf{q} + \mathbf{r}}{3}$$

Since, G divides OC in 2 : 1

$$\therefore \frac{2(\alpha\mathbf{p} + \beta\mathbf{q} + \gamma\mathbf{r}) + \frac{\mathbf{p} + \mathbf{q} + \mathbf{r}}{4}}{(2+1)} = \frac{\mathbf{p} + \mathbf{q} + \mathbf{r}}{3}$$

$\Rightarrow \quad 8(\alpha\mathbf{p} + \beta\mathbf{q} + \gamma\mathbf{r}) = 3(\mathbf{p} + \mathbf{q} + \mathbf{r})$

$$\Rightarrow \quad \left.\begin{matrix} 8\alpha = 3 \\ 8\beta = 3 \\ 8\gamma = 3 \end{matrix}\right\} \Rightarrow \alpha = \beta = \gamma = 3/8$$

$\therefore \alpha + 2\beta + 5\gamma = \frac{3}{8} + \frac{6}{8} + \frac{15}{8} = \frac{24}{8} = 3$

12. *(a)*

Key idea

Coefficient of rth term in $(1 + x)^{2n}$ is ${}^{2n}C_{r-1}$

Given, A = Coefficient of 30th term in the expansion of $(1+x)^{2n-1}$

and B = Coefficient of 12th term in the expansion of $(1+x)^{2n-1}$

$\therefore \quad A = {}^{2n-1}C_{29}$ and $B = {}^{2n-1}C_{11}$

Given, $2A = 5B$

$\Rightarrow \quad 2 \cdot {}^{2n-1}C_{29} = 5\ {}^{2n-1}C_{11}$

$\Rightarrow 2 \cdot \dfrac{(2n-1)!}{29!\,(2n-30)!} = 5 \cdot \dfrac{(2n-1)!}{11!\,(2n-12)!}$

$\Rightarrow \dfrac{1}{29 \times 28 \times 27 \times \ldots \times 12 \times 5}$

$= \dfrac{1}{(2n-12)(2n-13)(2n-14) \ldots (2n-29) \cdot 2}$

$\Rightarrow \dfrac{1}{30 \times 29 \times \ldots \times 12}$

$= \dfrac{1}{(2n-12)(2n-13)\ldots(2n-29) \cdot 12}$

$\Rightarrow \quad 2n - 12 = 30 \Rightarrow n = 21$

13. *(d)*

Key idea

Projection of $\mathbf{a}$ on $\mathbf{b}$ is $\mathbf{a} \cdot \hat{\mathbf{b}} = \dfrac{\mathbf{a} \cdot \mathbf{b}}{|\mathbf{b}|}$

Given, $\mathbf{a} = 3\hat{i} - \hat{j} + 2\hat{k}$

$\mathbf{b} = \mathbf{a} \times (\hat{i} - 2\hat{k})$

$\mathbf{c} = \mathbf{b} \times \hat{k}$

$\because \mathbf{b} = \mathbf{a} \times (\hat{i} - 2\hat{k}) = \begin{vmatrix} \hat{i} & \hat{j} & \hat{k} \\ 3 & -1 & 2 \\ 1 & 0 & -2 \end{vmatrix}$

$= 2\hat{i} + 8\hat{j} + \hat{k}$

and $\mathbf{c} = \mathbf{b} \times \hat{k} = 8\hat{i} - 2\hat{j}$

$\therefore \quad \mathbf{c} - 2\hat{j} = 8\hat{i} - 4\hat{j}$

Hence, projection of $(\mathbf{c} - 2\hat{j})$ on $\mathbf{a} = (\mathbf{c} - 2\hat{j}) \cdot \mathbf{a}$

$= \dfrac{(8\hat{i} - 4\hat{j}) \cdot (3\hat{i} - \hat{j} + 2\hat{k})}{\sqrt{3^2 + (-1)^2 + (2)^2}}$

$= \dfrac{28}{\sqrt{14}} = 2\sqrt{14}$

14. *(c)* Given,

$f(x) = \dfrac{2^{2x} - 1}{2^{2x} + 1} = \dfrac{2^{2x} + 1 - 2}{2^{2x} + 1}$

$= \dfrac{2^{2x} + 1}{2^{2x} + 1} - \dfrac{2}{2^{2x} + 1}$

$\Rightarrow \quad f(x) = 1 - \dfrac{2}{2^{2x} + 1} \quad \ldots \text{(i)}$

On differentiating Eq. (i) w.r.t x, we get

$f'(x) = 0 + \dfrac{2}{(2^{2x} + 1)^2} \times 2^{2x} \cdot \ln 2 \cdot 2$

as $2^{2x} > 0, \forall x \in R$ and $\ln 2 > 0$ always

So, $\quad f'(x) > 0, \forall x \in R$

So, $f(x)$ is strictly increasing function, i.e. $f(x)$ is one-one.

Now, $f(-\infty) = -1$ and $f(\infty) = 1$

$\therefore f(x) \in (-1, 1) \neq$ codomain of f (as $c_f : (-\infty, 1)$)

i.e. $f(x)$ is not onto.

Hence, $f(x)$ is one-one but not onto.

15. *(c)* Given, vertex of the parabola is $V\left(\dfrac{3}{2}, 3\right)$ and

equation of directrix is $x + 2y = 0 \ \ldots \text{(i)}$

So, equation of axis is

$y - 3 = 2\left(x - \dfrac{3}{2}\right)$

[$\because$ axis is $\perp$ to directrix and passes through focus]

$\Rightarrow \quad y - 2x = 0 \quad \ldots \text{(ii)}$

On solving Eqs. (i) and (ii) we, get $x = 0$ and $y = 0$ which is coordinates of foot of directrix, i.e. (0, 0).

and thus focus $\equiv (3, 6)$

Now, $\quad PS^2 = PM^2$

$\Rightarrow (x-3)^2 + (y-6)^2 = \left(\dfrac{x+2y}{\sqrt{5}}\right)^2$

$\Rightarrow 4x^2 + y^2 - 4xy - 30x - 60y + 225 = 0$

Comparing this with

$\alpha x^2 + \beta y^2 - \gamma xy - 30x - 60y + 225 = 0,$

we get $\quad \alpha = 4, \beta = 1$ and $\gamma = 4$

So, $\quad \alpha + \beta + \gamma = 9$

16. *(c)* Given,

$f(x) = 2\log_e(x-2) - x^2 + ax + 1$ is increasing in (2, 3)

$\therefore \quad f'(x) = \dfrac{2}{x-2} - 2x + a \geq 0$

and $\quad f''(x) = \dfrac{-2}{(x-2)^2} - 2 < 0$

then we can say that $f'(x)$ is strictly decreasing.

$\therefore \quad f'(3) \geq 0$

$\Rightarrow \quad 2 - 6 + a \geq 0 \Rightarrow a \geq 4$

$\therefore \quad a_{\min} = 4$

Also given, $g(x) = (x-1)^3\,(x + 2 - a)^2$ is strictly decreasing.

$\Rightarrow \quad g(x) = (x-1)^3\,(x+2-4)^2$

$= (x-1)^3\,(x-2)^2$

$\Rightarrow g'(x) = (x-1)^3\,2\,(x-2) + (x-2)^2 \cdot 3(x-1)^2$

$= (x-1)^2\,(x-2)(2x - 2 + 3x - 6)$

$= (x-1)^2\,(x-2)(5x-8) < 0$ in $x \in \left(\dfrac{8}{5}, 2\right)$

$\therefore \quad b = 8/5$ and $c = 2$.

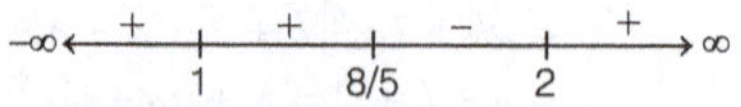

$\therefore 100\,(a + b - c) = 100\,(4 + 8/5 - 2)$

$= 360$

17. *(b)*

Key idea

Sum of first n terms of an AP is $\dfrac{n}{2}[2a + (n-1)d]$

Let a = first term

d = common difference of an AP

Also, $S_{40} = 1030$

$\Rightarrow \quad \dfrac{40}{2}(2a + 39d) = 1030 \quad \ldots \text{(i)}$

and $S_{12} = 57$

$\Rightarrow \quad \dfrac{12}{2}(2a + 11d) = 57 \quad \ldots \text{(ii)}$

On solving Eqs. (i) and (ii), we get

$a = -\dfrac{7}{2}$

and $\quad d = \dfrac{3}{2}$

$\therefore S_{30} - S_{10}$

$= \dfrac{30}{2}(2a + 29d) - \dfrac{10}{2}(2a + 9d)$

$= 20a + 390d$

$= 20 \times (-7/2) + 390\,(3/2) = 515$

18. *(d)* $\cot^{-1}\left(\beta+\frac{1+\beta^2}{\alpha-\beta}\right)$

$$+\cot^{-1}\left(\gamma+\frac{1+\gamma^2}{\beta-\gamma}\right)$$

$$+\cot^{-1}\left(\alpha+\frac{1+\alpha^2}{\gamma-\alpha}\right)$$

$$=\cot^{-1}\left(\frac{\alpha\beta+1}{\alpha-\beta}\right)+\cot^{-1}\left(\frac{\beta\gamma+1}{\beta-\gamma}\right)$$

$$+\cot^{-1}\left(\frac{\alpha\gamma+1}{\gamma-\alpha}\right)$$

$$=\tan^{-1}\left(\frac{\alpha-\beta}{1+\alpha\beta}\right)+\tan^{-1}\left(\frac{\beta-\gamma}{1+\beta\gamma}\right)$$

$$+\tan^{-1}\left(\frac{\gamma-\alpha}{1+\gamma\alpha}\right)$$

$$=\tan^{-1}\alpha-\tan^{-1}\beta$$

$$+\tan^{-1}\beta-\tan^{-1}\gamma$$

$$+(\pi+\tan^{-1}\gamma-\tan^{-1}\alpha)$$

$$[\because \alpha>\beta>\gamma>0 \therefore \gamma-\alpha<0]$$

$$=\pi$$

19. *(d)* $\log_{2/\pi}|\sin x|+\log_{2/\pi}|\cos x|=2$

$$\Rightarrow \log_{2/\pi}|\sin x\cos x|=2$$

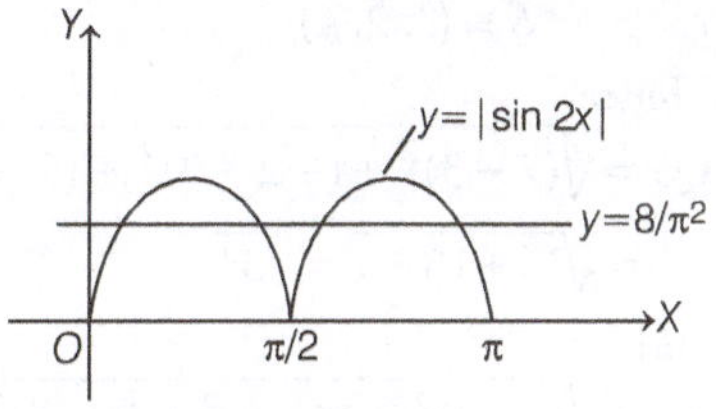

$$\Rightarrow |\sin 2x|=\frac{8}{\pi^2}$$

$$\left[\because \sin x\cos x=\frac{\sin 2x}{2}\right]$$

$\therefore$ Number of solution = 4,

$\therefore n(A)=4$

Now,

$\sqrt{x}(\sqrt{x}-4)-3|\sqrt{x}-2|+6=0$ ...(i)

Case I Let $\sqrt{x}=y<2$

Then, Eq. (i) becomes

$$y(y-4)+3|y-2|+6=0$$

$$\Rightarrow y^2-4y+3y-6+6=0$$

$$\Rightarrow y^2-y=0 \Rightarrow y=0,1$$

$$\Rightarrow \sqrt{x}=0,1 \Rightarrow x=0,1$$

Case II Again, let $\sqrt{x}=y>2$

Then, $y^2-4y-3y+6+6=0$

$$\Rightarrow y^2-7y+12=0$$

$$\Rightarrow y=3,4$$

$$\Rightarrow \sqrt{x}=3,4$$

$$\Rightarrow x=9,16$$

So, total number of solutions = 4

$\therefore n(B)=4$

Hence, $n(A\cup B)=4+4=8$

20. *(c)* Given equation of ellipse,

$$\frac{x^2}{25}+\frac{y^2}{16}=1$$

$\therefore$ Equation of chord with given middle point (3, 1) is $T=S_1$

$$\Rightarrow \frac{x\cdot x_1}{25}+\frac{y\cdot y_1}{16}-1=\frac{x_1^2}{25}+\frac{y_1^2}{16}-1$$

$$\Rightarrow \frac{3x}{25}+\frac{y}{16}=\frac{9}{25}+\frac{1}{16}$$

$$\Rightarrow 48x+25y=144+25$$

$$\Rightarrow 48x+25y=169$$

21. *(16)* Let $I=\int\frac{2x^2+5x+9}{\sqrt{x^2+x+1}}dx$ which

is of the form $\int\frac{ax^2+bx+c}{\sqrt{px^2+qx+r}}dx$

So, $2x^2+5x+9=A(x^2+x+1)$

$$+B\,(x^2+x+1)+C$$

$$\Rightarrow 2x^2+5x+9$$

$$=A(x^2+x+1)+B(2x+1)+C$$

On comparing the coefficient from both sides and then solving, we get

$$A=2,\ B=3/2$$

and $C=11/2$

So,

$$I=\int\frac{2(x^2+x+1)+\frac{3}{2}(2x+1)+\frac{11}{2}}{\sqrt{x^2+x+1}}dx$$

$$=2\int\frac{x^2+x+1}{\sqrt{x^2+x+1}}dx$$

$$+\frac{3}{2}\int\frac{2x+1}{\sqrt{x^2+x+1}}dx$$

$$+\frac{11}{2}\int\frac{1}{\sqrt{x^2+x+1}}dx$$

$$=2\int\sqrt{x^2+x+1}\,dx$$

$$+\frac{3}{2}\times 2\sqrt{x^2+x+1}$$

$$+\frac{11}{2}\int\frac{1}{\sqrt{x^2+x+1}}dx$$

$$=2\int\sqrt{\left(x+\frac{1}{2}\right)^2+\left(\frac{\sqrt{3}}{2}\right)^2}\,dx$$

$$+3\sqrt{x^2+x+1}$$

$$+\frac{11}{2}\int\frac{dx}{\sqrt{\left(x+\frac{1}{2}\right)^2+\left(\frac{\sqrt{3}}{2}\right)^2}}$$

$$=2\left[\frac{x+1/2}{2}\sqrt{x^2+x+1}+\frac{3}{8}\ln\left(x+\frac{1}{2}+\sqrt{x^2+x+1}\right)\right]$$

$$+3\sqrt{x^2+x+1}+\frac{11}{2}\ln\left(x+\frac{1}{2}+\sqrt{x^2+x+1}\right)+C,$$

where C is constant of integration.

$$=x\sqrt{x^2+x+1}+\frac{7}{2}\sqrt{x^2+x+1}$$

$$+\frac{25}{4}\ln\left(x+\frac{1}{2}+\sqrt{x^2+x+1}\right)+C$$

So, $\alpha=\frac{7}{2},\beta=\frac{25}{4}$

Hence, $\alpha+2\beta=16$

> **Concept Enhancer**
>
> To evaluate $\int\frac{px+q}{\sqrt{ax^2+bx+c}}dx$,
>
> Put $px+q=\lambda\frac{d}{dx}(ax^2+bx+c)+\mu$, then find values of λ, μ and then integrate it.

22. *(1)* Given differential equation is

$$2\cos x\frac{dy}{dx}=\sin 2x-4y\sin x,$$

$$x\in\left(0,\frac{\pi}{2}\right)$$

$$\Rightarrow \frac{dy}{dx}=\frac{2\sin x\cos x}{2\cos x}-\frac{4y\sin x}{2\cos x}$$

$$\Rightarrow \frac{dy}{dx}+(2\tan x)\,y=\sin x,$$

which is linear differential equation.

So, $\text{IF} = e^{\int 2\tan x\, dx} = e^{2\log_e \sec x}$

$$= e^{\log_e \sec^2 x} = \sec^2 x$$

Its solution is

$$y \times \sec^2 x = \int \sin x \times \sec^2 x\, dx$$

$$= \int \frac{\sin x}{\cos^2 x} dx = \int \tan x \cdot \sec x\, dx$$

$$= \sec x + C$$

Where, C is constant of integration.

$\because \quad y(\pi/3) = 0 \quad$ [given]

$\therefore \quad 0 \times \sec^2 \pi/3 = \sec \pi/3 + C$

$\Rightarrow \quad C = -2$

Hence, $y = \cos x - 2\cos^2 x$

$\Rightarrow \quad y' = -\sin x + 4\cos x \sin x$

Now, $y(\pi/4) = \frac{1}{\sqrt{2}} - 1$ and

$y'(\pi/4) = -\frac{1}{\sqrt{2}} + 2$

So, $y(\pi/4) + y'(\pi/4)$

$$= \frac{1}{\sqrt{2}} - 1 - \frac{1}{\sqrt{2}} + 2 = 1$$

23. *(392)*

(ways to connect one of {1,2,3,---, 98}to 1) × (99 can have image 1 or 0)

↓ ↓

98 × 2

× (100 can have image 0 or 1)

↓

× 2

= 392

24. *(55)* Given, $H_1 : \frac{x^2}{a^2} - \frac{y^2}{b^2} = 1 \quad$ ...(i)

and $\quad H_2 : \frac{-x^2}{A^2} + \frac{y^2}{B^2} = 1 \quad$ ...(ii)

Now, length of latus-rectum of Eq. (i), we get

$$\frac{2b^2}{a} = 15\sqrt{2} \text{ [given]} \quad \text{...(iii)}$$

and, length of latus-rectum of Eq. (ii)

$$\frac{2A^2}{B} = 12\sqrt{5} \text{ [given]} \quad \text{...(iv)}$$

Also, eccentricity of Eq. (i), we get

$$= e_1 = \sqrt{5/2} \quad \text{[given]}$$

$\Rightarrow \quad \sqrt{1 + \frac{b^2}{a^2}} = \frac{\sqrt{5}}{\sqrt{2}}$

$\Rightarrow \quad \sqrt{1 + \left(\frac{b^2}{a}\right) \times \frac{1}{a}} = \frac{\sqrt{5}}{\sqrt{2}}$

On squaring both sides, we get

$\Rightarrow \quad 1 + \frac{15\sqrt{2}}{2} \times \frac{1}{a} = \frac{5}{2}$

$\Rightarrow \quad \frac{1}{a} = \left(\frac{5}{2} - 1\right) \times \frac{2}{15\sqrt{2}} = \frac{1}{5\sqrt{2}}$

$\therefore \quad a = 5\sqrt{2}$

and hence $b = 5\sqrt{3}$

Also, given product of the lengths of transverse axis of Eqs. (i) and (ii) $= 100\sqrt{10}$

i.e., $\quad (2a)(2B) = 100\sqrt{10}$

$\Rightarrow \quad 2 \times 5\sqrt{2} \times 2 \times B = 100\sqrt{10}$

$\Rightarrow \quad B = 5\sqrt{5}$

$\therefore \quad \frac{2A^2}{5\sqrt{5}} = 12\sqrt{5}$

$\Rightarrow \quad A^2 = 6\sqrt{5} \times 5\sqrt{5} = 150$

$\Rightarrow \quad A = 5\sqrt{6} \quad [\because A > 0]$

Now, $e_2^2 = 1 + \frac{A^2}{B^2} = 1 + \frac{150}{125}$

$$= 1 + \frac{30}{25} = \frac{55}{25}$$

$\Rightarrow \quad 25e_2^2 = 55$

25. *(957)* Given equation of line,

$$\frac{x-1}{2} = \frac{y+1}{3} = \frac{z}{4} = \lambda \text{ (say)} \quad \text{...(i)}$$

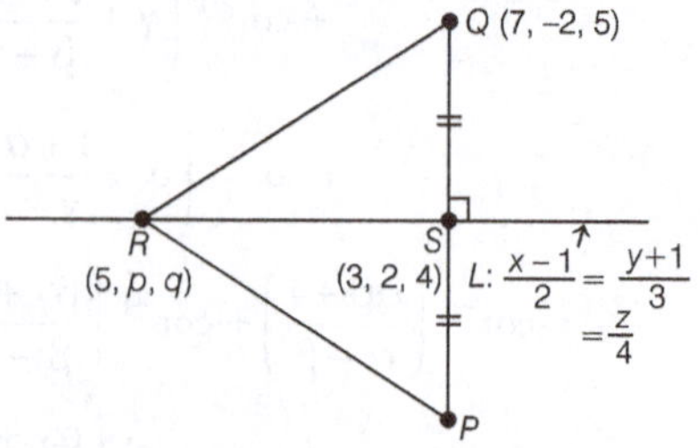

Any point taken on Eq. (i) is

$R \equiv (2\lambda + 1, 3\lambda - 1, 4\lambda) \equiv (5, p, q)$ [given]

$\therefore \quad 2\lambda + 1 = 5$

$\Rightarrow \quad \lambda = 2$

$\therefore R \equiv (5, 5, 8)$ and thus $p = 5$ and $q = 8$

Let $S \equiv (2\lambda + 1, 3\lambda - 1, 4\lambda)$

$\therefore$ **QS**

$$= (2\lambda - 6)\hat{i} + (3\lambda + 1)\hat{j} + (4\lambda - 5)\hat{k}$$

and let $\mathbf{b} = 2\hat{i} + 3\hat{j} + 4\hat{k}$

$\therefore \quad \mathbf{QS} \cdot \mathbf{b} = 0 \quad [\because = \mathbf{QS} \perp \mathbf{RS}]$

$\Rightarrow 4\lambda - 12 + 9\lambda + 3 + 16\lambda - 20 = 0$

$\Rightarrow \quad 29\lambda - 29 = 0 \Rightarrow \lambda = 1$

$\therefore \quad S \equiv (3, 2, 4)$

Hence,

$QS = \sqrt{(7-3)^2 + (-2-2)^2 + (5-4)^2}$

$= \sqrt{16 + 16 + 1} = \sqrt{33}$

and

$RS = \sqrt{(5-3)^2 + (5-2)^2 + (8-4)^2}$

$= \sqrt{4 + 9 + 16} = \sqrt{29}$

Now, (area of $\Delta PQR)^2$

$$= \left(\frac{1}{2} \times PQ \times RS\right)^2$$

$$= \left(\frac{1}{2} \times \sqrt{29} \times 2\sqrt{33}\right)^2 = 957$$

Physics

26. (d) $\mathbf{r} = 5t^2\hat{\mathbf{i}} - 5t\hat{\mathbf{j}}$

at $t = 2\text{s}$

$$\mathbf{v} = \frac{d\mathbf{r}}{dt} = 10t\hat{\mathbf{i}} - 5\hat{\mathbf{j}} = 10 \times 2\hat{\mathbf{i}} - 5\hat{\mathbf{j}} \quad (t = 2\text{s})$$

$$\mathbf{v} = 20\hat{\mathbf{i}} - 5\hat{\mathbf{j}}$$

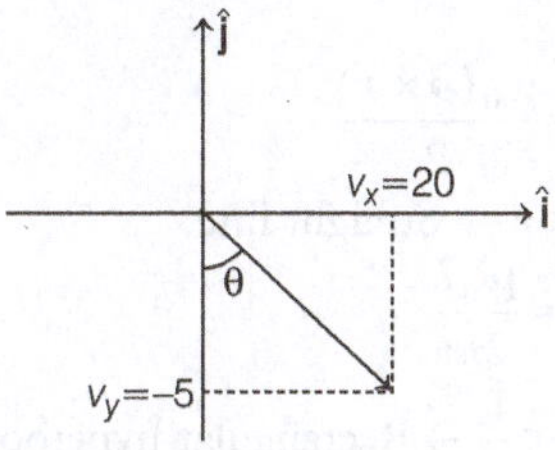

$$|\mathbf{v}| = \sqrt{(20)^2 + (5)^2} = \sqrt{400 + 25}$$
$$= \sqrt{425} = 5\sqrt{17} \text{ m/s}$$

$$\tan\theta = \frac{20}{5} = 4$$
$$\theta = \tan^{-1}(4)$$

From −ve Y-axis.

27. (a) $h\nu = \phi + \text{KE}_{\text{max}}$

$$\text{KE}_{\text{max}} = eV_0$$
$$h\nu = \phi + eV_0$$
$$V_0 = \frac{h\nu}{e} - \frac{\phi}{e}$$

It is in the form of $y = mx + c$

(A) V_0 versus ν is linear.

(B) Slope $= \dfrac{h}{e}$

(C) h = constant

(D) Incorrect

(E) $\phi = eV_0$

(A), (C) and (E) only are correct.

28. (a)

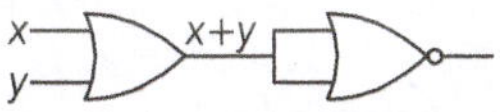

Combination of OR gate and NOT gate = NOR gate

x	y	z
0	0	1
0	1	0
1	0	0
1	1	0

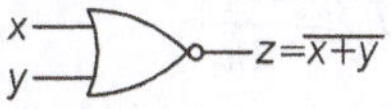

B, C and D are correct.

Concept Enhancer

Any type of gate can be constructed using NAND gate and NOR gate only. Hence, NAND gate and NOR gate are termed as universal gate.

29. (a)

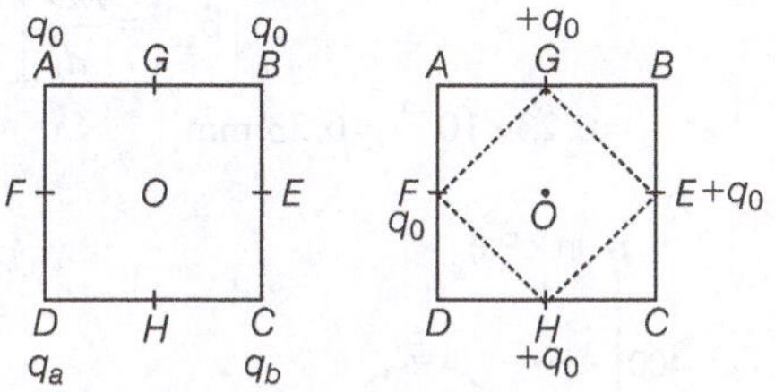

$$U_1 = \frac{4kq_0^2}{a} + \frac{2kq_0^2}{\sqrt{2}a} = \frac{kq_0^2}{a}(4 + \sqrt{2})$$

$$U_2 = \frac{\frac{kq_0^2}{a}(4+\sqrt{2})}{\sqrt{2}} = \frac{kq_0^2}{a}(4\sqrt{2} + 2)$$

$$\Delta U = U_2 - U_1 = \frac{kq_0^2}{a}(3\sqrt{2} - 2)$$

30. (d)

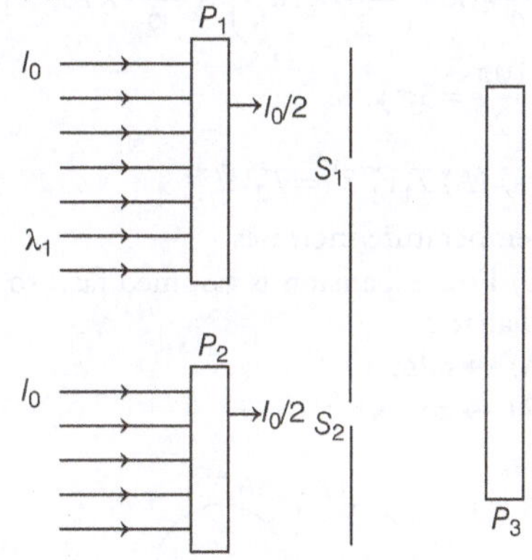

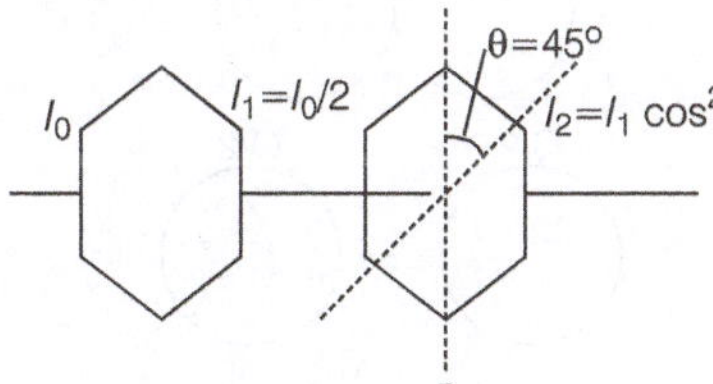

P_3
(Polariser)

$$I_2 = I_1\cos^2\theta = \frac{I_0}{2}\cos^2 45°$$

$$= \frac{I_0}{2}\left(\frac{1}{\sqrt{2}}\right)^2 = \frac{I_0}{4}$$

Now at a point where path difference is $\dfrac{\lambda}{3}$ phase difference

$$\Delta\phi = \frac{2\pi}{\lambda}\Delta x$$

$$\Delta\phi = \frac{2\pi}{\lambda}\cdot\frac{\lambda}{3} = \frac{2\pi}{3}$$

$$I_R = \frac{I_0}{4} + \frac{I_0}{4} + 2\sqrt{\frac{I_0}{4}\times\frac{I_0}{4}}\cos 120°$$

$$= \frac{I_0}{4}$$

31. (a)

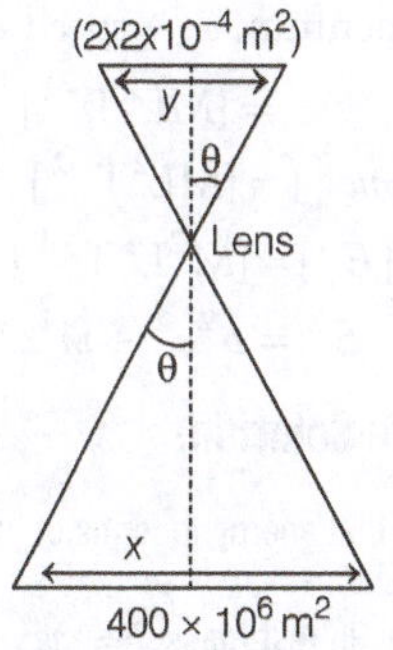

$H = 18$ km

Size of camera film = 2 cm × 2 cm

$$A_{\text{image}} = 400 \text{ km}^2$$
$$x = 20 \times 10^3 \text{ m}$$
$$x = 2 \times 10^4 \text{ m}$$

$$\frac{2\times10^4}{2\times10^{-2}} = \frac{x}{y} = 10^6 = \frac{18 \text{ km}}{f}$$

$$f = \frac{18 \text{ km}}{10^6}$$

$$f = \frac{18\times10^3 \text{ m}}{10^6} = 18\times10^{-3}$$

$$f = 18 \text{ mm} = 1.8 \text{ cm}$$

32. (d) $x(t) = x_0\sin^2\left(\dfrac{t}{2}\right)$, where $x_0 = 1$ m

$$x = x_0\left(\frac{1-\cos t}{2}\right)$$

$$\frac{x}{x_0} = \frac{1-\cos t}{2}$$

$$\frac{x}{x_0} = \frac{1}{2} - \frac{1}{2}\cos t$$

$$\frac{x}{x_0} - \frac{1}{2} = -\frac{1}{2}\cos t$$

where $x_0 = 1$ m

$$\frac{x}{1} - \frac{1}{2} = -\frac{1}{2}\cos t$$

$$x - \frac{1}{2} = -\frac{1}{2}\cos t$$

Particle is oscillating between $x = 0$ to $x = 1$

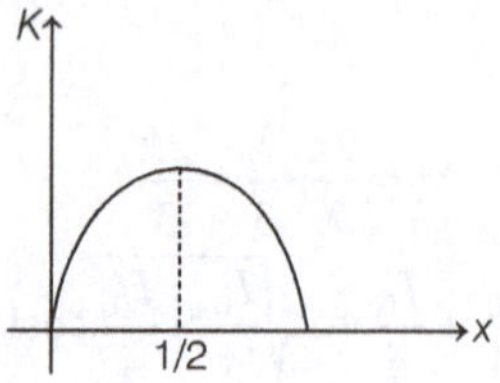

33. *(b)* $[E] = [M^1L^2T^{-2}]$

$[pc] = [MLT^{-1}L^1T^{-1}]$

$p \rightarrow$ momentum, $c \rightarrow$ speed of light

$= [ML^2T^{-2}]$

$[mc^2] = [ML^2T^{-2}]$

$[E^2] = [M^2L^4T^{-4}]$

$E^2 = p^2c^2 + m^2c^4$ is correct

Alternate Solution

We know that energy in terms of momentum in relativistic case, $E = pc$

and energy in rest mass, $E = mc^2$

$\therefore \quad E^2 = p^2c^2 + m^2c^4$

34. *(a)*

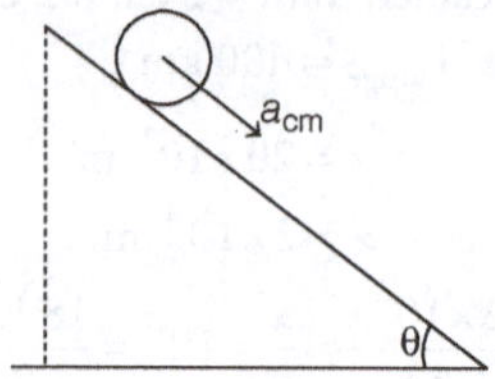

$$t = \sqrt{\frac{2l}{a_{cm}}}$$

$$a_{cm} = \frac{g\sin\theta}{1 + \frac{I_{cm}}{MR^2}}$$

$$a_1 = a_{cm_1} = \frac{5g\sin\theta}{7} \rightarrow \text{Solid} \quad \left[I = \frac{2}{5}MR^2\right]$$

$$a_2 = a_{cm_2} = \frac{3g\sin\theta}{5} \rightarrow \text{Hollow} \left[I = \frac{2}{3}MR^2\right]$$

$a_1 > a_2$

$\therefore \quad t_1 < t_2$

35. *(a)*

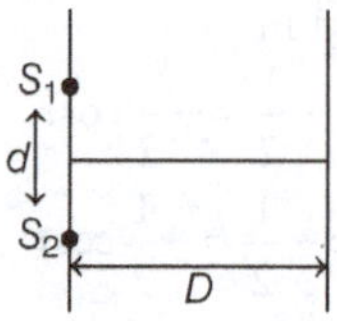

$d = 1.5 \text{ mm} = 1.5 \times 10^{-3}$ m

$\lambda = 690 \text{ nm} = 690 \times 10^{-9}$ m

$D = 0.72$ m

Fringe width, $\beta = ?$

$$= \frac{690 \times 10^{-9} \times 0.72}{1.44 \times 1.5 \times 10^{-3}} \left[\beta_m = \frac{\beta_{air}}{\mu}\right]$$

$$\left[\beta_{air} = \frac{\lambda D}{d}\right]$$

$= 0.23 \times 10^{-3} = 0.23$ mm

36. *(a)*

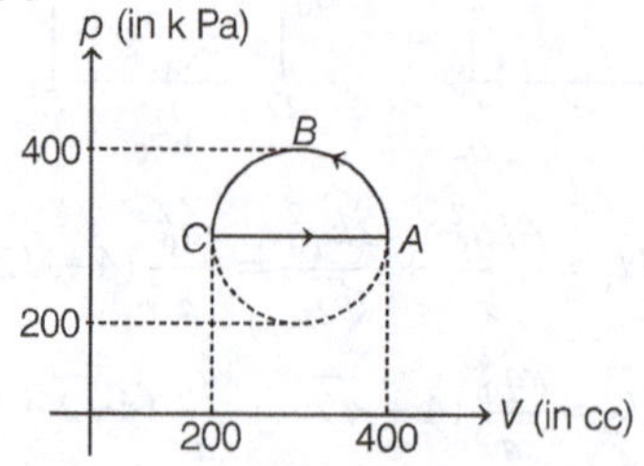

Work done = Area under ABC curve

$$= \frac{\pi R_1 R_2}{2}$$

$$= \frac{1}{2} \times \pi \times \left(\frac{200}{2} \times 10^3\right) \times \left(\frac{200}{2} \times 10^{-6}\right)$$

$$= \frac{10\pi}{2} = 5\pi \text{ J}$$

37. *(a)* (A) $T_1V_1^{\gamma-1} = T_2V_2^{\gamma-1}$

Temperature increases

(B) Free expansion is assumed fast, so adiabatic

(A) $\rightarrow$ false

(R) $\rightarrow$ correct

38. *(a)*

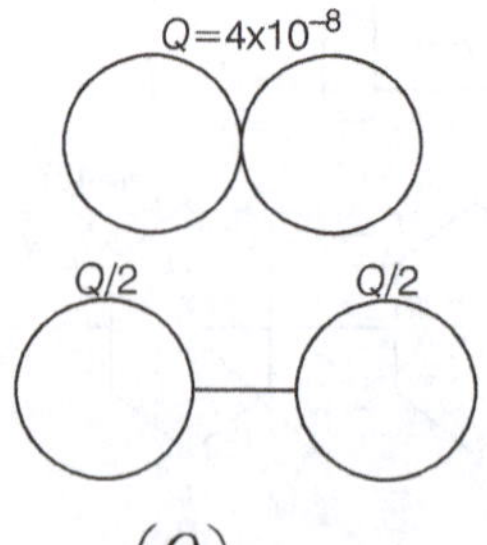

$$F = \frac{k\left(\frac{Q}{2}\right)(Q/2)}{r^2}$$

$$9 \times 10^{-3} = 9 \times 10^9 \times \frac{4 \times 10^{-8} \times 4 \times 10^{-8}}{4 \times r^2}$$

$$r^2 = \frac{9 \times 10^9 \times 16 \times 10^{-16}}{4 \times 9 \times 10^{-3}} = 4 \times 10^{-4}$$

$r = 2 \times 10^{-2}$ m

$\Rightarrow r = 2$ cm

39. *(a)*

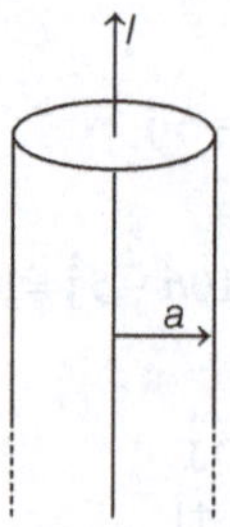

$$B_{in} = \frac{\mu_0(\mathbf{J} \times \mathbf{r})}{2}$$

$B_{in} \propto r \rightarrow$ Straight line

$$B_{out} = \frac{\mu_0 I}{2\pi r}$$

$B_{out} \propto \frac{1}{r} \rightarrow$ Rectangular hyperbola

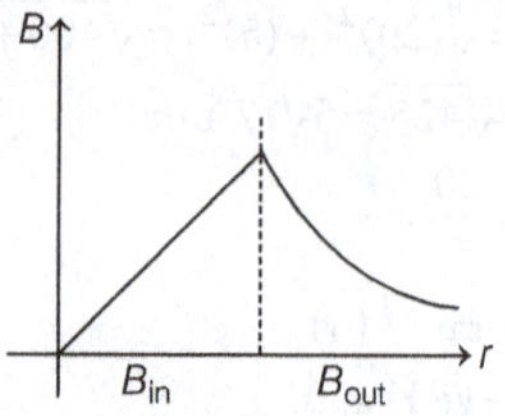

40. *(d)* $\frac{T_2 - T_1}{t} = k[T_{avg} - T_s]$

$T_1 = 24°C, T_2 = 40°C, t = 4$ min,

$T_s = 16°C$ given

$$\frac{40 - 24}{4} = k[32 - 16]$$

$$k = \frac{4}{16} = \frac{1}{4}$$

$$\text{Now, } \frac{24 - T}{4} = k\left[\frac{T + 24}{2} - 16\right]$$

$$\frac{3T}{2} = 28$$

$$T = \frac{56}{3} °C$$

41. *(b)*

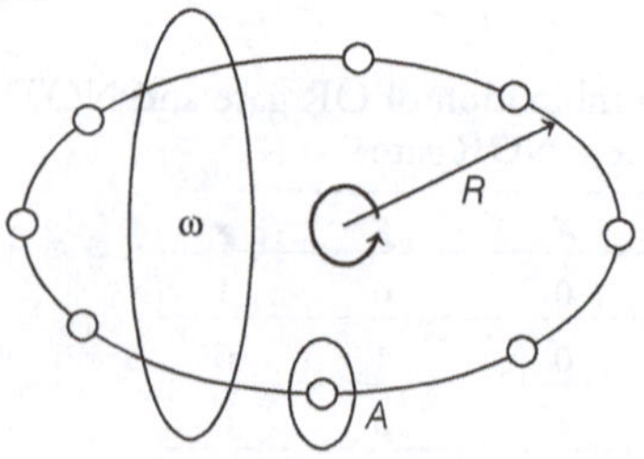

$$I_A = \frac{Nq}{\frac{2\pi}{\omega}}$$

$$I_A = \frac{Nq\omega}{2\pi}$$

$$I_B = 0$$

$$I_A - I_B = \frac{Nq\omega}{2\pi}$$

42. *(b)* $\mathbf{F} = q(\mathbf{v} \times \mathbf{B})$

$$\mathbf{F} = 0$$

$$\mathbf{v} \parallel \mathbf{B}$$

$$\theta = 0° \text{ or } 180°$$

Both A and R are true.

43. *(b)* γ-rays, X-rays, UV-rays, visible rays, IR, micro wave, radio wave

$\lambda \uparrow$

$\lambda_4 < \lambda_2 < \lambda_3 < \lambda_1$

44. *(d)* $$\frac{\text{Linear KE}}{\text{Rotational KE}} = \frac{\frac{1}{2}mv_{cm}^2}{\frac{1}{2}I\omega^2}$$

$$= \frac{\frac{1}{2}mv_{cm}^2}{\frac{1}{2} \times \frac{2}{5}mR^2\left(\frac{v}{R}\right)^2} \quad (v = R\omega)$$

$$= \frac{mv_{cm}^2}{\frac{2}{5}mR^2\frac{v^2}{R^2}}$$

$$= \frac{1}{\frac{2}{5}} = \frac{5}{2}$$

45. *(d)* $$C = \frac{5}{9}(F - 32)$$

$$C = \frac{5F}{9} - \frac{32 \times 5}{9}$$

$$C = \frac{5F}{9} - \frac{160}{9}$$

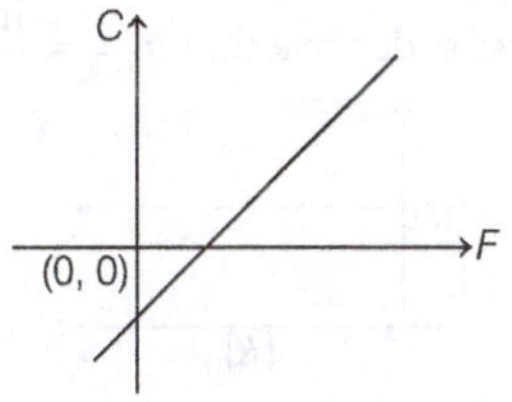

46. *(250)* Since time period of revolving charge is $\frac{2\pi m}{qB}$.

where B = magnetic field

Due to solenoid $B = \mu_0 nI$

$$T = \frac{2\pi m}{q(\mu_0 nI)}$$

$$75 \times 10^{-9} = \frac{2\pi \times 9 \times 10^{-31}}{1.6 \times 10^{-19} \times 4\pi \times 10^{-7} \times n \times 1.5}$$

$$\Rightarrow \quad n = 250$$

47. *(9)* Acceleration due to gravity on surface is given by

$$g = \frac{GM}{R_e^2}$$

Now since diameter is reduced to 1/3rd, radius also reduces to 1/3rd, keeping mass constant.

New value of acceleration due to gravity on earth surface is

$$g' = \frac{GM}{\left(\frac{R_e}{3}\right)^2} = \frac{GM}{\frac{R_e^2}{9}}$$

$$g' = \frac{9\,GM}{R_e^2}$$

$$g' = 9g \quad \left[g = \frac{GM}{R_e^2}\right]$$

48. *(36)*

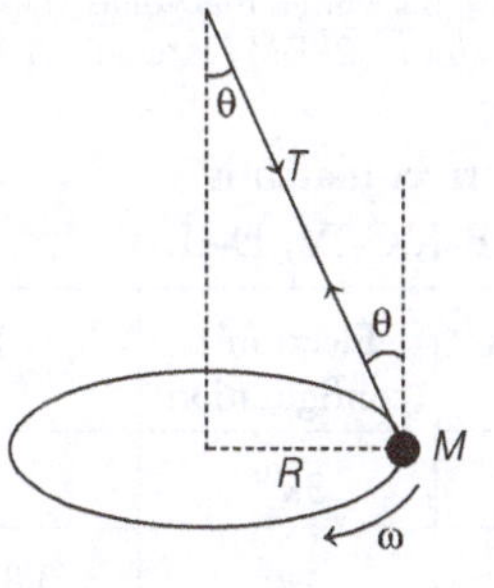

$$T\cos\theta = Mg \quad \ldots\text{(i)}$$

$$T\sin\theta = M\omega^2 R \quad \ldots\text{(ii)}$$

Using Eqs. (i) and (ii)

$$T\sin\theta = M\omega^2(L\sin\theta) \quad \left[\sin\theta = \frac{R}{L}\right]$$

$$T = M\omega^2 L$$

$$= M\left(\frac{3}{\pi} \times 2\pi\right)^2 \times L \quad \left(\omega = 2\pi \times \frac{3}{\pi}\right)$$

$$T = 36\,ML$$

49. *(5)* Since power emitted by source is given as

$$= \frac{\text{Total energy emitted}}{\text{Time}}$$

$$= \frac{(E_1 \text{ photon}) \times \text{Number of photons } (N)}{t}$$

$$P_1 = (E_1)n,\; n = \frac{N}{t} = \frac{\text{Number of photon}}{\text{second}}$$

$$\frac{P_1}{P_2} = \frac{(E_1)(n_1)}{(E_2)(n_2)} = \frac{\left(\frac{hc}{\lambda_1}\right)(n_1)}{\left(\frac{hc}{\lambda_2}\right)(n_2)}$$

$$\frac{P_1}{P_2} = \left(\frac{\lambda_2}{\lambda_1}\right)\left(\frac{n_1}{n_2}\right)$$

Substituting the given values, we get

$$2 = \frac{300}{600} \times \frac{2 \times 10^{15}}{n_2}$$

$$n_2 = \frac{1}{2} \times 10^{15} = 5 \times 10^{14} \text{ photon/second}$$

50. *(43)* Bulk modulus is given as

$$B = \frac{-\Delta p}{\left(\frac{\Delta V}{V}\right)}$$

$$2.15 \times 10^9 = -\frac{\Delta p}{-\left(\frac{0.2}{100}\right)}$$

$$\Delta p = 2.15 \times 10^9 \times 2 \times 10^{-3} = 4.3 \times 10^6$$

$$\Delta p = 43 \times 10^5 \text{ N/m}^2$$

Chemistry

51. *(d)* C : H : O

$$\frac{54.2}{12}:\frac{9.2}{1}:\frac{36.6}{16}$$

$$4.516:9.2:2.287$$

$$\frac{4.516}{2.287}:\frac{9.2}{2.287}:\frac{2.287}{2.287}$$

$$1.97:4.02:1\approx 2:4:1$$

$C_2H_4O \Rightarrow$ empirical formula

Empirical formula mass = 24 + 4 + 16 = 44 g/mol

and molar mass = 132 g/mol

Hence, molecular formula is

$$(C_2H_4O)_3 = C_6H_{12}O_3$$

52. *(a)* For hydrogen atom, the orbital(s) with lowest energy are given in B, C and D.

In hydrogen atom the orbitals in a shell are degenerate, means energy depends only on n.

$$E_{3p_x} = E_{3d_{x^2-y^2}} = E_{3d_{z^2}}$$

53. *(b)* Among the given options the value of E° for Cr^{3+}/Cr is lowest. i.e,

$$E^\circ_{Cr^{3+}/Cr} = -0.74\ V$$

Hence, Cr is the strongest reducing agent.

Time Saver Tech

Look for the option containing lowest value of reduction potential.
For strongest reducing agent, reduction potential should be lowest.

54. *(a)* The heat of formation of $SO_2(g)$.

$$SO_2(g) + \frac{1}{2}O_2(g) \longrightarrow SO_3(g),\quad \Delta H = -y$$

$$\Delta_r H = (\Delta_f H)_{SO_3} - (\Delta_f H)_{SO_2}$$

$$-y = -2x - (\Delta H_f)_{SO_2}$$

$$(\Delta_f H)_{SO_2} = y - 2x \text{ kcal}$$

Hence, option (a) is correct.

55. *(c)* The correct match is,

A-III, B-IV, C-II, D-I.

56. *(c)* The complete reaction is as follows

$$\underset{(A)}{CoS} \xrightarrow{\text{Aqua-regia}} \underset{(B)}{CoCl_2}$$

$$\xrightarrow[\text{(ii) AcOH}]{\text{(i) } KNO_2/NH_4OH} \underset{\text{Yellow ppt.}}{K_3[Co(NO_2)_6]\downarrow}$$

57. *(d)* The ionisation enthalpy for (IE_4) is suddenly very high, therefore element must have 3 valence electrons and it belongs to group 13.

58. *(c)* Statement I is true but II is false.

The correct form of statement II is, Due to resonance, bond lengths of ozone are identical.

59. *(b)* When ethane -1, 2-diamine is added progressively to an aqueous solution of nickel (II) chloride.

- $\underset{\text{(Green)}}{[Ni(H_2O)_6]^{2+}(aq)} + en(aq) \longrightarrow [Ni(H_2O)_4(en)]^{2+}(aq) + 2H_2O$
- $\underset{\text{(Pale blue)}}{[Ni(H_2O)_4 en]^{2+}(aq)} + en(aq) \longrightarrow [Ni(H_2O)_2(en)_2]^{2+}(aq) + 2H_2O$
- $\underset{\text{(Blue/purple)}}{[Ni(H_2O)_2 en]^{+2}(aq)} + en(aq) \longrightarrow [Ni(en)_3]^{2+}(aq) + 2H_2O$

60. *(b)*

Key idea

To solve this question, simply find out the number of unpaired electrons. Then, substitute this value in formula $\mu = \sqrt{n(n+2)}$ BM for required value.

The correct match is

A-III, B-I, C-IV, D-II.

Ions	Electronic configuration	μ_s
Sc^{3+}	$3d^0$	0
V^{2+}	$3d^3$	3.87 BM
Ni^{2+}	$3d^8$	2.84 BM
Ti^{3+}	$3d^1$	1.73 BM

61. *(b)* Statement I is true but statement II is false. The correct form of statement II is,

Order of ionisation enthalpy in kJ/mol

C	>	Si	>	Ge	>	Sn	>	Pb
1086		786		761		708		715

62. *(a)* The hybridisation of each carbon in the given compound is

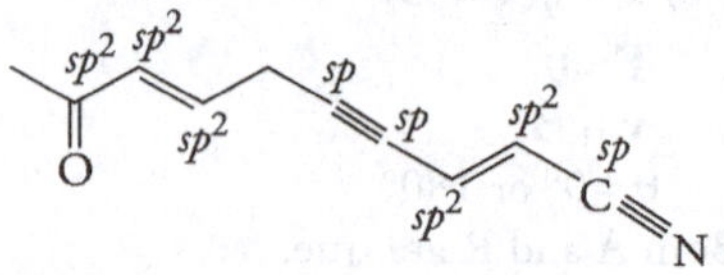

Hence, number of sp and sp^3 hybridised carbon atom are 3 and 5.

63. *(a)* The complete reaction is as follows

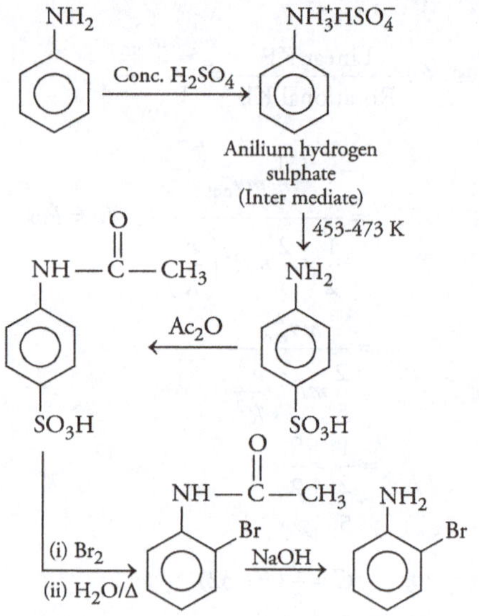

64. *(a)* Statement I is true but statement II is false.

The correct explanation for statement II is :

For first order reaction, $t_{1/2} = \frac{\ln 2}{k}$

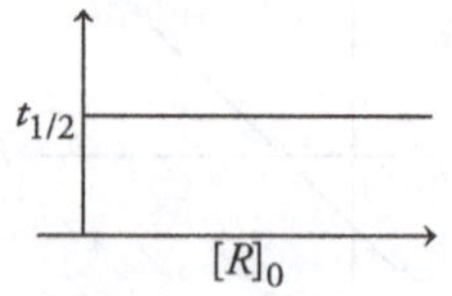

For first order reaction,

$$\log\frac{[R]_0}{[R]} = \frac{1}{2.303}kt$$

$$\log\frac{[R]_0}{[R]} = \left[\frac{k}{2.303}\right]t$$

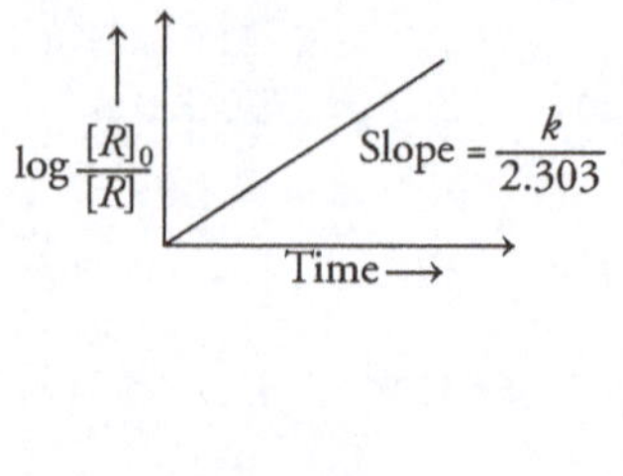

65. *(a)* Complete neutralisation of a strong acid by a strong base leads to the largest increase in temperature. Hence, option (a) is correct.

66. *(d)* For the reaction

$$H_2 + I_2 \rightleftharpoons 2HI$$

Concentration of H_2 and I_2 decrease and that of and concentration of HI increases till equilibrium condition and after equilibrium concentrations of all the reactants and the product remain constant.

67. *(d)* The correct match is

A-IV, B-II, C-I, D-II

(A) $RCN \xrightarrow[\text{(ii) } H_3O^+]{\text{(i) } SnCl_2, HCl} RCHO$

(Stephen reaction)

(B)

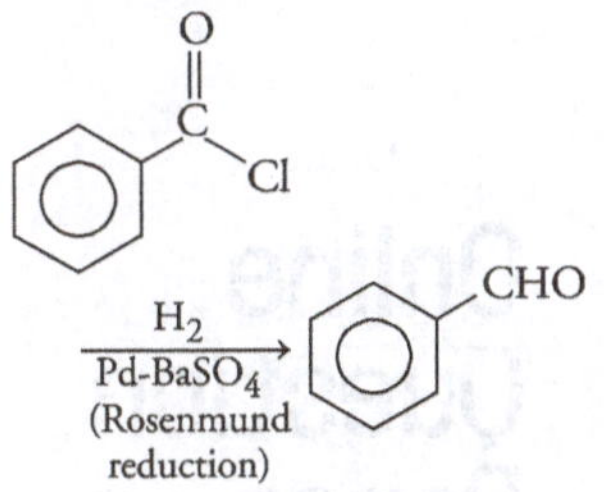

(C)

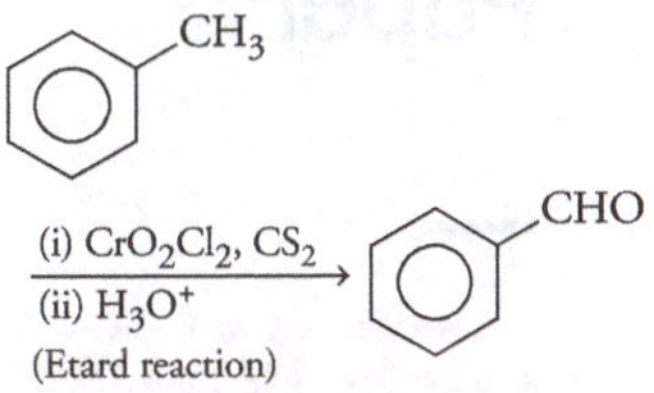

(D)

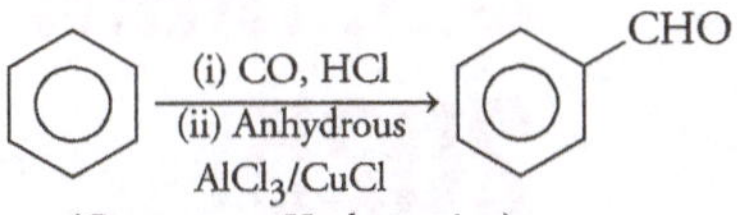

(Gattermann-Koch reaction)

68. *(c)* For $3d^4$,

if ligand is strong field ligand :

$t_{2g}^4 e_g^0$ (low spin)

If ligand is weak field ligand :

$t_{2g}^3 e_g^1$ (high spin)

Concept Enhancer

- Ligands for which $\Delta_o < P$ are called as weak field ligands, they form high spin complexes.
- Ligands for which $\Delta_o > P$ are called as strong filed ligands, they form low spin complexes.

69. *(b)* The complete reaction is as follows

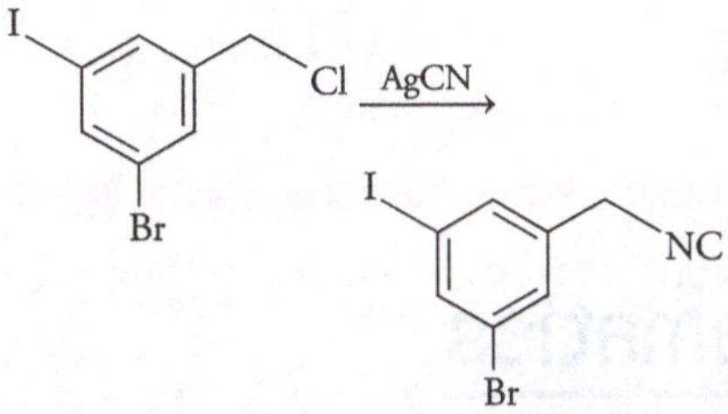

(Major)

70. *(b)* Among the given statements, A, C and D are correct. While B and E are incorrect.

Their correct forms are

- —CN is *meta* directing but —OH is *ortho*/*para* directing.
- Halides are deactivating group.

71. *(255)* Given, mass of organic compound = 0.25 g

Mass of AgBr = 0.15 g

% of bromine = ?

Percentage of bromine

$$= \frac{\text{Molar mass of bromine}}{\text{Molar mass of silver bromide}} \times \frac{\text{Weight of AgBr}}{\text{Weight of sample}} \times 100$$

$$= \frac{80}{188} \times \frac{0.15}{0.25} \times 100$$

$$= 25.5319 = 255.319 \times 10^{-1}$$

$$= 255 \times 10^{-1}$$

72. *(75)* The dissociation takes place as follows

$$MX_2 \longrightarrow M^{2+} + 2X^-$$

$$i = \frac{\text{Normal molar mass}}{\text{Observed molar mass}}$$

$$i = \frac{164}{65.6}$$

$$1 + (3-1)\alpha = \frac{164}{65.6}$$

$$2\alpha = \frac{98.4}{65.6}$$

$$\alpha = 0.75$$

Per cent degree of ionisation = 75%

73. *(4)*

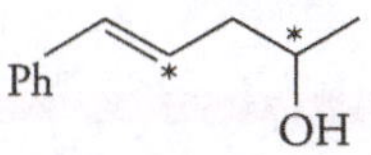

n (stereogenic unit) = 2, $2^2 = 4$

Stereoisomers are possible.

74. *(3)* Mass of carbon $= \frac{80 \times 90}{100} = 72$ g

Number of C-atoms $= \frac{72}{12} = 6$

Mass of hydrogen $= \frac{80 \times 10}{100} = 8$ g

Number of H-atoms $= \frac{8}{1} = 8$

So, molecular formula C_6H_8.

Hence, degree of unsaturation

$$= 6 + 1 - \frac{8}{2} = 7 - 4 = 3$$

75. *(20)* The resultant constant is given by

$$k = \sqrt{\frac{k_1 k_3}{k_2}}$$

$$Ae^{-E_a/RT} = \sqrt{\frac{A_1 e^{-E_{a_1}/RT} \times A_3 e^{-E_{a_3}/RT}}{A_2 e^{-E_{a_2}/RT}}}$$

By comparing exponential term

$$\frac{E_a}{RT} = \frac{1}{2} \times \left[\frac{E_{a_1}}{RT} + \frac{E_{a_3}}{RT} - \frac{E_{a_2}}{RT}\right]$$

$$E_a = \frac{(E_{a_1} + E_{a_3} - E_{a_2})}{2}$$

$$= \frac{(60 + 10 - 30)}{2} = 20 \text{ kJ/mol}$$

Time 9:00-12:00 **MM** : 300 Marks

28 January
Shift-I

Online Question Paper

JEE Main 2025

Mathematics

Section A
Objective Type Questions

01/ Let $< a_n >$ be a sequence such that $a_0 = 0, a_1 = \frac{1}{2}$ and $2a_{n+2} = 5a_{n+1} - 3a_n, n = 0, 1, 2, 3 \ldots\ldots$

Then, $\sum_{k=1}^{100} a_k$ is equal to

(a) $3a_{99} + 100$ (b) $3a_{100} - 100$
(c) $3a_{100} + 100$ (d) $3a_{99} - 100$

02/ The relation $R = \{(x, y) : x, y \in Z$ and $x + y$ is even$\}$

(a) symmetric and transitive but not reflexive
(b) reflexive and transitive but not symmetric
(c) an equivalence relation
(d) reflexive and symmetric but not transitive

03/ If $f(x) = \frac{2^x}{2^x + \sqrt{2}}, x \in R$, then $\sum_{k=1}^{81} f\left(\frac{k}{82}\right)$ is equal to

(a) 82 (b) $\frac{81}{2}$
(c) $81\sqrt{2}$ (d) 41

04/ Let the equation of the circle, which touches X-axis at the point $(a, 0)$, $a > 0$ and cuts off an intercept of length b on Y-axis be $x^2 + y^2 - \alpha x + \beta y + \gamma = 0$. If the circle lies below X-axis, then the ordered pair $(2a, b^2)$ is equal to

(a) $(\alpha, \beta^2 - 4\gamma)$ (b) $(\gamma, \beta^2 + 4\alpha)$
(c) $(\gamma, \beta^2 - 4\alpha)$ (d) $(\alpha, \beta^2 + 4\gamma)$

05/ The number of different 5-digit numbers greater than 50000 that can be formed using the digits 0, 1, 2, 3, 4, 5, 6, 7, such that the sum of their first and last digits should not be more than 8, is

(a) 4607 (b) 5720
(c) 5719 (d) 4608

06/ Let ${}^nC_{r-1} = 28, {}^nC_r = 56$ and ${}^nC_{r+1} = 70$. Let $A(4\cos t, 4\sin t)$, $B(2\sin t, -2\cos t)$ and $C(3r - n, r^2 - n - 1)$ be the vertices of a ΔABC, where t is a parameter. If $(3x - 1)^2 + (3y)^2 = \alpha$ is the locus of the centroid of ΔABC, then α equals

(a) 18 (b) 8 (c) 6 (d) 20

07/ If $\int_{-\frac{\pi}{2}}^{\frac{\pi}{2}} \frac{96x^2 \cos^2 x}{(1 + e^x)} dx = \pi(\alpha\pi^2 + \beta)$, $\alpha, \beta \in Z$, then $(\alpha + \beta)^2$ equals

(a) 144 (b) 196
(c) 100 (d) 64

08/ Let $ABCD$ be a trapezium whose vertices lie on the parabola $y^2 = 4x$. Let the sides AD and BC of the trapezium be parallel to Y-axis. If the diagonal AC is of length $\frac{25}{4}$ and it passes through the point (1, 0), then the area of $ABCD$ is

(a) $\frac{25}{2}$ (b) $\frac{125}{8}$ (c) $\frac{75}{4}$ (d) $\frac{75}{8}$

09/ Let T_r be the rth term of an AP. If for some m, $T_m = \frac{1}{25} \cdot T_{25} = \frac{1}{20}$ and $20\sum_{r=1}^{25} T_r = 13$, then $5m\sum_{r=m}^{2m} T_r$ is equal to

(a) 112 (b) 126 (c) 98 (d) 142

Imp.

10/ Let O be the origin, the point A be $z_1 = \sqrt{3} + 2\sqrt{2}i$, the point $B(z_2)$ be such that $\sqrt{3}|z_2| = |z_1|$ and $\arg(z_2) = \arg(z_1) + \frac{\pi}{6}$. Then,

(a) area of ΔABO is $\frac{11}{4}$

(b) ABO is a scalene triangle

(c) area of ΔABO is $\frac{11}{\sqrt{3}}$

(d) ABO is an obtuse angled isosceles triangle

Argument of Complex Number

11/ The sum of the squares of all the roots of the equation $x^2 + |2x - 3| - 4 = 0$, is

(a) $6(3-\sqrt{2})$ (b) $3(2-\sqrt{2})$ (c) $3(3-\sqrt{2})$ (d) $6(2-\sqrt{2})$

12/ Three defective oranges are accidently mixed with seven good ones and on looking at them, it is not possible to differentiate between them. Two oranges are drawn at random from the lot. If x denote the number of defective oranges, then the variance of x is

(a) $\frac{14}{25}$ (b) $\frac{18}{25}$

(c) $\frac{26}{75}$ (d) $\frac{28}{75}$

13/ If the image of the point (4, 4, 3) in the line $\frac{x-1}{2} = \frac{y-2}{1} = \frac{z-1}{3}$ is (α, β, γ), then $\alpha + \beta + \gamma$ is equal to

(a) 12 (b) 8

(c) 9 (d) 7

14/ The sum of all local minimum values of the function

$$f(x) = \begin{cases} 1-2x, & x < -1 \\ \frac{1}{3}(7+2|x|), & -1 \le x \le 2 \\ \frac{11}{18}(x-4)(x-5), & x > 2 \end{cases}$$ is

(a) $\frac{131}{72}$ (b) $\frac{167}{72}$ (c) $\frac{157}{72}$ (d) $\frac{171}{72}$

15/ $\cos\left(\sin^{-1}\frac{3}{5} + \sin^{-1}\frac{5}{13} + \sin^{-1}\frac{33}{65}\right)$ is equal to

(a) 1 (b) $\frac{32}{65}$ (c) 0 (d) $\frac{33}{65}$

16/ Two numbers k_1 and k_2 are randomly chosen from the set of natural numbers. Then, the probability that the value of $i^{k_1} + i^{k_2}$, $(i = \sqrt{-1})$ is non-zero equals

(a) $\frac{3}{4}$ (b) $\frac{1}{4}$ (c) $\frac{2}{3}$ (d) $\frac{1}{2}$

17/ The area (in sq. units) of the region $\{(x, y): 0 \le y \le 2|x| + 1, 0 \le y \le x^2 + 1, |x| \le 3\}$ is

(a) $\frac{80}{3}$ (b) $\frac{64}{3}$ (c) $\frac{17}{3}$ (d) $\frac{32}{3}$

18/ Let for some function $y = f(x)$, $\int_0^x t f(t)\,dt = x^2 f(x)$, $x > 0$ and $f(2) = 3$. Then, $f(6)$ is equal to

(a) 6 (b) 3 (c) 2 (d) 1

19/ Let $f: R \to R$ be a function defined by

$$f(x) = (2+3a)x^2 + \left(\frac{a+2}{a-1}\right)x + b, a \ne 1.$$

If $f(x+y) = f(x) + f(y) + 1 - \frac{2}{7}xy$, then the value of $28\sum_{i=1}^{5} |f(i)|$ is

(a) 735 (b) 675

(c) 545 (d) 715

20/ Let $A(x, y, z)$ be a point in XY plane, which is equidistant from three points (0, 3, 2), (2, 0, 3) and (0, 0, 1). Let $B = (1, 4, -1)$ and $C = (2, 0, -2)$. Then, among the statements $(S_1): \Delta ABC$ is an isosceles right angled triangle and (S_2) : the area of ΔABC is $\frac{9\sqrt{2}}{2}$,

(a) only $(S2)$ is true (b) both are false

(c) only $(S1)$ is true (d) both are true

Section B
Numerical Type Questions

21/ If $\alpha = 1 + \sum_{r=1}^{6} (-3)^{r-1}\ {}^{12}C_{2r-1}$, then the distance of the point $(12, \sqrt{3})$ from the line $\alpha x - \sqrt{3}y + 1 = 0$ is ……

22/ Let

$$f(x) = \begin{cases} 3x, & x < 0 \\ \min\{1 + x + [x], x + 2[x]\}, & 0 \le x \le 2, \text{ where} \\ 5, & x > 2 \end{cases}$$

[.] denotes greatest integer function. If α and β are the number of points, where f is not continuous and is not differentiable, respectively, then $\alpha + \beta$ equals ……

23/ Let $E_1 : \frac{x^2}{9} + \frac{y^2}{4} = 1$ be an ellipse. Ellipses E_1s are constructed such that their centres and eccentricities are same as that of E_1 and the length of minor axis of E_1 is the length of major axis of $E_{i+1} (i \ge 1)$. If A_i is the area of the ellipse E_i, then $\frac{5}{\pi}\left(\sum_{i=1}^{\infty} A_i\right)$ is equal to ……

24/ Let $\mathbf{a} = \hat{i} + \hat{j} + \hat{k}$, $\mathbf{b} = 2\hat{i} + 2\hat{j} + \hat{k}$ and $\mathbf{d} = \mathbf{a} \times \mathbf{b}$. If $\mathbf{c}$ is a vector such that $\mathbf{a} \cdot \mathbf{c} = |\mathbf{c}|, |\mathbf{c} - 2\mathbf{a}|^2 = 8$ and the angle between $\mathbf{d}$ and $\mathbf{c}$ is $\frac{\pi}{4}$, then $|10 - 3\mathbf{b} \cdot \mathbf{c}| + |\mathbf{d} \times \mathbf{c}|^2$ is equal to ………

Imp.

25/ Let M denote the set of all real matrices of order 3×3 and let $S = \{-3, -2, -1, 1, 2\}$. Let

$S_1 = \{A = [a_{ij}] \in M : A = A^T \text{ and } a_{ij} \in S, \forall i, j\}$,

$S_2 = \{A = [a_{ij}] \in M : A = -A^T \text{ and } a_{ij} \in S, \forall i, j\}$,

$S_3 = \{A = [a_{ij}] \in M : a_{11} + a_{22} + a_{33} = 0 \text{ and } a_{ij} \in S, \forall i, j\}$.

If $n(S_1 \cup S_2 \cup S_3) = 125\alpha$, then α equals ……

Symmetric and Skew-Symmetric Matrix

Physics

Section A
Objective Type Questions

26/ Three infinitely long wires with linear charge density λ are placed along the X-axis, Y-axis and Z-axis, respectively. Which of the following denotes an equipotential surface ?

(a) xyz = constant
(b) $(x + y)(y + z)(z + x)$ = constant
(c) $xy + yz + zx$ = constant
(d) $(x^2 + y^2)(y^2 + z^2)(z^2 + x^2)$ = constant

27/ A wire of resistance R is bent into an equilateral triangle and an identical wire is bent into a square. The ratio of resistance between the two end points of an edge of the triangle to that of the square is

(a) $\frac{27}{32}$ (b) $\frac{9}{8}$ (c) $\frac{8}{9}$ (d) $\frac{32}{27}$

28/ A hemispherical vessel is completely filled with a liquid of refractive index μ. A small coin is kept at the lowest point O of the vessel as shown in figure. The minimum value of the refractive index of the liquid, so that a person can see the coin from point E (at the level of the vessel) is ………

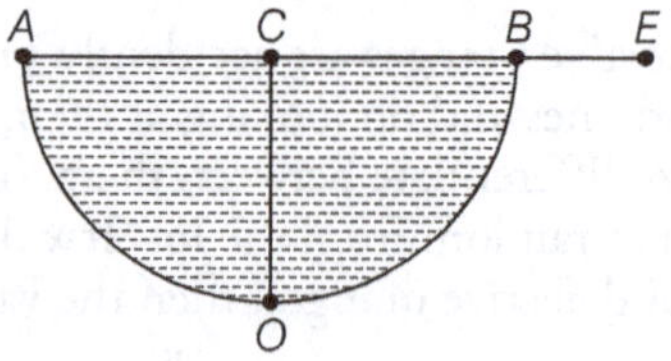

(a) $\frac{\sqrt{3}}{2}$ (b) $\frac{3}{2}$
(c) $\sqrt{3}$ (d) $\sqrt{2}$

29/ For a particular ideal gas which of the following graphs represents the variation of mean squared velocity of the gas molecules with temperature?

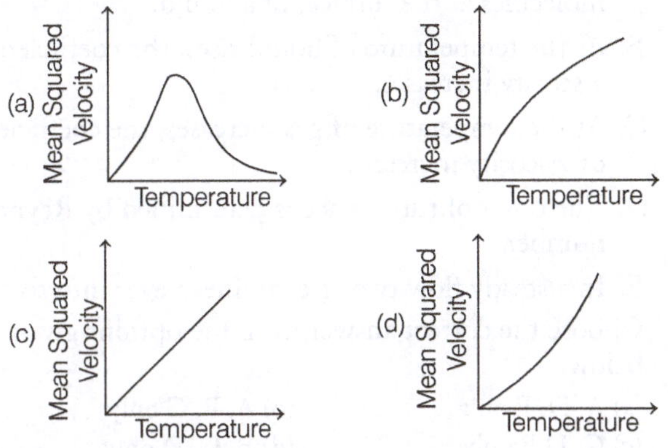

30/ Which of the following circuits has the same output as that of the given circuit?

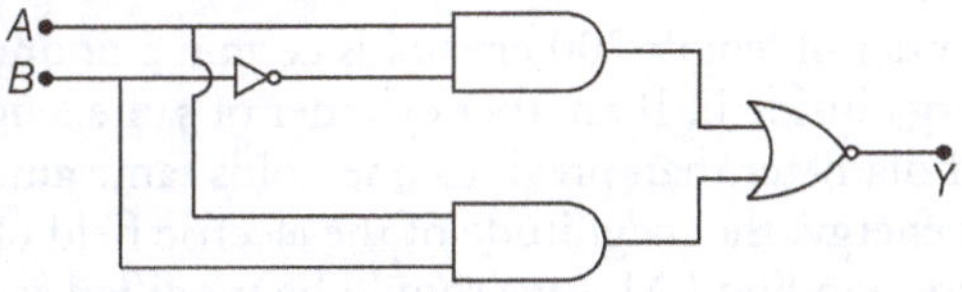

(a) 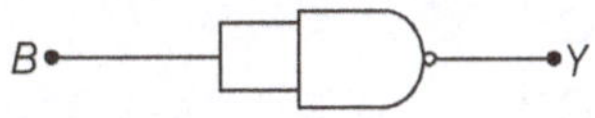

(b) 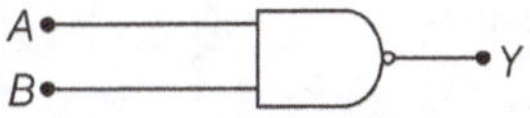

(c) 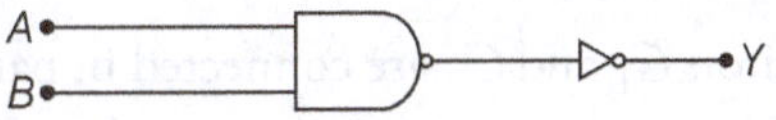

(d) 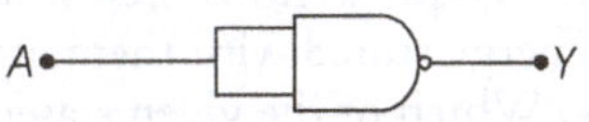

31/ In the experiment for measurement of viscosity η of given liquid with a ball having radius R, consider following statements.

A. Graph between terminal velocity v and R will be a parabola.

B. The terminal velocities of different diameter balls are constant for a given liquid.

C. Measurement of terminal velocity is dependent on the temperature.

D. This experiment can be utilised to assess the density of a given liquid.

E. If balls are dropped with some initial speed, the value of η will change.

Choose the correct answer from the options given below.

(a) C, D and E only (b) B, D and E only
(c) A, C and D only (d) A, B and E only

32/ The centre of mass of a thin rectangular plate with sides of length a and b, whose mass per unit area (σ) varies as $\sigma = \frac{\sigma_0 x}{ab}$ (where σ_0 is a constant), would be ......

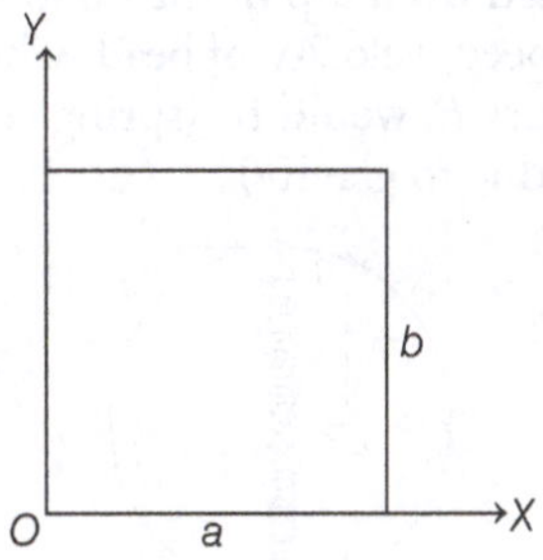

(a) $\left(\frac{1}{3}a, \frac{b}{2}\right)$ (b) $\left(\frac{2}{3}a, \frac{b}{2}\right)$
(c) $\left(\frac{a}{2}, \frac{b}{2}\right)$ (d) $\left(\frac{2}{3}a, \frac{2}{3}b\right)$

33/ A particle of mass m and charge q is fastened to one end A of a massless string having equilibrium length l, whose other end is fixed at point O. The whole system is placed on a frictionless horizontal plane and is initially at rest. If uniform electric field is switched on along the direction as shown in figure, then the speed of the particle when it crosses the X-axis is

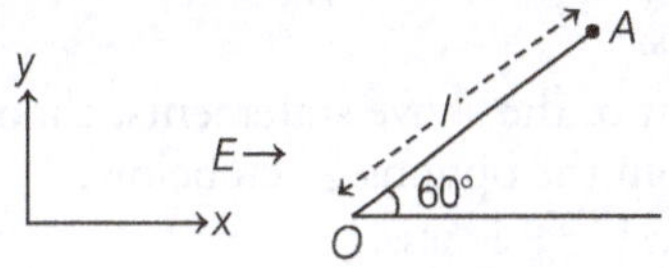

(a) $\sqrt{\frac{qEl}{m}}$ (b) $\sqrt{\frac{qEl}{2m}}$ (c) $\sqrt{\frac{2qEl}{m}}$ (d) $\sqrt{\frac{qEl}{4m}}$

Imp.

34/ Find the equivalent resistance between two ends of the following circuit

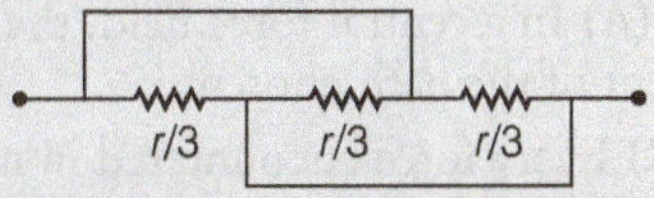

(a) $\frac{r}{9}$ (b) $\frac{r}{6}$ (c) r (d) $\frac{r}{3}$

Combination of Resistor

35/ Choose the correct nuclear process from the below options [p: proton, n: neutron, e^-: electron, e^+: positron, ν: neutrino, $\bar{\nu}$: anti-neutrino]

(a) $n \rightarrow p + e^- + \nu$ (b) $n \rightarrow p + e^+ + \bar{\nu}$
(c) $n \rightarrow p + e^+ + \nu$ (d) $n \rightarrow p + e^- + \bar{\nu}$

36/ A bead of mass m slides without friction on the wall of a vertical circular hoop of radius R as shown in figure. The bead moves under the combined action of gravity and a massless spring (k) attached to the bottom of the hoop. The equilibrium length of the spring is R. If the bead is released from top of the hoop with (negligible) zero initial speed, velocity of bead, when the length of spring becomes R, would be (spring constant is k, g is acceleration due to gravity)

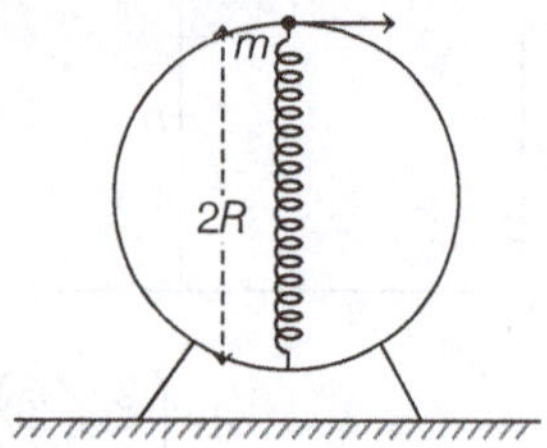

(a) $\sqrt{3Rg + \frac{kR^2}{m}}$ (b) $\sqrt{2Rg + \frac{4kR^2}{m}}$

(c) $2\sqrt{gR + \frac{kR^2}{m}}$ (d) $\sqrt{2Rg + \frac{kR^2}{m}}$

37/ Given below are two statements. One is labelled as Assertion (A) and the other is labelled as Reason (R).

Assertion (A) A sound wave has higher speed in solids than gases.

Reason (R) Gases have higher value of Bulk modulus than solids.

In the light of the above statements, choose the correct answer from the options given below.

(a) A is true but R is false.
(b) Both A and R are true and R is the correct explanation of A.
(c) A is false but R is true.
(d) Both A and R are true but R is not the correct explanation of A.

38/ Given below are two statements. One is labelled as Assertion (A) and the other is labelled as Reason (R).

Assertion (A) In a central force field, the work done is independent of the path chosen.

Reason (R) Every force encountered in mechanics does not have an associated potential energy.

In the light of the above statements, choose the most appropriate answer from the option given below.

(a) A is true but R is false.
(b) Both A and R are true and R is the correct explanation of A.
(c) A is false but R is true.
(d) Both A and R are true but R is not the correct explanation of A.

39/ Consider following statements.

A. Surface tension arises due to extra energy of the molecules at the interior as compared to the molecules at the surface, of a liquid.

B. As the temperature of liquid rises, the coefficient of viscosity increases.

C. As the temperature of gas increases, the coefficient of viscosity increases.

D. The onset of turbulence is determined by Reynold's number.

E. In a steady flow two stream lines never intersect.

Choose the correct answer from the options given below.

(a) A, D, E only (b) A, B, C only
(c) C, D E only (d) B, C, D only

40/ Due to presence of an EM wave whose electric component is given by $E = 100\sin(\omega t - kx)\text{NC}^{-1}$ a cylinder of length 200 cm holds certain amount of em energy inside it. If another cylinder of same length but half diameter than previous one holds same amount of em energy, the magnitude of the electric field of the corresponding EM wave should be modified as

(a) $25\sin(\omega t - kx)\text{NC}^{-1}$
(b) $50\sin(\omega t - kx)\text{NC}^{-1}$
(c) $200\sin(\omega t - kx)\text{NC}^{-1}$
(d) $400\sin(\omega t - kx)\text{NC}^{-1}$

41/ Two capacitors C_1 and C_2 are connected in parallel to a battery. Charge-time graph is shown below for the two capacitor. The energy stored with them are U_1 and U_2, respectively. Which of the given statements is true?

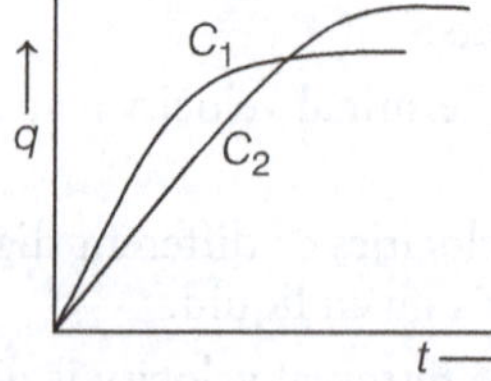

(a) $C_2 > C_1, U_2 < U_1$ (b) $C_1 > C_2, U_1 < U_2$
(c) $C_1 > C_2, U_1 > U_2$ (d) $C_2 > C_1, U_2 > U_1$

42/ Consider a long thin conducting wire carrying a uniform current I. A particle having mass M and charge q is released as a distance a from the wire with a speed v_0 along the direction of current in the wire. The particle gets attracted to the wire due to magnetic force. The particle turns round when it is at distance x from the wire. The value of x is

[μ_0 is vacuum permeability]

(a) $a\left[1-\frac{mv_0}{q\mu_0 I}\right]$ (b) $\frac{a}{2}$

(c) $ae^{-\frac{4\pi mv_0}{q\mu_0 I}}$ (d) $a\left[1-\frac{mv_0}{2q\mu_0 I}\right]$

43/ A proton of mass m_p has same energy as that of a photon of wavelength λ. If the proton is moving at non-relativistic speed, then ratio of its de-Broglie wavelength to the wavelength of photon is

(a) $\frac{1}{c}\sqrt{\frac{E}{2m_p}}$ (b) $\frac{1}{2c}\sqrt{\frac{E}{m_p}}$ (c) $\frac{1}{c}\sqrt{\frac{E}{m_p}}$ (d) $\frac{1}{c}\sqrt{\frac{2E}{m_p}}$

44/ A Carnot engine E is working between two temperatures 473 K and 273 K. In a new system, two engines-engine E_1 works between 473 K to 373 K and engine E_2 works between 373 K to 273 K. If $\eta_{12} \cdot \eta_1$ and η_2 are the efficiencies of the engines E, E_1 and E_2, respectively, then

(a) $\eta_{12} = \eta_1\eta_2$

(b) $\eta_{12} < \eta_1 + \eta_2$

(c) $\eta_{12} \geq \eta_1 + \eta_2$

(d) $\eta_{12} = \eta_1 + \eta_2$

45/ A thin prism P_1 with angle 4° made of glass having refractive index 1.54, is combined with another thin prism P_2 made of glass having refractive index 1.72 to get dispersion without deviation. The angle of the prism P_2 in degrees is

(a) 3 (b) 16/3

(c) 4 (d) 1.5

Section B

Numerical Type Questions

46/ Two iron solid discs of negligible thickness have radii R_1 and R_2 and moment of inertia I_1 and I_2, respectively. For $R_2 = 2R_1$, the ratio of I_1 and I_2 would be $\frac{1}{x}$. Where, $x = \ldots\ldots\ldots$

47/ In a measurement, it is asked to find modulus of elasticity per unit torque applied on the system. The measured quantity has dimension of $[M^a L^b T^c]$. If $b = 3$, the value of c is ………

48/ A tiny metallic rectangular sheet has length and breadth of 5 mm and 2.5 mm, respectively. Using a specially designed screw gauge which has pitch of 0.75 mm and 15 divisions in the circular scale, you are asked to find the area of the sheet. In this measurement, the maximum fractional error will be $\frac{x}{100}$, where x is …… .

49/ A double slit interference experiment performed with a light of wavelength 600 nm forms an interference fringe pattern on a screen with 10th bright fringe having its centre at a distance of 10 mm from the central maximum. Distance of the centre of the same 10th bright fringe from the central maximum when the source of light is replaced by another source of wavelength 660 nm would be ……… mm.

Imp.

50/ The moment of inertia of a solid disc rotating along its diameter is 2.5 times higher than the moment of inertia of a ring rotating in similar way. The moment of inertia of a solid sphere which has same radius as the disc and rotating in similar way, is n times higher than the moment of inertia of the given ring. Here, $n = \ldots\ldots$.

Consider all the bodies have equal masses.

Moment of Inertia

Chemistry

Section A

Objective Type Questions

51/ Consider the following elements In, Tl, Al, Pb, Sn and Ge. The most stable oxidation states of elements with highest and lowest first ionisation enthalpies, respectively, are

(a) +4 and +3 (b) +4 and +1 (c) +1 and +4 (d) +2 and +3

52/ The molecules having square pyramidal geometry are

(a) SbF_5 and PCl_5 (b) BrF_5 and PCl_5

(c) SbF_5 and $XeOF_4$ (d) BrF_5 and $XeOF_4$

53/

$[A]_0$ / mol L^{-1}	$t_{1/2}$ / min
0.100	200
0.025	100

For a given reaction $R \rightarrow P$, $t_{1/2}$ is related to $[A]_0$ as given in table.

Given, $\log 2 = 0.30$

Which of the following is true?

A. The order of the reaction is 1/2

B. If $[A]_0$ is 1M, then $t_{\frac{1}{2}}$ is $200\sqrt{10}$ min.

C. The order of the reaction changes to 1 if the concentration of reactant changes from 0.100 M to 0.500 M.

D. $t_{1/2}$ is 800 min for $[A]_0 = 1.6$ M

Choose the correct answer from the options given below.

(a) C and D only (b) A and C only
(c) A, B and D only (d) A and B only

54/ Given below are two statements.

Statement I D-glucose penta acetate reacts with 2, 4-dinitrophenylhydrazine.

Statement II Starch, on heating with concentrated sulphuric acid at 100°C and 2-3 atmosphere pressure produces glucose.

In the light of the above statements, choose the correct answer from the options given below

(a) Statement I is true but Statement II is false.
(b) Statement I is false but Statement II is true.
(c) Both Statement I and Statement II are false.
(d) Both Statement I and Statement II are true.

55/ The incorrect decreasing order of atomic radii is

(a) Be > Mg > Al > Si (b) Si > P > Cl > F
(c) Mg > Al > C > O (d) Al > B > N > F

56/ Consider '*n*' is the number of lone pair of electrons present in the equatorial position of the most stable structure of ClF_3. The ions from the following with '*n*' number of unpaired electrons are

A. V^{3+} B. Ti^{3+} C. Cu^{2+} D. Ni^{2+}
E. Ti^{2+}

Choose the correct answer from the options given below

(a) B and C only (b) A, D and E only
(c) B and D only (d) A and C only

57/ A molecule ("*P*") on treatment with acid undergoes rearrangement and gives ("*Q*"). ("*Q*") on ozonolysis followed by reflux under alkaline condition gives ("*R*"). The structure of ("*R*") is given below.

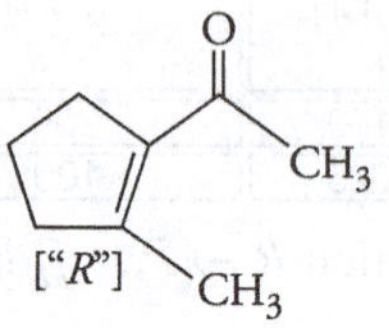

["*R*"]

The structure of ("*P*") is

(a) 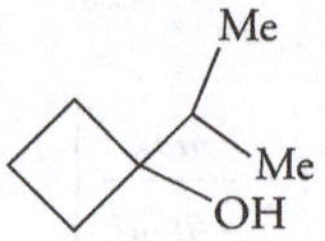 (b) 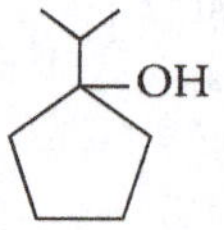

(c) 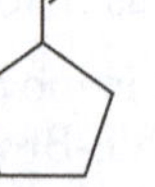 (d) 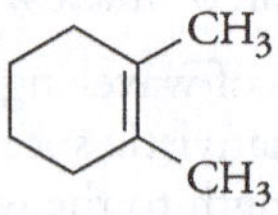

58/ Ice and water are placed in a closed container at a pressure of 1 atm and temperature 273.15 K. If pressure of the system is increased 2 times, keeping temperature constant, then identify correct observation from following.

(a) The amount of ice decreases.
(b) The solid phase (ice) disappears completely.
(c) Volume of system increases.
(d) Liquid phase disappears completely.

Imp.

59/ Which of the following oxidation reactions are carried out by both $K_2Cr_2O_7$ and $KMnO_4$ in acidic medium ?

A. $I^- \rightarrow I_2$ B. $S^{2-} \rightarrow S$ C. $Fe^{2+} \rightarrow Fe^{3+}$
D. $I^- \rightarrow IO_3^-$ E. $S_2O_3^{2-} \rightarrow SO_4^{2-}$

Choose the correct answer from the options given below

(a) C, D and E only (b) A, B and C only
(c) B, C and D only (d) A, D and E only

Oxidation Reaction of $K_2Cr_2O_7$ and $KMnO_4$

60/ Given below are two statements:

Statement I 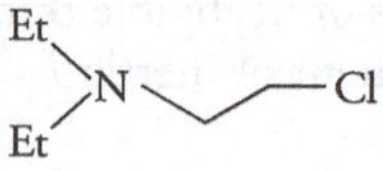 will undergo alkaline hydrolysis at a faster rate than 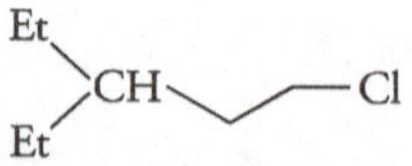

Statement II In 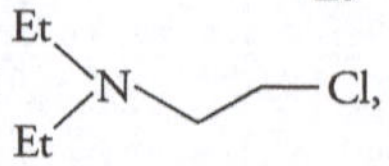, intramolecular substitution takes place first by involving lone pair of electrons on nitrogen.

In the light of the above statements, choose the most appropriate answer from the options given below

(a) Statement I is incorrect but Statement II is correct.
(b) Statement I is correct but Statement II is incorrect.
(c) Both Statement I and Statement II are incorrect.
(d) Both Statement I and Statement II are correct.

61/ Given below are two statements:

Statement I In the oxalic acid *vs* $KMnO_4$ (in the presence of dil. H_2SO_4) titration the solution needs to be heated initially to 60°C, but no heating is required in ferrous ammonium sulphate (FAS) *vs* $KMnO_4$ titration (in the presence of dil. H_2SO_4).

Statement II In oxalic acid *vs* $KMnO_4$ titration, the initial formation of $MnSO_4$ takes place at high temperature, which then acts as catalyst for further reaction. In the case of FAS *vs* $KMnO_4$, heating oxidises Fe^{2+} into Fe^{3+} by oxygen of air and error may be introduced in the experiment.

In the light of the above statements, choose the correct answer from the options given below

(a) Statement I is true but Statement II is false.
(b) Statement I is false but Statement II is true.
(c) Both Statement I and Statement II are false.
(d) Both Statement I and Statement II are true.

62/ Match the List-I with List-II.

	List-I (Redox reaction)		List-II (Type of redox reaction)
A.	$CH_4(g) + 2O_2(g) \xrightarrow{\Delta} CO_2(g) + 2H_2O(l)$	I.	Disproportionation reaction
B.	$2NaH(s) \xrightarrow{\Delta} 2Na(s) + H_2(g)$	II.	Combination reaction
C.	$V_2O_5(s) + 5Ca(s) \xrightarrow{\Delta} 2V(s) + 5CaO(s)$	III.	Decomposition reaction
D.	$2H_2O_2(aq) \xrightarrow{\Delta} 2H_2O(l) + O_2(g)$	IV.	Displacement reaction

Choose the correct answer from the options given below

(a) A-II, B-III, C-IV, D-I
(b) A-III, B-IV, C-I, D-II
(c) A-IV, B-I, C-II, D-III
(d) A-II, B-III, C-I, D-IV

63/ In a multielectron atom, which of the following orbitals described by three quantum numbers will have same energy in absence of electric and magnetic fields?

A. $n = 1, l = 0, m_l = 0$
B. $n = 2, l = 0, m_l = 0$
C. $n = 2, l = 1, m_l = 1$
D. $n = 3, l = 2, m_l = 1$
E. $n = 3, l = 2, m_l = 0$

Choose the correct answer from the options given below.

(a) A and B only
(b) D and E only
(c) C and D only
(d) B and C only

64/ Both acetaldehyde and acetone (individually) undergo which of the following reactions?

A. Iodoform reaction
B. Cannizzaro reaction
C. Aldol condensation
D. Tollen's test
E. Clemmensen reduction

Choose the correct answer from the options given below.

(a) A, B and D only
(b) A, C and E only
(c) B, C and D only
(d) C and E only

65/ What is the freezing point depression constant of a solvent, 50 g of which contain 1 g non-volatile solute (molar mass 256 g mol^{-1}) and the decrease in freezing point is 0.40 K?

(a) 5.12 K kg mol^{-1}
(b) 3.72 K kg mol^{-1}
(c) 4.43 K kg mol^{-1}
(d) 1.86 K kg mol^{-1}

66/ The compounds that produce CO_2 with aqueous $NaHCO_3$ solution are

(A) 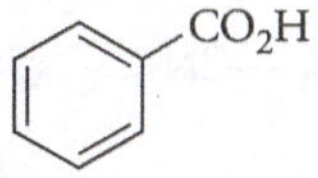

(B) 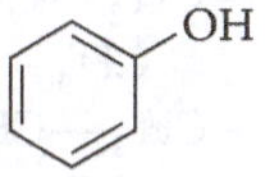

(C) 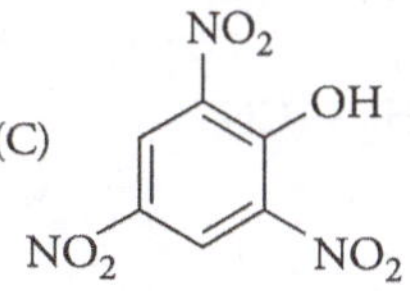

(D) 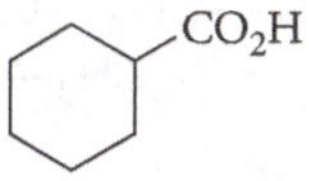

(E) 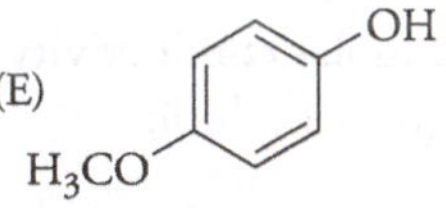

Choose the correct answer from the options given below.

(a) A and B only
(b) A, B and E only
(c) A, C and D only
(d) A and C only

67/ A weak acid H*A* has degree of dissociation *x*. Which option gives the correct expression of (pH- pK_a)?

(a) 0
(b) $\log(1 + 2x)$
(c) $\log\left(\frac{1-x}{x}\right)$
(d) $\log\left(\frac{x}{1-x}\right)$

68/ The metal ion whose electronic configuration is not affected by the nature of the ligand and which gives a violet colour in non-luminous flame under hot condition in borax bead test is

(a) Ti^{3+}
(b) Ni^{2+}
(c) Mn^{2+}
(d) Cr^{3+}

Imp.

69/ The correct order of stability of following carbocations is

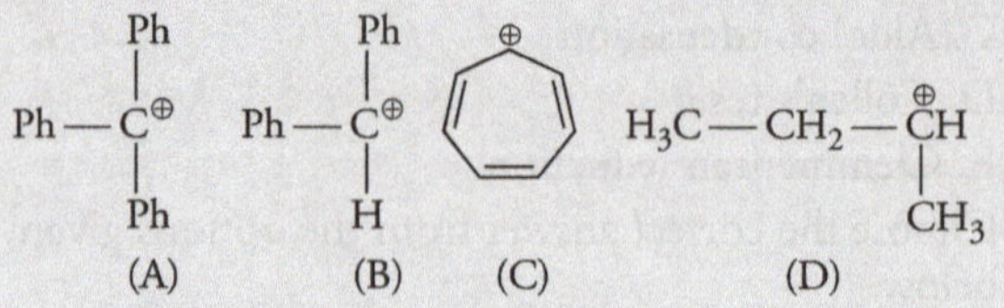

(a) B > C > A > D (b) C > B > A > D
(c) C > A > B > D (d) A > B > C > D

Stability of Carbocation

70/ The products A and B in the following reactions, respectively are

$$A \xleftarrow{Ag-NO_2} CH_3-CH_2-CH_2-Br \xrightarrow{AgCN} B$$

(a) $CH_3-CH_2-CH_2-NO_2$, $CH_3-CH_2-CH_2-NC$
(b) $CH_3-CH_2-CH_2-ONO$, $CH_3-CH_2-CH_2-CN$
(c) $CH_3-CH_2-CH_2-ONO$, $CH_3-CH_2-CH_2-NC$
(d) $CH_3-CH_2-CH_2-NO_2$, $CH_3-CH_2-CH_2-CN$

Section B
Numerical Type Questions

71/ Given below is the plot of the molar conductivity *vs* $\sqrt{\text{concentration}}$ for KCl in aqueous solution.

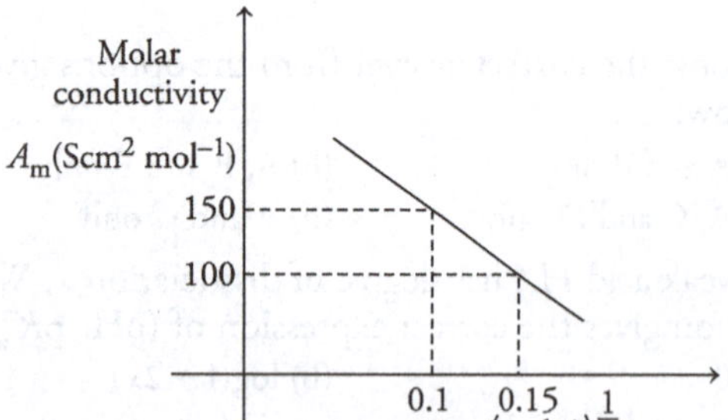

If, for the higher concentration of KCl solution, the resistance of the conductivity cell is 100 Ω, then the resistance of the same cell with the dilute solution is 'x' Ω

The value of x is .... (Nearest integer)

72/ The molarity of a 70% (mass/mass) aqueous solution of a monobasic acid (X) is ..... $\times 10^{-1}$ M (Nearest integer)
[Given, density of aqueous solution of (X) is 1.25 g mL^{-1}
Molar mass of the acid is 70 g mol^{-1}]

73/ The formation enthalpies, $\Delta H_f^\ominus$ for H(g) and O(g) are 220.0 and 250.0 kJ mol^{-1}, respectively, at 298.15 K, and $\Delta H_f^\ominus$ for $H_2O(g)$ is −242.0 kJ mol^{-1} at the same temperature. The average bond enthalpy of the O—H bond in water at 298.15 K is .... kJ mol^{-1} (Nearest integer).

74/ Quantitative analysis of an organic compound (X) shows following % composition.
C : 14.5% Cl : 64.46% H : 1.8%
(Empirical formula mass of the compound (X) is ..... $\times 10^{-1}$
(Given, molar mass in g mol^{-1} of C : 12, H : 1, O : 16, Cl : 35.5)

Imp.

75/ Consider the following sequence of reactions

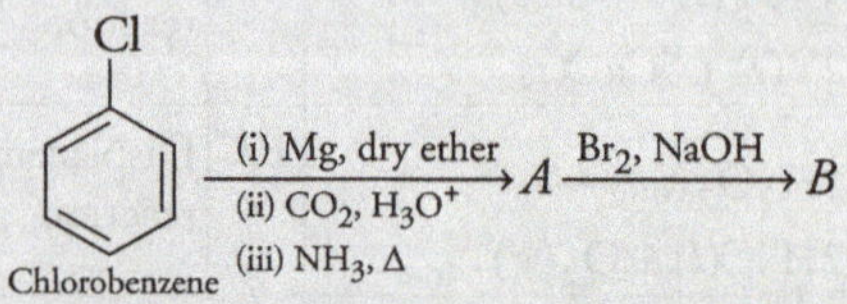

11.25 mg of chlorobenzene will produce .... $\times 10^{-1}$ mg of product B.
(Consider the reactions result in complete conversion.)
[Given, molar mass of C, H, O, N and Cl as 12, 1, 16, 14 and 35.5 g mol^{-1} respectively.]

Reaction of Chlorobenzene in Blend with Stoichiometry

Answer Key

Mathematics

1	(b)	2	(c)	3	(b)	4	(a)	5	(a)	6	(d)	7	(c)	8	(c)	9	(b)	10	(d)
11	(d)	12	(d)	13	(c)	14	(c)	15	(c)	16	(a)	17	(b)	18	(d)	19	(b)	20	(c)
21	(5)	22	(5)	23	(54)	24	(6)	25	(1613)										

Physics

26	(d)	27	(d)	28	(d)	29	(c)	30	(d)	31	(c)	32	(b)	33	(a)	34	(a)	35	(d)
36	(a)	37	(a)	38	(d)	39	(c)	40	(c)	41	(d)	42	(c)	43	(a)	44	(b)	45	(a)
46	(16)	47	(6)	48	(3)	49	(11)	50	(4)										

Chemistry

51	(a)	52	(d)	53	(c)	54	(b)	55	(a)	56	(b)	57	(b)	58	(b)	59	(b)	60	(d)
61	(d)	62	(d)	63	(b)	64	(b)	65	(a)	66	(c)	67	(d)	68	(b)	69	(c)	70	(a)
71	(150)	72	(125)	73	(466)	74	(1655)	75	(93)										

***Note** (*) None of the option is correct.*

Detailed Explanations

28 January
Shift-I

Mathematics

1. (b)

> **Key idea**
> Sum of first n terms of a GP,
> $$S_n = \frac{a(1-r^n)}{1-r}$$

$$2a_{n+2} = 5a_{n+1} - 3a_n$$
$$\Rightarrow \quad 2a_{n+2} - 5a_{n+1} + 3a_n = 0$$
$$\Rightarrow \quad 2x^2 - 5x + 3 = 0$$

have roots α and β.

Then, $a_n = p\alpha^n + q\beta^n$

and $\quad 2x^2 - 5x + 3 = 0$

$$\Rightarrow \quad 2x^2 - 2x - 3x + 3 = 0$$
$$\Rightarrow \quad 2x(x-1) - 3(x-1) = 0$$
$$\Rightarrow \quad (2x-3)(x-1) = 0$$
$$x = 1, 3/2$$
$$\therefore \quad a_n = p(1)^n + q(3/2)^n$$

Let $\alpha = 1$ and $\beta = \frac{3}{2}$

Put $n = 0$

$$\Rightarrow a_0 = p + q \Rightarrow p + q = 0 \qquad \ldots\text{(i)}$$

Put $n = 1 \Rightarrow a_1 = p + q\left(\frac{3}{2}\right) = \frac{1}{2}$

$$\Rightarrow -q + \frac{3}{2}q = \frac{1}{2} \Rightarrow \frac{q}{2} = \frac{1}{2} \Rightarrow q = 1$$
$$\Rightarrow \quad p = -1$$
$$\therefore \quad a_n = -4 + \left(\frac{3}{2}\right)^n \Rightarrow S_n = \Sigma a_n$$
$$\therefore \quad \sum_{k=1}^{100} a_k = a_1 + a_2 + \ldots + a_{100}$$
$$= -100 + \frac{3}{2} + \left(\frac{3}{2}\right)^2 + \ldots + \left(\frac{3}{2}\right)^{100}$$
$$= -100 + \frac{3}{2}\left[1.\frac{(1-(3/2)^{100})}{1-3/2}\right]$$
$$= -100 - \frac{3/2}{1/2}(1-(3/2)^{100})$$
$$= -100 - 3(1-(3/2)^{100})$$
$$= -100 + 3a_{100} = 3a_{100} - 100$$

2. (c) $R = \{(x, y) : x, y \in Z$ and $x + y$ is even$\}$

Reflexive Let $x \in Z$.

Then, $\quad x + x = 2x$, which is given, $\forall\, x \in Z$.

$\Rightarrow R$ is reflexive.

Symmetric Let $x, y \in Z$, Then,

$(x, y) \in R$

$\Rightarrow \quad x + y$ is an even number.

Therefore, $y + x$ will also be an even number because $x + y = y + x$, $\forall\, x, y \in Z$

$\Rightarrow \quad R$ is symmetric.

Transitive Let $x, y, z \in Z$ and $(x, y) \in R, (y, z) \in R$

$\Rightarrow \quad x + y = 2n$ for some $n \in Z$

and $\quad y + z = 2m$ for some $m \in Z$,

then $(x + y) + (y + z) = 2n + 2m$

$$x + 2y + z = 2n + 2m$$
$$x + z = 2(n + m - y)$$

$\Rightarrow x + z$ is an even number $\forall\, n, m, y \in Z$.

Therefore, $(x, z) \in R$.

$\Rightarrow R$ is transitive.

Hence, R is an equivalence relation.

3. (b) $f(x) = \dfrac{2^x}{2^x + \sqrt{2}}$

$$\Rightarrow f(1-x) = \frac{2^{1-x}}{2^{1-x} + \sqrt{2}}$$
$$= \frac{2 \cdot 2^{-x}}{2 \cdot 2^{-x} + \sqrt{2}}$$
$$\Rightarrow f(1-x) = \frac{2}{2 + \sqrt{2} \cdot 2^x} = \frac{\sqrt{2}}{\sqrt{2} + 2^x}$$

Then,

$$f(x) + f(1-x) = \frac{2^x}{2^x + \sqrt{2}} + \frac{\sqrt{2}}{\sqrt{2} + 2^x}$$
$$= \frac{2^x + \sqrt{2}}{2^x + \sqrt{2}} = 1 \Rightarrow \sum_{k=1}^{81} f\left(\frac{k}{82}\right)$$
$$= f\left(\frac{1}{82}\right) + f\left(\frac{2}{82}\right) + f\left(\frac{3}{82}\right) + \ldots + f\left(\frac{41}{82}\right) + \ldots + f\left(\frac{80}{82}\right) + f\left(\frac{81}{82}\right)$$
$$= f\left(\frac{1}{81}\right) + f\left(\frac{81}{82}\right) + f\left(\frac{2}{82}\right) + f\left(\frac{80}{82}\right) + \ldots + f\left(\frac{41}{82}\right)$$
$$= (1 + 1 + \ldots + 1)_{40\text{ times}} + f\left(\frac{1}{2}\right)$$
$$= 40 + \frac{\sqrt{2}}{\sqrt{2} + \sqrt{2}} = 40 + \frac{1}{2}$$
$$= \frac{81}{2}$$

4. (a) $x^2 + y^2 - \alpha x + \beta y + \gamma = 0$

$$\text{Centre} = \left(\frac{\alpha}{2}, -\frac{\beta}{2}\right)$$
$$\text{Radius} = \sqrt{\left(\frac{\alpha}{2}\right)^2 + \left(\frac{\beta}{2}\right)^2 - \gamma}$$
$$\text{Clearly, radius} = \frac{\beta}{2}$$

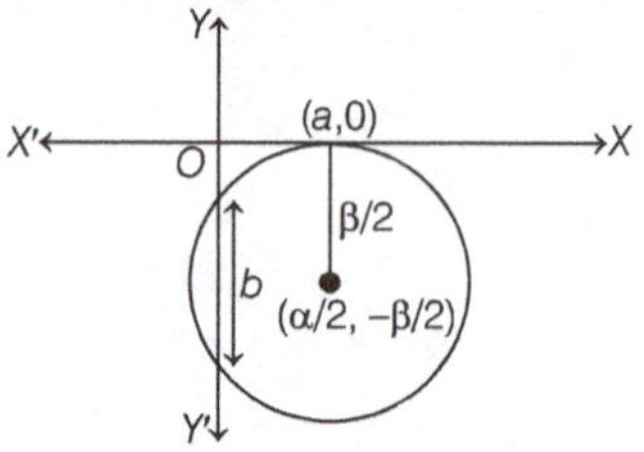

$$\Rightarrow \quad \sqrt{\frac{\alpha^2}{4} + \frac{\beta^2}{4} - \gamma} = \frac{\beta}{2}$$

On squaring both sides, we get

$$\frac{\alpha^2}{4} + \frac{\beta^2}{4} - \gamma = \frac{\beta^2}{4}$$
$$\frac{\alpha^2}{4} = \gamma \qquad \ldots\text{(i)}$$

$\because (a, 0)$ satisfies the given equation of circle

$$\Rightarrow \quad a^2 - a\alpha + \gamma = 0$$
$$a^2 - a\alpha + \frac{\alpha^2}{4} = 0 \Rightarrow \left(a - \frac{\alpha}{2}\right)^2 = 0$$
$$\Rightarrow \quad a = \frac{\alpha}{2} \Rightarrow 2a = \alpha$$

Length of y-intercept $= 2\sqrt{g^2 - c}$

$$= 2\sqrt{\left(\frac{\beta}{2}\right)^2 - \gamma} = b$$

$$\Rightarrow \quad 4\left(\frac{\beta^2}{4} - \gamma\right) = b^2$$

$$\beta^2 - 4\gamma = b^2$$

$$\Rightarrow \quad (2a, b^2) = (\alpha, \beta^2 - 4\gamma)$$

5. *(a)* 0, 1, 2, 3, 4, 5, 6, 7,

$\Rightarrow$ Total digits = 8

□□□□□ Number of digits

5 → 0, 1, 2, 3 → 8×8×8×4

6 → 0, 1, 2 → 8×8×8×3

7 → 0, 1 → 8×8×8×2

Total $= 8 \times 8 \times 8\,(4 + 3 + 2) - 1$

$= 4608 - 1 = 4607$

6. *(d)* $\dfrac{{}^nC_{r-1}}{{}^nC_r} = \dfrac{28}{56}$

$$\frac{r}{n - r + 1} = \frac{1}{2} \Rightarrow n - 3r = -1 \quad \ldots(i)$$

and $\dfrac{{}^nC_r}{{}^nC_{r+1}} = \dfrac{56}{70}$

$$\frac{r+1}{n-r} = \frac{4}{5} \Rightarrow 4n - 9r = 5 \ldots(ii)$$

From Eqs. (i) and (ii), we get

$n = 8$ and $r = 3$

$C\,(3r - n, r^2 - n - 1)$

$= (9 - 8, 9 - 8 - 1) = (1, 0)$

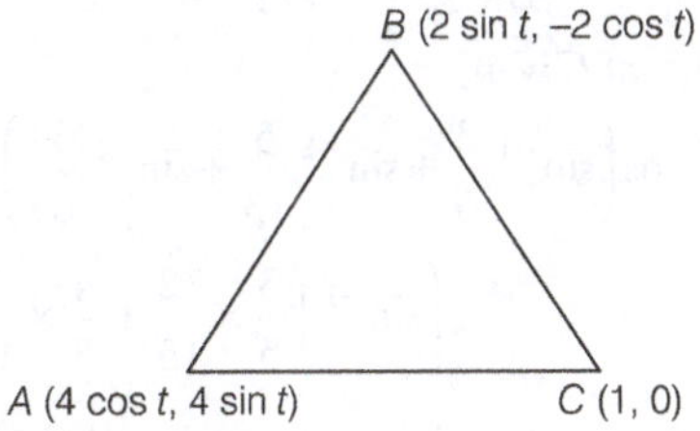

Centroid $(h, k) =$

$$\left(\frac{2\sin t + 4\cos t + 1}{3}, \frac{4\sin t - 2\cos t}{3}\right)$$

$$\Rightarrow \quad 3h - 1 = 2\sin t + 4\cos t$$

$$\Rightarrow \quad 3k = 4\sin t - 2\cos t$$

$$\Rightarrow \quad (3h - 1)^2 + (3k)^2 = 16 + 4$$

$\Rightarrow$ Locus : $(3x - 1)^2 + (3y)^2 = 20$

$$\Rightarrow \quad \alpha = 20$$

7. *(c)*

Key idea

$\int_{-a}^{a} f(x)\,dx = \int_0^a [f(x) + f(-x)]\,dx$

Let $I = \displaystyle\int_{-\pi/2}^{\pi/2} \frac{96x^2 \cos^2 x}{1 + e^x}\,dx$

$$I = \int_0^{\pi/2} \left(\frac{96x^2\cos^2 x}{1 + e^x} + \frac{96x^2 \cos^2 x}{1 + e^{-x}}\right) dx$$

$$= \int_0^{\pi/2} 96x^2 \cos^2 x\,dx$$

$$= 96\int_0^{\pi/2} x^2 \left(\frac{1 + \cos 2x}{2}\right) dx$$

$$= 48\left[\frac{x^3}{3} + \frac{x^2}{2}\sin 2x - \int 2x \cdot \frac{\sin 2x}{2}\,dx\right]_0^{\pi/2}$$

$$= 48\left[\frac{x^3}{3} + \frac{x^2}{2}\sin 2x + \frac{x}{2}\cos 2x - \frac{\sin 2x}{4}\right]_0^{\pi/2}$$

$$= 48\left[\frac{\pi^3}{24} + 0 + \frac{\pi}{4}(-1) - 0\right]$$

$$= 2\pi^3 - 12\pi = \pi\,(2\pi^2 - 12)$$

$$\Rightarrow \quad \alpha = 2, \beta = -12$$

$$\therefore \quad (\alpha + \beta)^2 = (2 - 12)^2 = 100$$

8. *(c)* $y^2 = 4x \Rightarrow a = 1$

$\Rightarrow$ Focus $= (1, 0)$

Hence, AC is a focal chord.

Length of focal chord

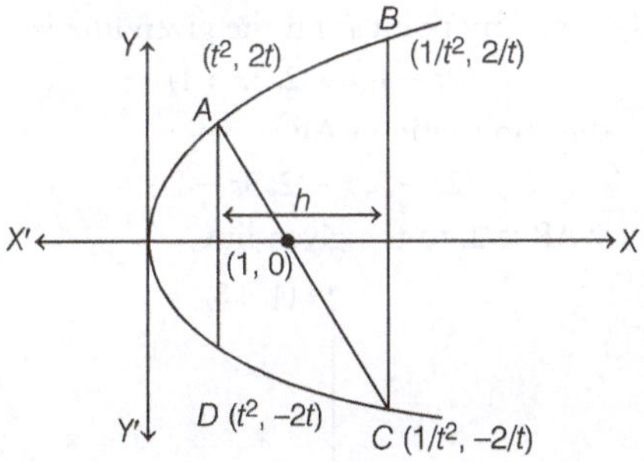

$$= a\left(t + \frac{1}{t}\right)^2 = \frac{25}{4}$$

$$\left(t + \frac{1}{t}\right)^2 = \frac{25}{4} \Rightarrow t + \frac{1}{t} = \frac{5}{2}$$

$$\Rightarrow \quad t^2 + \frac{1}{t^2} + 2 = \frac{25}{4}$$

$$\Rightarrow \quad t^2 + \frac{1}{t^2} = \frac{17}{4}$$

$$\Rightarrow \quad \left|t - \frac{1}{t}\right| = \sqrt{t^2 + \frac{1}{t^2} - 2}$$

$$= \sqrt{\frac{17}{4} - 2} = \frac{3}{2}$$

Area of $ABCD = \dfrac{1}{2} \times h\,(AD + BC)$

$$= \frac{1}{2} \times \left(\frac{1}{t^2} - t^2\right)\left(4t + \frac{4}{t}\right)$$

$$= \frac{4}{2}\left(\frac{1}{t} - t\right)\left(\frac{1}{t} + t\right)\left(t + \frac{1}{t}\right)$$

$$= 2\left(\frac{3}{2}\right)\left(\frac{5}{2}\right)\left(\frac{5}{2}\right) = \frac{75}{4}$$

9. *(b)* $T_{25} = a + 24d = \dfrac{1}{20}$ $\ldots$(i)

$$20 \sum_{r=1}^{25} T_r = 13$$

$$20 \times \frac{25}{2}[2a + 24d] = 13$$

$$a + a + 24d = \frac{13}{250}$$

$$a + \frac{1}{20} = \frac{13}{250}$$

$$\Rightarrow a = \frac{13}{250} - \frac{1}{20} = \frac{26 - 25}{500}$$

$$a = \frac{1}{500}$$

$$\Rightarrow \quad \frac{1}{500} + 24d = \frac{1}{20}$$

$$\Rightarrow \quad 24d = \frac{1}{20} - \frac{1}{500}$$

$$24d = \frac{25 - 1}{500}$$

$$d = \frac{1}{500}$$

$$T_m = a + (m - 1)\,d = \frac{1}{25}$$

$$\frac{1 + m - 1}{500} = \frac{1}{25}$$

$$\Rightarrow \quad m = 20$$

$$5m \sum_{r=m}^{2m} T_r = 5 \times 20 \sum_{r=20}^{40} T_r$$

$= 100\,[T_{20} + T_{21} + T_{22} + \ldots + T_{40}]$

$= 100\,[S_{40} - S_{19}]$

$$= 100\left[\frac{40}{2}\left(\frac{2}{500}+\frac{39}{500}\right) - \frac{19}{2}\left(\frac{2}{500}+\frac{18}{500}\right)\right]$$

$$= \frac{100}{2\times 500}[40\times 41 - 19\times 20]$$

$$= \frac{1}{10}\times 20\,[82-19] = 126$$

10. *(d)* $z_1 = \sqrt{3} + 2\sqrt{2}i$

$$\sqrt{3}|z_2| = |z_1| \Rightarrow \left|\frac{z_2}{z_1}\right| = \frac{1}{\sqrt{3}}$$

Also, $\arg(z_2) = \arg(z_1) + \frac{\pi}{6}$

$$\arg\left(\frac{z_2}{z_1}\right) = \frac{\pi}{6}$$

$$\therefore \quad z_2 = \left|\frac{z_2}{z_1}\right|\cdot z_1 e^{i(\pi/6)}$$

$$z_2 = \frac{(\sqrt{3}+2\sqrt{2}i)(\sqrt{3}+i)}{2\sqrt{3}}$$

$$z_2 = \frac{(3-2\sqrt{2}) + i(2\sqrt{6}+\sqrt{3})}{2\sqrt{3}}$$

$$z_1 - z_2 = \frac{3+2\sqrt{2}+i(2\sqrt{6}-\sqrt{3})}{2\sqrt{3}}$$

$$|z_1 - z_2| = |z_2|$$

$\therefore$ ΔABO = isosceles triangle and angles are $\frac{\pi}{6}, \frac{\pi}{6}, \frac{2\pi}{3}$

$\therefore$ Obtuse angle triangle.

11. *(d)* If $x \geq 3/2$

$$\Rightarrow x^2 + 2x - 3 - 4 = 0$$

$$\Rightarrow x^2 + 2x - 7 = 0$$

$$\Rightarrow \alpha = \frac{-2 \pm \sqrt{4+28}}{2} = -1 \pm 2\sqrt{2}$$

$$\therefore \quad x = 2\sqrt{2} - 1 \qquad [\text{as } x \geq 3/2]$$

and $x \leq 3/2 \Rightarrow x^2 - 2x + 3 - 4 = 0$

$$\Rightarrow x^2 - 2x - 1 = 0$$

$$\Rightarrow x = \frac{2\pm\sqrt{4+4\times 1}}{2} = \frac{2\pm 2\sqrt{2}}{2}$$

$$= 1 \pm \sqrt{2}$$

$$\Rightarrow x = 1 - \sqrt{2} [\text{as } x \leq 3/2]$$

$\therefore$ Sum of the squares of all roots

$$= (2\sqrt{2}-1)^2 + (1-\sqrt{2})^2$$

$$= 8 + 1 - 2\cdot 2\sqrt{2} + 1 + 2 - 2\sqrt{2}$$

$$= 12 - 6\sqrt{2} = 6(2-\sqrt{2})$$

12. *(d)* Since, two oranges are drawn.

Then, $x = 0$; both oranges are good.

$x = 1$; one orange defective and one good.

$x = 2$; both oranges are defective.

Total number of ways to choose 2 oranges from 10 is

$$= {}^{10}C_2 = \frac{10\times 9}{2} = 45$$

$$X = 0 \Rightarrow {}^7C_2 = \frac{7\times 6}{2} = 21$$

$$\therefore \quad P(X=0) = \frac{21}{45} = \frac{7}{15}$$

$$X = 1 \Rightarrow {}^7C_1 \times {}^3C_1 = 21$$

$$\therefore \quad P(X=1) = \frac{21}{45} = \frac{7}{15}$$

$$X = 2 \Rightarrow {}^3C_2 = \frac{3\times 2}{2} = 3$$

$$\therefore \quad P(X=2) = \frac{3}{45} = \frac{1}{15}$$

$$\therefore \quad E(X) = \Sigma x\, P(X=x)$$

$$= 0\times\frac{7}{15} + 1\times\frac{7}{15} + 2\times\frac{1}{15}$$

$$= \frac{7+2}{15} = \frac{9}{15} = \frac{3}{5}$$

$$E(X^2) = \Sigma x^2 P(X=x)$$

$$= 0 + 1^2\times\frac{7}{15} + 2^2\times\frac{1}{15} = \frac{11}{15}$$

$$\therefore Var(X) = E(X^2) - [(E(X)]^2$$

$$= \frac{11}{15} - \left(\frac{3}{5}\right)^2 = \frac{55-27}{75} = \frac{28}{75}$$

13. *(c)* Any point P on the given line is

$$(2t+1, t+2, 3t+1)$$

direction ratio of AP

$$2t-3, t-2, 3t-2$$

$\because$ AP is $\perp$ to the given line,

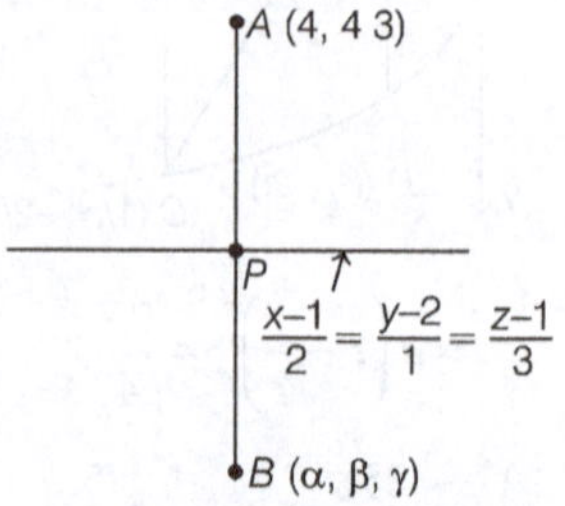

$$2(2t-3) + 1(t-2) + 3(3t-2) = 0$$

$$4t - 6 + t - 2 + 9t - 6 = 0$$

$$14t = 14 \Rightarrow t = 1$$

$$P : (3, 3, 4)$$

$\because P$ is the mid-point of AB, then

$$(3, 3, 4) = \left(\frac{4+\alpha}{2}, \frac{4+\beta}{2}, \frac{3+\gamma}{2}\right)$$

$$\alpha = 2, \quad \beta = 2$$

and $\quad \gamma = 5$

$$\therefore \quad \alpha + \beta + \gamma = 2 + 2 + 5 = 9$$

14. *(c)* Given,

$$f(x) = \begin{cases} 1 - 2x, & x < -1 \\ \frac{1}{3}(7 + 2|x|), & -1 \leq x \leq 2 \\ \frac{11}{18}(x-4)(x-5), & x > 2 \end{cases}$$

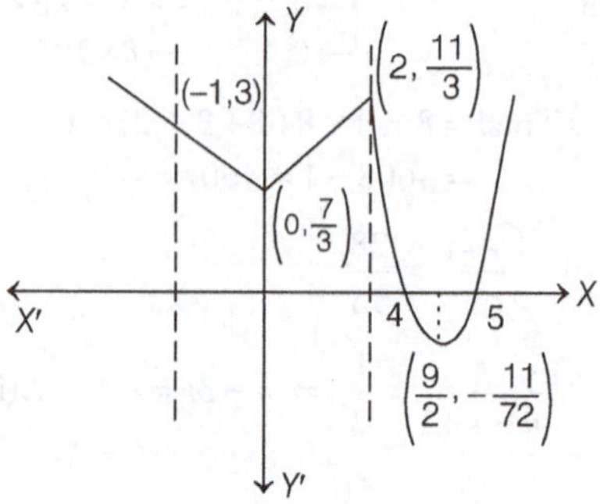

Local minimum at $\left(0, \frac{9}{2}\right)$

$$f(0) = \frac{7}{3}$$

$$f\left(\frac{9}{2}\right) = \frac{11}{18}\left(\frac{1}{2}\right)\left(-\frac{1}{2}\right) = -\frac{11}{72}$$

$$\text{Sum} = \frac{7}{3} - \frac{11}{72} = \frac{168-11}{72} = \frac{157}{72}$$

15. *(c)* Given,

$$\cos\left(\sin^{-1}\frac{3}{5} + \sin^{-1}\frac{5}{13} + \sin^{-1}\frac{33}{65}\right)$$

$$= \cos\left(\sin^{-1}\left(\frac{3}{5}\times\frac{12}{13} + \frac{4}{5}\times\frac{5}{13}\right) + \sin^{-1}\frac{33}{65}\right)$$

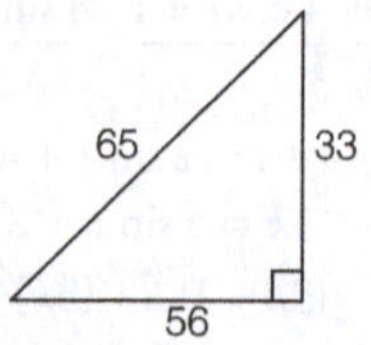

$$= \cos\left(\sin^{-1}\left(\frac{56}{65}\right) + \sin^{-1}\left(\frac{33}{65}\right)\right)$$

$$= \cos\left(\cos^{-1}\left(\frac{33}{65}\right) + \sin^{-1}\left(\frac{33}{65}\right)\right)$$

$$= \cos\left(\frac{\pi}{2}\right) = 0$$

Concept Enhancer

$\sin^{-1} x + \sin^{-1} y$

$$= \sin^{-1}[x\sqrt{1-y^2} + y\sqrt{1-x^2}],$$

if $x, y \in [-1, 1]$ and $x^2 + y^2 \le 1$

16. (a) $i^{k_1} + i^{k_2} \ne 0,\ k_1, k_2 \in N$

k_1	k_2	$i^{k_1} + i^{k_2}$
$4n$	$4m+2$	0
$4n+2$	$4m$	0
$4n+1$	$4m+3$	0
$4n+3$	$4m+1$	0

$k_1 = 4n, 4n+1, 4n+2, 4n+3$

$k_2 = 4m, 4m+1, 4m+2, 4m+3$

P (probability sum is zero) $= \dfrac{4}{16} = \dfrac{1}{4}$

$\therefore$ Required probability $= 1 - \dfrac{1}{4} = \dfrac{3}{4}$

17. (b) $R = \{(x, y) : 0 \le y \le 2|x| + 1,\ 0 \le y \le x^2 + 1, |x| \le 3\}$

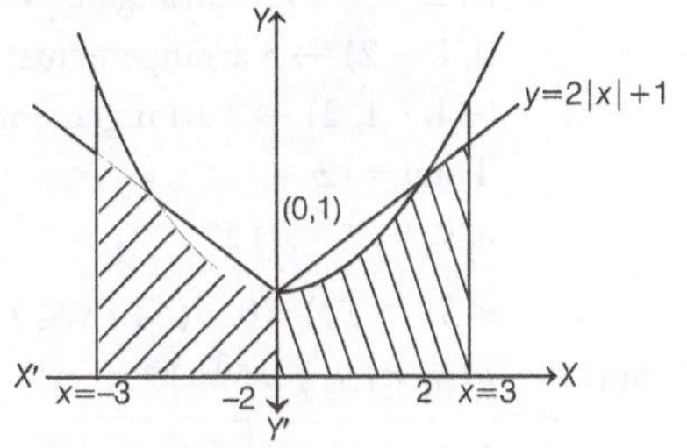

Area = 2 × area in first quadrant

$$= 2\int_0^2 (x^2+1)\,dx + 2\int_2^3 (2x+1)\,dx$$

$$= 2\left[\frac{x^3}{3} + x\right]_0^2 + 2[x^2 + x]_2^3$$

$$= \frac{64}{3}$$

18. (d)

Key idea

If $f(x) = \int_{g(x)}^{h(x)} F(x)\,dx$

$\Rightarrow f'(x) = h'(x)\,F(h(x)) - g'(x)\,F(g(x))$

$$\int_0^x t\,f(t)\,dt = x^2 f(x)$$

On differentiating w.r.t x, wet get

$$xf(x) = 2xf(x) + x^2 f'(x)$$

$$\Rightarrow \quad x^2 f'(x) = -xf(x)$$

$$\Rightarrow \quad \frac{f'(x)}{f(x)} = -\frac{1}{x}$$

$$\Rightarrow \quad \int \frac{f'(x)}{f(x)}\,dx = -\int \frac{1}{x}$$

$$\Rightarrow \quad \log(f(x)) = -\log(x) + \log(c)$$

$$f(x) = \frac{c}{x}$$

$$\because \quad f(2) = 3 \Rightarrow 3 = \frac{c}{2} \Rightarrow c = 6$$

$$\Rightarrow \quad f(x) = \frac{6}{x}$$

$$\therefore \quad f(6) = \frac{6}{6} = 1$$

19. (b) $f(x) = (2+3a)x^2 + \left(\dfrac{a+2}{a-1}\right)x + b$

$$f(x+y) = f(x) + f(y) + 1 - \frac{2}{7}xy$$

$$\Rightarrow (2+3a)(x+y)^2 + \left(\frac{a+2}{a-1}\right)(x+y) + b$$

$$= (2+3a)(x^2+y^2) + \left(\frac{a+2}{a-1}\right)(x+y) + 2b + 1 - \frac{2}{7}xy$$

$$\Rightarrow (2+3a)(2xy) = b + 1 - \frac{2}{7}xy$$

$$\left(2 + 3a + \frac{1}{7}\right)2xy - (b+1) = 0$$

$$\Rightarrow \quad 3a = -\frac{15}{7} \text{ and } b = -1$$

$$\therefore \quad a = -\frac{5}{7} \text{ and } b = -1$$

$$f(x) = -\frac{1}{7}x^2 + \left(\frac{-\frac{5}{7}+2}{-\frac{5}{7}-1}\right)x - 1$$

$$= -\frac{1}{7}x^2 - \frac{9}{12}x - 1$$

$$= -\frac{1}{7}x^2 - \frac{3}{4}x - 1$$

$$28\sum_{i=1}^{5} |f(i)|$$

$$= 28\left[\sum_{i=1}^{5} \frac{1}{7}i^2 + \frac{3}{4}\sum_{i=1}^{5} i + 5\right]$$

$$= 28\left[\frac{1}{7} \times \frac{5 \times 6 \times 11}{6} + \frac{3}{4} \times \frac{5 \times 6}{2} + 5\right]$$

$$= 4 \times 5 \times 11 + 7 \times 45 + 140$$

$$= 220 + 315 + 140 = 675$$

20. (c) Let point $A(x, y, 0)$

Let $P(0, 3, 2)$, $Q(2, 0, 3)$ and $R(0, 0, 1)$

$$AP = AQ = AR$$

$$x^2 + (y-3)^2 + 4 = (x-2)^2 + y^2 + 9$$

$$= x^2 + y^2 + 1$$

$$AQ = AR \Rightarrow -4x + 13 = 1$$

$$x = 3$$

$$AP = AR \Rightarrow y = 2$$

$$A \equiv (3, 2, 0)$$

$$AB = 3$$

$$BC = 3\sqrt{2}$$

$$AC = 3$$

$\because AB^2 + AC^2 = BC^2$ and $AB = AC$

Area of ΔABC is

$$= \frac{1}{2} \times 3 \times 3 = \frac{9}{2} \text{ sq unit}$$

21. (5) $\alpha = 1 + \sum_{r=1}^{6} (-3)^{r-1}\ {}^{12}C_{2r-1}$

$$\because C_1 - C_3 + C_5 - \ldots = \frac{(1+i)^n - (1-i)^n}{2} = 2^{n/2}\sin\frac{n\pi}{4}$$

$$C_1\sqrt{3} - C_3(\sqrt{3})^3 + C_5(\sqrt{3})^5 - \ldots$$

$$= \frac{(1+i\sqrt{3})^{12} - (1-i\sqrt{3})^{12}}{2}$$

$$= \frac{2^{12}[(e^{i\pi/3})^{12} - (e^{-i\pi/3})^{12}]}{2}$$

$$= 2^{11}(e^{i4\pi} - e^{-i4\pi})$$

$$= 2^{11}(1-1) = 0$$

$$\Rightarrow \quad \alpha = 1$$

Line : $\alpha x - \sqrt{3}y + 1 = 0$

$$\text{Distance} = \left|\frac{12 - 3 + 1}{\sqrt{1+3}}\right| = \left|\frac{10}{2}\right| = 5$$

22. *(5)* $f(x)=\begin{cases} 3x, & x<0 \\ \min\{1+x,[x]\}, & 0\le x<1 \\ \min\{2+x, x+2\}, & 1\le x<2 \\ 5, & x>2 \end{cases}$

$$f(x)=\begin{cases} 3x, & x<0 \\ x, & 0\le x<1 \\ x+2, & 1\le x<2 \\ 5, & x>2 \end{cases}$$

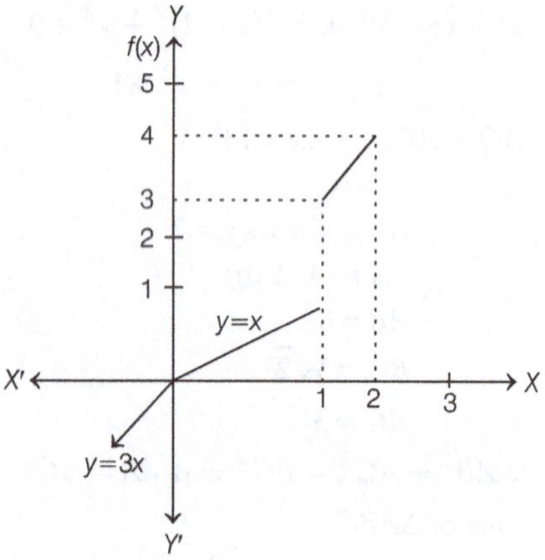

Not continuous at $x=1, 2 \Rightarrow \alpha=2$

Not differentiable at $x=0, 1, 2 \Rightarrow \beta=3$

$\therefore \quad \alpha+\beta=5$

23. *(54)* Given,

$$e_1 = e_i$$

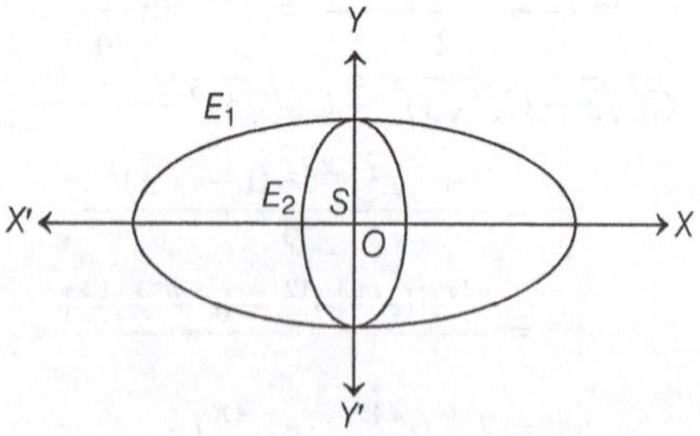

$$b_i = a_{i+1}$$

$$e_1 = \sqrt{1-4/9} = \sqrt{5}/3$$

Area of $E_i = A_i = \pi a_i b_i$

and area of $E_{i+1} = A_{i+1}$

$= \pi a_{i+1} + b_{i+1}$

$$e_i^2 = e_{i+1}^2$$

$$\Rightarrow \quad 1-\frac{b_i^2}{a_i^2} = 1-\frac{b_{i+1}^2}{e_{i+1}^2}$$

$$\Rightarrow \quad \frac{b_i^4}{a_i^2} = b_{i+1}^2$$

$$\Rightarrow \quad b_{i+1} = \frac{b_i^2}{a_i}$$

$$\Rightarrow \quad A_{i+1} = \pi b_i \left(\frac{b_i^2}{a_i}\right)$$

$$= \pi a_i b_i \left(\frac{b_i}{a_i}\right)^2$$

$$= A_i\,[1-e_i^2]$$

$$= A_i\,(1-5/9) = A_i.4/9$$

$$\Rightarrow \quad \frac{5}{\pi}\sum_{i=1}^{\infty} A_i = \frac{5}{\pi}\cdot S_1 \times \left[1-\frac{4}{9}\right]$$

$$= \frac{5}{\pi}\times \pi \times 3 \times 2 \times \frac{9}{5}$$

$(S_1 = \pi \times 2 \times 3 = 6\pi)$

$= 54$

24. *(6)* $\mathbf{d} = \mathbf{a}\times\mathbf{b} = -\hat{\mathbf{i}}+\hat{\mathbf{j}}$

$\because \; |\mathbf{c}-2\mathbf{a}|^2 = 8$

$\Rightarrow |\mathbf{c}|^2 + 4|\mathbf{a}|^2 - 4(\mathbf{a}\cdot\mathbf{c}) = 8$

$\Rightarrow |\mathbf{c}| = 2 \qquad [\because |\mathbf{a}| = \sqrt{3},\ \mathbf{a}\cdot\mathbf{c} = c]$

We have,

$\quad \mathbf{d} = \mathbf{a}\times\mathbf{b}$

Or $\quad \mathbf{d}\times\mathbf{c} = (\mathbf{a}\times\mathbf{b})\times\mathbf{c}$

$$\left[|\mathbf{d}||\mathbf{c}|\sin\frac{\pi}{4}\right]^2 = [(\mathbf{a}\cdot\mathbf{c})\mathbf{b} - (\mathbf{b}\cdot\mathbf{c})\mathbf{a}]^2$$

$$4 = 4b^2 + (\mathbf{b}\cdot\mathbf{c})^2(\mathbf{a})^2 - 2(\mathbf{a}\cdot\mathbf{c})(\mathbf{a}\cdot\mathbf{b})(\mathbf{b}\cdot\mathbf{c})$$

Let $\mathbf{b}\cdot\mathbf{c} = x$

$$4 = 36 + 3x^2 - 20x$$

$$\Rightarrow \quad 3x^2 - 20x + 32 = 0$$

$$\Rightarrow \quad x = \frac{8}{3}, 4$$

$$\Rightarrow \quad \mathbf{b}\cdot\mathbf{c} = \frac{8}{3}$$

Now, $|10 - 3\mathbf{b}\cdot\mathbf{c}| + |\mathbf{b}\times\mathbf{c}|^2$

$= |10-8| + (2)^2 = 6$

> **Concept Enhancer**
>
> For vectors **a** and **b**, $|\mathbf{a}\times\mathbf{b}| = |\mathbf{a}||\mathbf{b}|\sin\theta$, θ is angle between **a** and **b**
>
> and for vector **a**, **b** and **c**,
>
> $\mathbf{a}\times(\mathbf{b}\times\mathbf{c}) = (\mathbf{a}\cdot\mathbf{c})\mathbf{b} - (\mathbf{a}\cdot\mathbf{b})\mathbf{c}$

25. *(1613)* $S_1 = \{A = [a_{ij}] \in M : A = A^T$ and $a_{ij} \in S, \forall\, i, j\}$

$$n(S_1) = 5^6$$

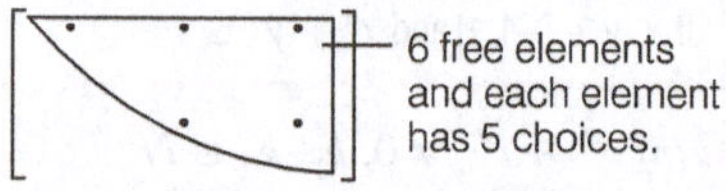

$S_2 = \{A = [a_{ij}] \in M : A = -A^T$ and $a_{ij} \in S, \forall\, i, j\}$

Skew-symmetric matrix has diagonal entries as zero.

But $\quad 0 \notin S$

Then, $\quad n(S_2) = 0$

$S_3 = \{A = [a_{ij}] \in M, :$

$a_{11} + a_{22} + a_{33} = 0$ and $a_{ij} \in S, \forall\, i, j]$

$$\begin{bmatrix} a_{11} & \cdot & \cdot \\ \cdot & a_{22} & \cdot \\ \cdot & \cdot & a_{33} \end{bmatrix} \text{6 free}$$

elements and each has 5 choices.

$a_{11} + a_{22} + a_{33} = 0$

$(1, 2, -3) \to 3!$ arrangements

$(1, 1, -2) \to 3$ arrangements.

$(-1, -1, 2) \to 3$ arrangements

Total $= 12$

$n(S_3) = 5^6 \times 12$

$n(S_1 \cap S_2) = 0,\ n(S_2 \cap S_3) = 0$

and $\quad n(S_1 \cap S_3) = 5^3 \cdot 12$

$$\begin{bmatrix} a_{11} & & \\ & a_{22} & \\ & & a_{33} \end{bmatrix}$$

$a_{11} + a_{22} + a_{33} = 0$

Clearly, $n(S_1 \cap S_2 \cap S_3) = 0$

$\Rightarrow n(S_1 \cup S_2 \cup S_3) = n(S_1) + n(S_2) + n(S_3) - n(S_1 \cap S_2) - n(S_2 \cap S_3) - n(S_1 \cap S_3) + n(S_1 \cap S_2 \cap S_3) = 0$

$= 5^6 + 5^6 \cdot 12 - 5^3 \cdot 12$

$= 5^3\,[125 + 125 \times 12 - 12]$

$= 5^3\,(1613)$

$\Rightarrow \quad \alpha = 1613$

Physics

26. *(d)* $V = -\int \mathbf{E} \cdot dr = \int \frac{2k\lambda}{r} dr$

$$= 2k\lambda \ln r + c$$

Net potential due to all wires

$$V = -2k\lambda \ln \sqrt{x^2 + y^2}$$

$$-2k\lambda \ln \sqrt{y^2 + z^2} - 2k\lambda \ln \sqrt{z^2 + x^2} + c$$

For V = constant

$$\sqrt{(x^2 + y^2)(y^2 + z^2)(z^2 + x^2)} = c$$

$$\therefore \quad (x^2 + y^2)(y^2 + z^2)(z^2 + x^2) = c$$

where c = constant

27. *(d)* Let the resistance per unit length be λ, then

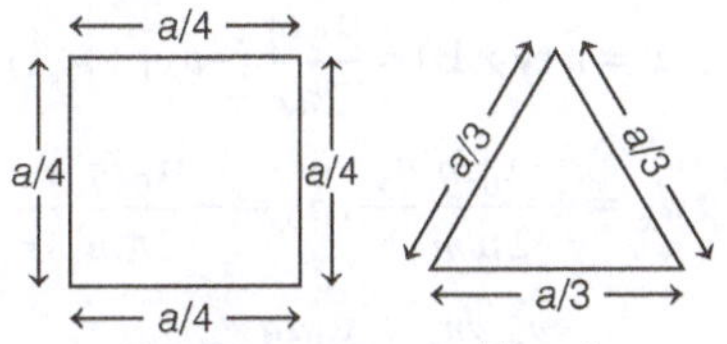

Square Triangle

$$R_{\text{Triangle}} = \frac{\left(\frac{2\lambda a}{3}\right)\left(\frac{\lambda a}{3}\right)}{\frac{2\lambda a}{3} + \frac{\lambda a}{3}} = \frac{2\lambda a}{9} \quad \ldots \text{(i)}$$

$$R_{\text{Square}} = \frac{\left(\frac{3\lambda a}{4}\right)\left(\frac{\lambda a}{4}\right)}{\frac{3\lambda a}{4} + \frac{\lambda a}{4}} = \frac{3\lambda a}{16} \quad \ldots \text{(ii)}$$

From Eqs. (i) and (ii), we get

$$\frac{R_{\text{Triangle}}}{T_{\text{Square}}} = \frac{2\lambda a}{9} \times \frac{16}{3\lambda a} = \frac{32}{27}$$

Alternate Solution

For a polygon of n sides, the equivalent resistance across two adjacent vertices is given by

$$R_{eq} = \frac{(n-1)R}{n^2}$$

where R is the total resistance.

So, according to question.

For triangle

$n = 3$ and $R = R$

$$(R_{eq})_t = \frac{(n-1)R}{n^2} = \frac{(3-1) \times R}{3^2} = \frac{2R}{9}$$

For square

$n = 4$ and $R = R$

$$(R_{eq})_{sq} = \frac{(n-1)R}{n^2} = \frac{(4-1)R}{4^2} = \frac{3R}{16}$$

$$\frac{(R_{eq})_t}{(R_{eq})_{sq}} = \frac{2R}{9 \times \frac{3R}{16}} = \frac{32}{27}$$

28. *(d)*

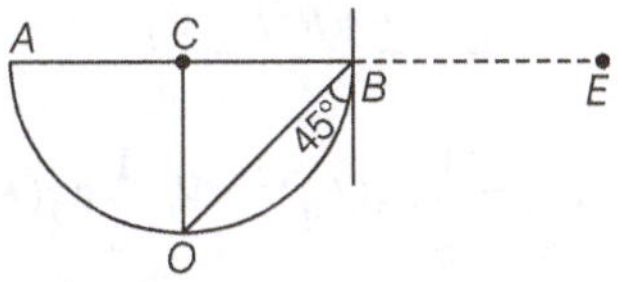

In this case,

Critical angle, $\theta_c = 45°$

Using the condition, we have

$$\sin \theta_c = \frac{1}{\mu}$$

$$\mu = \frac{1}{\sin 45°} = \sqrt{2}$$

29. *(c)* We know that, relation between energy of gas molecules with temperature

$$\frac{1}{2} M v_{\text{rms}}^2 = \frac{3}{2} RT \Rightarrow v_{\text{rms}}^2 \propto T$$

30. *(d)* From the given circuit, the expression of Boolean algebra is

$$\overline{Y} = \overline{A\overline{B} + AB}$$

$$= \overline{A(B + \overline{B})} \Rightarrow Y = \overline{A}$$

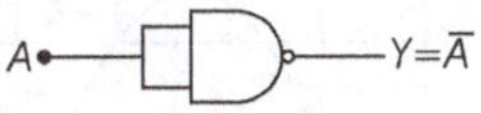

31. *(c)*

Key idea

Viscosity of fluid mainly depends upon the temperature.

For a liquid, viscosity decreases with increase of temperature.

For a gas, viscosity increases with increase of temperature.

We know,

$$v_T = \frac{2}{9} R^2 \frac{g}{\eta} (d - \rho)$$

Statements A, C and D are correct.

32. *(b)* As there is no dependence of mass density in y-direction

So, $\quad y_{\text{cm}} = \frac{b}{2}$

For x-direction

$$dm = \left(\frac{\sigma_0 x}{ab}\right) b \cdot dx$$

For centre of mass in x-direction

$$x_{\text{cm}} = \frac{\int_0^a x \, dm}{\int_0^a dm}$$

$$= \frac{\frac{\sigma_0}{a} \int_0^a x^2 \, dx}{\frac{\sigma_0}{a} \int_0^a x \, dx}$$

$$x_{\text{cm}} = \frac{a^3 \times 2}{3 \times a^2} = \frac{2}{3} a$$

So, coordinates of $r_{\text{cm}} = \left(\frac{2}{3} a, \frac{b}{2}\right)$

33. *(a)* Using the work-energy theorem, we have

$$W = \Delta \text{KE}$$

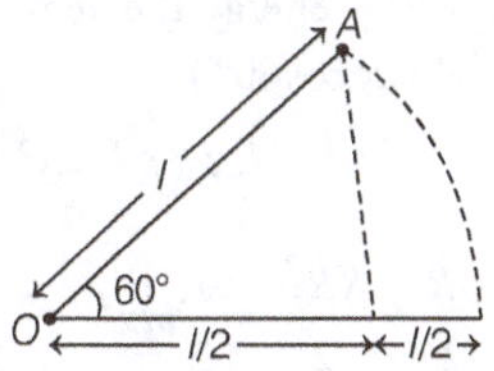

$$qE(l - l\cos 60°) = \frac{1}{2} mv^2 - 0$$

$$qE\left(l - \frac{l}{2}\right) = \frac{1}{2} mv^2$$

$$qE \frac{l}{2} = \frac{1}{2} mv^2$$

$$\frac{qE \cdot l}{m} = v^2 \Rightarrow v = \sqrt{\frac{qEl}{m}}$$

34. *(a)* From the following circuit, all the resistances are parallel to each other, so it can be redrawn as

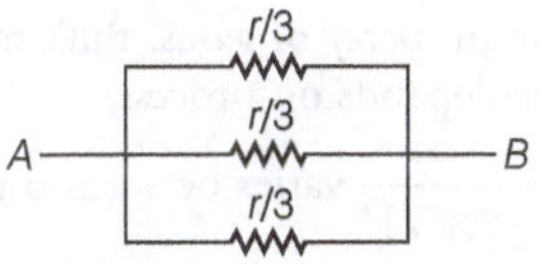

$$R_{eq} = \frac{1}{3}\left(\frac{r}{3}\right) = \frac{r}{9}$$

35. *(d)*

Key idea

A neutron outside the nucleus is unstable having mean life of 1000 s. So, if breaks into proton and electron given by the relation

$$n \rightarrow p + e^- + \nu^-$$

A proton outside the nucleus is stable but inside the nucleus it can break into neutron and position for the stablity of nucleus.

$$p \rightarrow n + e^+ + \nu$$

In the following options, (b) and (c) are not following charge conservation and in neutron decay, anti-neutrino ($\overline{\nu}$) is released

$$n \rightarrow p + e^- + \overline{\nu}$$

36. *(a)*

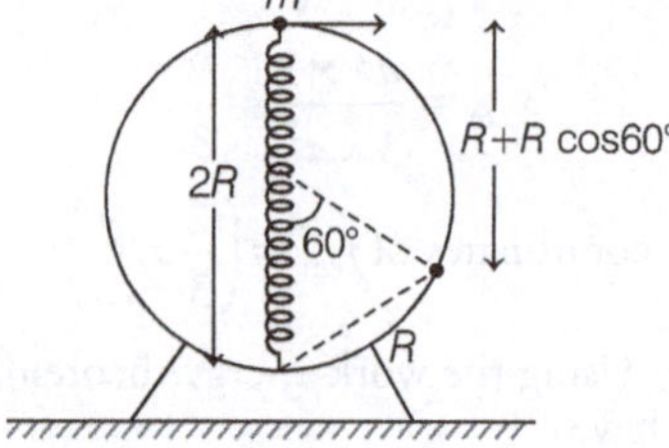

Using work - energy theorem

$$mg(R + R\cos 60^\circ) + \frac{1}{2}K(R^2 - 0^2) = \frac{1}{2}mv^2$$

$$mg\frac{3R}{2} + \frac{KR^2}{2} = \frac{1}{2}mv^2$$

$$v = \sqrt{3gR + \frac{KR^2}{m}}$$

37. *(a)* Speed of sound in medium depends on elastic and inertial property of medium.

For gas, $v = \sqrt{\frac{B}{\rho}}$

For solids, $v = \sqrt{\frac{Y}{\rho}}$

The elastic property of solids happens to be many fold greater than that of elastic property of gases. Bulk modulus of gas depends on process,

$B = -\frac{\Delta p}{\Delta V / V}$ varies between 0 to ∞.

Therefore, Reason is incorrect.

38. *(d)* Both A and R are true but R is not the correct explanation of A.

39. *(c)* A. Surface tension is due to less stability or unbalanced forces on the molecules at the surface.

B. For liquid, coefficient of viscosity decreases with temperature

Statements C, D and E are correct.

40. *(c)* For EM wave,

$$E = 100\sin(\omega t - kx)$$

Energy density $P_{av} = \frac{1}{2}\varepsilon_0 (E_0)^2$

Initial volume, $V_0 = \pi R^2 l$

Volume in next situation

$$= V = \frac{\pi R^2 l}{4} = \frac{V_0}{4}$$

$$U_0 = \frac{1}{2}\varepsilon_0 (E_0)^2 \times V_0 \; \frac{1}{2}\varepsilon_0 (E_1{}^2) \cdot \frac{V_0}{4}$$

$$E_0{}^2 = \frac{E_1{}^2}{4} \Rightarrow E_1 = 2E_0 = 200 \text{ V/m}$$

So, the electric field of the corresponding EM-wave is modified as $200\sin(\omega t - kx)$ N/C.

41. *(d)* Here, in this case both the capacitors C_1 and C_2 are connected in parallel to a battery.

In steady state,

$$\frac{C_1}{C_2} = \frac{Q_1}{Q_2} < 1 \Rightarrow C_2 > C_1$$

For energy stored in capacitor,

$$\frac{C_1}{C_2} = \frac{U_1}{U_2} < 1 \Rightarrow U_2 > U_1$$

42. *(c)*

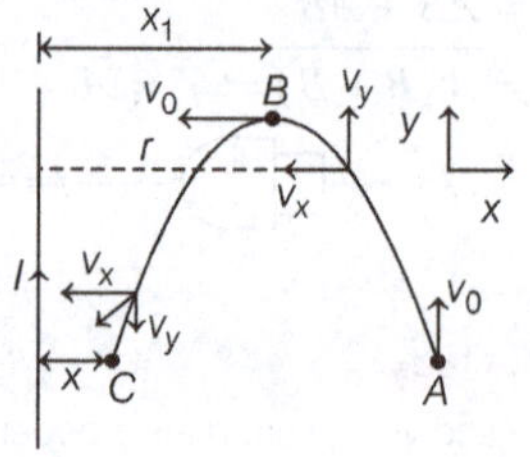

$A \rightarrow B$

$$\mathbf{v} = -v_x\hat{\mathbf{i}} + v_y\hat{\mathbf{j}}$$

$$\mathbf{B} = \frac{\mu_0 I}{2\pi r}(-\hat{\mathbf{k}})$$

$$\mathbf{F} = q(\mathbf{v} \times \mathbf{B}) = \frac{\mu_0 Iq}{2\pi r}(-v_x\hat{\mathbf{j}} - v_y\hat{\mathbf{i}})$$

$$a_x = -\frac{\mu_0 Iq}{2\pi m} \cdot \frac{v_y}{r}$$

$$a_y = -\frac{\mu_0 Iq}{2\pi m} \cdot \frac{v_x}{r}$$

$$\frac{v_x dv_x}{dr} = -\frac{\mu_0 Iq}{2\pi m}\frac{v_y}{r}$$

$$\frac{v_x dv_x}{v_y} = -\frac{\mu_0 Iq}{2\pi m}\frac{dr}{r}$$

$$\int_0^{v_0} \frac{v_x dv_x}{\sqrt{v_0^2 - v_x^2}} = -\frac{\mu_0 I_q}{2\pi m}\int_a^{x_1} \frac{dr}{r}$$

Let $z^2 = v_0^2 - v_x^2$

$$2z\,dz = -2v_x dv_x$$

$$z\,dz = -v_x dv_x$$

$$\frac{v_x dv_x}{\sqrt{v_0^2 - v_x^2}} = -\frac{z\,dz}{z} = -dz$$

Then integral becomes

$$-\int_{v_0}^{0} dz = -\frac{\mu_0 Iq}{2\pi m}\ln\frac{x_1}{a}$$

$$v_0 = -\frac{\mu_0 Iq}{2\pi m}\ln\frac{x_1}{a}$$

$$x_1 = ae^{\frac{-2\pi m v_0}{\mu_0 Iq}} \quad \ldots\text{(i)}$$

For $B \rightarrow C$, $\mathbf{v} = -v_x\mathbf{i} - v_y\hat{\mathbf{j}}$

$$\mathbf{B} = \frac{\mu_0 I}{2\pi r}(-\hat{\mathbf{k}})$$

$$\mathbf{F} = q(\mathbf{v} \times \mathbf{B}) = \frac{\mu_0 Iq}{2\pi r}(-v_x\hat{\mathbf{j}} + v_y\hat{\mathbf{i}})$$

$$a_x = +\frac{\mu_0 Iq}{2\mu m}\frac{v_y}{r}, \; a_y = -\frac{\mu_0 Iq}{2\pi m} \cdot \frac{v_x}{r}$$

$$\frac{v_x\,dv_x}{dr} = \frac{\mu_0 Iq}{2\pi m}\frac{v_y}{r}$$

$$\Rightarrow \int_{v_0}^{0} \frac{v_x dv_x}{\sqrt{v_0^2 - v_x^2}} = \frac{\mu_0 Iq}{2\pi m}\int_{x_1}^{x} \frac{dr}{r}$$

$$\frac{\mu_0 Iq}{2\pi m}\ln\frac{x}{x_1} = -\int_0^{v_0} dt = -v_0$$

$$x = x_1 e^{\frac{-2\pi m v_0}{\mu_0 Iq}} \quad \ldots\text{(ii)}$$

From Eqs. (i) and (ii), $x = a\,e^{\frac{-4\pi m v_0}{\mu_0 Iq}}$

43. *(a)* Given,

Energy of photon = Kinetic energy of proton of mass m_p

$$\frac{1}{2}m_p v^2 = E_0$$

$$\Rightarrow \quad p = \sqrt{2m_p E_0}$$

So, $\quad \lambda_p = \frac{h}{\sqrt{2m_p E_0}} \quad \ldots\text{(i)}$

For photon, $\frac{hc}{\lambda} = E_0$

$$\Rightarrow \quad \lambda_{\text{Ph}} = \frac{hc}{E_0} \quad \ldots\text{(ii)}$$

Using Eqs. (i) and (ii), we get

$$\frac{\lambda_p}{\lambda_{\text{ph}}} = \frac{h}{\sqrt{2m_p E}} \times \frac{E}{hc} = \frac{1}{c}\sqrt{\frac{E}{2m_p}}$$

44. *(b)* Using efficiency of engine, we have

$$\eta = 1 - \frac{T_0}{T_1}, (T_2 < T_1)$$

For engine -E

$$\eta_{12} = 1 - \frac{273}{473} = \frac{200}{473}$$

For engine - E_1

$$\eta_1 = 1 - \frac{373}{473} = \frac{100}{473}$$

For engine - E_2

$$\eta_2 = 1 - \frac{273}{373} = \frac{100}{373}$$

So, $\eta_{12} - \eta_1 = \frac{100}{473} < \eta_2$

Which gives $\eta_{12} < \eta_1 + \eta_2$

45. *(a)* Angle of deviation for thin prism

$$\delta = A(n-1)$$

For no deviation, $\delta_1 = \delta_2$

$$4° (1.54 - 1) = (1.72 - 1) A$$

$$\frac{4 \times 0.54}{0.72} = A \Rightarrow A = 3°$$

46. *(16)* Given, $R_2 = 2R_1$

$$M_1 = \sigma \times \pi R_1^2 = M_0$$

$$M_2 = \sigma \times \pi R_2^2 = \sigma \times \pi [2R_1]^2$$

$$= \sigma \times 4\pi R_1^2 = 4M_0$$

$$\frac{I_1}{I_2} = \frac{\frac{M_1 R_1^2}{2}}{\frac{M_2 R_2^2}{2}} = \frac{M_1 R_1^2}{M_2 R_2^2} = \frac{1}{4} \times \frac{1}{4} = \frac{1}{16}$$

$$x = 16$$

47. *(0)* Modulus of elasticity $= \frac{\text{Force}}{\text{Area}}$

$$= \frac{[MLT^{-2}]}{[L^2]} = [ML^{-1}T^{-2}] \quad \ldots(i)$$

Torque = Force × perpendicular distance

$$= [MLT^{-2}][L^1]$$

$$= [ML^2T^{-2}] \quad \ldots(ii)$$

$$\frac{\text{Modulus of elasticity}}{\text{Torque}} = \frac{[ML^{-1}T^{-2}]}{[ML^2T^{-2}]} = [M^0L^{-3}T^0]$$

Measured quantity

$$= [M^a L^b T^c] = [M^0 L^{-3} T^0]$$

For $b = -3, c = 0$

If $b = 3$, then also $c = 0$

48. *(3)* Least count of screw gauge

$$= \frac{0.75}{15} = 0.05 \text{ mm}$$

Now, area of rectangular sheet $= l \times b$

$$\frac{\Delta S}{S} = \frac{\Delta l}{l} + \frac{\Delta b}{b}$$

$$\frac{\Delta S}{S} = \frac{0.05}{5} + \frac{0.05}{2.5} = \frac{1}{100} + \frac{2}{100} = \frac{3}{100}$$

So, value of $x = 3$

49. *(11)* Distance of 10th bright fringe from central maxima is

$$y = 10\frac{\lambda D}{d}$$

$$\frac{y_2}{y_1} = \frac{\lambda_2}{\lambda_1}$$

(For different wavelengths)

$$y_2 = \frac{660\text{nm}}{600\text{nm}} \times 10 \text{ mm}$$

$$y_1 = 11 \text{ mm}$$

50. *(4)* $I_{\text{disc}} = \frac{MR_1^2}{4}, I_{\text{ring}} = \frac{MR_2^2}{2},$

$$I_{\text{sphere}} = \frac{2MR_1^2}{5}$$

According to the question,

$$\frac{I_{\text{disc}}}{I_{\text{ring}}} = 2.5 \Rightarrow \frac{\frac{MR_1^2}{4}}{\frac{MR_2^2}{2}} = \frac{5}{2}$$

$$\Rightarrow \frac{R_1^2}{R_2^2} = 5 \quad \ldots(i)$$

Also, we are given that

$$\frac{I_{\text{sphere}}}{I_{\text{ring}}} = n \Rightarrow \frac{\frac{2MR_1^2}{5}}{\frac{MR_2^2}{2}} = n$$

$$\Rightarrow \frac{4R_1^2}{5R_2^2} = n \quad \ldots(ii)$$

From Eqs. (i) and (ii)

$$n = 4$$

Chemistry

51. *(a)* Highest ionisation enthalpy among the given elements = Ge

Most stable oxidation state of Ge = + 4

Lowest first ionisation enthalpy among the given element = In

Most stable oxidation state of In = +3

52. *(d)* Among the given options, BrF_5 and $XeOF_4$ has square pyramidal geometry.

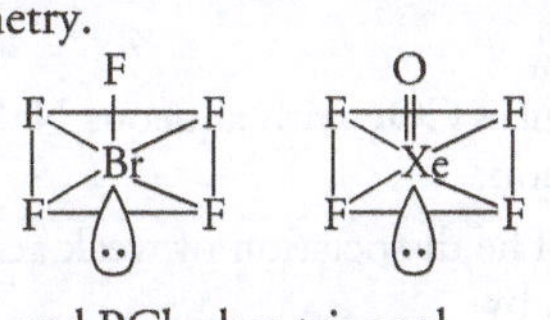

SbF_5 and PCl_5 has trigonal bipyramidal geometry.

53. *(c)* Among the given statements, A, B and D are correct, while statement C is incorrect.

The correct form of statement C is : 1st order kinetics have half-life ($t_{1/2}$) independent of concentration. So, upon changing the concentration $t_{1/2}$ should not change for 1st order reaction.

54. *(b)* Statement I is false but statement II is true.

The correct form of statement I is Glucose does not respond to 2, 4-DNP test, as it does not contain any free aldehydic group in its cyclic form.

55. *(a)* Atomic radii increases on moving down the group while it decreases along the period.

So, the incorrect decreasing order of atomic radii is given in option (a). Its correct decreasing order is

Be < Mg > Al > Si

56. *(b)* The most stable structure of ClF_3 is

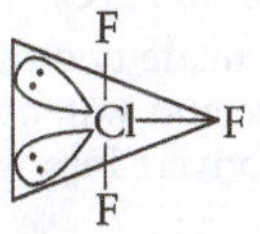

Number of lone pair present in equitorial plane,

$$n = 2$$

Ions	Configuration	Unpaired electrons
V^{3+}	$[Ar] 3d^2$	2
Ti^{3+}	$[Ar] 3d^1$	1
Cu^{2+}	$[Ar] 3d^9$	1
Ni^{2+}	$[Ar] 3d^8$	2
Ti^{2+}	$[Ar] 3d^2$	2

Hence, V^{3+}, Ni^{2+} and Ti^{2+} have same n number of unpaired electrons i.e. 2.

57. *(b)* The given reaction involved in the given question is as follows

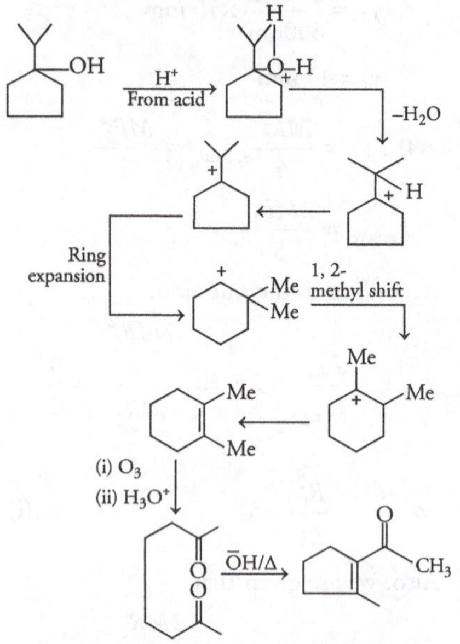

So, option (b) is the correct answer.

58. *(b)* The correct answer is option (b). The solid phase (ice) disappears completely.

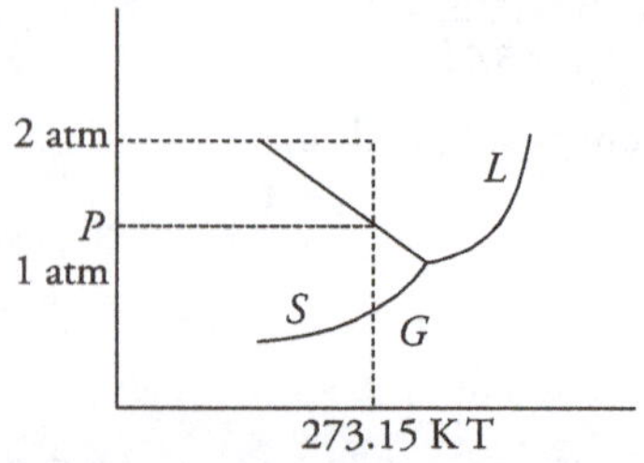

Phase diagram of H_2O.

If pressure is made two times, then mixture of ice and water will completely convert into water (liquid) form.

59. *(b)* The conversions given in $I^- \longrightarrow I_2$, $S^{2-} \longrightarrow S$ and $Fe^{2+} \longrightarrow Fe^{3+}$ occurs in acidic medium, while $I^- \longrightarrow IO_3^-$, $S_2O_3^{2-} \longrightarrow SO_4^{2-}$ occurs in neutral or faintly alkaline solution. Thus, oxidation reactions given in A, B and C can be carried out by both $K_2Cr_2O_7$ and $KMnO_4$ in acidic medium.

60. *(d)* Both statements I and II are correct.

61. *(d)* Both statement I and statement II are true.

$$2MnO_4^- + 5(COO)_2^{2-} + 16H^+$$
$$\downarrow$$
$$10CO_2 + 2Mn^{2+} + 8H_2O$$

This reaction is slow at room temperature but become fast at 60°C. Manganese (II) ions catalyse the reaction, thus the reaction is autocatalytic reaction.

62. *(d)* The correct match is

A-II, B-III, C-I, D-IV

A : Combustion of hydrocarbon is combination reaction.

B : Decomposition into gaseous product.

C : Displacement of V by Ca atom.

D : Disproportionation of H_2O_2.

Time Saver Tech

Use elimination technique as time saver D-IV is given only is option (d).

63. *(b)*

Set of quantum number	Orbitals
$A: n=1, l=0, m_l=0$	$1s$
$B: n=2, l=0, m_l=0$	$2s$
$C: n=2, l=1, m_l=1$	$2p$
$D: n=3, l=2, m_l=1$	$3d$
$E: n=3, l=2, m_l=1$	$3d$

In absence of electric and magnetic fields, all orbitals of $3d$ are degenerate.

64. *(b)* Both acetaldehyde and acetone (individually) undergoes

(A) Iodoform reaction

(C) Aldol condensation

(E) Clemmensen reduction

Name of reaction	Acetaldehyde	Acetone
Iodoform	(+ ve)	(+ ve)
Cannizzaro	(– ve)	(– ve)
Aldol condensation	(+ ve)	(+ ve)
Tollen's test	(+ ve)	(– ve)
Clemmensen reduction	(+ ve)	(+ ve)

Concept Enhancer

- Positive **Iodoform test** can be given by those compounds that have $CH_3-\overset{O}{\overset{\|}{C}}-$ and $H_3C-\overset{OH}{CH}$ group.
- **Cannizzaro reaction** Aldehydes that do not have an α-hydrogen undergoes self oxidation and reduction reaction of heating with concentrated alkali.
- **Aldol condensation** Aldehydes and ketones with at least one α-hydrogen undergo a reaction in the presence of dilute alkali as catalyst to form β-hydroxy aldehyde.
- **Tollen's test** It is a mild oxidising agent that converts aldehydes into corresponding carboxylate anion.
- **Clemmensen reduction** In this reaction, carbonyl group of aldehydes and ketones is reduced to CH_2 group on treatment with zinc amalgam and concentrated hydrochloric acid.

65. *(a)* Given molar mass of non-volatile,

Solute = 256 g/mol

Mass of solvent = 50 g

Mass of solute = 1 g

$$\Delta T_f = 0.40 \text{ g}$$

The depression in freezing point is given by the formula,

$$\Delta T_f = K_F m$$

On substituting the values, we get

$$0.4 = \frac{K_F \times \frac{1}{256}}{50 \times 10^{-3}}$$

$$\Rightarrow \quad K_F = 5.12 \text{ K kg/mol}$$

66. *(c)* Among the given options,

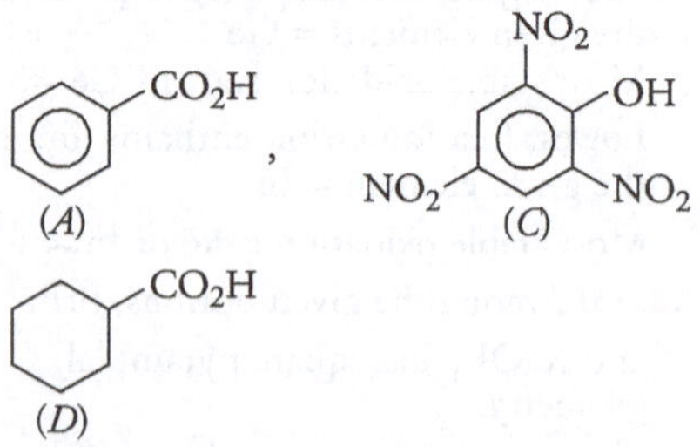

produces CO_2 with aqueous $NaHCO_3$ solution.

67. *(d)* The dissociation of weak acid is given by

$$HA \rightleftharpoons H^+ + A^-$$

	HA	H^+	A^-
at $t=0$	a	–	
at $t=t$	$a(1-x)$	ax	ax

$$K_a = (ax)\frac{(x)}{(1-x)},\ [\text{H}^+] = ax$$

$$-\log(K_a) = -\log(ax) - \log\left(\frac{x}{1-x}\right)$$

$$\text{p}K_a = \text{pH} - \log\left(\frac{x}{1-x}\right)$$

$$\therefore \text{pH} - \text{p}K_a = \log\left(\frac{x}{1-x}\right)$$

68. *(b)* Among the given ionic species. Ni^{2+} gives violet coloured bead in non-luminous flame under hot conditions. Ni^{2+} has d^8 configuration which does not depend on nature of ligand present in octahedral complex.

$Ni^{2+} : t_{2g}^6 e_g^2$

69. *(c)*

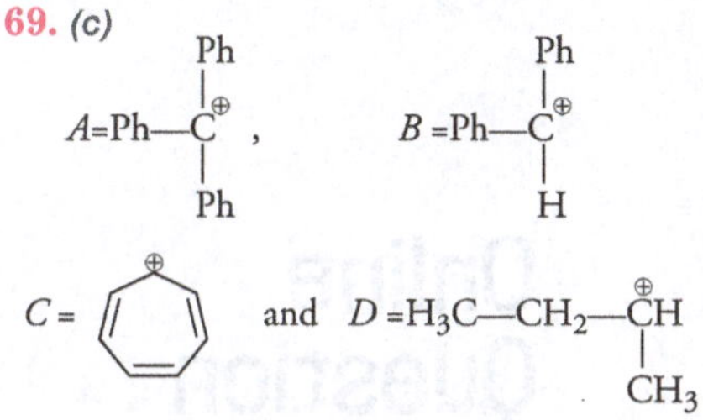

Now, C is aromatic due to positive charge and hence it is most stable.

A has more resonance structure.

B has less resonance structure

D has only hyperconjugation.

$\therefore$ Stability is in order,

Aromaticity > Resonance > Hyper conjugation

Hence, the correct order of stability is $C > A > B > D$.

70. *(a)* The complete reaction involved is as follows.

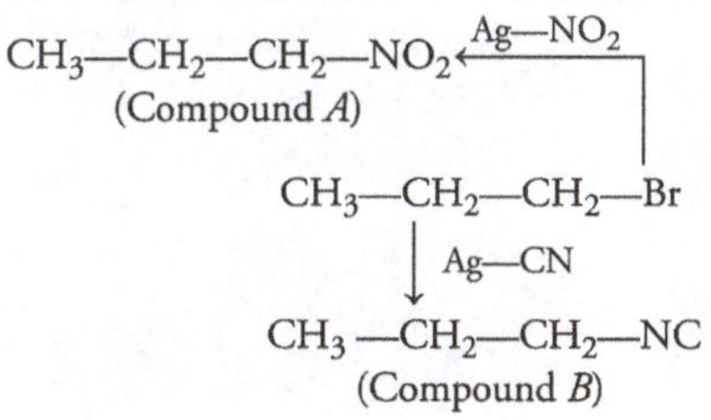

71. *(150)* Given, $R = 100\ \Omega$

Resistance of same cell with dilute solution $x = ?$

The resistance is given by the formule

$$R = \rho\frac{l}{A}$$

$$\kappa = G \times G^*\ (\text{where, } G = \frac{1}{R})$$

$$\text{and } G^*\ (\text{Cell constant}) = \frac{l}{A}$$

$$\frac{\kappa_C}{\kappa_d} = \frac{R_d}{R_C};$$

$$\lambda_m = \frac{\kappa \times 1000}{C}$$

$$\frac{\kappa_C}{\kappa_d} = \frac{(\lambda_m . C)_c}{(\lambda_m . C)_d} = \frac{R_d}{R_C}$$

$C \to$ concentrated solution,

$d \to$ dilute solution.

$$\frac{100 \times (0.15)^2}{150 \times (0.1)^2} = \frac{R_d}{100}$$

$$\Rightarrow \quad R_d = 150\Omega$$

So, the value of x is 150.

72. *(125)* Given, $\frac{\text{Mass}}{\text{Mass}}\% = 70\%$

Molarity = 2

Let 100 g solution contain 70 g solute.

Volume of 100 g solution will be $\frac{100}{1.25}$ mL

$$\text{Molarity} = \frac{70/70}{100/1.25} \times 1000$$

$$= 12.5 \text{ or } 125 \times 10^{-1}$$

73. *(466)* The reaction involved is as follows

$$\frac{1}{2}H_2(g) \longrightarrow H(g);$$

$$\Delta_f H(H(g)) = 220 \text{ kJ/ mol} \quad \ldots(i)$$

$$\frac{1}{2}O_2(g) \longrightarrow O(g);$$

$$\Delta_f H(O_g) = 250 \text{ kJ/mol} \quad \ldots(ii)$$

[Eq (i) × 2 + Eq (ii)] −2 [BE(O—H)]

$$H_2(g) + \frac{1}{2}O_2(g) \longrightarrow H_2O(g)$$

$$\Delta H_f(H_2O(l)) = -242 = 440 + 250 - 2(\text{BE(O—H)})$$

BE(O—H) = 466 kJ/ mol

74. *(1655)*

Key idea To solve this problem, first of all find the molar ratio of all the elements, then find the minimum integral ratio and put this value as a subscript with respective element to obtain empirical formula.

	C :	Cl :	H :	O
% mass	14.5	64.46	1.8	19.24

Molar ratio

$$= \frac{14.5}{12}, \frac{64.46}{35.5}, \frac{1.8}{1}, \frac{19.24}{16}$$

$$= 1.2 \quad 1.8 \quad 1.8 \quad 1.2$$

Minimum integral ratio 2 3 3 2

$\therefore$ Empirical formulae $= C_2H_3Cl_3O_2$

So, the empirical mass is

$$= 165.5$$

$$\text{Mass} = 1655 \times 10^{-1}$$

75. *(93)* The complete reaction is as follows

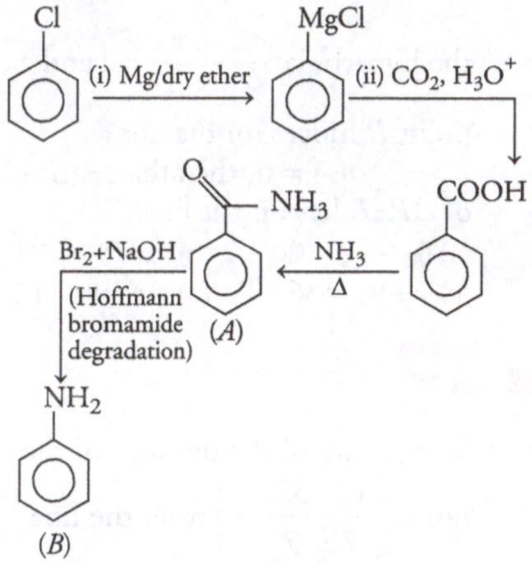

Number of moles of 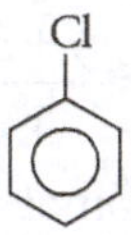

= Number of moles of 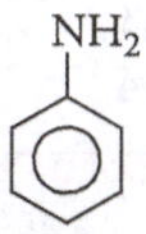

$$\frac{11.25 \times 10^{-3}}{112.5} = \frac{x \times 10^{-1} \times 10^{-3}}{93}$$

$$x = 93 \text{ mg}$$

Time 3:00-6:00 | MM : 300 Marks

28 January
Shift-II

Online Question Paper

—

JEE Main 2025

Mathematics

Section A
Objective Type Questions

01/ If A and B are points of intersection of the circle $x^2+y^2-8x=0$ and the hyperbola $\frac{x^2}{9}-\frac{y^2}{4}=1$ and a point P moves on the lines $2x-3y+4=0$, then the centroid of ΔPAB lies on the line :

(a) $6x-9y=20$ (b) $9x-9y=32$
(c) $x+9y=36$ (d) $4x-9y=12$

Imp.

02/ The square of the distance of the point $\left(\frac{15}{7},\frac{32}{7},7\right)$ from the line $\frac{x+1}{3}=\frac{y+3}{5}=\frac{z+5}{7}$ in the direction of the vector $\hat{i}+4\hat{j}+7\hat{k}$ is

(a) 44 (b) 41
(c) 54 (d) 66

Distance of a Point from the Line

03/ If the mid-point of a chord of the ellipse $\frac{x^2}{9}+\frac{y^2}{4}=1$ is $\left(\sqrt{2},\frac{4}{3}\right)$ and the length of the chord is $\frac{2\sqrt{\alpha}}{3}$, then α is

(a) 26 (b) 18
(c) 20 (d) 22

04/ If the components of $a=\alpha\hat{i}+\beta\hat{j}+\gamma\hat{k}$ along and perpendicular to respectively, are $\frac{16}{11}(3\hat{i}+\hat{j}-\hat{k})$ and $\frac{1}{11}(-4\hat{i}-5\hat{j}-17\hat{k})$, then $\alpha^2+\beta^2+\gamma^2$ is equal to

(a) 23 (b) 16 (c) 26 (d) 18

05/ Let f be a real valued continuous function defined on the positive real axis such that $g(x)=\int_0^x tf(t)\,dt$

If $g(x^3)=x^6+x^7$, then value of $\sum_{r=1}^{15} f(r^3)$ is

(a) 270 (b) 320 (c) 340 (d) 310

06/ Let $[x]$ denote the greatest integer less than or equal to x. Then, the domain of $f(x)=\sec^{-1}(2[x]+1)$ is

(a) $(-\infty,-1]\cup[1,\infty)$
(b) $(-\infty,-1]\cup[0,\infty)$
(c) $(-\infty,\infty)-\{0\}$
(d) $(-\infty,\infty)$

07/ Let S be the set of all the words that can be formed by arranging all the letters of the words GARDEN. From the set S, one word is selected at random. The probability that the selected word will not have vowels in alphabetical order is.

(a) 1/3 (b) 1/2 (c) 1/4 (d) 2/3

08/ If $f(x)=\int \frac{1}{x^{1/4}(1+x^{1/4})}dx, f(0)=-6$, then $f(1)$ is equal to

(a) $2-\log_e 2$ (b) $\log_e 2+2$
(c) $4(\log_e 2+2)$ (d) $4(\log_e 2-2)$

09/ Let $f:R-\{0\}\to(-\infty,1)$ be a polynomial of degree 2, satisfying $f(x)f\left(\frac{1}{x}\right)=f(x)+f\left(\frac{1}{x}\right)$.

If $f(K)=-2K$, then the sum of squares of all possible values of K is

(a) 7 (b) 6 (c) 1 (d) 9

10/ Let $A=\begin{bmatrix}\frac{1}{\sqrt{2}} & -2\\ 0 & 1\end{bmatrix}$ and $P=\begin{bmatrix}\cos\theta & -\sin\theta\\ \sin\theta & \cos\theta\end{bmatrix}, \theta>0$. If $B=PAP^T, C=P^TB^{10}P$ and the sum of the diagonal elements of C is $\frac{m}{n}$, where $gcd(m,n)=1$, then $m+n$ is

(a) 127 (b) 65 (c) 258 (d) 2049

11/ Let A, B, C be three points in XY-plane, whose position vector are given by $\sqrt{3}\hat{i}+\hat{j}, \hat{i}+\sqrt{3}\hat{j}$ and $a\hat{i}+(1-a)\hat{j}$, respectively with respect to the origin O. If the distance of the point C from the line bisecting the angle between the vectors **OA** and **OB** is $\frac{9}{\sqrt{2}}$, then the sum of all the possible values of a is

(a) 1 (b) $\frac{9}{2}$ (c) 0 (d) 2

12/ Bag B_1 contains 6 white and 4 blue balls, Bag B_2 contains 4 white and 6 blue balls and Bag B_3 contains 5 white and 5 blue balls. One of the bags is selected at random and a ball is drawn from it, If the ball is white, then the probability that the ball is drawn from Bag B_2 is

(a) $\frac{2}{5}$ (b) $\frac{2}{3}$ (c) $\frac{1}{3}$ (d) $\frac{4}{15}$

13/ Two equal sides of an isosceles triangle are along $-x+2y=4$ and $x+y=4$. If m is the slope of its third side, then the sum of all possible distinct values of m is

(a) 6 (b) $-2\sqrt{10}$ (c) 12 (d) -6

14/ For positive integers n, if $4a_n=(n^2+5n+6)$ and $S_n=\sum_{k=1}^{n}\left(\frac{1}{a_k}\right)$, then the value of $507\,S_{2025}$ is

(a) 675 (b) 1350 (c) 540 (d) 135

15/ Let $f:R\to R$ be a twice differentiable function such that $f(2)=1$. If $f(x)=xf(x)\ \forall x\in R$,

$\int_0^2 xF'(x)dx=6$ and $\int_0^2 x^2F''(x)dx=40$, then

$F'(2)+\int_0^2 F(x)dx$ is equal to

(a) 13 (b) 11 (c) 15 (d) 9

16/ If $\sum_{r=1}^{13}\left\{\frac{1}{\sin\left(\frac{\pi}{4}+(r-1)\frac{\pi}{6}\right)\sin\left(\frac{\pi}{4}+\frac{r\pi}{6}\right)}\right\}$ $=a\sqrt{3}+b, a, b\in Z$, then a^2+b^2 is equal to

(a) 2 (b) 8 (c) 10 (d) 4

17/ The area of the region bounded by the curves $x(1+y^2)=1$ and $y^2=2x$ is

(a) $\frac{\pi}{4}-\frac{1}{3}$ (b) $2\left(\frac{\pi}{2}-\frac{1}{3}\right)$ (c) $\frac{\pi}{2}-\frac{1}{3}$ (d) $\frac{1}{2}\left(\frac{\pi}{2}-\frac{1}{3}\right)$

18/ Let $f:[0,3]\to A$ be defined by $f(x)=2x^3-15x^2+36x+7$ and $g:[0,\infty)\to B$ be defined by $g(x)=\frac{x^{2025}}{x^{2025}+1}$. If both the functions are onto and $S=\{x\in Z: x\in A \text{ or } x\in B\}$, then $n(S)$ is equal to

(a) 36 (b) 29 (c) 31 (d) 30

19/ Let the coefficients of three consecutive terms T_r, T_{r+1} and T_{r+2} in the binomial expansion of $(a+b)^{12}$ be in a GP and let p be the number of all possible values of r. Let q be the sum of all rational terms in the binomial expansion of $(\sqrt[4]{3}+\sqrt[3]{4})^{12}$. Then, $p+q$ is equal to

(a) 295 (b) 283 (c) 287 (d) 299

20/ If $\alpha+i\beta$ and $\gamma+i\delta$ are the roots of $x^2-(3-2i)x-(2i-2)=0, i=\sqrt{-1}$, then $\alpha\gamma+\beta\delta$ is equal to

(a) 6 (b) -6 (c) -2 (d) 2

Section B

Numerical Type Questions

21/ The interior angles of a polygon with n sides are in an AP with common difference $6°$. If the largest interior angle of the polygon is $219°$, then n is equal to……

22/ Let $f(x) = \lim_{n \to \infty} \sum_{r=0}^{n} \left(\frac{\tan(x/2^{r+1}) + \tan^3(x/2^{r+1})}{1 - \tan^2(x/2^{r+1})} \right)$,

then $\lim_{x \to 0} \frac{e^x - e^{f(x)}}{(x - f(x))}$ is equal to

23/ If $y = y(x)$ is the solution of the differential equation,

$$\sqrt{4 - x^2} \frac{dy}{dx} = \left(\left(\sin^{-1}\left(\frac{x}{2}\right) \right)^2 - y \right) \sin^{-1}\left(\frac{x}{2}\right),$$

$-2 \le x \le 2$, $y(2) = \frac{\pi^2 - 8}{4}$, then $y^2(0)$ is equal to ……… .

Imp.

24/ The number of natural number between 212 and 999 such that the sum of their digits is 15, is ……… .

Binomial Expansion

25/ Let A and B be the two points of intersection of the line $y + 5 = 0$ and the mirror image of the parabola $y^2 = 4x$ with respect to the line $x + y + 4 = 0$. If d denotes the distance between A and B and a denotes the area of ΔSAB, where S is the focus of the parabola $y^2 = 4x$, then the value of $(a + d)$ is

Physics

Section A

Objective Type Questions

26/ A body of mass 4 kg is placed on a plane at a point P having coordinate (3,4) m. Under the action of force $\mathbf{F} = (2\hat{i} + 3\hat{j})$ N, it moves to a new point Q having coordinates (6,10) m in 4 s. The average power and instantaneous power at the end of 4 s are in the ratio of

(a) 4 : 3 (b) 6 : 13 (c) 13 : 6 (d) 1 : 2

27/ The frequency of revolution of the electron in Bohr's orbit varies with n, the principal quantum number is

(a) $\frac{1}{n^3}$ (b) $\frac{1}{n^2}$ (c) $\frac{1}{n}$ (d) $\frac{1}{n^4}$

28/ A 400 g solid cube having an edge of length 10 cm floats in water. How much volume of the cube is outside the water ?

(Given, density of water = 1000 kg m^{-3})

(a) 4000 cm^3 (b) 1400 cm^3 (c) 600 cm^3 (d) 400 cm^3

29/ The ratio of vapour densities of two gases at the same temperature is 4/25, then the ratio of rms velocities will be

(a) $\frac{5}{2}$ (b) $\frac{2}{5}$ (c) $\frac{25}{4}$ (d) $\frac{4}{25}$

30/ The kinetic energy of translation of the molecules in 50 g of CO_2 gas at 17°C is

(a) 4205.5 J (b) 3986.3 J (c) 4102.8 J (d) 3582.7 J

31/ A concave mirror produces an image of an object such that the distance between the object and image is 20 cm. If the magnification of the image is −3, then the magnitude of the radius of the curvature of the mirror is

(a) 7.5 cm (b) 30 cm (c) 3.75 cm (d) 15 cm

32/ In a long glass tube, mixture of two liquids A and B with refractive indices 1.3 and 1.4 respectively, forms a convex refractive meniscus towards A. If an object placed at 13 cm from the vertex of the meniscus in A forms an image with a magnification of −2, then the radius of curvature of meniscus is

(a) 2/3 cm (b) 4/3 cm (c) 1/3 cm (d) 1 cm

33/ Given below are two statements. One is labelled as Assertion (A) and the other is labelled as Reason (R).

Assertion(A) Knowing initial position x_0 and initial momentum p_0 is enough to determine the position and momentum at any time for a simple harmonic motion with a given angular frequency ω.

Reason (R) The amplitude and phase can be expressed in terms of x_0 and p_0.

In the light of the above statements, choose the correct answer from the options given below.

(a) (A) is false but (R) is true.
(b) (A) is true but (R) is false.
(c) Both (A) and (R) are true and (R) is not the correct explanation of (A).
(a) Both (A) and (R) are true and (R) is the correct explanation of (A).

34/

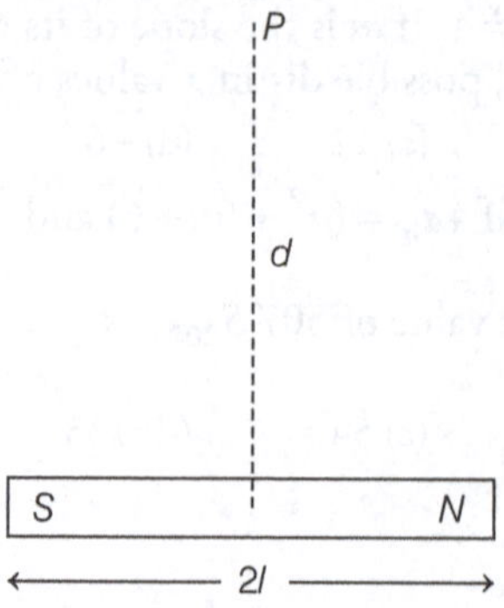

A bar magnet has total length $2l = 20$ units and the field point P is at a distance $d = 10$ units from the centre of the magnet. If the relative uncertainly of length measurement is 1%, then uncertainty of the magnetic field at point P is

(a) 5% (b) 3% (c) 4% (d) 10%

35/ A uniform rod of mass 250 g having length 100 cm is balanced on a sharp edge at 40 cm mark. A mass of 400 g is suspended at 10 cm mark. To maintain the balance of the rod, the mass to be suspended at 90 cm mark is

(a) 290 g (b) 190 g (c) 300 g (d) 200 g

36/

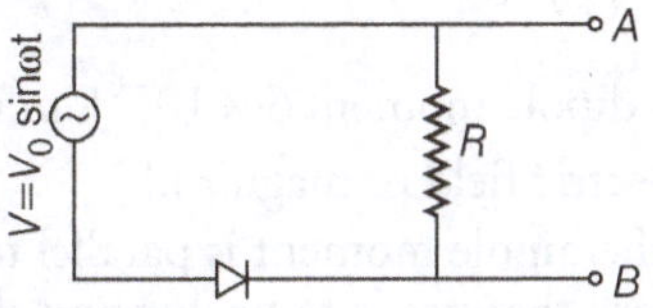

In the circuit shown here, assuming threshold voltage of diode is negligibly small, then voltage V_{AB} is correctly represented by

(a) V_{AB} would be zero at all time

(b)
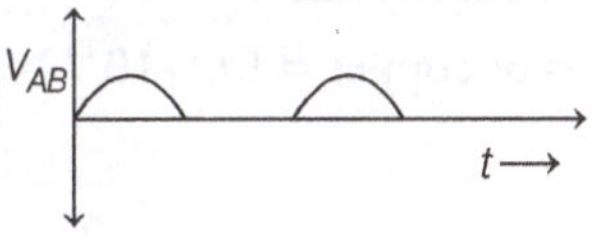

(c)
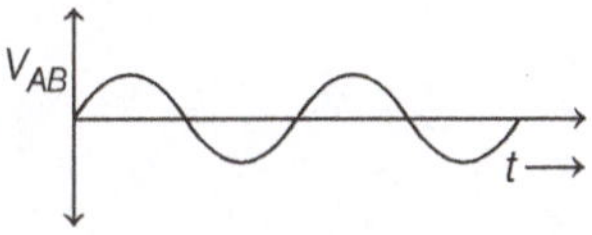

(d)
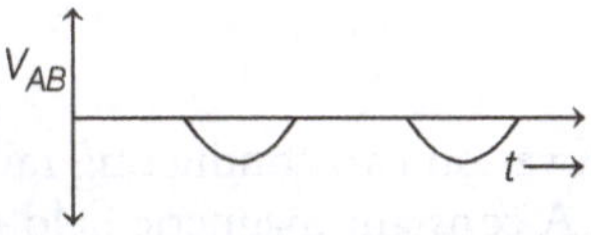

37/ A balloon and its content having mass M is moving up with an acceleration a. The mass that must be released from the content, so that the balloon starts moving up with an acceleration $3a$ will be (Take, g as acceleration due to gravity)

(a) $\dfrac{3Ma}{2a+g}$ (b) $\dfrac{3Ma}{2a-g}$

(c) $\dfrac{2Ma}{3a+g}$ (d) $\dfrac{2Ma}{3a-g}$

38/ Which of the following phenomena can not be explained by wave theory of light ?

(a) Refraction of light
(b) Diffraction of light
(c) Reflection of light
(d) Compton effect

Imp.

39/ The magnetic field of an EM wave is given by

$$\mathbf{B} = \left(\frac{\sqrt{3}}{2}\hat{i} + \frac{1}{2}\hat{j}\right) 30 \sin\left[\omega\left(t - \frac{z}{c}\right)\right] \text{ (SI Units).}$$

The corresponding electric field in SI unit is

(a) $\mathbf{E} = \left(\frac{\sqrt{3}}{2}\hat{i} - \frac{1}{2}\hat{j}\right) 30c \sin\left[\omega\left(t + \frac{z}{c}\right)\right]$

(b) $\mathbf{E} = \left(\frac{1}{2}\hat{i} - \frac{\sqrt{3}}{2}\hat{j}\right) 30c \sin\left[\omega\left(t - \frac{z}{c}\right)\right]$

(c) $\mathbf{E} = \left(\frac{1}{2}\hat{i} + \frac{\sqrt{3}}{2}\hat{j}\right) 30c \sin\left[\omega\left(t + \frac{z}{c}\right)\right]$

(d) $\mathbf{E} = \left(\frac{3}{4}\hat{i} + \frac{1}{4}\hat{j}\right) 30c \cos\left[\omega\left(t - \frac{z}{c}\right)\right]$

Characteristic of EM Wave

40/ Match List-I with List-II.

	List-I		List-II
(A)	Angular impulse	(I)	$[M^0L^2T^{-2}]$
(B)	Latent heat	(II)	$[ML^2T^{-3}A^{-1}]$
(C)	Electrical resistivity	(III)	$[ML^2T^{-1}]$
(D)	Electromotive force	(IV)	$[ML^3T^{-3}A^{-2}]$

Choose the correct answer from the options given below.

(a) (A)-(I), (B)-(III), (C)-(IV), (D)-(II)
(b) (A)-(III), (B)-(I), (C)-(II), (D)-(IV)
(c) (A)-(III), (B)-(I), (C)-(IV), (D)-(II)
(d) (A)-(II), (B)-(I), (C)-(IV), (D)-(III)

41/ The velocity-time graph of an object moving along a straight line is shown in figure. What is the distance covered by the object between $t = 0$ to $t = 4$ s?

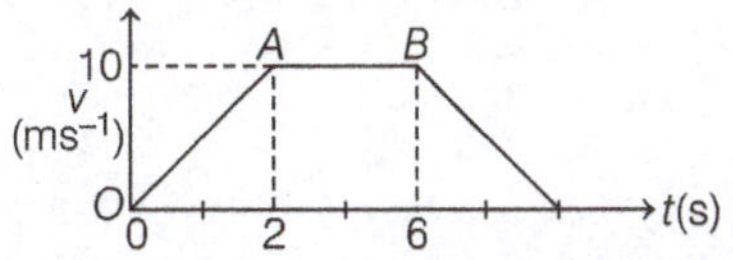

(a) 30 m (b) 13 m (c) 11 m (d) 10 m

42/ A uniform magnetic field of 0.4 T acts perpendicular to a circular copper disc 20 cm in radius. The disc is having a uniform angular velocity of 10π rad s^{-1} about an axis through its centre and perpendicular to the the disc. What is the potential difference developed between the axis of the disc and the rim ? ($\pi = 3.14$)

(a) 0.2512 V (b) 0.0628 V (c) 0.1256 V (d) 0.5024 V

43/ 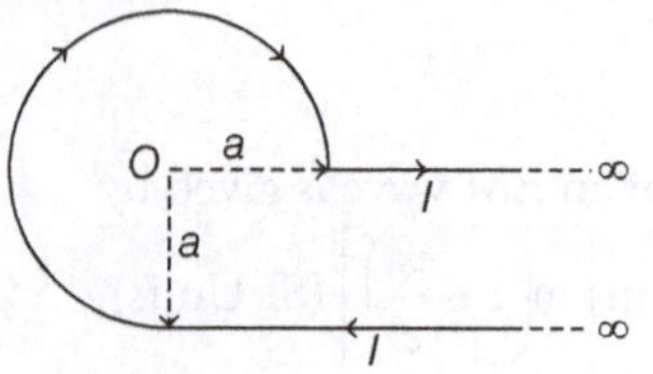

An infinite wire has a circular bend of radius a and carrying a current I as shown in figure. The magnitude of magnetic field at the origin O of the arc is given by

(a) $\frac{\mu_0}{4\pi}\frac{I}{a}\left[\frac{3\pi}{2}+1\right]$ (b) $\frac{\mu_0}{4\pi}\frac{I}{a}\left[\frac{\pi}{2}+1\right]$

(c) $\frac{\mu_0}{4\pi}\frac{I}{a}\left[\frac{3\pi}{2}+2\right]$ (d) $\frac{\mu_0}{2\pi}\frac{I}{a}\left[\frac{\pi}{2}+2\right]$

44/ Earth has mass 8 times and radius 2 times that of a planet. If the escape velocity from the earth is 11.2 km/s the escape velocity in km/s from the planet will be

(a) 2.8 (b) 5.6

(c) 8.4 (d) 11.2

Imp.

45/ A parallel plate capacitor of capacitance $1\mu F$ is charged to a potential difference of 20 V. The distance between plates is $1\ \mu m$. The energy density between plates of capacitor is

(a) $1.8\times10^3\ J/m^3$ (b) $2\times10^2\ J/m^3$

(c) $1.8\times10^5\ J/m^3$ (d) $2\times10^{-4}\ J/m^3$

Energy Density

Section B

Numerical Type Questions

46/ A thin transparent film with refractive index 1.4, is held on circular ring of radius 1.8 cm. The fluid in the film evaporates such that transmission through the film at wavelength 560 nm goes to a minimum every 12 s. Assuming that the film is flat on its two sides, the rate of evaporation is ....$\pi\times10^{-13}\ m^3/s$.

47/ The value of current I in the electrical circuit as given below, when potential at A is equal to the potential at B, will be......... A.

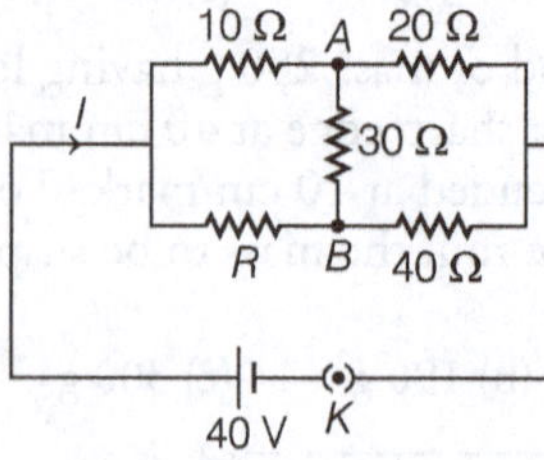

48/ An electric dipole of dipole moment 6×10^{-6}Cm is placed in uniform electric field of magnitude 10^6 V/m. Initially, the dipole moment is parallel to electric field. The work that needs to be done on the dipole to make its dipole moment opposite to the field, will be.......J.

49/ The volume contraction of a solid copper cube of edge length 10 cm, when subjected to a hydraulic pressure of 7×10^6, Pa, would be......mm^3.

(Given, bulk modulus of copper $= 1.4\times10^{11}\ Nm^{-2}$)

50/ 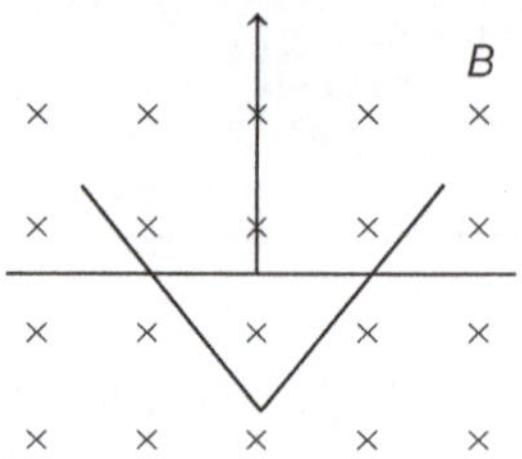

A conducting bar moves on two conducting rails as shown in the figure. A constant magnetic field B exists into the page. The bar starts to move from the vertex at time $t=0$ with a constant velocity. If the induced EMF is $E\propto t^n$, then value of n is..........

Chemistry

Section A

Objective Type Questions

51/ Match List-I with List-II.

	List-I (Complex)		List-II (Hybridisation of central metal ion)
(A)	$[CoF_6]^{3-}$	I.	d^2sp^3
(B)	$[NiCl_4]^{2-}$	II.	sp^3
(C)	$[Co(NH_3)_6]^{3+}$	III.	sp^3d^2
(D)	$[Ni(CN)_4]^{2-}$	IV.	dsp^2

Choose the correct answer from the options given below.

(a) (A)-(III), (B)-(II), (C)-(I), (D)-(IV)
(b) (A)-(I), (B)-(II), (C)-(III), (D)-(IV)
(c) (A)-(I), (B)-(IV), (C)-(III), (D)-(II)
(d) (A)-(III), (B)-(IV), (C)-(I), (D)-(II)

52/ The product *B* formed in the following reaction sequence is

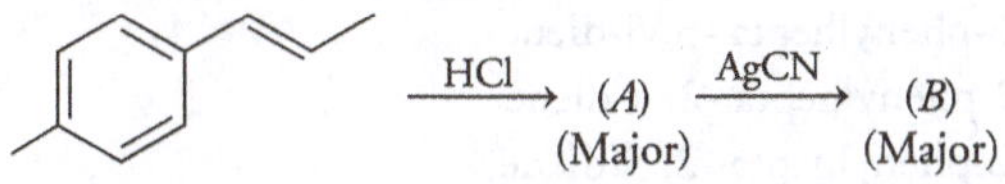

(a) 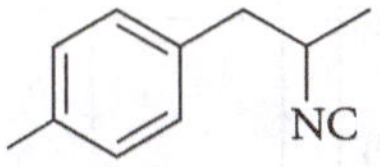

(b) 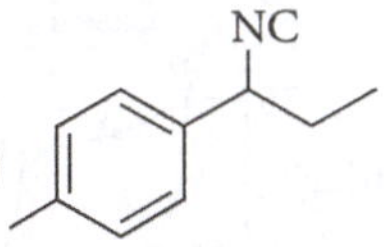

(c) 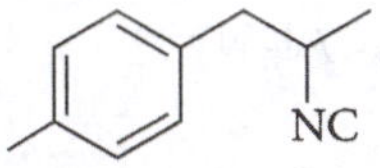

(d) 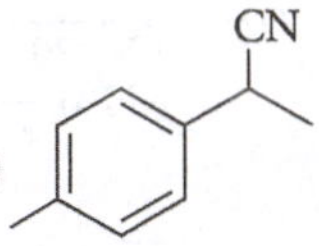

53/ The amphoteric oxide among V_2O_3, V_2O_4 and V_2O_5, upon reaction with alkali leads to formation of an oxide anion. The oxidation state of V in the oxide anion is

(a) +4 (b) +7
(c) +5 (d) +3

54/ For bacterial growth in a cell culture, growth law is very similar to the law of radioactive decay. Which of the following graphs is most suitable to represent bacterial colony growth?

Where N-number of bacteria at any time, N_0 –Initial number of bacteria.

(a) 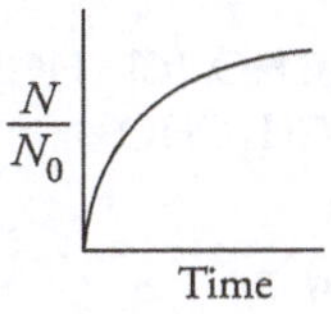

(b) 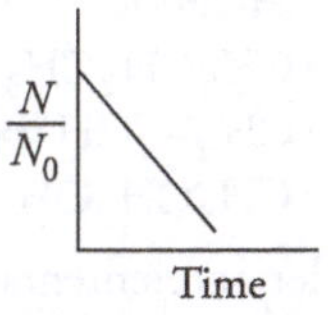

(c) 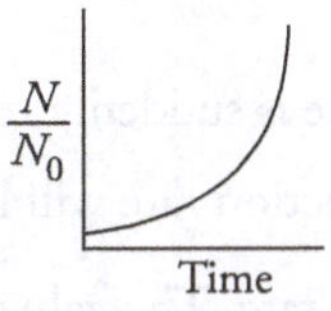

(d) 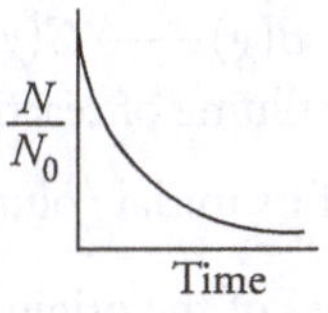

55/ Identify correct statements.

A. Primary amines do not give diazonium salts when treated with $NaNO_2$ in acidic condition.
B. Aliphatic and aromatic primary amines on heating with $CHCl_3$ and ethanolic KOH form carbylamines.
C. Secondary and tertiary amines also give carbylamine test.
D. Benzenesulphonyl chloride is known as Hinsberg's reagent.
E. Tertiary amines reacts with benzenesulphonyl chloride very easily.

Choose the correct answer from the options given below.

(a) (D) and (E) only (b) (B) and (C) only
(c) (B) and (D) only (d) (A) and (B) only

Imp.

56/ Identify correct conversion during acidic hydrolysis from the following.

A. Starch gives galactose.
B. Cane sugar gives equal amount of glucose and fructose.
C. Milk sugar gives glucose and galactose.
D. Amylopectin gives glucose and fructose.
E. Amylose gives only glucose.

Choose the correct answer from the options given below

(a) (A), (B) and (C) only (b) (B), (C) and (D) only
(c) (C), (D) and (E) only (d) (B), (C) and (E) only

Acidic Hydrolysis of Carbohydrates

57/ Identify product [*A*], [*B*] and [*C*] in the following reaction sequence

$$CH_3—C\equiv CH \xrightarrow[H_2]{Pd/C} [A] \xrightarrow[(ii)\ Zn,\ H_2O]{(i)\ O_3} [B]+[C]$$

(a) $[A]: CH_3—CH=CH_2$, $[B]: CH_3CHO$, $[C]: HCHO$

(b) $[A]: CH_2=CH_2$, $[B]: H_3C—\overset{\overset{\large O}{\|}}{C}—CH_3$, $[C]: HCHO$

(c) $[A]: CH_3CH_2CH_3$, $[B]: CH_3CHO$, $[C]: HCHO$

(d) $[A]: CH_3—CH=CH_2$, $[B]\ CH_3CHO$, $[C]: CH_3CH_2OH$

58/ Consider an elementary reaction,

$A(g) + B(g) \longrightarrow C(g) + D(g)$

If the volume of reaction mixture is suddenly reduced to $\frac{1}{3}$ of its initial volume, the reaction rate will become 'x' times of the original reaction rate. The value of x is

(a) 1/9 (b) 1/3 (c) 9 (d) 3

Imp.

59/ Arrange the following in increasing order of solubility product $Ca(OH)_2$, AgBr, PbS, HgS

(a) $Ca(OH)_2$ < AgBr < HgS < PbS

(b) HgS < PbS < AgBr < $Ca(OH)_2$

(c) PbS < HgS < $Ca(OH)_2$ < AgBr

(d) HgS < AgBr < PbS < $Ca(OH)_2$

Order of Solubility Product

60/ The purification method based on the following physical transformation is

$$\underset{(X)}{\text{Solid}} \xrightarrow{\text{Heat}} \underset{(Y)}{\text{Vapour}} \xrightarrow{\text{Cool}} \underset{(Z)}{\text{Solid}}$$

(a) Extraction (b) Sublimation

(c) Distillation (d) Crystallisation

61/ The total number of compounds from below when treated with hot $KMnO_4$ giving benzoic acid is

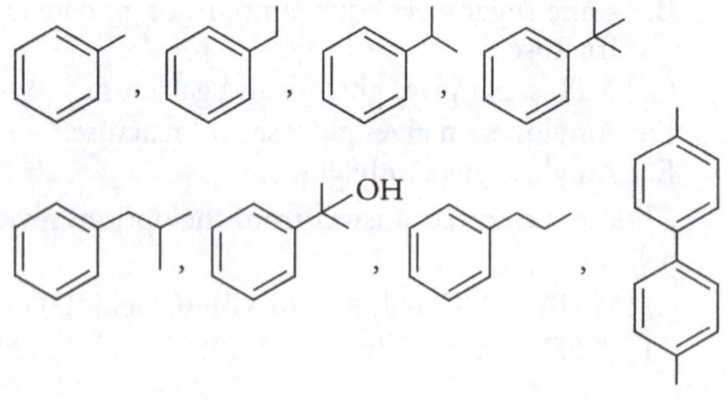

(a) 5 (b) 6 (c) 3 (d) 4

62/ Assume a living cell with 0.9% (ω / ω) of glucose solution (aqueous). This cell is immersed in another solution having equal mole fraction of glucose and water.

(Consider the data upto first decimal place only)

The cell will :

(a) Show no change in volume since solution is 0.9% (ω/ ω)

(b) Shrink since solution is 0.45% (ω/ ω) as a result of association of glucose molecules (due to hydrogen bonding)

(c) Swell up since solution is 1% (ω/ ω)

(d) Shrink since solution is 0.5% (ω/ ω)

63/ Which of the following is/are not correct with respect to energy of atomic orbitals of hydrogen atom?

(A) $1s < 2p < 3d < 4s$

(B) $1s < 2s = 2p < 3s = 3p$

(C) $1s < 2s < 2p < 3s < 3p$

(D) $1s < 2s < 4s < 3d$

Choose the correct answer from the options given below

(a) (A) and (B) only (b) (C) and (D) only

(c) (B) and (D) only (d) (A) and (C) only

64/ The major product of the following reaction is

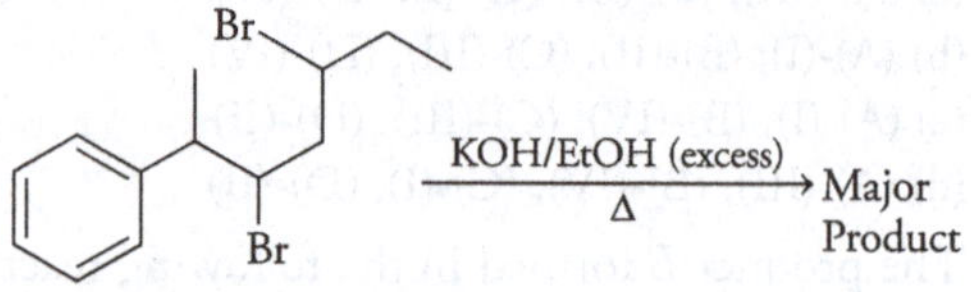

(a) 6-phenylhepta -2, 4-diene

(b) 6-phenylhepta -3, 5-diene

(c) 2-phenylhepta -2, 4-diene

(d) 2-phenylhepta -2, 5-diene

65/

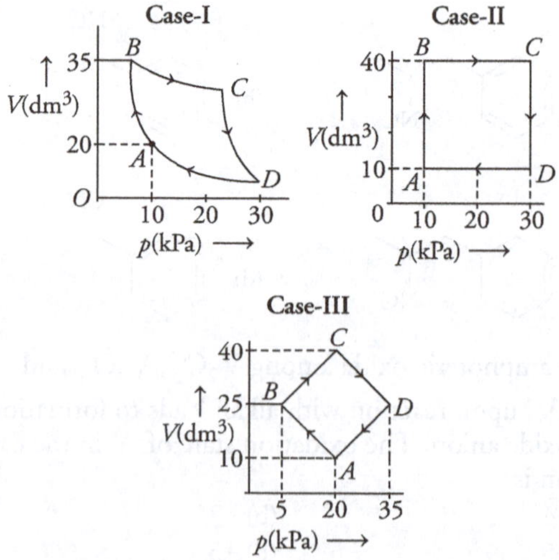

An ideal gas undergoes a cyclic transformation starting from the point A and coming back to the same point by tracing the path $A \rightarrow B \rightarrow C \rightarrow D \rightarrow A$ as shown in the three cases above.

Choose the correct option regarding ΔU :
(a) ΔU (Case-I) = ΔU (Case-II) = ΔU (Case-III)
(b) ΔU (Case-I) > ΔU (Case-II) > ΔU (Case-III)
(c) ΔU (Case-III) > ΔU (Case-II) > ΔU (Case-I)
(d) ΔU (Case-I) > ΔU (Case-III) > ΔU (Case-II)

66/ Given below are two statements :

Statement I 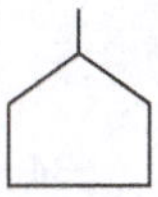 and 

are isomeric compounds.

Statement II 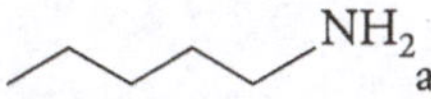

and 

are functional group isomers.
In the light of the above statements, choose the correct answer from the options given below :
(a) Both Statement I and Statement II are true.
(b) Statement I is true but Statement II is false.
(c) Statement I is false but Statement II is true.
(d) Both Statement I and Statement II are false.

67/ Concentrated nitric acid is labelled as 75% by mass. The volume in mL of the solution which contains 30g of nitric acid is......... .
(Given, density of nitric acid solution is 1.25 g/mL.)
(a) 32 (b) 55
(c) 45 (d) 40

Imp.

68/ Identify the inorganic sulphides that are yellow in colour
(A) $(NH_4)_2S$ (B) PbS
(C) CuS (D) As_2S_3
(E) As_2S_5
Choose the correct answer from the options given below :
(a) (A) and (C) only (b) (D) and (E) only
(c) (A) and (B) only (d) (A), (D) and (E) only

Colour of Salt (Salt Analysis)

69/ Given below are two statements.
Statement I According to the law of octaves, the elements were arranged in the increasing order of their atomic number.
Statement II Meyer observed a periodically repeated pattern upon plotting physical properties of certain elements against their respective atomic numbers.
In the light of the above statements, choose the correct answer from the options given below
(a) Statement I is true but Statement II is false.
(b) Both Statement I and Statement II are true.
(c) Statement I is false but Statement II is true.
(d) Both Statement I and Statement II are false.

70/ Match List-I with List-II.

	List-I (Saccharides)		List-II (Glycosidic-linkages found)
(A)	Sucrose	I.	$\alpha 1-4$
(B)	Maltose	II.	$\alpha\, 1-4$ and $\alpha\, 1-6$
(C)	Lactose	III.	$\alpha\, 1-\beta\, 2$
(D)	Amylopectin	IV.	$\beta 1-4$

Choose the correct answer from the options given below :
(a) (A)-(III), (B)-(I), (C)-(IV), (D)-(II)
(b) (A)-(II), (B)-(IV), (C)-(III), (D)-(I)
(c) (A)-(IV), (B)-(II), (C)-(I), (D)-(III)
(d) (A)-(I), (B)-(II), (C)-(III), (D)-(IV)

Section B
Numerical Type Questions

71/ Electrolysis of 600 mL aqueous solution of NaCl for 5 min changes the pH of the solution to 12.
The current in Amperes used for the given electrolysis is.......(Nearest integer).

72/ Consider the following data :
Heat of formation of $CO_2(g) = -393.5$ kJ mol^{-1}
Heat of formation of $H_2O(l) = -286.0$ kJ mol^{-1}
Heat of combustion of benzene $= -3267.0$ kJ mol^{-1}
The heat of formation of benzene is ....kJ mol^{-1}
(Nearest integer)

73/ Total number of molecules/species from following which will be paramagnetic is....
O_2, O_2^+, O_2^-, NO, NO_2, CO,
$K_2[NiCl_4], [Co(NH_3)_6]Cl_3, K_2[Ni(CN)_4]$

74/ The spin only magnetic moment (μ) value (BM) of the compound with strongest oxidising power among Mn_2O_3, TiO and VO is .... BM (Nearest integer).

75/ A group 15 element forms $d\pi - d\pi$ bond with transition metals. It also forms hydride, which is a strongest base among the hydrides of other group members that form $d\pi - d\pi$ bond. The atomic number of the element is.......

Answer Key

Mathematics

1 (a)	2 (d)	3 (d)	4 (c)	5 (d)	6 (d)	7 (b)	8 (d)	9 (b)	10 (b)
11 (a)	12 (d)	13 (a)	14 (a)	15 (d)	16 (b)	17 (c)	18 (d)	19 (b)	20 (d)
21 (20)	22 (1)	23 (4)	24 (64)	25 (14)					

Physics

26 (b)	27 (a)	28 (c)	29 (a)	30 (c)	31 (d)	32 (a)	33 (d)	34 (b)	35 (b)
36 (d)	37 (c)	38 (d)	39 (b)	40 (c)	41 (a)	42 (a)	43 (a)	44 (b)	45 (a)
46 (54)	47 (2)	48 (12)	49 (50)	50 (1)					

Chemistry

51 (a)	52 (b)	53 (c)	54 (c)	55 (c)	56 (d)	57 (a)	58 (c)	59 (b)	60 (b)
61 (a)	62 (*)	63 (b)	64 (c)	65 (a)	66 (a)	67 (a)	68 (d)	69 (d)	70 (a)
71 (2)	72 (48)	73 (6)	74 (5)	75 (15)					

***Note** (*) None of the option is correct.*

Detailed Explanations

Mathematics

1. *(a)* Put $y^2 = 8x - x^2$

$$\frac{x^2}{9} - \frac{8x - x^2}{4} = 1$$

$$\Rightarrow 4x^2 - 72x + 9x^2 = 36$$

$$\Rightarrow 13x^2 - 72x - 36 = 0$$

$$\Rightarrow 13x^2 - 78x + 6x - 36 = 0$$

$$\Rightarrow x = 6 \text{ or } x = \frac{-6}{13} \text{ (not possible)}$$

If $x = 6$, then $y = \pm 2\sqrt{3}$

$\Rightarrow A \equiv (6, 2\sqrt{3})$ and $B \equiv (6, -2\sqrt{3})$

and $P \equiv (\alpha, \beta)$, then $2\alpha - 3\beta + 4 = 0$

Centroid of ΔPAB is

$$\left(\frac{\alpha + 12}{3}, \frac{\beta}{3}\right) \equiv (h, k)$$

$$\alpha = 3h - 12, \beta = 3k.$$

$$\Rightarrow \quad 6h - 24 - 9k + 4 = 0$$

$$\Rightarrow \quad 6h - 9k = 20$$

Locus is $\quad 6x - 9y = 20$

2. *(d)*

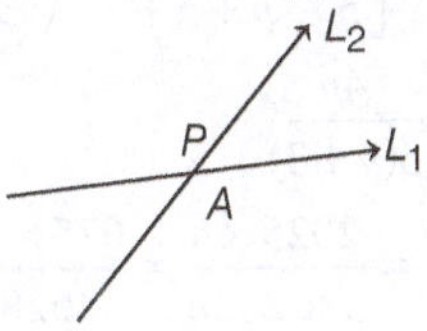

Let $L_1 : \dfrac{x+1}{3} = \dfrac{y+3}{5} = \dfrac{z+5}{7} = t$ (say)

$L_2 : \dfrac{x - \frac{15}{7}}{1} = \dfrac{y - \frac{32}{7}}{4} = \dfrac{z-7}{7} = r$ (say)

$$\Rightarrow P(3t - 1, 5t - 3, 7t - 5)$$

$$= P\left(\frac{15}{7} + r, \frac{32}{7} + 4r, 7 + 7r\right)$$

$$\Rightarrow 3t - 1 = \frac{15}{7} + r \quad \text{...(i)}$$

$$5t - 3 = \frac{32}{7} + 4r \quad \text{...(ii)}$$

Multiplying by 4 in Eq. (i) and subtract Eq. (ii), we get

$$12t - 4 = \frac{60}{7} + 4r$$

$$5t - 3 = \frac{32}{7} + 4r$$

$$- \quad + \quad - \quad -$$

$$7t - 1 = 4$$

$$\Rightarrow \quad 7t = 5 \Rightarrow t = \frac{5}{7}$$

$\therefore$ Coordinate of point

$$P \equiv \left(3 \times \frac{5}{7} - 1, 5 \times \frac{5}{7} - 3, 7 \times \frac{5}{7} - 5\right)$$

$$= \left(\frac{8}{7}, \frac{4}{7}, 0\right)$$

and coordinate of point $A \equiv \left(\dfrac{15}{7}, \dfrac{32}{7}, 7\right)$

$$\therefore \quad AP = \sqrt{\left(\frac{15}{7} - \frac{8}{7}\right)^2 + \left(\frac{32}{7} - \frac{4}{7}\right)^2 + (7 - 0)^2}$$

$$= \sqrt{1^2 + 4^2 + 7^2}$$

$$= \sqrt{1 + 16 + 49} = \sqrt{66}$$

$$\Rightarrow AP^2 = 66$$

3. *(d)* Equation of chord is $T = S_1$

$$\Rightarrow \frac{x\sqrt{2}}{9} + \frac{y}{3} = \frac{2}{3} \Rightarrow \sqrt{2}\,x + 3y = 6$$

$$\text{Now, } \frac{\left(\frac{6 - 3y}{\sqrt{2}}\right)^2}{9} + \frac{y^2}{4} = 1$$

$$\Rightarrow \quad (y - 2)(3y - 2) = 0$$

$$\Rightarrow \quad y = 2, \frac{2}{3}$$

If $y = 2$, then $x = 0$

If $y = \dfrac{2}{3}$, then $x = 2\sqrt{2}$

Length of chord is $\sqrt{8 + \dfrac{16}{9}} = \dfrac{2\sqrt{22}}{3}$

$$\frac{2\sqrt{\alpha}}{3} = \frac{2\sqrt{22}}{3}$$

$$\Rightarrow \quad \alpha = 22$$

4. *(c)* Given $\mathbf{a} = \mathbf{a}$ along $\mathbf{b} + \mathbf{a} \perp \mathbf{b}$

$$\Rightarrow \alpha\hat{i} + \beta\hat{j} + \gamma\hat{k} = \frac{16}{11}(3\hat{i} + \hat{j} - \hat{k}) + \frac{1}{11}(-4\hat{i} - 5\hat{j} - 17\hat{k})$$

$$= 4\hat{i} + \hat{j} - 3\hat{k}$$

$$\Rightarrow \alpha = 4, \beta = 1 \text{ and } \gamma = -3$$

$$\Rightarrow \alpha^2 + \beta^2 + \gamma^2 = 16 + 1 + 9 = 26$$

5. *(d)* $\quad g(x^3) = \displaystyle\int_0^{x^3} t\, f(t)\, dt$

$$\Rightarrow \quad x^7 + x^6 = \int_0^{x^3} t\, f(t)\, dt$$

$$\Rightarrow 7x^6 + 6x^5 = 3x^2 \cdot x^3 f(x^3)$$

$$\Rightarrow \quad f(x^3) = \frac{7}{3}x + 2$$

$$\text{Now, } \sum_{r=1}^{15} f(r^3) = \frac{7}{3}\sum_{r=1}^{15} r + 2 \times 15$$

$$= \frac{7}{3} \cdot \frac{15 \cdot 16}{2} + 30 = 310$$

> **Concept Enhancer**
>
> Let $\quad f(x) = \displaystyle\int_{g(x)}^{h(x)} f(x)dx$, then
>
> $$f'(x) = h'(x)f(h(x)) - g'(x)f(g(x))$$

6. *(d)* Given, $f(x) = \sec^{-1}(2[x] + 1)$

$2[x] + 1 \geq 1$ or $2[x] + 1 \leq -1$

$$\Rightarrow \quad [x] \geq 0 \text{ or } [x] \leq -1$$

$$\Rightarrow \quad x \in [0, \infty) \text{ or } x \in (-\infty, 0]$$

$$\Rightarrow \quad x \in (-\infty, \infty)$$

7. *(b)* There are 2 vowels A and E only

So, probability the selected word will not have vowels in alphabetical order

$$1 - \frac{\left(\frac{6!}{2!}\right)}{6!} = \frac{1}{2}$$

8. *(d)* Let $x^{1/4} = t$

$$\Rightarrow \quad x = t^4 \Rightarrow dx = 4t^3\, dt$$

$$\Rightarrow \quad I = 4\int \frac{t^3\, dt}{t\,(t + 1)}$$

$$= 4\int\left(\frac{t^2 - 1}{t + 1} + \frac{1}{t + 1}\right) dt$$

$$I = 4\left[\frac{t^2}{2} - t + \log_e(t + 1)\right] + C$$

Given, $f(x) = 2x^{1/2} - 4x^{1/4} + 4\log_e(x^{1/4} + 1) + C$

$f(0) = -6 \Rightarrow C = -6$

$f(1) = 2 - 4 + 4\log_e 2 - 6$

$= 4(\log_e 2 - 2)$

9. (b) $f(x)\, f\left(\frac{1}{x}\right) = f(x) + f\left(\frac{1}{x}\right)$

$\Rightarrow f(x) = 1 \pm x^2$ [degree 2 is given]

$\Rightarrow f(x) = 1 + x^2$ [not possible]

Since, range of $f(x)$ is $(-\infty, 1)$ [given]

$\Rightarrow f(x) = 1 - x^2 \Rightarrow 1 - K^2 = -2K$

$\Rightarrow K^2 - 2K - 1 = 0$

$\Rightarrow K_1 + K_2 = 2$

and $K_1 K_2 = -1$

$\Rightarrow K_1^2 + K_2^2 = (K_1 + K_2)^2 - 2K_1K_2$

$= 4 + 2 = 6$

10. (b) $P = \begin{bmatrix} \cos\theta & -\sin\theta \\ \sin\theta & \cos\theta \end{bmatrix}$

$P^T = \begin{bmatrix} \cos\theta & \sin\theta \\ -\sin\theta & \cos\theta \end{bmatrix}$

$PP^T = P^T P = I$

Now, $C = P^T B^{10} P$

$= P^T (PAP^T)(PAP^T) \ldots\ldots (PAP^T) P$

$= A^{10}$

Now, $A^2 = \begin{bmatrix} \left(\frac{1}{\sqrt{2}}\right)^2 & -\sqrt{2} - 2 \\ 0 & 1 \end{bmatrix}$

$A^3 = \begin{bmatrix} \left(\frac{1}{\sqrt{2}}\right)^3 & -\sqrt{2} - 3 \\ 0 & 1 \end{bmatrix}$

$A^n = \begin{bmatrix} \left(\frac{1}{\sqrt{2}}\right)^n & -\sqrt{2} - n \\ 0 & 1 \end{bmatrix}$

$tr(A^n) = \left(\frac{1}{\sqrt{2}}\right)^n + 1$

$tr(A^{10}) = \left(\frac{1}{\sqrt{2}}\right)^{10} + 1 = \frac{33}{32}$

$m = 33$ and $n = 32$

$\therefore m + n = 65$

11. (a)

Key idea

Distance of point $P(x_1, y_1)$ from $ax + by + c = 0$ is $\left|\frac{ax_1 + by_1 + c}{\sqrt{a^2 + b^2}}\right|$

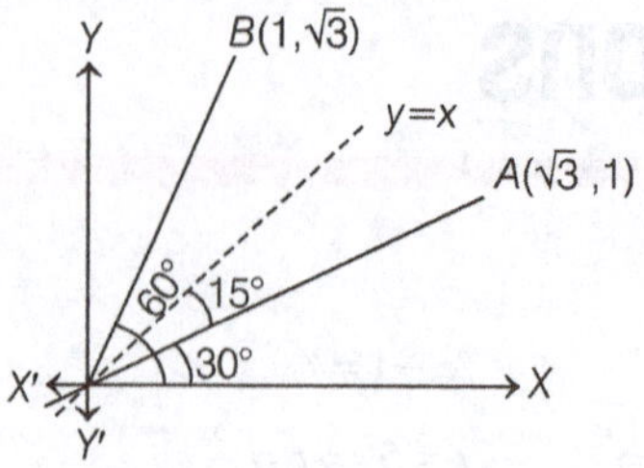

Distance of point $(a, 1 - a, 0)$ from line $y - x = 0$ is $\frac{9}{\sqrt{2}}$

$\Rightarrow \left|\frac{1 - a - a}{\sqrt{2}}\right| = \frac{9}{\sqrt{2}}$

$\Rightarrow 1 - 2a = \pm 9 \Rightarrow 1 \pm 9 = 2a$

$\Rightarrow -8, 10 = 2a \Rightarrow a = -4, 5$

$\therefore$ Sum $= -4 + 5 = 1$

12. (d)

Bag B_1	Bag B_2	Bag B_3
6W 4B	4W 6B	5W 5B

$P(B_1) = P(B_2) = P(B_3) = \frac{1}{3}$

$P\left(\frac{W}{B_1}\right) = \frac{6}{10}, P\left(\frac{W}{B_2}\right) = \frac{4}{10},$

$P\left(\frac{W}{B_3}\right) = \frac{5}{10}$

$\therefore P\left(\frac{B_2}{W}\right) = \frac{P(B_2)\, P\left(\frac{W}{B_2}\right)}{P(W)}$

$= \frac{\frac{1}{3} \times \frac{4}{10}}{\frac{1}{3} \times \frac{6}{10} + \frac{1}{3} \times \frac{4}{10} + \frac{1}{3} \times \frac{5}{10}}$

$= \frac{\frac{4}{30}}{\frac{15}{30}} = \frac{4}{30} \times \frac{30}{15} = \frac{4}{15}$

13. (a) Given, lines $L_1: -x + 2y = 4$ and $L_2: x + y = 4$

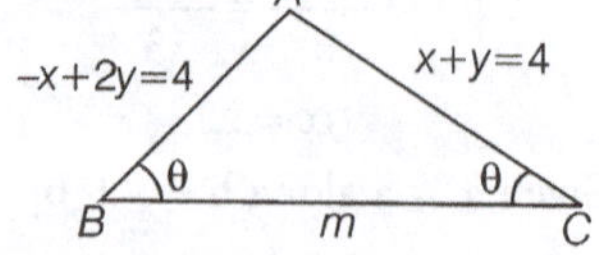

Slope $m_1 = \frac{1}{2}$ and $m_2 = -1$

In ΔABC $\tan\alpha = \tan\beta$

$\frac{m - m_1}{1 + mm_1} = \frac{-m_2 - m}{1 + mm_2}$

$\Rightarrow \frac{m - \frac{1}{2}}{1 + \frac{1}{2}m} = \frac{-1 - m}{1 - m}$

$\Rightarrow \frac{2m - 1}{2 + m} = \frac{-m - 1}{1 - m}$

$\Rightarrow m^2 - 6m - 1 = 0$

$\Rightarrow m = \frac{6 \pm \sqrt{40}}{2} \Rightarrow m = 3 \pm \sqrt{10}$

$\therefore$ Sum $= 6$

14. (a) Given,

$S_n = \sum_{k=1}^{n} \frac{1}{a_k} = \sum_{k=1}^{n} \frac{4}{k^2 + 5k + 6}$

$= \sum_{k=1}^{n} \frac{4}{(k+2)(k+3)}$

$= 4\sum_{k=1}^{n} \left(\frac{1}{k+2} - \frac{1}{k+3}\right)$

$= 4\left[\left(\frac{1}{3} - \frac{1}{4}\right) + \left(\frac{1}{4} - \frac{1}{5}\right) + \ldots + \left(\frac{1}{n+2} - \frac{1}{n+3}\right)\right]$

$= 4\left[\frac{1}{3} - \frac{1}{n+3}\right] = 4\left(\frac{n+3-3}{3(n+3)}\right)$

$= \frac{4n}{3(n+3)}$

$\therefore S_{2025} = \frac{2025 \times 4}{3 \times 2028} = \frac{675 \times 4}{2028}$

$\Rightarrow 507 S_{2025} = \frac{507 \times 675 \times 4}{2028}$

$= \frac{675}{4} \times 4 = 675$

15. (d) Given, $\int_0^2 x^2 F''(x)dx = 40$

$\Rightarrow [x^2 F'(x)]_0^2 - 2\int_0^2 xF'(x)dx = 40$

$\Rightarrow 4F'(2) - 0 - 2(6) = 40$

(given $\int_0^2 xF'(x)dx = 6$)

$\Rightarrow F'(2) = \frac{52}{4} = 13$

Also, $\int_0^2 xF'(x)dx = 6$

$\Rightarrow [xF(x)]_0^2 - \int_0^2 F(x)dx = 6$

$\Rightarrow 2F(2) - 0 - \int_0^2 F(x)dx = 6$

$\Rightarrow 2(1) - \int_0^2 F(x)dx = 6$

$$\Rightarrow \int_0^2 F(x)dx = -4$$

$$\therefore F'(2) + \int_0^2 F(x)dx = 13 - 4 = 9$$

16. *(b)*

$$\sum_{r=1}^{13} \frac{1}{\sin\left(\frac{\pi}{4} + (r-1)\frac{\pi}{6}\right)\sin\left(\frac{\pi}{4} + \frac{r\pi}{6}\right)}$$

$$= 2\sum_{r=1}^{13} \frac{\sin\left(\left(\frac{\pi}{4} + \frac{r\pi}{6}\right) - \left(\frac{\pi}{4} + (r-1)\frac{\pi}{6}\right)\right)}{\sin\left(\frac{\pi}{4} + (r-1)\frac{\pi}{6}\right)\sin\left(\frac{\pi}{4} + \frac{r\pi}{6}\right)}$$

$$= 2\sum_{r=1}^{13} \left[\cot\left(\frac{\pi}{4} + (r-1)\frac{\pi}{6}\right) - \cot\left(\frac{\pi}{4} + \frac{r\pi}{6}\right)\right]$$

$$= 2\left[\left(\cot\frac{\pi}{4} - \cot\left(\frac{\pi}{4} + \frac{\pi}{6}\right)\right)\right.$$

$$+ \left\{\left(\cot\frac{\pi}{4} + \frac{\pi}{6}\right) - \cot\left(\frac{\pi}{4} + \frac{2\pi}{6}\right)\right\} + \ldots$$

$$\ldots + \left\{\cot\left(\frac{\pi}{4} + \frac{12\pi}{6}\right) - \cot\left(\frac{\pi}{4} + \frac{13\pi}{6}\right)\right\}$$

$$= 2\left[1 - \cot\left(\frac{\pi}{4} + \left(2\pi + \frac{\pi}{6}\right)\right)\right]$$

$$= 2\left[1 - \cot\left(\frac{\pi}{4} + \frac{\pi}{6}\right)\right]$$

$$= 2\,[1 - 2 + \sqrt{3}] = 2(\sqrt{3} - 1) = a\sqrt{3} + b$$

[given]

$$\Rightarrow \quad a = 2 \text{ and } b = -2$$

$$\Rightarrow a^2 + b^2 = 4 + 4 = 8$$

17. (c) Given, $x(1 + y^2) = 1$ and $y^2 = 2x$

$$\Rightarrow \; x(1 + 2x) = 1 \quad \Rightarrow x + 2x^2 - 1 = 0$$

$$\Rightarrow 2x^2 + x - 1 = 0$$

$$\Rightarrow 2x^2 + 2x - x - 1 = 0$$

$$\Rightarrow 2x(x + 1) - (x + 1) = 0$$

$$\Rightarrow (2x - 1)(x + 1) = 0 \Rightarrow x = -1, \frac{1}{2}$$

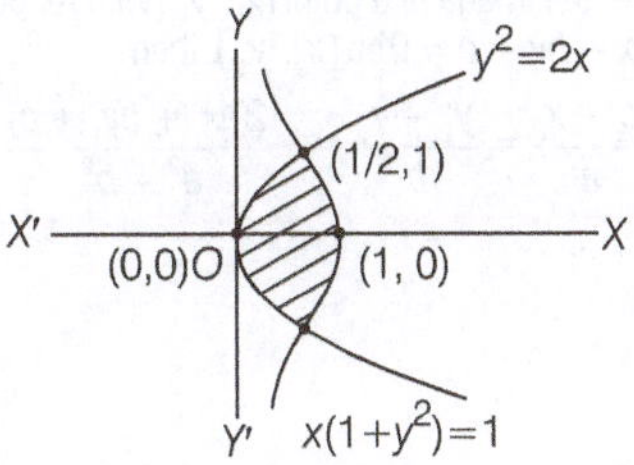

$$\therefore \text{Area} = 2\int_0^1 (x_2 - x_1)dy$$

$$= 2\int_0^1 \left(\frac{1}{y^2 + 1} - \frac{y^2}{2}\right)dy$$

$$= 2\left[\tan^{-1} y - \frac{y^3}{6}\right]_0^1$$

$$= 2\left[\left(\frac{\pi}{4} - \frac{1}{6}\right) - 0\right] = \frac{\pi}{2} - \frac{1}{3}$$

18. *(d)* $f(x) = 2x^3 - 15x^2 + 36x + 7$

$$\Rightarrow f'(x) = 6x^2 - 30x + 36$$

Put $f'(x) = 0 \Rightarrow 6(x^2 - 5x + 6) = 0$

$$\Rightarrow 6(x - 2)(x - 3) = 0 \Rightarrow x = 2, 3$$

$\therefore$ $f(2) = 35$, $f(3) = 34$ and $f(0) = 7$

$$\Rightarrow \quad A = [7, 35]$$

$\Rightarrow$ Total number of integer in $A = 29$

Now, $g(x) = \dfrac{x^{2025}}{x^{2025} + 1} = 1 - \dfrac{1}{x^{2025} + 1}$

$$\Rightarrow B = [0, 1)$$

$\Rightarrow$ Number of integer in $B = 1$

So, total number of integers in the range $A \cup B = 30$

(because no element is common in A and B)

19. *(b)*

Key idea

General term in the expansion of $(x + a)^n$ is

$$T_{r+1} = {}^nC_r x^{n-r} a^r$$

Given, $T_r, T_{r+1}, T_{r+2} \ldots$ are in GP

$$\Rightarrow (T_{r+1})^2 = T_r \cdot T_{r+2}$$

$$\Rightarrow ({}^{12}C_r\, a^{12-r} b^r)^2 = {}^{12}C_{r-1}\, a^{13-r} \cdot b^{r-1} \times {}^{12}C_{r+1}\, a^{11-r}\, b^{r+1}$$

$$\Rightarrow \quad ({}^{12}C_r)^2 = {}^{12}C_{r-1} \times {}^{12}C_{r+1}$$

$$\Rightarrow \quad \frac{{}^{12}C_{r-1}}{{}^{12}C_r} = \frac{{}^{12}C_r}{{}^{12}C_{r+1}}$$

$$\Rightarrow \frac{\dfrac{12!}{(r-1)!\,(13-r)!}}{\dfrac{12!}{r!\,(12-r)!}} = \frac{\dfrac{12!}{r!\,(12-r)!}}{\dfrac{12!}{(r+1)!\,(11-r)!}}$$

$$\Rightarrow -r = 13 - r \quad \Rightarrow \quad 13 = 0$$

Not possible

(no three consecutive binomial coefficients are in GP or HP but AP is possible) $\Rightarrow$ $p = 0$

Now, $T_{k+1} = {}^{12}C_k\, (3^{1/4})^{12-k}\, (4^{1/3})^k$

$$= {}^{12}C_k\, 3^{3-\frac{k}{4}}\, 4^{\frac{k}{3}}$$

For terms to be rational (4, 3) divides k

$\Rightarrow$ 12 divides k

$\Rightarrow$ 0, 12 $\rightarrow$ Two terms

$\therefore$ Sum of rational terms

$$= {}^{12}C_0 3^3 4^0 + {}^{12}C_{12}\, 4^4 3^0 = 3^3 + 4^4$$

$$= 27 + 256 = 283 = q$$

$$\therefore \; p + q = 0 + 283 = 283$$

20. *(d)* $x^2 - (3 - 2i)x - (2i - 2) = 0$

$$x = \frac{(3 - 2i) \pm \sqrt{(3 - 2i)^2 + 4 \times 1\,(2i - 2)}}{2 \times 1}$$

$$= \frac{(3 - 2i) \pm \sqrt{9 - 12i + 4i^2 + 8i - 8}}{2}$$

$$= \frac{(3 - 2i) \pm \sqrt{-3 - 4i}}{2}$$

$$= \frac{(3 - 2i) \pm \sqrt{(1 - 2i)^2}}{2}$$

$$= \frac{3 - 2i + 1 - 2i}{2}, \frac{3 - 2i - 1 + 2i}{2}$$

$$= \frac{4 - 4i}{2}, \frac{2}{2} = 2 - 2i, 1$$

Let $\alpha + i\beta = 2 - 2i$ and $\gamma + i\delta = 1$

$$\Rightarrow \quad \alpha = 2, \beta = -2, \gamma = 1, \delta = 0$$

$$\therefore \quad \alpha\gamma + \beta\delta = 2 \times 1 + 0 = 2$$

21. *(20)* $a + (n - 1)\,d = 219°$

$$\Rightarrow \quad a + (n - 1)6 = 219° \quad \ldots(i)$$

$$\Rightarrow \quad \frac{n}{2}[2a + (n - 1)6] = (n - 2)180°$$

$$\Rightarrow \quad \frac{n}{2}[a + 219] = (n - 2)180°$$

$$\Rightarrow \quad \frac{n}{2}[219 + 219 - (n - 1)6] = (n - 2)180°$$

$$\Rightarrow \quad n[219 - 3(n - 1)] = (n - 2)180°$$

$$\Rightarrow \quad n[222 - 3n] = 180n - 360$$

$$\Rightarrow \quad 3n^2 - 222n + 180n - 360 = 0$$

$$\Rightarrow \quad 3n^2 - 42n - 360 = 0$$

$$\Rightarrow n^2 - 14n - 120 = 0$$

$$\Rightarrow (n + 6)(n - 20) = 0 \Rightarrow n = 20$$

22. *(1)* Since,

$$\frac{\tan\theta + \tan^3\theta}{1 - \tan^2\theta} = \frac{2\tan\theta - \tan\theta + \tan^3\theta}{1 - \tan^2\theta}$$

$$= \frac{2\tan\theta}{1 - \tan^2\theta} - \frac{\tan\theta\,(1 - \tan^2\theta)}{1 - \tan^2\theta}$$

$= \tan 2\theta - \tan\theta$

$\therefore f(x) = \lim_{n\to\infty} \sum_{r=0}^{n} \left(\dfrac{\tan \dfrac{x}{2^{r+1}} + \tan^3 \dfrac{x}{2^{r+1}}}{1 - \tan^2 \dfrac{x}{2^{r+1}}} \right)$

$= \lim_{n\to\infty} \sum_{r\to 0}^{n} \tan\left(\dfrac{2x}{2^{r+1}}\right) - \tan\left(\dfrac{x}{2^{r+1}}\right)$

$= \lim_{n\to\infty} \left[\left(\tan\dfrac{x}{1} - \tan\dfrac{x}{2}\right) + \left(\tan\dfrac{x}{2} - \tan\dfrac{x}{2^2}\right) + \ldots + \left(\tan\dfrac{x}{2^n} - \tan\dfrac{x}{2^{n+1}}\right) \right]$

$= \lim_{n\to\infty} \left(\tan x - \tan\dfrac{x}{2^{n+1}}\right)$

$= \tan x - 0 = \tan x$

$\therefore \lim_{x\to 0} \dfrac{e^x - e^{f(x)}}{(x - f(x))} = \lim_{x\to 0} \dfrac{e^x - e^{\tan x}}{x - \tan x}$

$= \lim_{x\to 0} e^{\tan x} \lim_{x\to 0} \dfrac{e^{x-\tan x} - 1}{x - \tan x} = e^0 \cdot 1 = 1$

23. *(4)* Since, $\sqrt{4-x^2}\dfrac{dy}{dx}$

$= \left\{\left(\sin^{-1}\dfrac{x}{2}\right)^2 - y\right\}\sin^{-1}\dfrac{x}{2}$

$= \left(\sin^{-1}\dfrac{x}{2}\right)^3 - \sin^{-1}\dfrac{x}{2}\cdot y$

$\Rightarrow \dfrac{dy}{dx} + \dfrac{\sin^{-1}\dfrac{x}{2}\cdot y}{\sqrt{4-x^2}} = \dfrac{\left(\sin^{-1}\dfrac{x}{2}\right)^3}{\sqrt{4-x^2}}$

which is liner differential equation

$\therefore \text{IF} = e^{\int \frac{\sin^{-1}\frac{x}{2}}{\sqrt{4-x^2}}dx}$

Put $t = \sin^{-1}\dfrac{x}{2} \Rightarrow dt = \dfrac{1}{\sqrt{4-x^2}}dx$

$\Rightarrow e^{\int t\,dt} = e^{\frac{t^2}{2}} = e^{\frac{\left(\sin^{-1}\frac{x}{2}\right)^2}{2}} \Rightarrow y\cdot e^{\frac{\left(\sin^{-1}\frac{x}{2}\right)^2}{2}}$

$= \int \dfrac{\left(\sin^{-1}\dfrac{x}{2}\right)^3}{\sqrt{4-x^2}}\cdot e^{\frac{\left(\sin^{-1}\frac{x}{2}\right)^2}{2}} dx + C$

Put $\dfrac{\left(\sin^{-1}\dfrac{x}{2}\right)^2}{2} = z$

$\Rightarrow ye^z = 2\int z e^z dz + C$

$= 2[e^z(z-1)] + C$

$\Rightarrow 2\sin^{-1}\dfrac{x}{2} \times \dfrac{1}{\sqrt{4-x^2}} dx = 2dz$

$ye^{\frac{\left(\sin^{-1}\frac{x}{2}\right)^2}{2}} = 2\left[e^{\frac{\left(\sin^{-1}\frac{x}{2}\right)^2}{2}} \left(\dfrac{\left(\sin^{-1}\dfrac{x}{2}\right)^2}{2} - 1 \right) \right] + c$

Given, $y(2) = \dfrac{\pi^2 - 8}{4}$

$\Rightarrow \left(\dfrac{\pi^2-8}{4}\right)e^{\frac{\pi^2}{8}} = 2e^{\frac{\pi^2}{8}}\left[\dfrac{\pi^2-8}{8}\right] + C$

$\Rightarrow C = 0$

$\therefore ye^{\frac{\left(\sin^{-1}\frac{x}{2}\right)^2}{2}}$

$= 2e^{\frac{\left[\sin^{-1}\left(\frac{x}{2}\right)^2\right]}{2}} \left[\dfrac{\left(\sin^{-1}\dfrac{x}{2}\right)^2}{2} - 1 \right] + C$

Put $x = 0 \Rightarrow ye^0 = 2e^0[0-1]$

$\Rightarrow y(0) = -2 \Rightarrow y^2(0) = 4$

24. *(64)* Coefficient of x^{15} in $(1 + x + x^2 + \ldots + x^9)^3$

$=$ Coefficient of x^{15} in $\left(\dfrac{1-x^{10}}{1-x}\right)^3$

$=$ Coefficient of x^{15} in $(1-x^{10})^3(1-x)^{-3}$

$=$ Coefficient of x^{15} in $(1 - 3x^{10} + {}^3C_2 x^{20} + \ldots)(1-x)^{-3}$

$= {}^{3+15-1}C_{15} - 3\,{}^{3+5-1}C_5$

$= {}^{17}C_{15} - 3\,{}^{7}C_5$

$= \dfrac{17\times 16}{2} - 3\times\dfrac{7\times 6}{2} = 73$

$\therefore$ Number of less than 212 whose sum of digits is (078, 087, 069, 096, 195, 159, 186, 177, 168)

$\therefore$ Required number $= 73 - 9 = 64$

25. *(14)*

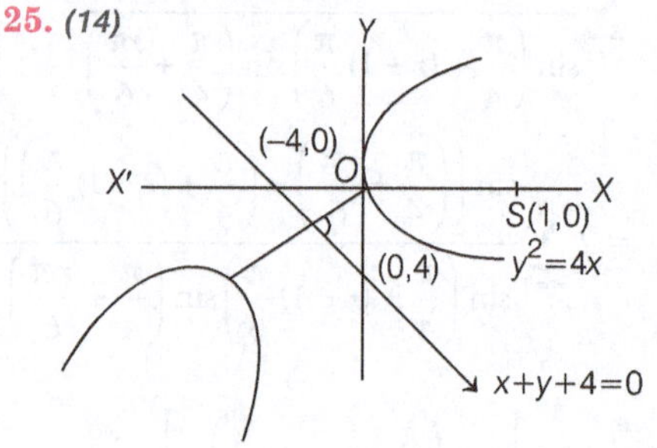

Let $L_1 \equiv x + y + 4 = 0$

image of point $(t^2, 2t)$ in $x + y + 4 = 0$ is

$\dfrac{x - t^2}{1} = \dfrac{y - 2t}{1}$

$= \dfrac{-2(t^2 + 2t + 4)}{2}$

$= -(t^2 + 2t + 4)$

$\Rightarrow x = -2t - 4,\ y = -t^2 - 4$

$\therefore$ Image of point is $y = -\left(\dfrac{x+4}{-2}\right)^2 - 4$

$\Rightarrow \quad (x+4)^2 = -4(y+4)$

Given, $\quad y + 5 = 0$

$\Rightarrow \quad y = -5$

$\therefore (x+4)^2 = -4(-5+4) = 4$

$x + 4 = \pm 2 \Rightarrow x = \pm 2 - 4 = -2, -6$

$\therefore$ Points are $(-2, -5)$ and $(-6, -5)$

$\Rightarrow d = 4$

$\therefore$ Area

$\Delta SAB = \dfrac{1}{2}\left\| \begin{matrix} 1 & 0 & 1 \\ -2 & -5 & 1 \\ -6 & -5 & 1 \end{matrix} \right\| = 10 = a$

$\therefore a + d = 4 + 10 = 14$

Concept Enhancer

Let the image of a point (x_1, y_1) with respect to $ax + by + c = 0$ be (x_2, y_2), then

$\dfrac{x_2 - x_1}{a} = \dfrac{y_2 - y_1}{b} = -\dfrac{2(ax_1 + by_1 + c)}{a^2 + b^2}$

Physics

26. (b) $m = 4\text{ kg}$

$F = 2\hat{i} + 3\hat{j}$

$m\mathbf{a} = 2\hat{i} + 3\hat{j}$

$(4)\mathbf{a} = 2\hat{i} + 3\hat{j}$

$\mathbf{a} = \frac{2\hat{i}}{4} + \frac{3\hat{j}}{4} = \frac{1}{2}\hat{i} + \frac{3}{4}\hat{j}$

$\frac{dv}{dt} = \frac{1}{2}\hat{i} + \frac{3}{4}\hat{j}$

$\mathbf{v} = \left(\frac{1}{2}\hat{i} + \frac{3}{4}\hat{j}\right)t$

at $t = 4$ s

$\mathbf{v} = \left(\frac{1}{2}\hat{i} + \frac{3}{4}\hat{j}\right)4 = 2\hat{i} + 3\hat{j}$

$P_{\text{inst}} = \mathbf{F} \cdot \mathbf{v} = (2\hat{i} + 3\hat{j}) \cdot (2\hat{i} + 3\hat{j})$

$= 4 + 9 = 13$

$s = ut + \frac{1}{2}at^2$

$s = 0 + \frac{1}{2}\frac{(2\hat{i} + 3\hat{j})}{4}(4)^2$

$= 4\hat{i} + 6\hat{j}$

$r_i = 3\hat{i} + 4\hat{j}$

If $r_f = 6\hat{i} + 10\hat{j}$, then

Final position will be (6, 10)

$P_{\text{avg}} = \frac{\Delta W}{\Delta t} = \frac{F\Delta s}{\Delta t}$

$\Delta s = r_f - r_i$

$= (6\hat{i} + 10\hat{j}) - (3\hat{i} + 4\hat{j})$

$= (3\hat{i} + 6\hat{j})$

$= \frac{(2\hat{i} + 3\hat{j})(3\hat{i} + 6\hat{j})}{4}$

$= \frac{6 + 18}{4} = \frac{24}{4} = 6$

$< P_{\text{avg}} > = 6$

$\frac{P_{\text{avg}}}{P_{\text{inst}}} = \frac{6}{13} = 6:13$

27. (a) n = Principal quantum number

$$f_n = \frac{4\pi^2 k^2 e^4 m}{h^3} \times \frac{Z^2}{n^3}$$

$$f_n \propto \frac{Z^2}{n^3}$$

28. (c) $Mg = F_B$

$M = 400\text{ g}, l = 10\text{ cm}, \rho = 10^3\text{ kg/m}^3$

$M = 400 \times 10^{-3}\text{ kg}$

$Mg = V_d \rho g \qquad \left[\begin{array}{l}\rho = \frac{M}{V} \\ M = \rho V\end{array}\right]$

$400 \times 10^{-3} = V_d \times 10^3$

$V_d = 400 \times 10^{-6} = 400 \times 10^{-6}\text{ m}^3$

$(\text{Volume})_{\text{out}} = (10 \times 10^{-2})^3 - 400 \times 10^{-6}$

$= 1000 \times 10^{-6} - 400 \times 10^{-6}$

$= 600 \times 10^{-6} = 600\text{ cm}^3$

Time Saver Tech

Volume of cube outside the water is given by

$V_{\text{outside}} = h^3\left[1 - \frac{\rho_b}{\rho_f}\right]$

where h is edge length

According to question

$\rho_b = \frac{\text{mass}}{h^3} = \frac{400 \times 10^{-3}}{(10 \times 10^{-2})^3}$

$= \frac{4 \times 10^{-1}}{10^{-3}} = 4 \times 10^2$

$V_{\text{outside}} = h^3\left[1 - \frac{\rho_b}{\rho_f}\right]$

$= (10 \times 10^{-2})^3\left[1 - \frac{400}{1000}\right]$

$= \frac{10^{-3} \times 600}{1000} = 600\text{cm}^2$

29. (a) Molecular mass of a gas is directly proportional to its vapour density

$\Rightarrow \quad \frac{M_1}{M_2} = \frac{\rho_1}{\rho_2} = \frac{4}{25}$

Now, $v_{\text{rms}} = \sqrt{\frac{3RT}{M}}$

As temperature is constant

$\Rightarrow \quad \frac{v_{\text{rms}1}}{v_{\text{rms}2}} = \sqrt{\frac{M_2}{M_1}}$

$= \sqrt{\frac{25}{4}} = \frac{5}{2}$

30. (c) $m = 50\text{ g}$ at $T = 17°\text{C}$

$\text{KE}_{\text{Translational}} = \left[\frac{3}{2}k_B T\right] \times$ Number of molecule

Number of molecule $= \frac{50N_0}{12 + 32} = \frac{50N_0}{44}$

$\text{KE}_T = \frac{3}{2}k_B T \times \frac{50}{40}N_0$

$= \frac{3}{2}RT \times \frac{50}{40}$

$= \frac{3}{2} \times 8.3 \times (273 + 17) \times \frac{50}{44}$

$= 4102.84\text{ J}$

Alternate Solution

$\text{KE} = \frac{1}{2}mv^2 = \frac{1}{2} \times m \times \frac{3RT}{M \cdot w}$

$= \frac{1}{2} \times \frac{50 \times 3 \times 8.3 \times 290}{44}$

$= 4102.84\text{ J}$

31. (d)

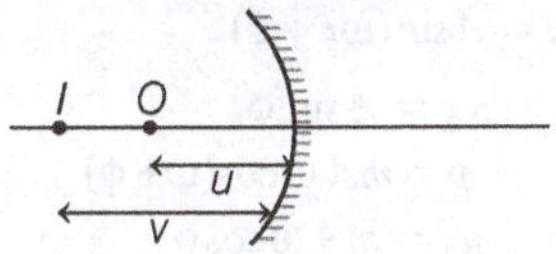

$m = -3 = -\frac{v}{u}$

and $v - u = 20\text{ cm}$

$f = \frac{vu}{v + u} = \frac{(-30)(-10)}{-30 - 10} = \frac{300}{-40}$

$\therefore f = -\frac{30}{4} = -\frac{15}{2}$

$R = -15\text{ cm} \qquad [2f = R]$

$\therefore$ Magnitude of R is 15 cm.

32. (a) $u = -13\text{ cm}$

$n_1 = 1.3, \; n_2 = 1.4$

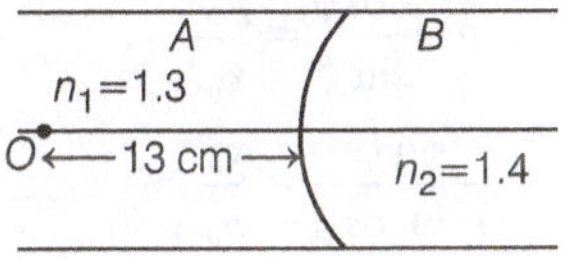

$m = \frac{\frac{v}{n_2}}{\frac{u}{n_1}}$

$-2 = \frac{v}{1.4} \times \frac{1.3}{(-13)}$

$\frac{v}{1.4} = +\frac{2 \times 13}{1.3}$

$\frac{1.4}{v} = \frac{1.3}{2 \times 13} \qquad \ldots(\text{i})$

$\frac{n_2}{v} - \frac{n_1}{u} = \frac{n_2 - n_1}{R}$

$\frac{1.4}{v} - \frac{1.3}{(-13)} = \frac{1.4 - 1.3}{R}$

$\frac{1.4}{v} + \frac{13}{130} = \frac{0.1}{R}$

$$\frac{1.4}{v} = \frac{1}{10R} - \frac{1}{10}$$

$$\frac{1.4}{v} = \frac{1-R}{10R} \quad \ldots\text{(ii)}$$

Put Eq. (i) in Eq. (ii)

$$\frac{13}{2\times13} = \frac{1-R}{10R}$$

$$+13R = 26 - 26R$$

$$39R = 26$$

$$R = \frac{26}{39} = \frac{2}{3} = \frac{2}{3}\ \text{cm}$$

33. *(d)* $x = A\sin(\omega t + \phi)$

$$x_0 = A\sin\phi \quad \ldots\text{(i)}$$

$$p = mA\,\omega\cos(\omega + \phi)$$

$$p_0 = mA\,\omega\cos\phi \quad \ldots\text{(ii)}$$

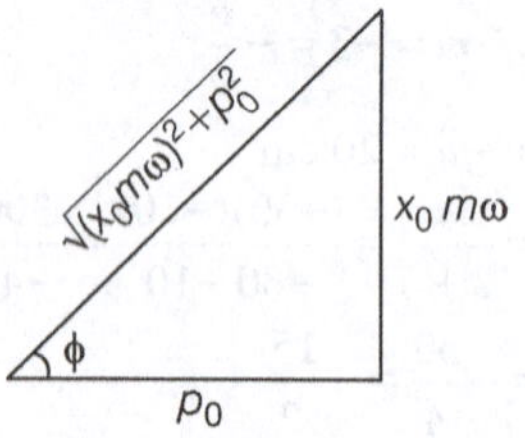

Dividing Eq. (ii) by Eq. (i), we get

$$\frac{mA\omega\cos\phi}{A\sin\phi} = \frac{p_0}{x_0}$$

$$m\omega\frac{\cos\phi}{\sin\phi} = \frac{p_0}{x_0}$$

$$\frac{\sin\phi}{m\omega\cos\phi} = \frac{x_0}{p_0}$$

$$\frac{\tan\phi}{m\omega} = \frac{x_0}{p_0}$$

$$\Rightarrow \quad \tan\phi = \left(\frac{x_0}{p_0}\right)m\omega$$

$$\sin\phi = \frac{x_0 m\omega}{\sqrt{(x_0 m\omega)^2 + p_0^2}}$$

$$A = \frac{x_0}{\sin\phi} = \frac{x_0}{\dfrac{x_0 m\omega}{\sqrt{(x_0 m\omega)^2 + p_0^2}}}$$

34. *(b)* $B = \frac{\mu_0}{4\pi}\frac{m}{r^3}$

$$B \propto \frac{1}{r^3}$$

$$\frac{\Delta r}{r}\times100 = 1\%$$

$$100\times\frac{\Delta B}{B} = 3\times\left(\frac{\Delta r}{r}\right)\times100$$

% uncertainty in $B = 3\%$

35. *(b)* $\tau_{\text{net}} = 0$

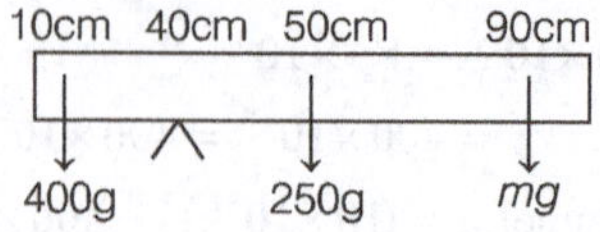

$$\tau_{\text{net}} = 0$$

$$\Rightarrow (400g\times30) = (250g\times10) + mg\times50$$

$$m = 190\ \text{g}$$

36. *(d)*

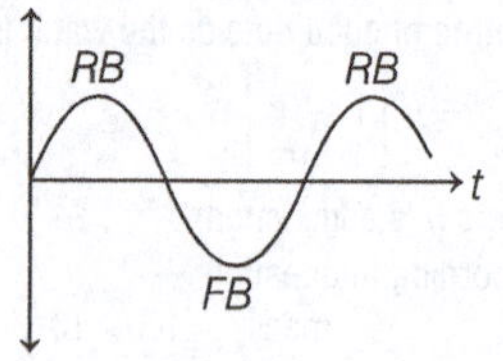

$V = V_0 \sin\omega t$

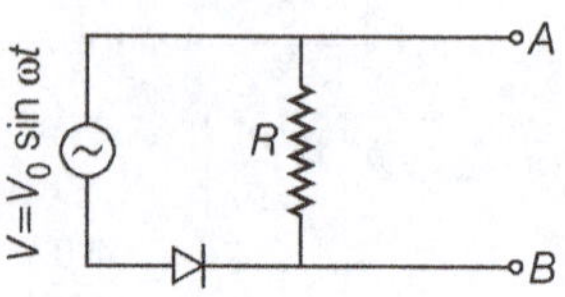

For first half cycle $\rightarrow RB$

For second half cycle $\rightarrow FB$

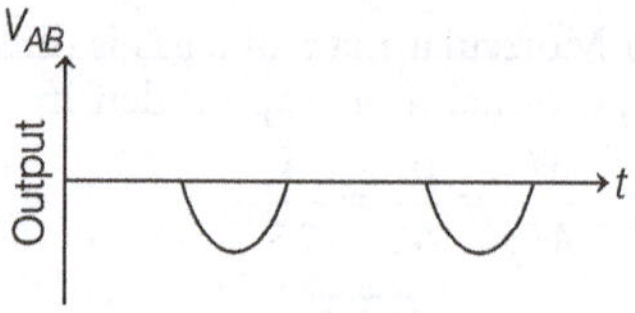

37. *(c)*

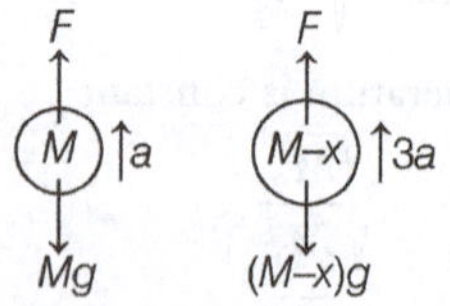

$$F - Mg = Ma$$

$$F = Mg + Ma \quad \ldots\text{(i)}$$

$$F - (M-x)g = (M-x)3a \quad \ldots\text{(ii)}$$

Put Eq. (i) in Eq. (ii)

$$Mg + Ma - Mg + xg = 3Ma - 3xa$$

$$3xa + xg = 3Ma - Ma$$

$$x(3a+g) = 2Ma$$

$$x = \frac{2Ma}{3a+g}$$

38. *(d)* Wave theory of light cannot explain Compton effect, or particle nature of light.

Wave theory of light can explain

$\rightarrow$ Reflection of light

$\rightarrow$ Diffraction of light

$\rightarrow$ Refraction of light

> **Concept Enhancer**
> When a high frequency photon scatters due to an interaction with a charged particle, there is a decrease in the energy of the photon and thus, an increase in its wavelength. This trade off between wavelength and energy in response to collision is the Compton effect.

39. *(b)*

$$\mathbf{B} = \left(\frac{\sqrt{3}}{2}\hat{\mathbf{i}} + \frac{1}{2}\hat{\mathbf{j}}\right)30\sin\left[\omega\left(t - \frac{z}{c}\right)\right]$$

$$B_0 = 30\sqrt{\frac{3}{4} + \frac{1}{4}} = 30$$

$$E_0 = C|B_0| = 30C$$

Now **B** lies in *XY*-plane and wave travels in +*z*-direction by the righ hand rule, *E* must lie in *XY*-plane and perpendicular to **B**.

$$\mathbf{E} = \mathbf{B}\times\mathbf{C}$$

$$= \left(\frac{\sqrt{3}}{2}\hat{\mathbf{i}} + \frac{1}{2}\hat{\mathbf{j}}\right)\times(\hat{\mathbf{k}})$$

$$= \left(-\frac{\sqrt{3}}{2}\hat{\mathbf{j}} + \frac{1}{2}\hat{\mathbf{k}}\right)$$

$$= \left(\frac{1}{2}\hat{\mathbf{i}} - \frac{\sqrt{3}}{2}\hat{\mathbf{j}}\right)$$

$$\therefore \mathbf{E} = \left(\frac{1}{2}\hat{\mathbf{i}} - \frac{\sqrt{3}}{2}\hat{\mathbf{j}}\right)30C\sin\left[w\left(t - \frac{z}{c}\right)\right]$$

40. *(c)* Angular impulse $= [ML^2T^{-1}]$

Latent heat $= [M^0L^2T^{-2}]$

Electrical resistivity $= [ML^3T^{-3}A^{-2}]$

Electromotive force $= [ML^2T^{-3}A^{-1}]$

$A \rightarrow$ III, $B \rightarrow$ I, $C \rightarrow$ IV, $D \rightarrow$ II

41. *(a)* $t = 0$ to $t = 4$s

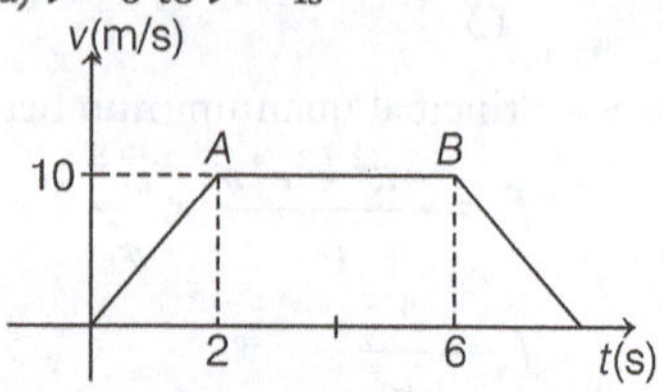

Distance = Area under *v*-*t* graph

$$= \frac{1}{2}\times2\times10 + 2\times10$$

$$= 10 + 20 = 30\ \text{m}$$

42. *(a)* $B = 0.4\ \text{T},\ R = 20\text{cm}$

and $\omega = 10\pi$ rad/s

$$E = \frac{1}{2}B\omega R^2$$

$$= \frac{1}{2}\times 0.4\times 10\times 3.14\times(20\times10^{-2})^2$$

$$= 0.2512\ \text{V}$$

43. *(a)*

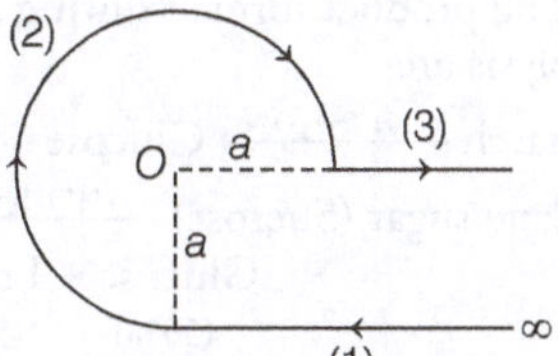

$$\mathbf{B}_1 = \frac{\mu_0 I}{4\pi a}\otimes$$

$$\mathbf{B}_2 = \frac{\mu_0}{4\pi}\frac{I}{a}\left(\frac{3\pi}{2}\right)\otimes$$

$$\mathbf{B}_3 = 0$$

$$\mathbf{B}_{\text{net}} = \mathbf{B}_1 + \mathbf{B}_2 + \mathbf{B}_3$$

$$= \frac{\mu_0 I}{4\pi a} + \frac{\mu_0}{4\pi}\frac{I}{a}\left(\frac{3\pi}{2}\right) + 0$$

$$= \frac{\mu_0 I}{4\pi a}\left[1 + \frac{3\pi}{2}\right]$$

44. *(b)* $v_{\text{escape}} = \sqrt{\dfrac{2GM}{R}}$

$$\frac{(v_{\text{esp}})_{\text{planet}}}{(v_{\text{esp}})_{\text{earth}}} = \sqrt{\frac{M_P}{M_E}\times\frac{R_E}{R_P}}$$

$$(v_{\text{esp}})_{\text{planet}} = \frac{1}{2}(v_{\text{esp}})_{\text{earth}}$$

$$= \frac{1}{2}\times 11.2 = 5.6\ \text{km/s}$$

45. *(a)* $C = 1\mu\text{F},\ V = 20\text{V},\ d = 1\,\mu\text{m}$

$$\text{Energy density} = \frac{1}{2}\varepsilon_0 E^2$$

$$V = Ed \Rightarrow E = \frac{V}{d}$$

$$\text{Energy density} = \frac{1}{2}\varepsilon_0\left(\frac{V}{d}\right)^2$$

$$= \frac{1}{2}\varepsilon_0\frac{V^2}{d^2} \qquad \left[\frac{V}{d} = 20\times10^{+6}\right]$$

$$= \frac{1}{2}\times 8.85\times10^{-12}\times(20\times10^{+6})^2$$

$$= 1.77\times10^3\ \text{J/m}^3 \cong 1.8\times10^3\ \text{J/m}^3$$

46. *(54)* Maxima condition,

$$2\mu t = n\lambda$$

$$t = \frac{n\lambda}{2\mu},\ n = 1, 2, \ldots\ldots$$

$$t = \frac{\lambda}{2\mu}, \frac{2\lambda}{2\mu}$$

Minima condition

$$2\mu t = (2n-1)\frac{\lambda}{2}, \ldots\ldots$$

$$t = \frac{(2n-1)\lambda}{2\mu\times 2} = \frac{(2n-1)\lambda}{4\mu}$$

$$t = \frac{\lambda}{4\mu}, \frac{3\lambda}{4\mu}\ [n = 1, 2, \ldots\ldots]$$

$$\Delta t = \frac{3\lambda}{4\mu} - \frac{\lambda}{4\mu}$$

$$= \frac{2\lambda}{4\mu} = \frac{\lambda}{2\mu}$$

$$\text{Rate of evaporation} = \frac{A\cdot\Delta t}{\text{time}}$$

$$= \frac{\pi r^2\times\Delta t}{t}$$

$$= \frac{\pi\times(1.8\times10^{-2})^2\times 2\times 560\times10^{-9}}{4\times1.4\times2}$$

$$= 54\pi\times10^{-13}\ \text{m}^3/\text{s}$$

47. *(2)*

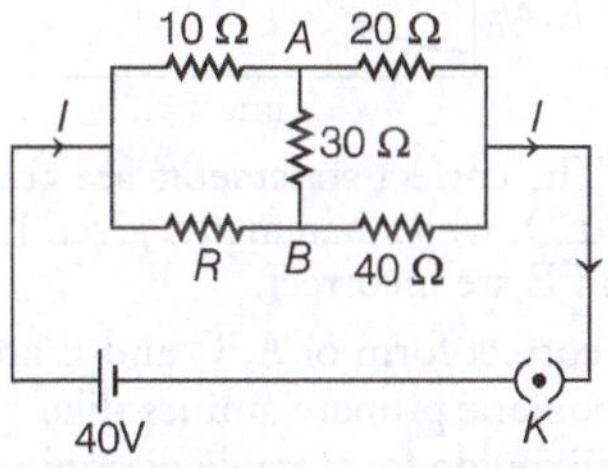

$V_A = V_B$, the bridge is in balanced condition

$$\frac{10}{R} = \frac{20}{40}$$

$$\Rightarrow \quad R = 20\Omega$$

$$I = \frac{V}{R} = \frac{40}{20} = 2\text{A}$$

48. *(12)* $p = 6\times10^{-6}$ C-m

$$E = 10^6\ \text{V/m}$$

$$W = \Delta U = -pE(\cos\theta_f - \cos\theta_i)$$

$$W = -pE(\cos180° - \cos0°)$$

$$= -pE(-1-1)$$

$$= 2pE = 12\text{J}$$

49. *(50)*

$$B = 1.4\times10^{11}\ \text{N/m}^2,\ p = 7\times10^6\ \text{Pa}$$

$$l = (10\times10^{-2})^3$$

$$B = \frac{\Delta p}{\frac{\Delta V}{V}}$$

$$B = \frac{\Delta p\cdot V}{\Delta V}$$

$$\Delta V = \frac{\Delta p\cdot V}{B}$$

$$= \frac{7\times10^6}{1.4\times10^{11}}\times(10\times10^{-2})^3$$

$$\Delta V = 50\text{mm}^3$$

50. *(1)*

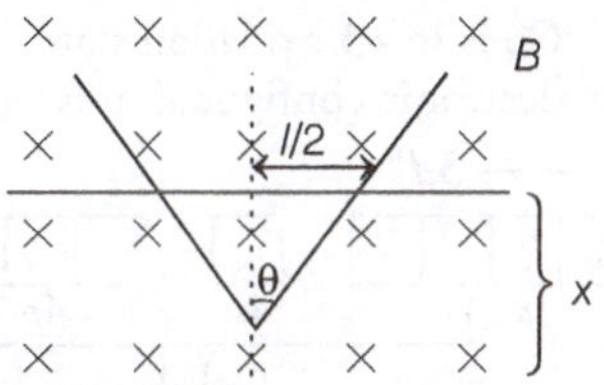

From the figure,

$$\tan\theta = \frac{l}{2x}$$

$$l = 2x\tan\theta$$

$$E = Blv = 2x\tan\theta vB \qquad (x = vt)$$

$$E = 2\tan\theta v^2 tB$$

$$E \propto t$$

Chemistry

51. *(a)* The correct match is

A-III, B-II, C-I, D-IV

$A \rightarrow [CoF_6]^{3-}$

Here, Co is in +3 oxidation state. Thus, outer electronic configuration is

$Co^{3+} \longrightarrow 3d^6$

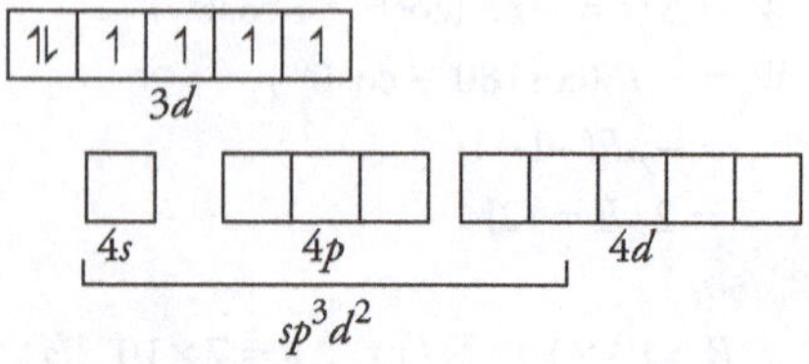

(No pairing will take place as F^- is a weak field ligand.)

$B \longrightarrow [NiCl_4]^{2-}$

Here, Ni is in +2 oxidation state. Thus, outer electronic configuration is

$Ni^{2+} \longrightarrow 3d^8$

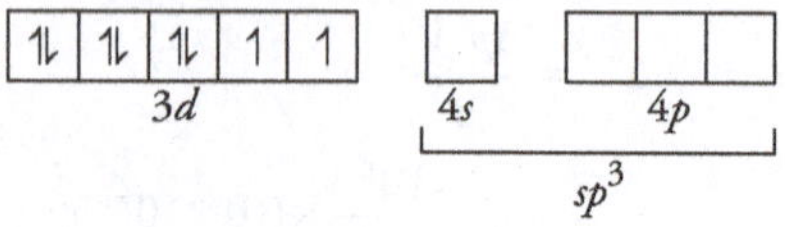

(No pairing will take place as Cl^- is a weak field ligand.)

$C \longrightarrow [Co(NH_3)_6]^{3+}$

Here, Co is in +3 oxidation state. Thus, outer electronic configuration is

$Co^{3+} \longrightarrow 3d^6$

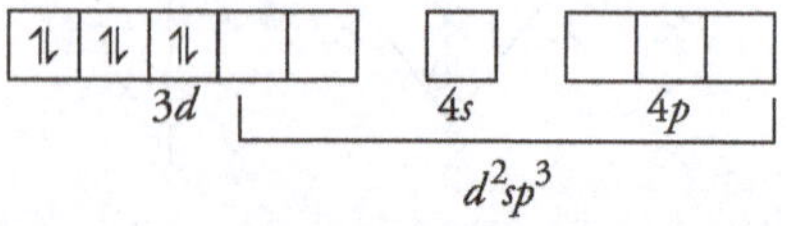

(Pairing will take place as NH_3 will act as strong field ligand as Co is in +3 oxidation state.)

$D \longrightarrow [Ni(CN)_4]^{2-}$

Here, Ni is in +2 oxidation state, thus outer electronic configuration is

$Ni^{2+} \longrightarrow 3d^8$

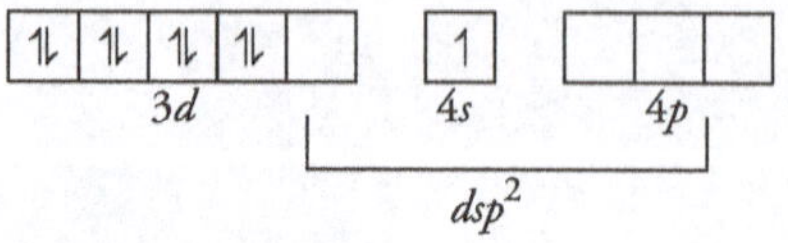

(Pairing will take place as CN is a strong field ligand.)

52. *(b)* The complete reaction is as follows.

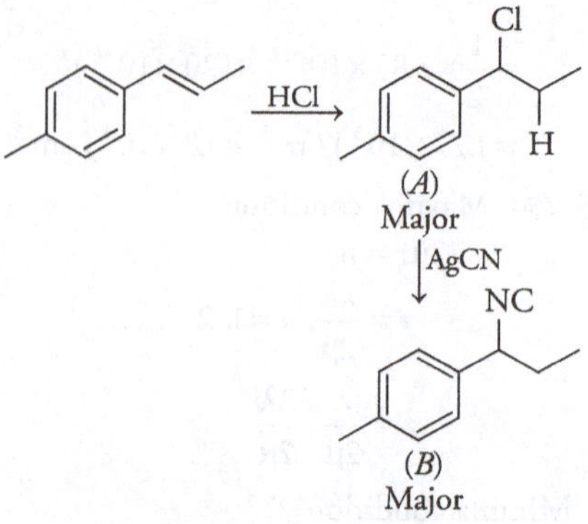

53. *(c)* Among the given oxides, V_2O_3, V_2O_4 and V_2O_5. The amphoteric oxide is V_2O_5. The reaction of V_2O_5 with alkali is given by

$$V_2O_5 + \text{Alkali} \longrightarrow VO_4^{3-}$$

In VO_4^{3-} ion, vanadium is in +5 oxidation state.

54. *(c)* Initial number of bacteria $= N_0$

Number of bacteria at any given time $= N$

Bacterial growth is given as

$$N = N_0 e^{kt}$$

Where, k = growth constant for bacterial growth.

The most suitable graph to show bacterial growth is

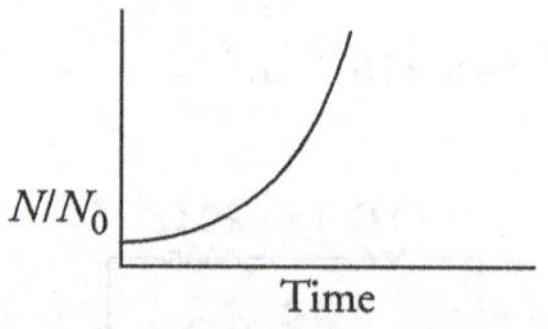

55. *(c)* The correct statements are given in B and D, while statements given is A , C and E are incorrect.

The correct form of A, C and E are

- Aromatic primary amines (like aniline) do form stable diazonium salts under the given conditions. However, aliphatic primary amines form diazonium salts that are highly unstable and decompose rapidly.
- The carbylamine reaction is specific to primary amines. Secondary and tertiary amines do not give this test.
- Tertiary amines do not react with benzenesulphonyl chloride, as they lack a hydrogen atom to participate in the reaction.

Concept Enhancer

Benzene sulphonyl chloride ($C_6H_5SO_2Cl$) is an organosulphur compound. It reacts with compounds that have reactive O—H and N—H bonds. This reagent is used as a test for amines.

56. *(d)* The product formed during acidic hydrolysis are

(A) Starch $\xrightarrow{H^+/H_2O}$ Glucose

(B) Cane sugar (Sucrose) $\xrightarrow{H^+/H_2O}$ Glucose + Fructose (50%) (50%)

(C) Milk sugar (Lactose) $\xrightarrow{H^+/H_2O}$ Glucose + Galactose

(D) Amylopectin $\xrightarrow{H^+/H_2O}$ Glucose

(E) Amylose $\xrightarrow{H^+/H_2O}$ Glucose

So, correct conversions are given in B, C and E only.

57. *(a)* The complete reaction is as follows.

$$\underset{}{CH_3—C \equiv CH} \xrightarrow[H_2]{Pd/C} \underset{(A)}{CH_3—CH{=}CH_2} \xrightarrow[\text{(ii) Zn, } H_2O]{\text{(i) } O_3} \underset{(B)}{CH_3—CH{=}O} + \underset{(C)}{HCHO}$$

58. *(c)* The rate of reaction is given as

$$R_1 = K[A]^1[B]^1$$

$$R_1 = K\left[\frac{n_A}{V}\right]^1\left[\frac{n_B}{V}\right]^1$$

Now, for R_2 rate is given as

$$R_2 = K\left[\frac{3n_A}{V}\right]^1\left[\frac{3n_B}{V}\right]^1 = 9R_1$$

59. *(b)* Based on K_{sp} values and salt analysis cation identification, we can say that order of k_{sp} value is

$HgS < PbS < AgBr < Ca(OH)_2$

K_{sp} values for given salts are

$$HgS \rightarrow 4 \times 10^{-53},$$
$$PbS \rightarrow 8 \times 10^{-28}$$
$$AgBr \rightarrow 5 \times 10^{-13},$$
$$Ca(OH)_2 \rightarrow 5.5 \times 10^{-6}$$

60. *(b)* When solid is heated and it is directly converted to vapour phase without going through liquid phase, then it is called as sublimation.

Same process happen when vapour is directly cooled down to solid phase bypassing liquid phase

$$\text{Vapour} \xrightarrow[\text{Sublimation}]{\text{Cool}} \text{Solid}$$

$$\text{Solid} \xrightarrow[\text{Sublimation}]{\text{Hot}} \text{Vapour}$$

61. *(a)* Compounds which have at least one α-hydrogen will react with $KMnO_4$ and give benzoic acid.

There are five compounds with at least one α-hydrogen, these are

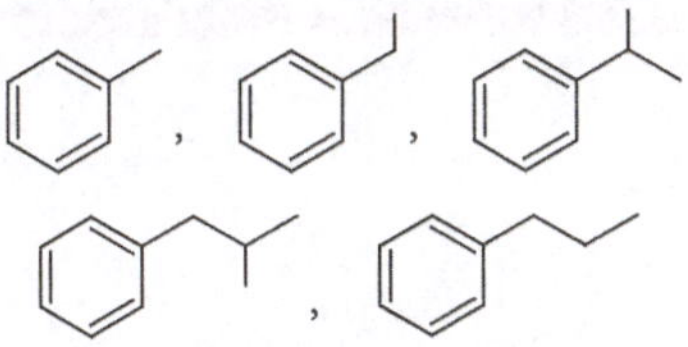

62. *(*)* Living cell = 0.9 g in 100 g of solution

$\% \, w/w = 0.9$

Solution is having equal moles of glucose and water = 0.5

Weight of solution

$= 0.5 \times 180 + 0.5 \times 18 = 99$ g

$\% \, w/w \simeq 90\%$

Concentrated solution means cell will shrink.

63. *(b)* Energy of atomic orbitals of hydrogen atom is given incorrect in C and D.

For single electron species energy only depends on 'n' (principal quantum number)

So, energy of $2s = 2p$ and energy of $3d < 4s$

64. *(c)* The complete reactions is as follow

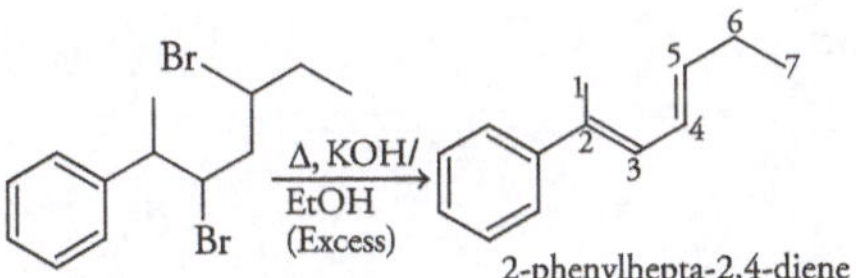

2-phenylhepta-2,4-diene (Major product)

65. *(a)* As internal energy 'U' is a state function, its cyclic integral must be zero in a cyclic process.

$\therefore \Delta U$(case I) = ΔU (case II)

$= \Delta U$ (case III)

66. *(a)* Both statement I and statement II are true.

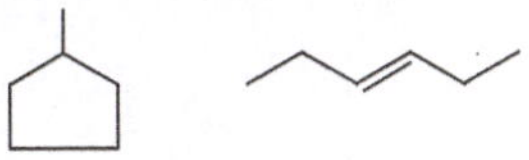

Both are ring chain isomers.

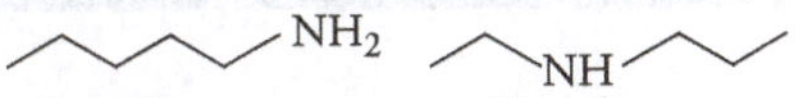

1° amine 2° amine

1° amine and 2° amine are different functional groups, hence both are functional group isomers.

67. *(a)* % w/w of $HNO_3 = 75\%$,

Means that 100 g, of solution contains 75 g of HNO_3

and $\left(\frac{g}{mL}\right)_{\text{Solution}} = 1.25 = \frac{100}{V}$ g

$\because$ 75 g of HNO_3 present in $\frac{100}{1.25}$ mL solution.

$\therefore$ 30 g of HNO_3 present in

$\frac{100}{1.25 \times 75} \times 30 = 32$ mL solution

68. *(d)* $(NH_4)_2S$, As_2S_3 and As_2S_5 are yellow colour sulphides, while PbS and CuS are black in colour.

69. *(d)* Both statement I and statement II are false. Their correct forms are

- Law of octaves was arranged in the increasing order of their atomic weight.
- Lothar Meyer plotted the physical properties such as atomic volume, melting point and boiling point against atomic weight.

70. *(a)* The correct match is

A-III, B-I, C-IV, D-II

Sucrose $\rightarrow \alpha_1 - \beta_2$ Glycosidic linkage

Maltose $\rightarrow \alpha 1 - 4$ Glycosidic linkage

Lactose $\rightarrow \beta 1 - 4$ Glycosidic linkage

Amylopectin $\rightarrow \alpha 1 - 4$ and $\alpha 1 - 6$ Glycosidic linkage.

Time Saver Tech

Use elimination technique as time saver A-III is given is option (a) only.

71. *(2)*

Key idea

To solve this problem, first of all find the value of $[OH^-]$ by using the pH value. After that use Faraday's law to calculate the value of current in ampere.

Electrolysis of NaCl is

$$NaCl + H_2O(aq) \longrightarrow NaOH(aq) + \frac{1}{2}Cl_2(g) + \frac{1}{2}H_2(g)$$

Since during electrolysis pH changes to 12.

So, $[OH^-] = 10^{-2}$ and $[H^+] = 10^{-12}$

So, according to Faraday law,

Gram amount of substance deposited = Amount of electricity passed

$10^{-2} \times \frac{600}{1000} \times 96500 = I \times t$

$\frac{10^{-2} \times 600}{1000} \times 96500 = I \times 5 \times 60$

$I = 1.93 \text{ A} \simeq 2$ ampere.

72. *(48)* Given, $\Delta H_f[CO_2(g)]$

$= -393.5$ kJ/mole

$\Delta H_f[H_2O(l)] = -286$ kJ/mole

$\Delta H_C[C_6H_6] = -3267.0$ kJ/mole

$\Delta H_f C_6H_6 = ?$

$$C_6H_6 + \frac{15}{2}O_2(g) \rightarrow 6CO_2(g) + 3H_2O(l)$$

$\Delta H_R = \Delta H_C = \Sigma \Delta H_f(P) - \Sigma \Delta H_f(R)$

$-3267 = 6 \times (-393.5) + 3 \times (-286) - \Delta H_f(C_6H_6)$

$\therefore \Delta H_f(C_6H_6) = 48$ kJ/mole

73. *(6)* According to molecular orbital theory.

$O_2 \rightarrow 2$ unpaired electron

$O_2^+ \rightarrow 1$ unpaired electron

$O_2^- \rightarrow 1$ unpaired electron

$NO \rightarrow$ odd electron species.

$NO_2 \rightarrow$ odd electron species.

$K_2[NiCl_4] \rightarrow Ni^{2+} = 3d^8$

(Weak field ligand)

Coordination number = 4

$\therefore$ Tetrahedral, paramagnetic with 2 unpaired electrons.

So, total 6 paramagnetic species are present.

74. *(5)* Mn_2O_3 has strongest oxidising power among the given compounds. This is due to value of $E°$.

$E°_{Mn^{3+}/Mn^{2+}} = +1.57$ V

$Mn^{3+} \longrightarrow d^4$ configuration

(So, 4 unpaired e^-)

$\mu = \sqrt{24}$ BM = 4.89 BM $\simeq$ 5 BM

75. *(15)* Phosphorus belong to 15th group and forms $d\pi - d\pi$ bond with transition metal and PH_3 is strongest base among the other group member except NH_3.

So, atomic number of element is 15.

Time 9:00-12:00 **MM** : 300 Marks

29 January
Shift-I

Online Question Paper — JEE Main 2025

Mathematics

Section A
Objective Type Questions

01/ Let the line $x+y=1$ meet the circle $x^2+y^2=4$ at the points A and B. If the line perpendicular to AB and passing through the mid-point of the chord AB intersects the circle at C and D, then the area of the quadrilateral $ADBC$ is equal to

(a) $\sqrt{14}$ (b) $5\sqrt{7}$
(c) $3\sqrt{7}$ (d) $2\sqrt{14}$

02/ Let $x_1, x_2 \dots, x_{10}$ be ten observations such that $\sum_{i=1}^{10}(x_i-2)=30$, $\sum_{i=1}^{10}(x_i-\beta)^2=98, \beta>2$ and their variance is $\frac{4}{5}$. If μ and σ^2 are respectively the mean and the variance of $2(x_1-1)+4\beta$, $2(x_2-1)+4\beta, \dots, 2(x_{10}-1)+4\beta$, then $\frac{\beta\mu}{\sigma^2}$ is equal to

(a) 120 (b) 110
(c) 100 (d) 90

03/ The number of solutions of the equation $\left(\frac{9}{x}-\frac{9}{\sqrt{x}}+2\right)\left(\frac{2}{x}-\frac{7}{\sqrt{x}}+3\right)=0$ is

(a) 4 (b) 1
(c) 3 (d) 2

Imp.

04/ Let

$$A=[a_{ij}]=\begin{bmatrix}\log_5 128 & \log_4 5\\ \log_5 8 & \log_4 25\end{bmatrix}$$

If A_{ij} is the cofactor of a_{ij},

$C_{ij}=\sum_{k=1}^{2} a_{ik}\,A_{jk}, 1\le i, j\le 2$ and

$C=|C_{ij}|$, then $8|C|$ is equal to

(a) 262 (b) 222 (c) 288 (d) 242

Minor and Cofactor of Matrices

05/ Define a relation R on the interval $[0, \pi/2)$ by $x\,R\,y$ if and only if $\sec^2 x-\tan^2 y=1$. Then, R is

(a) both reflexive and transitive but not symmetric
(b) reflexive but neither symmetric not transitive
(c) both reflexive and symmetric but not transitive
(d) an equivalence relation

06/ Let $L_1: \frac{x-1}{1}=\frac{y-2}{-1}=\frac{z-1}{2}$ and $L_2: \frac{x+1}{-1}=\frac{y-2}{2}=\frac{z}{1}$ be two lines.

Let L_3 be a line passing through the point (α, β, γ) and be perpendicular to both L_1 and L_2. If L_3 intersects L_1, then $|5\alpha-11\beta-8\gamma|$ equals

(a) 16 (b) 20 (c) 18 (d) 25

07/ Let $\mathbf{a} = \hat{i} + 2\hat{j} + \hat{k}$ and $\mathbf{b} = 2\hat{i} + 7\hat{j} + 3\hat{k}$.

Let $L_1 : \mathbf{r} = (-\hat{i} + 2\hat{j} + \hat{k}) + \lambda\mathbf{a}, \lambda \in R$ and $L_2 : \mathbf{r} = (\hat{j} + \hat{k}) + \mu\mathbf{b}, \mu \in R$ be two lines. If the line L_3 passes through the point of intersection of L_1 and L_2 and is parallel to $\mathbf{a} + \mathbf{b}$, then L_3 passes through the point

(a) (5, 17, 4) (b) (8, 26, 12)
(c) (2, 8, 5) (d) (−1, −1, 1)

08/ Let the area of the region $\{(x, y) : 2y \le x^2 + 3, y + |x| \le 3, y \ge |x - 1|\}$ be A. Then, 6A is equal to

(a) 18 (b) 12 (c) 16 (d) 14

09/ Let $y = y(x)$ be the solutions of the differential equation $\cos x\,(\log_e(\cos x))^2\,dy + (\sin x - 3y\sin x\,\log_e(\cos x))\,dx = 0, x \in \left(0, \frac{\pi}{2}\right)$. If $y\left(\frac{\pi}{4}\right) = \frac{-1}{\log_e 2}$, then $y\left(\frac{\pi}{6}\right)$ is equal to

(a) $\frac{2}{\log_e(3) - \log_e(4)}$ (b) $\frac{1}{\log_e(3) - \log_e(4)}$

(c) $\frac{1}{\log_e(4) - \log_e(3)}$ (d) $-\frac{1}{\log_e(4)}$

Imp.

10/ Consider an AP of positive integers, whose sum of the first three terms is 54 and the sum of the first twenty terms lies between 1600 and 1800. Then, its 11th term is

(a) 108 (b) 122 (c) 84 (d) 90

Arithmetic Progression

11/ Let P be the set of seven-digit numbers with sum of their digits equal to 11. If the numbers in P are formed by using the digits 1, 2 and 3 only, then the number of elements in the set P is

(a) 164 (b) 158 (c) 161 (d) 173

12/ Let M and m, respectively be the maximum and the minimum values of

$$f(x) = \begin{vmatrix} 1+\sin^2 x & \cos^2 x & 4\sin 4x \\ \sin^2 x & 1+\cos^2 x & 4\sin 4x \\ \sin^2 x & \cos^2 x & 1+4\sin 4x \end{vmatrix},$$

$x \in R$. Then, $M^4 - m^4$ is equal to

(a) 1040 (b) 1215 (c) 1295 (d) 1280

13/ Let $|z_1 - 8 - 2i| \le 1$ and $|z_2 - 2 + 6i| \le 2, z_1, z_2 \in C$. Then, the minimum value of $|z_1 - z_2|$ is

(a) 10 (b) 3 (c) 7 (d) 13

14/ The integral $80 \int_0^{\pi/4} \left(\frac{\sin\theta + \cos\theta}{9 + 16\sin 2\theta}\right) d\theta$ is equal to

(a) $2\log_e 3$ (b) $3\log_e 4$ (c) $6\log_e 4$ (d) $4\log_e 3$

15/ Two parabolas have the same focus (4, 3) and their directrices are the X-axis and the Y-axis, respectively. If these parabolas intersects the points A and B, then $(AB)^2$ is equal to

(a) 392 (b) 384 (c) 96 (d) 192

16/ Let ABC be a triangle formed by the lines $7x - 6y + 3 = 0, x + 2y - 31 = 0$ and $9x - 2y - 19 = 0$. Let the point (h, k) be the image of the centroid of ΔABC in the line $3x + 6y - 53 = 0$. Then, $h^2 + k^2 + hk$ is equal to

(a) 37 (b) 36 (c) 47 (d) 40

17/ Let $\mathbf{a} = 2\hat{i} - \hat{j} + 3\hat{k}, \mathbf{b} = 3\hat{i} - 5\hat{j} + \hat{k}$ and $\mathbf{c}$ be a vector such that $\mathbf{a} \times \mathbf{c} = \mathbf{c} \times \mathbf{b}$ and $(\mathbf{a} + \mathbf{c}) \cdot (\mathbf{b} + \mathbf{c}) = 168$. Then, the maximum value of $|\mathbf{c}|^2$ is

(a) 154 (b) 308 (c) 462 (d) 77

18/ Let the ellipse $E_1 : \frac{x^2}{a^2} + \frac{y^2}{b^2} = 1, a > b$ and $E_2 : \frac{x^2}{A^2} + \frac{y^2}{B^2} = 1, A < B$ have same eccentricity $1/\sqrt{3}$. Let the product of their lengths of latus rectums be $32/\sqrt{3}$ and the distance between the foci of E_1 be 4. If E_1 and E_2 meets at A, B, C and D, then the area of the quadrilateral $ABCD$ equals

(a) $\frac{24\sqrt{6}}{5}$ (b) $\frac{12\sqrt{6}}{5}$ (c) $\frac{18\sqrt{6}}{5}$ (d) $6\sqrt{6}$

19/ The least value of n for which the number of integral terms in the binomial expansion of $(\sqrt[3]{7} + \sqrt[12]{11})^n$ is 183, is

(a) 2148 (b) 2184 (c) 2172 (d) 2196

20/ The value of $\lim_{n \to \infty} \left(\sum_{k=1}^{n} \frac{k^3 + 6k^2 + 11k + 5}{(k+3)!}\right)$ is

(a) 4/3 (b) 5/3 (c) 7/3 (d) 2

Section B

Numerical Type Questions

Imp.

21/ The number of 6-letter words with or without meaning that can be formed using the letters of the word MATHS such that any letter that appears in the word must appear at least twice, is ……… .

Permutation and Combination

22/ Let $[t]$ be the greatest integer less than or equal to t. Then, the least value of $p \in N$ for which

$$\lim_{x \to 0^+}\left(x\left(\left[\frac{1}{x}\right]+\left[\frac{2}{x}\right]+\ldots+\left[\frac{p}{x}\right]\right)-x^2\left(\left[\frac{1}{x^2}\right]+\left[\frac{2^2}{x^2}\right]+\ldots+\left[\frac{9^2}{x^2}\right]\right)\right) \geq 1 \text{ is equal}$$

to ………… .

23/ Let $S=\{x:\cos^{-1}x=\pi+\sin^{-1}x+\sin^{-1}(2x+1)\}$. Then, $\sum_{x \in S}(2x-1)^2$ is equal to ………… .

24/ Let $S=\{m \in Z: A^{m^2}+A^m=3I-A^{-6}\}$, where $A=\begin{bmatrix}2 & -1\\ 1 & 0\end{bmatrix}$. Then, $n(S)$ is equal to ………… .

25/ Let $f:(0,\infty)\to R$ be a twice differentiable function. If for some $a \neq 0$, $\int_0^1 f(\lambda x)\,d\lambda = af(x)$, $f(1)=1$ and $f(16)=\frac{1}{8}$, then $16-f'\left(\frac{1}{16}\right)$ is equal to ………… .

Physics

Section A

Objective Type Questions

26/ Given below are two statements. One is labelled as Assertion (A) and the other is labelled as Reason (R).

Assertion (A) Emission of electrons in photoelectric effect can be suppressed by applying a sufficiently negative electron potential to the photoemissive substance.

Reason (R) A negative electric potential, which stops the emission of electrons from the surface of a photoemissive substance, varies linearly with frequency of incident radiation.

In the light of the above statements, choose the most appropriate answer from the options given below.

(a) Both A and R are true and R is the correct explanation of A.
(b) Both A and R are true but R is not the correct explanation of A.
(c) (A) is false but R is true.
(d) (A) is true but R is false.

27/ Let u and v be the distances of the object and the image from a lens of focal length f. The correct graphical representation of u and v for a convex lens when $|u|>f$, is

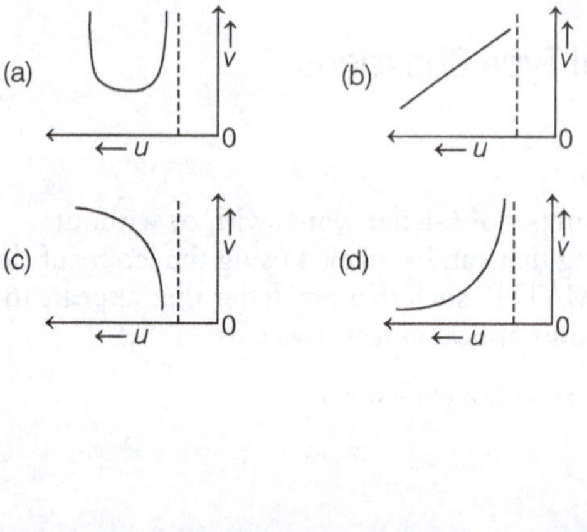

28/ A coil of area A and N turns is rotating with angular velocity ω in a uniform magnetic field **B** about an axis perpendicular to **B**. Magnetic flux ϕ and instant emf ε across it, at an instant when **B** is parallel to the plane of coil, are

(a) $\phi=AB$, $\varepsilon=NAB\omega$
(b) $\phi=0$, $\varepsilon=NAB\omega$
(c) $\phi=AB$, $\varepsilon=0$
(d) $\phi=0$, $\varepsilon=0$

29/ An electric dipole of mass m, charge q and length l is placed in a uniform electric field $\mathbf{E}=E_0\hat{\mathbf{i}}$. When the dipole is rotated slightly from its equilibrium position and released, the time period of its oscillations will be

(a) $\frac{1}{2\pi}\sqrt{\frac{2ml}{qE_0}}$ (b) $2\pi\sqrt{\frac{ml}{2qE_0}}$ (c) $2\pi\sqrt{\frac{ml}{qE_0}}$ (d) $\frac{1}{2\pi}\sqrt{\frac{ml}{2qE_0}}$

30/ At the interface between two materials having refractive indices n_1 and n_2, the critical angle for reflection of an em wave is θ_{1C}. The n_2 material is replaced by another material having refractive index n_3 such that the critical angle at the interface between n_1 and n_3 materials is θ_{2C}. If $n_3>n_2>n_1$, $\frac{n_2}{n_3}=\frac{2}{5}$ and $\sin\theta_{2C}-\sin\theta_{1C}=1/2$, then θ_{1C} is

(a) $\sin^{-1}\left(\frac{5}{6n_1}\right)$
(b) $\sin^{-1}\left(\frac{2}{3n_1}\right)$
(c) $\sin^{-1}\left(\frac{1}{3n_1}\right)$
(d) $\sin^{-1}\left(\frac{1}{6n_1}\right)$

31/ A body of mass m connected to a massless and unstretchable string goes in verticle circle of radius R under gravity g. The other end of the string is fixed at the centre of circle. If velocity at top of circular path is $n\sqrt{gR}$, where, $n \geq 1$, then ratio of kinetic energy of the body at bottom to that at top of the circle is

(a) $\frac{n+4}{n}$ (b) $\frac{n^2}{n^2+4}$ (c) $\frac{n}{n+4}$ (d) $\frac{n^2+4}{n^2}$

32/ As shown below, bob A of a pendulum having massless string of length R is released from 60° to the vertical. It hits another bob B of half the mass that is at rest on a frictionless table in the centre. Assuming elastic collision, the magnitude of the velocity of bob A after the collision will be (take g as acceleration due to gravity)

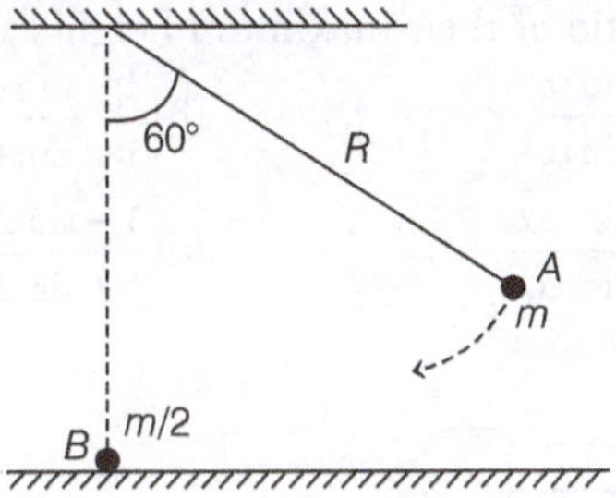

(a) $\sqrt{Rg}$ (b) $\frac{1}{3}\sqrt{Rg}$

(c) $\frac{2}{3}\sqrt{Rg}$ (d) $\frac{4}{3}\sqrt{Rg}$

33/ Consider I_1 and I_2 are the currents flowing simultaneously in two nearby coils 1 and 2, respectively. If L_1 = self-inductance of coil 1, M_{12} = mutual inductance of coil 1 with respect to coil 2, then the value of induced emf in coil 1 will be

(a) $\varepsilon_1 = -L_1\frac{dI_1}{dt} - M_{12}\frac{dI_1}{dt}$

(b) $\varepsilon_1 = -L_1\frac{dI_1}{dt} - M_{12}\frac{dI_2}{dt}$

(c) $\varepsilon_1 = -L_1\frac{dI_1}{dt} + M_{12}\frac{dI_2}{dt}$

(d) $\varepsilon_1 = -L_1\frac{dI_2}{dt} - M_{12}\frac{dI_1}{dt}$

34/ Given below are two statements. One is labelled as Assertion (A) and the other is labelled as Reason (R).

Assertion (A) Choke coil is simply a coil having a large inductance but a small resistance. Choke coils are used with fluorescent mercury-tube fittings. If household electric power is directly connected to a mercury tube, the tube will be damaged.

Reason (R) By using the choke coil, the voltage across the tube is reduced by a factor $(R/\sqrt{R^2+\omega^2L^2})$, where ω is frequency of the supply across resistor R and inductor L. If the choke coil were not used, the voltage across the resistor would be the same as the applied voltage.

In the light of the above statements, choose the most appropriate answer from the options given below.

(a) A is false but R is true.
(b) Both A and R are true and R is the correct explanation of A.
(c) Both A and R are true but R is not the correct explanation of A.
(d) A is true but R is false.

35/ Match List-I with List-II.

	List-I		List-II
(A)	Electric field inside (distance $r > 0$ from centre) of a uniformly charged spherical shell with surface charge density σ and radius R.	(I)	σ/ε_0
(B)	Electric field at distance $r > 0$ from a uniformly charged infinite plane sheet with surface charge density σ.	(II)	$\sigma/2\varepsilon_0$
(C)	Electric field outside (distance $r > 0$ from centre) of a uniformly charged spherical shell with surface charge density σ and radius R.	(III)	0
(D)	Electric field between two oppositely charged infinite plane parallel sheets with uniform surface charge density σ.	(IV)	$\frac{\sigma}{\varepsilon_0 r^2}$

Choose the correct answer from the options given below.

(a) (A)-(II), (B)-(I), (C)-(IV), (D)-(III)
(b) (A)-(III), (B)-(II), (C)-(IV), (D)-(1)
(c) (A)-(IV), (B)-(II), (C)-(III), (D)-(I)
(d) (A)-(IV), (B)-(I), (C)-(III), (D)-(II)

Imp.

36/ If λ and K are de-Broglie wavelength and kinetic energy, respectively, of a particle with constant mass. The correct graphical representation for the particle will be

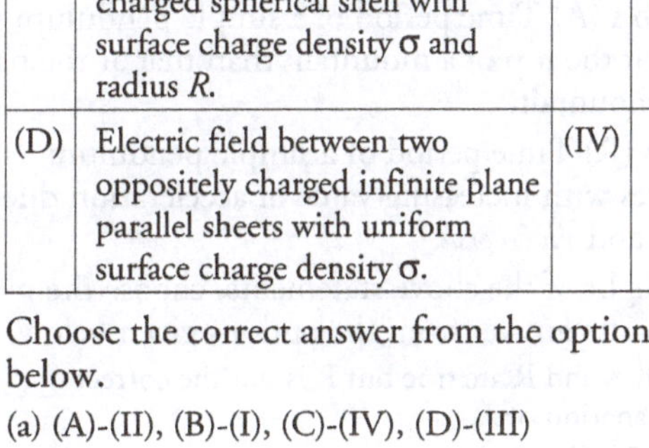

de-Broglie Wavelength

37/ The pair of physical quantities not having same dimensions is
(a) surface tension and impulse
(b) angular momentum and Planck's constant
(c) pressure and Young's modulus
(d) torque and energy

38/ Given below are two statements. One is labelled as Assertion (A) and the other is labelled as Reason (R).
Assertion (A) Electromagnetic waves carry energy but not momentum.
Reason (R) Mass of a photon is zero.
In the light of the above statements, choose the most appropriate answer from the options given below.
(a) A is false but R is true.
(b) A is true but R is false.
(c) Both A and R are true but R is not the correct explanation of A.
(d) Both A and R are true and R is the correct explanation of A.

39/ Given below are two statements. One is labelled as Assertion (A) and the other is labelled as Reason (R).
Assertion (A) Time period of a simple pendulum is longer at the top of a mountain than that at the base of the mountain.
Reason (R) Time period of a simple pendulum decreases with increasing value of acceleration due to gravity and *vice-versa*.
In the light of the above statements, choose the most appropriate answer from the options given below.
(a) Both A and R are true but R is not the correct explanation of A.
(b) A is false but R is true.
(c) Both A and R are true and R is the correct explanation of A.
(d) A is true but R is false.

40/ The expression given below shows the variation of velocity (v) with time (t),
$v = At^2 + \frac{Bt}{C+t}$. The dimension of ABC is
(a) $[M^0L^1T^{-3}]$ (b) $[M^0L^2T^{-2}]$
(c) $[M^0L^2T^{-3}]$ (d) $[M^0L^1T^{-2}]$

41/ The fractional compression $\left(\frac{\Delta V}{V}\right)$ of water at the depth of 2.5 km below the sea level is .... %. Given, the Bulk modulus of water $= 2 \times 10^9$ Nm^{-2}, density of water $= 10^3$ kg m^{-3} acceleration due to gravity $= 10$ ms^{-2}.
(a) 1.5 (b) 1.0 (c) 1.75 (d) 1.25

42/ The work done in an adiabatic change in an ideal gas depends upon only
(a) change in its pressure (b) change in its temperature
(c) change in its specific heat (d) change in its volume

Imp.

43/ Two projectiles fired with same initial speed from same point on ground at angles of $(45° - \alpha)$ and $(45° + \alpha)$, respectively, with the horizontal direction. The ratio of their maximum heights attained is
(a) $\frac{1+\sin\alpha}{1-\sin\alpha}$ (b) $\frac{1-\tan\alpha}{1+\tan\alpha}$
(c) $\frac{1+\sin 2\alpha}{1-\sin 2\alpha}$ (d) $\frac{1-\sin 2\alpha}{1+\sin 2\alpha}$

Projectile Motion

44/ 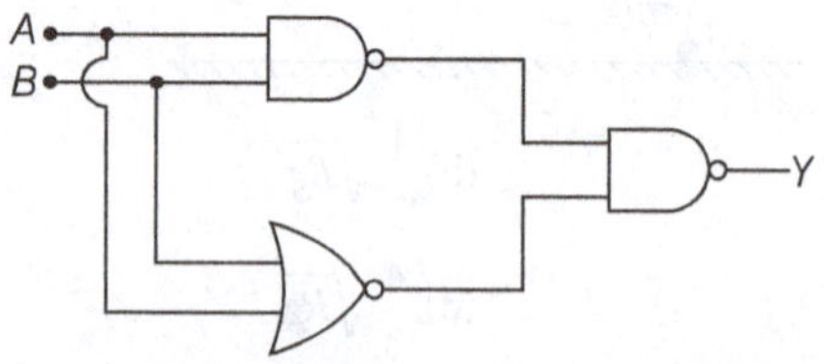

For the circuit shown above, equivalent gate is
(a) AND gate (b) NOT gate
(c) NAND gate (d) OR gate

45/ Consider a long straight wire of a circular cross-section (radius a) carrying a steady current I. The current is uniformly distributed across this cross-section. The distances from the centre of the wire's cross-section at which the magnetic field (inside the wire, outside the wire) is half of the maximum possible magnetic field, anywhere due to the wire, will be
(a) $[a/4, 3a/2]$ (b) $[a/2, 3a]$ (c) $[a/4, 2a]$ (d) $[a/2, 2a]$

Section B

Numerical Type Questions

46/ The coordinates of a particle with respect to origin in a given reference frame is (1, 1, 1) m. If a force of $\mathbf{F} = \hat{i} - \hat{j} + \hat{k}$ acts on the particle, then the magnitude of torque (with respect to origin) in z-direction is ....

47/ A container of fixed volume contains a gas at 27°C. To double the pressure of the gas, the temperature of gas should be raised to .... °C.

48/ Two light beams fall on a transparent material block at point 1 and 2 with angle θ_1 and θ_2, respectively, as shown in figure. After refraction, the beams intersect at point 3 which is exactly on the interface at other end of the block. Given the distance between 1 and 2,

$d = 4\sqrt{3}$ cm and $\theta_1 = \theta_2 = \cos^{-1}(n_2/2n_1)$, where refractive index of the block n_2 > refractive index of the outside medium n_1, then the thickness of the block is .... cm.

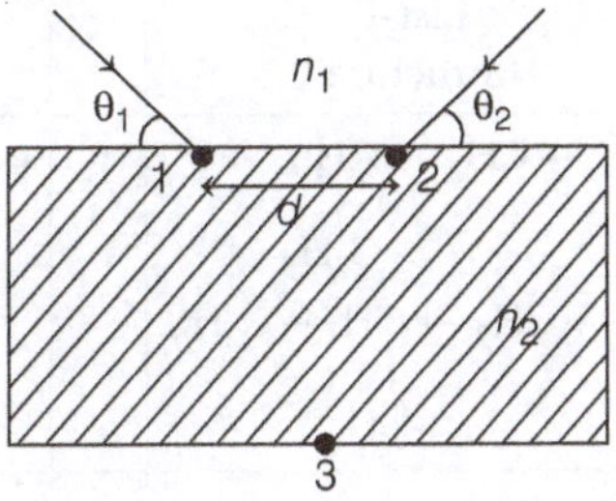

49/ The maximum speed of a boat in still water is 27 km/h. Now, this boat is moving downstream in a river flowing at 9 km/h. A man in the boat throws a ball vertically upwards with speed of 10 m/s. Range of the ball as observed by an observer at rest on the river bank, is ...... cm. (Take, $g = 10\ m/s^2$)

50/ In a hydraulic lift, the surface area of the input piston is 6 cm^2 and that of the output piston is 1500 cm^2. If 100 N force is applied to the input piston to raise the output piston by 20 cm, then the work done is .... kJ.

Chemistry

Section A

Objective Type Questions

51/ Match List-I with List-II.

	List-I (Carbohydrate)		List-II (Linkage source)
(A)	Amylose	I.	$\beta-C_1-C_4$, plant
(B)	Cellulose	II.	$\alpha-C_1-C_4$, animal
(C)	Glycogen	III.	$\alpha-C_1-C_4, \alpha-C_1-C_6$, plant
(D)	Amylopectin	IV.	$\alpha-C_1-C_4$, plant

Choose the correct answer from the options given below :

(a) (A)-(IV), (B)-(I), (C)-(III), (D)-(II)
(b) (A)-(IV), (B)-(I), (C)-(II), (D)-(III)
(c) (A)-(III), (B)-(II), (C)-(I), (D)-(IV)
(d) (A)-(II), (B)-(III), (C)-(I), (D)-(IV)

52/ For a Mg $|Mg^{2+}(aq)||Ag^+(aq)|$ Ag the correct Nernst equation is

(a) $E_{cell} = E^\circ_{cell} - \frac{RT}{2F} \ln \frac{[Mg^{2+}]}{[Ag^+]}$

(b) $E_{cell} = E^\circ_{cell} + \frac{RT}{2F} \ln \frac{[Ag^+]^2}{[Mg^{2+}]}$

(c) $E_{cell} = E^\circ_{cell} - \frac{RT}{2F} \ln \frac{[Ag^+]}{[Mg^{2+}]}$

(d) $E_{cell} = E^\circ_{cell} - \frac{RT}{2F} \ln \frac{[Ag^+]^2}{[Mg^{2+}]}$

53/ In the following substitution reaction,

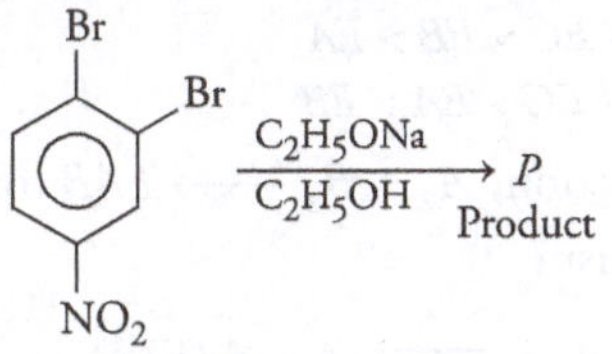

Product P formed is

(a) 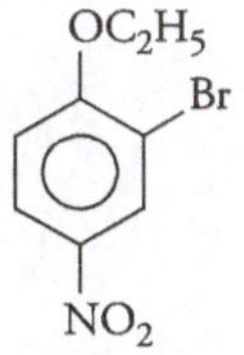

(b) 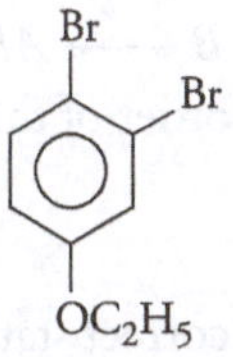

(c) 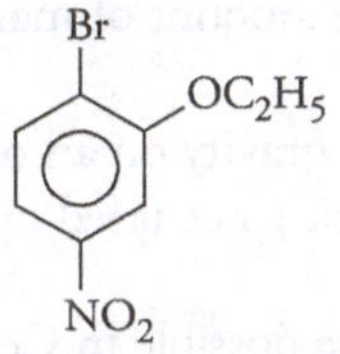

(d) 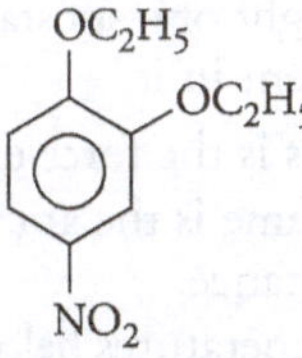

54/ The molar conductivity of a weak electrolyte when plotted against the square root of its concentration, which of the following is expected to be observed?

(a) Molar conductivity decreases sharply with increase in concentration.
(b) A small decrease in molar conductivity is observed at infinite dilution.
(c) A small increase in molar conductivity is observed at infinite dilution.
(d) Molar conductivity increases sharply with increase in concentration.

55/ If a_0 is denoted as the Bohr radius of hydrogen atom, then what is the de-Broglie wavelength (λ) of the electron present in the second orbit of hydrogen atom? [n : any integer]

(a) $\frac{2a_0}{n\pi}$ (b) $\frac{8\pi a_0}{n}$

(c) $\frac{4\pi a_0}{n}$ (d) $\frac{4n}{\pi a_0}$

56/ An element E has the ionisation enthalpy value of 374 kJ mol^{-1}, E reacts with elements A, B, C and D with electron gain enthalpy values of $-328, -349, -325$ and -295 kJ mol^{-1}, respectively. The correct order of the products EA, EB, EC and ED in terms of ionic character is

(a) $EB > EA > EC > ED$

(b) $EA > EB > EC > ED$

(c) $ED > EC > EB > EA$

(d) $ED > EC > EA > EB$

57/ The reaction, $A_2 + B_2 \longrightarrow 2AB$ follows the mechanism

$$A_2 \underset{k_{-1}}{\overset{k_1}{\rightleftharpoons}} A + A \text{ (Fast)}$$

$$A + B_2 \xrightarrow{k_2} AB + B \text{ (Slow)}$$

$$A + B \longrightarrow AB \text{ (Fast)}$$

The overall order of the reaction is

(a) 2 (b) 1.5

(c) 2.5 (d) 3

58/ Choose the correct statements.

A. Weight of a substance is the amount of matter present in it.

B. Mass is the force exerted by gravity on an object.

C. Volume is the amount of space occupied by a substance.

D. Temperatures below 0°C are possible in Celsius scale, but in Kelvin scale negative temperature is not possible.

E. Precision refers to the closeness of various measurements for the same quantity.

Choose the correct answer from the options given below

(a) C, D and E only

(b) B, C and D only

(c) A, B and C only

(d) A, D and E only

Imp.

59/ Match List-I with List-II.

	List-I (Structure)		List-II (IUPAC name)
(A)	$H_3C-CH_2-CH(C_2H_5)-CH_2-CH(CH_3)-C_2H_5$	I.	4-methylpent-1-ene
(B)	$(CH_3)_2C\,(C_3H_7)_2$	II.	3-ethyl-5-methylheptane
(C)	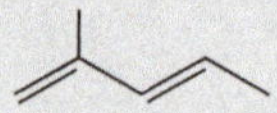	III.	4,4-dimethylheptane
(D)	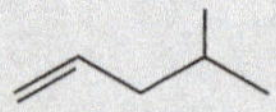	IV.	2-methyl-1,3-pentadiene

Choose the correct answer from the options given below :

(a) (A)-(III), (B)-(II), (C)-(I), (D)-(IV)

(b) (A)-(II), (B)-(III), (C)-(I), (D)-(IV)

(c) (A)-(II), (B)-(III), (C)-(IV), (D)-(I)

(d) (A)-(III), (B)-(II), (C)-(IV), (D)-(I)

IUPAC Nomenclature

60/ The correct option with order of melting points of the pairs (Mn, Fe), (Tc, Ru) and (Re, Os) is

(a) Mn < Fe, Tc < Ru and Os < Re

(b) Fe < Mn, Ru < Tc and Os < Re

(c) Mn < Fe, Tc < Ru and Re < Os

(d) Fe < Mn, Ru < Tc and Re < Os

61/ Given below are two statements.

Statement I The radii of isoelectronic species increases in the order.

$$Mg^{2+} < Na^+ < F^- < O^{2-}$$

Statement II The magnitude of electron gain enthalpy of halogen decreases in the order.

$$Cl > F > Br > I$$

In the light of the above statements, choose the most appropriate answer from the options given below.

(a) Statement I is incorrect but Statement II is correct.

(b) Both Statement I and Statement II are correct.

(c) Both Statement I and Statement II are incorrect.

(d) Statement I is correct but Statement II is incorrect.

62/ The correct increasing order of stability of the complexes based on Δ_o value is

I. $[Mn(CN)_6]^{3-}$ II. $[Co(CN)_6]^{4-}$
III. $[Fe(CN)_6]^{4-}$ IV. $[Fe(CN)_6]^{3-}$

(a) I < II < IV < III (b) II < III < I < IV
(c) IV < III < II < I (d) III < II < IV < I

63/ The standard reduction potential values of some of the *p*-block ions are given below. Predict the one with the strongest oxidising capacity.

(a) $E^{\ominus}_{Al^{3+}/Al} = -1.66$ V (b) $E^{\ominus}_{Pb^{4+}/Pb^{2+}} = +1.67$ V
(c) $E^{\ominus}_{Sn^{4+}/Sn^{2+}} = +1.15$ V (d) $E^{\ominus}_{Tl^{3+}/Tl} = +1.26$ V

64/ Total number of nucleophiles from the following is :

NH_3, PhSH, $(H_3C)_2S$, $H_2C=CH_2$
$\overset{\ominus}{O}H^{\ominus}$, $H_3O^{\oplus}$, $(CH_3)_2$ CO, 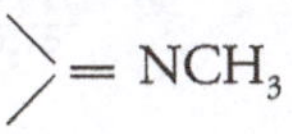

(a) 5 (b) 6
(c) 7 (d) 4

65/ The product (*P*) formed in the following reaction is

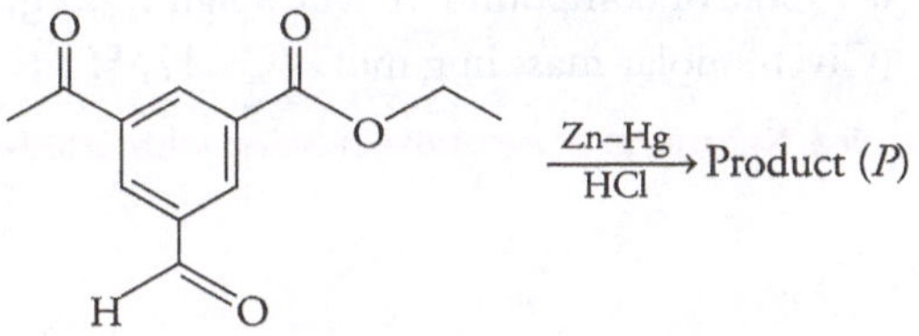

(a) 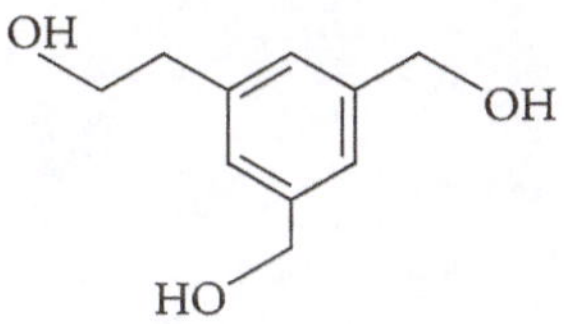

(b) 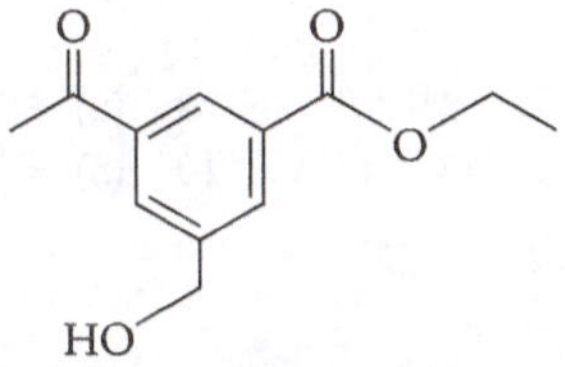

(c) 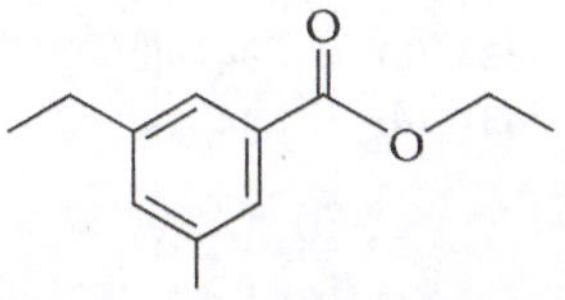

(d) 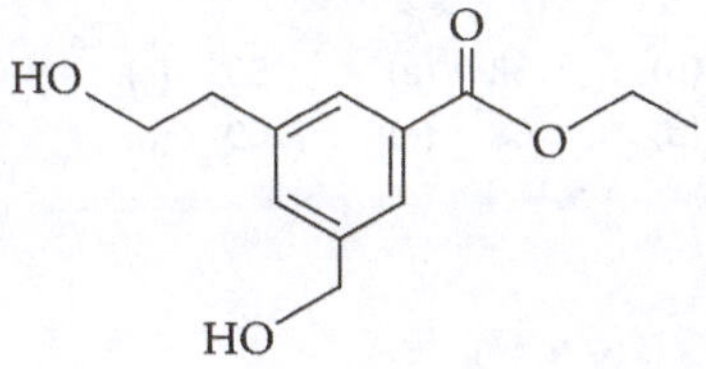

66/ 500 J of energy is transferred as heat to 0.5 mol of Argon gas at 298 K and 1.00 atm. The final temperature and the change in internal energy respectively are (Given : $R = 8.3$ J K^{-1} mol^{-1}

(a) 368 K and 500 J (b) 378 K and 300 J
(c) 378 K and 500 J (d) 348 K and 300 J

67/ 1.24 g of AX_2 (molar mass 124 g mol^{-1}) is dissolved in 1 kg of water to form a solution with boiling point of 100.0156°C, while 25.4 g of AY_2 (molar mass 250 g mol^{-1}) in 2 kg of water constitutes a solution with a boiling point of 100.0260°C.

$[K_b(H_2O) = 0.52 \text{ K kg mol}^{-1}]$

Which of the following is correct?

(a) AX_2 and AY_2 (both) are fully ionised.
(b) AX_2 is fully ionised while AY_2 is completely unionised.
(c) AX_2 is completely unionised while AY_2 is fully ionised.
(d) AX_2 and AY_2 (both) are completely unionised.

68/ At temperature *T*, compound $AB_2(g)$ dissociates as

$AB_2(g) \rightleftharpoons AB(g) + \frac{1}{2}B_2(g)$ having degree of dissociation *x* (small compared to unity). The correct expression for *x* in terms of K_p and *p* is

(a) $\sqrt{K_p}$ (b) $\sqrt[4]{\frac{2K_p}{p}}$
(c) $\sqrt[3]{\frac{2K_p^2}{p}}$ (d) $\sqrt[3]{\frac{2K_p}{p}}$

69/ Match List-I with List-II.

	List-I (Complex)		List-II (Hybridisation and Magnetic characters)
(A)	$[MnBr_4]^{2-}$	I.	d^2sp^3 and diamagnetic
(B)	$[FeF_6]^{3-}$	II.	sp^3d^2 and paramagnetic
(C)	$[Co(C_2O_4)_3]^{3-}$	III.	sp^3 and diamagnetic
(D)	$[Ni(CO)_4]$	IV.	sp^3 and paramagnetic

Choose the correct answer from the options given below

(a) (A)-(IV), (B)-(II), (C)-(I), (D)-(III)
(b) (A)-(III), (B)-(II), (C)-(I), (D)-(IV)
(c) (A)-(IV), (B)-(I), (C)-(II), (D)-(III)
(d) (A)-(III), (B)-(I), (C)-(II), (D)-(IV)

70/ The steam volatile compounds among the following are

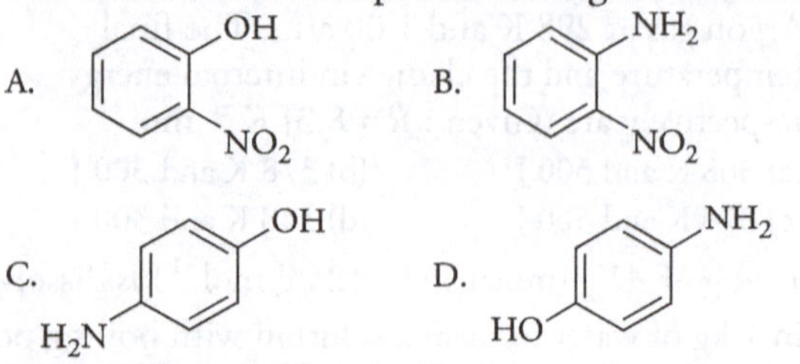

Choose the correct answer from the options given below

(a) (A) and (B) only
(b) (A), (B) and (C) only
(c) (A) and (C) only
(d) (B) and (D) only

Section B

Numerical Type Questions

71/ The molar mass of the water insoluble product formed from the fusion of chromite ore ($FeCr_2O_4$) with Na_2CO_3 in presence of O_2 is .... g mol^{-1}.

72/ Given below are some nitrogen containing compounds

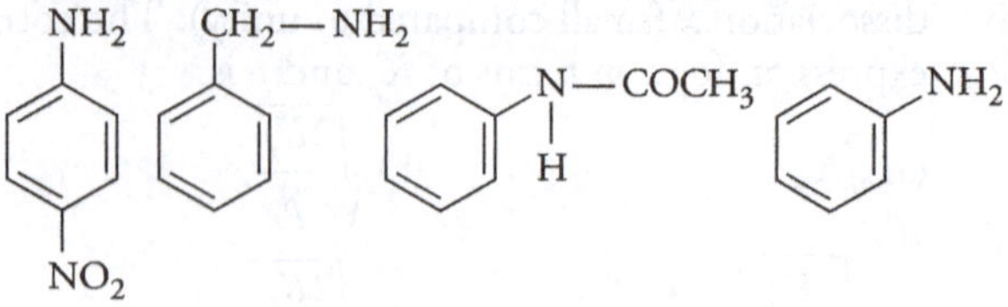

Each of them is treated with HCl separately. 1.0 g of the most basic compound will consume .... mg of HCl.

(Given, molar mass in g mol^{-1} C : 12, H : 1, O : 16, Cl : 35.5)

73/ The sum of sigma (σ) and pi (π) bonds in hex-1,3-dien-5-yne is ....

74/ If A_2B is 30% ionised in an aqueous solution, then the value of van't Hoff factor (i) is .... $\times 10^{-1}$

Imp.

75/

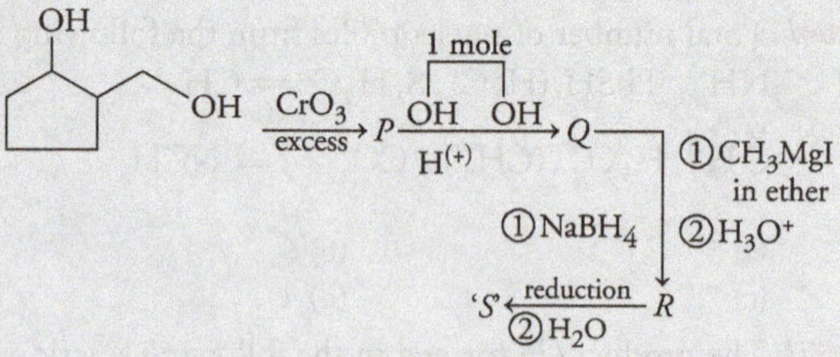

0.1 mole of compound 'S' will weigh ...... g.
(Given, molar mass in g mol^{-1} C : 12, H : 1, O : 16)

Chemical Reactions of Alcohols in Blend with Stoichiometry

Answer Key

Mathematics

1 (d)	2 (c)	3 (a)	4 (d)	5 (d)	6 (d)	7 (b)	8 (d)	9 (b)	10 (d)
11 (c)	12 (d)	13 (c)	14 (d)	15 (d)	16 (a)	17 (b)	18 (a)	19 (b)	20 (b)
21 (1405)	22 (24)	23 (5)	24 (2)	25 (112)					

Physics

26 (b)	27 (d)	28 (b)	29 (b)	30 (*)	31 (b)	32 (b)	33 (b)	34 (b)	35 (*)
36 (d)	37 (a)	38 (*)	39 (c)	40 (c)	41 (d)	42 (b)	43 (d)	44 (d)	45 (d)
46 (2)	47 (327)	48 (6)	49 (2000)	50 (5)					

Chemistry

51 (b)	52 (b)	53 (a)	54 (a)	55 (b)	56 (a)	57 (b)	58 (a)	59 (c)	60 (a)
61 (b)	62 (a)	63 (b)	64 (a)	65 (c)	66 (d)	67 (b)	68 (c)	69 (a)	70 (a)
71 (160)	72 (341)	73 (15)	74 (16)	75 (13)					

***Note** (*) None of the option is correct.*

Detailed Explanations

29 January
Shift-I

Mathematics

1. *(d)* First of all we have to write equation of line CD.

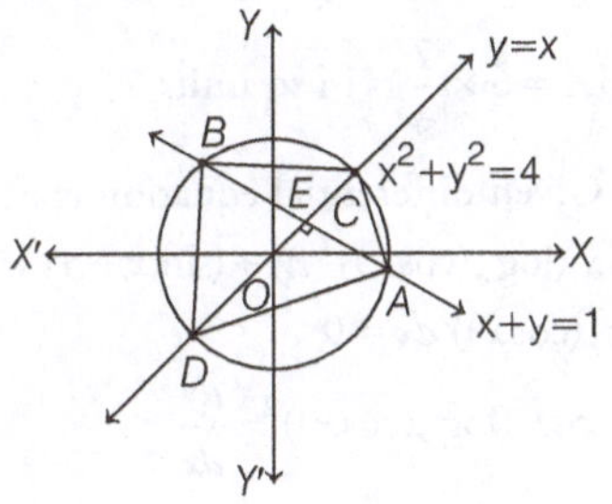

As $\quad CD \perp AB$

$\therefore \quad M_{CD} = \dfrac{-1}{m_{AB}} = \dfrac{-1}{-1} = 1$

and CD passes through mid-point of AB.
when we solve $x + y = 1$ with $x^2 + y^2 = 4$.
then we get coordinats of A and B

i.e. $A \equiv \left(\dfrac{(1+\sqrt{7})}{2}, \dfrac{1-\sqrt{7}}{2}\right)$ and

$B \equiv \left(\dfrac{1-\sqrt{7}}{2}, \dfrac{1+\sqrt{7}}{2}\right)$ and thus, we

find that mid-point of AB is $\left(\dfrac{1}{2}, \dfrac{1}{2}\right)$.

So, equation of line CD will be

$$y - \frac{1}{2} = 1\left(x - \frac{1}{2}\right) \Rightarrow y = x \qquad \ldots\text{(i)}$$

Now, solving $y = x$ with $x^2 + y^2 = 4$ to get coordinates of C and D. Then,

$x^2 + x^2 = 4 \Rightarrow x^2 = 2 \Rightarrow x = \pm\sqrt{2}$

$\therefore \quad y = \pm\sqrt{2}$

So, $C \equiv (\sqrt{2}, \sqrt{2})$ and
$D \equiv (-\sqrt{2}, -\sqrt{2})$ [$\because C$ lies in first quadrant and D lies in third quadrant]

Now, area of quadrilateral $ADBC$

$$= 2 \times (\text{area of } \Delta BCD)$$

$$= 2 \times \frac{1}{2} \begin{vmatrix} \sqrt{2} & \sqrt{2} & 1 \\ \dfrac{1-\sqrt{7}}{2} & \dfrac{1+\sqrt{7}}{2} & 1 \\ -\sqrt{2} & -\sqrt{2} & 1 \end{vmatrix}$$

$= 2\sqrt{14}$ sq unit

2. *(c)*

Key idea Mean and variance of n observations $x_1, x_2 \ldots, x_n$ are $\bar{x} = \dfrac{\Sigma x_i}{n}$, $\sigma^2 = \dfrac{1}{n}\sum_{i=1}^{n} x_i^2 - (\bar{x})^2$, respectively.

Given, variance $= 4/5$

$$\Rightarrow \quad \frac{\sum x_i^2}{10} - \left(\frac{\sum x_i}{10}\right)^2 = \frac{4}{5}$$

$$\Rightarrow \quad \frac{\sum x_i^2}{10} - 25 = \frac{4}{5}$$

$$\because \quad \sum_{i=1}^{10}(x_i - 2) = 30$$

$$\therefore \quad \sum_{i=1}^{10} x_i = 50$$

$$\therefore \quad \left(\frac{\sum x_i^2}{10}\right)^2 = (5)^2 = 25$$

$$\Rightarrow \quad \sum x_i^2 = 258$$

$$\text{Now,} \quad \sum_{i=1}^{10}(x_i - \beta)^2 = 98$$

$$\Rightarrow \quad \sum_{i=1}^{10}(x_i^2 - 2\beta \cdot x_i + \beta^2) = 98$$

$$\Rightarrow \quad 258 - 2\beta(50) + 10\beta^2 = 98$$

$$\Rightarrow \quad (\beta - 8)(\beta - 2) = 0 \Rightarrow \beta = 2, 8$$

but $\beta > 2$ [given]

$\therefore \ \beta = 8$ (only)

Now, $2(x_1 - 1) + 4\beta, 2(x_2 - 1) + 4\beta, \ldots$
$2(x_{10} - 1) + 4\beta$
$= 2x_1 + 30, 2x_2 + 30, \ldots\ldots, 2x_{10} + 30$

So, mean of above observation

$$\mu = 2(5) + 30 = 40$$

[$\because$ mean of $x_1, x_2, x_3 \ldots, x_{10}$ is 5]

and variance $= \sigma^2 = 2^2\left(\dfrac{4}{5}\right) = \dfrac{16}{5}$

Hence, $\dfrac{\beta\mu}{\sigma^2} = \dfrac{8 \times 40}{\frac{16}{5}} = 100$

3. *(a)* Given equation is

$$\left(\frac{9}{x} - \frac{9}{\sqrt{x}} + 2\right)\left(\frac{2}{x} - \frac{7}{\sqrt{x}} + 3\right) = 0 \quad \ldots\text{(i)}$$

Let $\dfrac{1}{\sqrt{x}} = t$, so that $t > 0$

From Eq. (i), we get

$$(9t^2 - 9t + 2)(2t^2 - 7t + 3) = 0$$

$$\Rightarrow (3t - 2)(3t - 1)(t - 3)(2t - 1) = 0$$

$$\Rightarrow \quad t = \frac{1}{3}, \frac{1}{2}, \frac{2}{3}, 3$$

$$\therefore \quad x = 9, 4, \frac{9}{4}, \frac{1}{9}$$

Hence, number of solutions = 4

4. *(d)* Given $A = \begin{bmatrix} \log_5 128 & \log_4 5 \\ \log_5 8 & \log_4 25 \end{bmatrix}$

$\therefore |A| = (\log_5 128 \times \log_4 25)$

$$- (\log_4 5 \times \log_5 8)$$

$$= (\log_5 128 \times 2 \times \log_4 5) - \left(\log_5 8 \times \frac{1}{\log_5 4}\right)$$

$$= 2 \times \log_4 128 - \log_4 8$$

$$= \log_4 (128)^2 - \log_4 8$$

$$= \log_4\left(\frac{128 \times 128}{8}\right)$$

$$= \log_4 (4^2 \times 4^3 \times 2)$$

$$= \log_4 (4^5) + \log_4 2$$

$$= 5 + \frac{1}{2} = \frac{11}{2}$$

Now,

$$C_{11} = \sum_{k=1}^{2} a_{1k} \cdot A_{1k} = a_{11}A_{11} + a_{12}A_{12}$$

$$= |A| = \frac{11}{2}$$

$$C_{12} = \sum_{k=1}^{2} a_{1k}A_{2k}$$

$$= a_{11}A_{21} + a_{12}A_{22} = 0$$

$$C_{21} = \sum_{k=1}^{2} a_{2k} \cdot A_{1k}$$

$$= a_{21}A_{11} + a_{22}A_{12} = 0$$

$$C_{22}=\sum_{k=1}^{2}a_{2k}A_{2k}=a_{21}A_{21}+a_{22}A_{22}$$

$$=|A|=\frac{11}{2}$$

$$\therefore C=\begin{bmatrix}11/2 & 0\\ 0 & 11/2\end{bmatrix}$$

$$\Rightarrow |C|=\frac{11}{2}\times\frac{11}{2}-0=\frac{121}{4}$$

Hence, $8|C|=8\times\frac{121}{4}=242$

5. *(d)* $xRy \Rightarrow \sec^2 x-\tan^2 y=1$

To check Reflexivity

$xRx \Rightarrow \sec^2 x-\tan^2 x=1$ which is true.

So, R is reflexive.

To check symmetricity

$xRy \Rightarrow yRx$

i.e. $\sec^2 x-\tan^2 y=1$

$\Rightarrow \tan^2 x+1-\sec^2 y+1=1$

$\Rightarrow \tan^2 x-\sec^2 y=-1$

$\Rightarrow \sec^2 y-\tan^2 x=1 \Rightarrow y\,R\,x.$

So, R is symmetric.

To check Transitivity

$x\,R\,y,\ y\,R\,z \Rightarrow x\,R\,z$

i.e. $\sec^2 x-\tan^2 y=1$ ...(i)

$\Rightarrow \sec^2 y-\tan^2 z=1$ ...(ii)

On adding Eqs. (i) and (ii), we get

$\sec^2 x+\sec^2 y-\tan^2 y-\tan^2 z=2$

$\Rightarrow \sec^2 x+1-\tan^2 z=2$

$\Rightarrow \sec^2 x-\tan^2 z=1 \Rightarrow x\,R\,z$

$\therefore$ R is transitive.

Hence, R is an equivalence relation.

Time Saver Tech

Given relation $xRy \Rightarrow \sec^2 x-\tan^2 y=1$

$\sec^2 x=1+\tan^2 y$

$\sec^2 x=\sec^2 y$

$x=y\ x,\ y\in\left(0,\frac{\pi}{2}\right)$

$\therefore$ $x=y$ is reflexive, symmetric and Transitive

$\therefore$ If is equivalence relation.

6. *(d)* Given

$L_1:\frac{x-1}{1}=\frac{y-2}{-1}=\frac{z-1}{2}=\lambda$ (say)

$\therefore$ Any point taken on L_1 is

$(\lambda+1,-\lambda+2,2\lambda+1)$

Let equation of L_3 be

$$\frac{x-\alpha}{a}=\frac{y-\beta}{b}=\frac{z-\gamma}{c}\quad\text{...(i)}$$

where, $(a\hat{i}+b\hat{j}+c\hat{k})$ is parallel to

$[(\hat{i}-\hat{j}+2\hat{k})\times(-\hat{i}+2\hat{j}+\hat{k})]$

So, $(a,b,c)\equiv(5,3,-2)$

From Eq. (i), we get

$$\Rightarrow \frac{x-\alpha}{5}=\frac{y-\beta}{3}=\frac{z-\gamma}{-1}=\mu\ \text{(say)}\quad\text{...(ii)}$$

Any point taken on Eq. (ii) i.e. L_3 is

$(5\mu+\alpha,3\mu+\beta,-\mu+\gamma)$

Since, L_3 intersects L_1

So, $5\mu+\alpha=\lambda+1 \Rightarrow \alpha=(\lambda+1-5\mu)$

$3\mu+\beta=-\lambda+2 \Rightarrow \beta=(-\lambda+2-3\mu)$

and $-\mu+\gamma=2\lambda+1 \Rightarrow \gamma=(2\lambda+1+\mu)$

Hence, $|5\alpha-11\beta-8\gamma|=|-25|=25$

7. *(b)*

$L_1:\mathbf{r}=(-\hat{i}+2\hat{j}+\hat{k})+\lambda(\hat{i}+2\hat{j}+\hat{k})$

$\Rightarrow \mathbf{r}=(\lambda-1)\hat{i}+2(\lambda+1)\hat{j}+(\lambda+1)\hat{k}$

$L_2:\mathbf{r}=(\hat{j}+\hat{k})+\mu(2\hat{i}+7\hat{j}+3\hat{k})$

$\Rightarrow \mathbf{r}=(2\mu)\hat{i}+(1+7\mu)\hat{j}+(1+3\mu)\hat{k}$

For point of intersection, we equating respective components

i.e. $\lambda-1=2\mu$ ...(i)

$2(\lambda+1)=1+7\mu$ ...(ii)

$\lambda+1=1+3\mu$ ...(iii)

On solving these equation simultaneously, we get

$\lambda=3$ and $\mu=1$

$\Rightarrow$ Point of intersection $=(2,8,4)$

Given, $\mathbf{a}+\mathbf{b}=3\hat{i}+9\hat{j}+4\hat{k}$

$L_3:\mathbf{r}=2\hat{i}+8\hat{j}+4\hat{k}+\alpha(3\hat{i}+9\hat{j}+4\hat{k})$

For $\alpha=2$, $\mathbf{r}=8\hat{i}+26\hat{j}+12\hat{k}$

Hence, L_3 passes through (8, 26, 12)

8. *(d)* $\therefore$ Required area $=A$

ar (rectangle $ABDE$) − ar (region EDC)

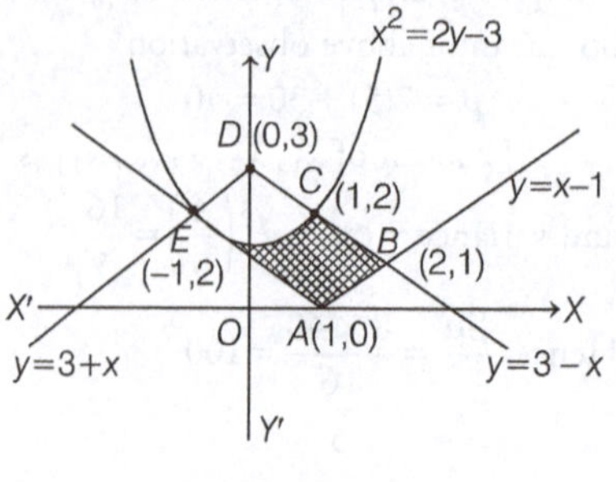

$$\Rightarrow A=4-2\int_0^1(3-x)-\left(\frac{x^2+3}{2}\right)dx$$

$$=4-2\left[3x-\frac{x^2}{2}-\frac{x^3}{6}-\frac{3}{2}x\right]_0^1$$

$$=4-2\left[3-\frac{1}{2}-\frac{1}{6}-\frac{3}{2}\right]=\frac{7}{3}$$

$\therefore$ $6A=6\times\frac{7}{3}=14$ sq units

9. *(b)* Given differential equation is

$\cos x(\log_e(\cos x))^2\,dy+(\sin x-3y(\sin x)\log_e(\cos x))\,dx=0$

$$\Rightarrow \cos x(\log_e(\cos x))^2\cdot\frac{dy}{dx}-3\sin x\cdot\log_e(\cos x)y=-\sin x$$

$$\Rightarrow \frac{dy}{dx}+\frac{3\tan x}{\log_e(\sec x)}y=\frac{-\tan x}{(\log_e(\sec x))^2}$$

$(\because \log_e(\sec x)=-\log_e(\cos x))$

The above differential equation is linear differential equation

So, integrating factor (IF)

$$=e^{\int\frac{3\tan x}{\log_e(\sec x)}dx}=(\log_e(\sec x))^3$$

Its solution is

$$y\times(\log_e(\sec x))^3=\int\frac{-\tan x}{(\log_e(\sec x))^2}(\log_e(\sec x))^3\,dx$$

$$\Rightarrow y\times(\log_e(\sec x))^3=-\frac{1}{2}(\log_e(\sec x))^2+C\quad\text{...(i)}$$

where C is constant of integration.

Given that when $x=\pi/4$, then

$y=-\frac{1}{\log_e 2}$

So, $\frac{-1}{\log_e 2}\times(\log_e\sqrt{2})^3=\frac{-1}{2}(\log_e\sqrt{2})^2+C$

$$\Rightarrow \frac{-1}{8\log_e 2}\times(\log_e 2)^3=\frac{-1}{2}\times\frac{1}{4}(\log_e 2)^2+C$$

$$\Rightarrow \frac{-1}{8}(\log_e 2)^2=\frac{-1}{8}(\log_e 2)^2+C$$

$\Rightarrow C=0$

So, Eq. (i), we get

$$\Rightarrow y \times (\log_e(\sec x))^3 = \frac{-1}{2}(\log_e(\sec x))^3$$

$$\Rightarrow y = \frac{-1}{2\log_e(\sec x)} = \frac{1}{2\log_e(\cos x)}$$

Hence,

$$y\left(\frac{\pi}{6}\right) = \frac{1}{2\log_e\left(\cos\frac{\pi}{6}\right)} = \frac{1}{2\log_e\left(\frac{\sqrt{3}}{2}\right)}$$

$$= \frac{1}{2\left(\frac{1}{2}\log_e 3 - \log_e 2\right)}$$

$$= \frac{1}{\log_e 3 - \log_e 4}$$

10. *(d)* Consider AP as $a, a+d, a+2d$, ... where $a, d \in N$.

Given sum of first three terms of AP

$$= S_3 = 3a + 3d = 54.$$

$$\Rightarrow a + d = 18 \quad \text{...(i)}$$

and $S_{20} = 10(2a + 19d)$

$$= 10(36 + 17d)$$

According to the question,

$1600 < S_{20} < 1800$

$$\Rightarrow 1600 < 10(36 + 17d) < 1800$$

$$\Rightarrow 160 < 36 + 17d < 180$$

$$\Rightarrow 124 < 17d < 144$$

$$\Rightarrow \frac{124}{17} < d < \frac{144}{17}$$

Since, common difference will be natural number

$\therefore \quad d = 8$

$\Rightarrow \quad a + d = 18$

$\Rightarrow \quad a + 8 = 18$

$\Rightarrow \quad a = 10$

Hence,

$$a_{11} = 10 + (10 \times 8) = 10 + 80 = 90$$

11. *(c)* Case (I) Number of numbers created using 3221111 is

$$n_1 = \frac{7!}{4!2!} = \frac{7\times6\times5\times4}{4!\times2\times1} = 105$$

Case (II) Number of numbers created using 2222111 is

$$n_2 = \frac{7!}{4!3!} = \frac{7\times6\times5\times4!}{4!\times3\times2} = 35$$

Case (III) Number of numbers created 3311111 is

$$n_3 = \frac{7!}{5!2!} = \frac{7\times6\times5!}{5!\,2\times1} = 21$$

Hence, total number of elements in the set

$$P = n_1 + n_2 + n_3 = 105 + 35 + 21 = 161$$

Concept Enhancer

Number of permutation of n things of which P_1 are alike of the one kind P_2 are alike of second kind......P_r are rth kind such that $P_1 + P_2 + \ldots + P_r = n$ is $\frac{n!}{P_1!P_2!\ldots P_r!}$

12. *(d)* Given,

$$f(x) = \begin{vmatrix} 1+\sin^2 x & \cos^2 x & 4\sin 4x \\ \sin^2 x & 1+\cos^2 x & 4\sin 4x \\ \sin^2 x & \cos^2 x & 1+4\sin 4x \end{vmatrix}$$

$$\Rightarrow f(x) = \begin{vmatrix} 1+\sin^2 x & \cos^2 x & 4\sin 4x \\ -1 & 1 & 0 \\ -1 & 0 & 1 \end{vmatrix}$$

(applying $R_2 \to R_2 - R_1$ and $R_3 \to R_3 - R_1$)

$$\Rightarrow \quad f(x) = 2 + 4\sin 4x$$

(expanding along R_1)

$[\because -1 \le \sin 4x \le 1 \Rightarrow -4 \le 4\sin 4x \le 4]$

$\Rightarrow -4 + 2 \le 2 + 4\sin 4x \le 4 + 2$

$\Rightarrow -2 \le f(x) \le 6$

$\therefore f(x)|_{\min} = m = -2$

and $f(x)|_{\max} = M = 6$

Hence,

$$M^4 - m^4 = (6)^4 - (-2)^4 = 1280$$

13. *(c)* Given $|z_1 - 8 - 2i| \le 1$

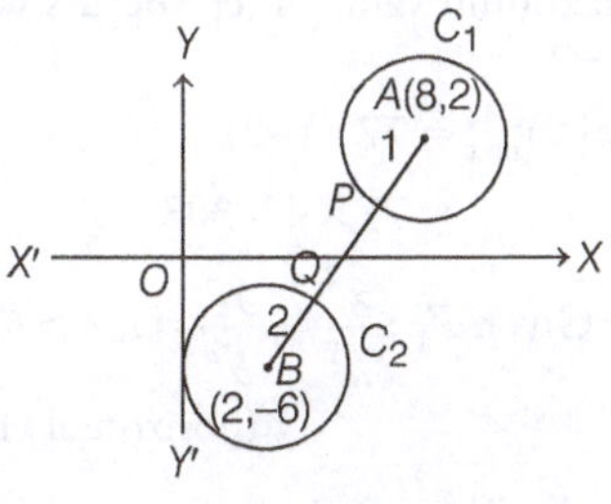

$\Rightarrow \quad |z_1 - (8 + 2i)| \le 1 = C_1$

and $\quad |z_2 - 2 + 6i| \le 2$

$\Rightarrow \quad |z_2 - (2 - 6i)| \le 2 = C_2$

$\therefore AB = \sqrt{(8-2)^2 + (2+6)^2}$

$= \sqrt{36 + 64}$

$= 10$

$\therefore |z_1 - z_2|_{\min} = PQ = AB - BQ - AP$

$= 10 - 2 - 1 = 7$

14. *(d)* Let $I = 80\int_0^{\frac{\pi}{4}} \frac{\sin\theta + \cos\theta}{9 + 16\sin 2\theta}\,d\theta$

$$= 80\int_0^{\frac{\pi}{4}} \frac{\sin\theta + \cos\theta}{9 + 16(2\sin\theta\cdot\cos\theta)}\,d\theta$$

$$= 80\int_0^{\frac{\pi}{4}} \frac{\sin\theta + \cos\theta}{9 + 16(1 + 2\sin\theta\cos\theta - 1)}\,d\theta$$

$$= 80\int_0^{\frac{\pi}{4}} \frac{\sin\theta + \cos\theta}{9 + 16 - 16(\sin\theta - \cos\theta)^2}\,d\theta$$

On putting $\sin\theta - \cos\theta = z$ so that $(\cos\theta + \sin\theta)\,d\theta = dz$

θ	0	π/4
z	−1	0

$$I = 80\int_{-1}^{0} \frac{dz}{25 - 16z^2}$$

$$= 80\int_{-1}^{0} \frac{dz}{16\left(\frac{25}{16} - z^2\right)}$$

$$= \frac{80}{16}\int_{-1}^{0} \frac{dz}{\left(\frac{5}{4}\right)^2 - (z)^2}$$

$$= 5 \times \frac{1}{2\times\frac{5}{4}}\log_e\left(\frac{\frac{5}{4} + z}{\frac{5}{4} - z}\right)_{-1}^{0}$$

$$= 2\log_e(1) + 4\log_e(3)$$

$$= 2\times 0 + 4\log_e(3) = 4\log_e(3)$$

15. *(d)* According to the question, scenario, diagram is created as below-

Now, let us consider that coordinates of A and B are (x_1, y_1) and (x_2, y_2), respectively.

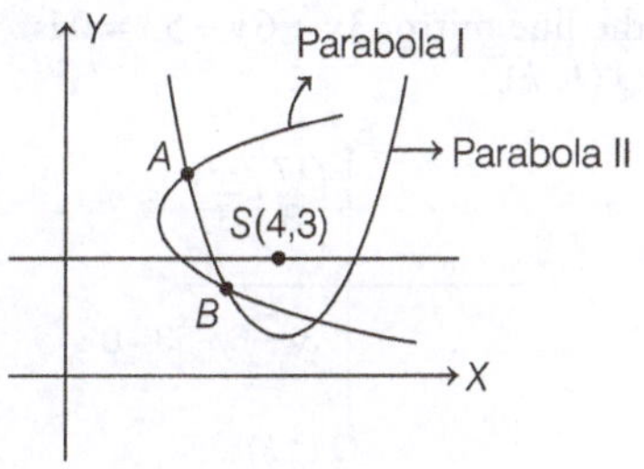

Now, we can easily write equation of parabolas I and II such as

$$(x-4)^2+(y-3)^2=x^2 \quad \ldots\text{(i)}$$

and $$(x-4)^2+(y-3)^2=y^2 \quad \ldots\text{(ii)}$$

Since, A and B will satisfy the Eqs. (i) and (ii) as well as

We observe that, $x=y$ ...(iii)

Put in Eq. (i) we get,

$$x^2-14x+25=0 \Rightarrow x_1+x_2=14$$

and $$x_1x_2=25$$

$$\therefore \quad AB^2=(x_1-x_2)^2+(y_1-y_2)^2$$

$$=2(x_1-x_2)^2$$

$$[\because x=y \therefore x_1=y_1 \text{ and } x_2=y_2]$$

$$=2[(x_1+x_2)^2-4x_1x_2]$$

$$=2[(14)^2-4\times 25]=2\times[196-100]$$

$$=2\times 96=192$$

16. *(a)* Given equation of sides of ΔABC are

$AB: 7x-6y+3=0$ ...(i)

$BC: 9x-2y-19=0$ ...(ii)

$CA: x+2y-31=0$ ...(iii)

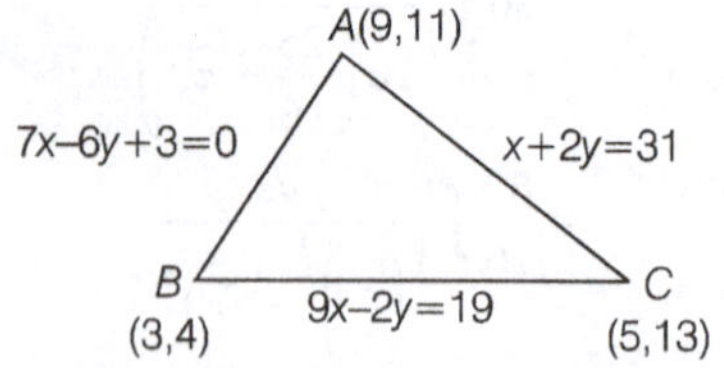

On solving these equations taken two at a time we get the coordinates of A, B and C such that

$A\equiv(9,11)$, $B\equiv(3,4)$ and $C\equiv(5,13)$

Now, centroid of

$$\Delta ABC=\left(\frac{9+3+5}{3},\frac{11+4+13}{3}\right)$$

$$=\left(\frac{17}{3},\frac{28}{3}\right)$$

Let image of the centroid $P\left(\frac{17}{3},\frac{28}{3}\right)$ in the line mirror $3x+6y-53=0$ is $Q(h,k)$.

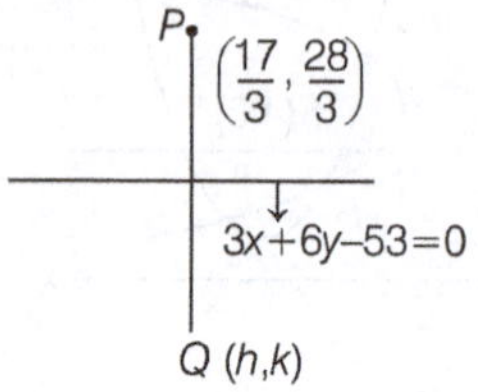

So, $m_{PQ}\times m_{\text{line}}=-1$

$$\Rightarrow \left(\frac{k-\frac{28}{3}}{h-\frac{17}{3}}\right)\times\left(-\frac{1}{2}\right)=-1 \quad \ldots\text{(iv)}$$

and $$3\left(\frac{h+\frac{17}{3}}{2}\right)+6\left(\frac{k+\frac{28}{3}}{2}\right)=55$$

[$\because$ mid-point of PQ lies on $3x+6y-53=0$]

On solving Eqs. (iv) and (v), we get

$h=3$, $k=4$

$$\therefore h^2+k^2+hk=9+16+12=37$$

17. *(b)* Given, $\mathbf{a}=2\hat{i}-\hat{j}+3\hat{k}$

and $\mathbf{b}=3\hat{i}-5\hat{j}+\hat{k}$

Also, $\mathbf{a}\times\mathbf{c}=\mathbf{c}\times\mathbf{b}\Rightarrow\mathbf{a}\times\mathbf{c}=-\mathbf{b}\times\mathbf{c}$

$\Rightarrow \mathbf{a}\times\mathbf{c}+\mathbf{b}\times\mathbf{c}=\mathbf{o}$

$\Rightarrow (\mathbf{a}+\mathbf{b})\times\mathbf{c}=\mathbf{o}\Rightarrow\mathbf{c}=\lambda(\mathbf{a}+\mathbf{b})$

where λ is a parameter

$$\Rightarrow \quad \mathbf{c}=\lambda(5\hat{i}-6\hat{j}+4\hat{k}) \quad \ldots\text{(i)}$$

$$\Rightarrow |\mathbf{c}|^2=\lambda^2(25+36+16)=77\lambda^2 \quad \ldots\text{(ii)}$$

Also, given, $(\mathbf{a}+\mathbf{c})\cdot(\mathbf{b}+\mathbf{c})=168$

$$\Rightarrow \mathbf{a}\cdot\mathbf{b}+\mathbf{a}\cdot\mathbf{c}+\mathbf{c}\cdot\mathbf{b}+|\mathbf{c}|^2=168$$

$$\Rightarrow 14+\mathbf{c}\cdot(\mathbf{a}+\mathbf{b})+77\lambda^2=168$$

[using Eq. (ii) $\mathbf{a}\cdot\mathbf{b}=14$]

$$\Rightarrow \lambda|5\hat{i}-6\hat{j}+4\hat{k}|^2+77\lambda^2=154$$

(using Eq. (i))

$$\Rightarrow 77\lambda+77\lambda^2-154=0$$

$$\Rightarrow \quad \lambda^2+\lambda-2=0\Rightarrow\lambda=-2,1$$

$\therefore$ Maximum value of $|\mathbf{c}|^2$ occurs when $\lambda=-2$

$$\therefore (|\mathbf{c}|^2)_{\max}=77\times(-2)^2$$

$$=77\times 4=308$$

18. *(a)* Given $E_1:\frac{x^2}{a^2}+\frac{y^2}{b^2}=1$, $a>b$

(horizontal ellipse)

$$E_2:\frac{x^2}{A^2}+\frac{y^2}{B^2}=1,\ A<B$$

(vertical ellipse)

both are having eccentricity $=\frac{1}{\sqrt{3}}=e$

Also, given, distance between focal of $E_1=4\Rightarrow 2ae=4$

$$\Rightarrow 2a\left(\frac{1}{\sqrt{3}}\right)=4\Rightarrow a=2\sqrt{3}$$

$$\because \quad b^2=a^2(1-e^2)$$

$$=(2\sqrt{3})^2\left(1-\left(\frac{1}{\sqrt{3}}\right)^2\right)$$

$$=12\left(1-\frac{1}{3}\right)=12\times\frac{2}{3}=8$$

According to the question, product of LR's of both ellipse $=\frac{32}{\sqrt{3}}$

i.e. $$\frac{2b^2}{a}\times\frac{2A^2}{B}=\frac{32}{\sqrt{3}}$$

$$\Rightarrow 2\left(\frac{8}{2\sqrt{3}}\right)\cdot\frac{2A^2}{B}=\frac{32}{\sqrt{3}}\Rightarrow A^2=2B$$

Also, $A^2=B^2(1-e^2)$

$$\Rightarrow 2B=B^2\left(1-\frac{1}{3}\right)\Rightarrow 2B=B^2\times\frac{2}{3}$$

$$\Rightarrow B^2-3B=0\Rightarrow B=0,3 \text{ but } B\neq 0$$

$$\therefore B=3 \quad \therefore A^2=6$$

So, $$E_1:\frac{x^2}{12}+\frac{y^2}{8}=1 \quad \ldots\text{(i)}$$

$$E_2:\frac{x^2}{6}+\frac{y^2}{9}=1 \quad \ldots\text{(ii)}$$

On solving Eqs. (i) and (ii), we get

$$(x,y)\equiv\left(\frac{\sqrt{6}}{\sqrt{5}},\frac{6}{\sqrt{5}}\right),\left(\frac{-\sqrt{6}}{\sqrt{5}},\frac{6}{\sqrt{5}}\right),$$

$$\left(\frac{\sqrt{6}}{\sqrt{5}},\frac{-6}{\sqrt{5}}\right),\left(\frac{-\sqrt{6}}{\sqrt{5}},\frac{-6}{\sqrt{5}}\right)$$

These coordinates shows that this forms a rectangle.

So, ar (quadrant $\cdot$ $ABCD$)

$$=\text{ar (rectangle } ABCD)=\frac{2\sqrt{6}}{\sqrt{5}}\times\frac{12}{\sqrt{5}}$$

$$=\frac{24\sqrt{6}}{5}$$

19. *(b)*

Key idea

$(r+1)^{\text{th}}$ term in expansion of $(a+b)^n$ is $T_{r+1}={}^nC_r\,a^{n-r}b^r$

General term

$$=T_{r+1}={}^nC_r\,(7^{\frac{1}{3}})^{n-r}(11^{\frac{1}{12}})^r$$

$$= {}^nC_r (7)^{\frac{n-r}{3}} (11)^{\frac{r}{12}}$$

For integral terms, r must be multiple of 12

$\therefore r = 12\lambda, \lambda \in W$

So, $r = 0, 12, 24, 36, 48, \ldots\ldots$

$\because$ Number of forms $= 183$

$\therefore r_{max} = 12 \times 182 = 2184$

$\therefore$ Minimum value of n will be 2184 as 3/2184 (i.e. 3 divides 2184).

20. *(b)* $\lim\limits_{n\to\infty} \sum\limits_{k=1}^{n} \dfrac{k^3 + 6k^2 + 11k + 5}{(k+3)!}$

$$= \lim_{n\to\infty} \sum_{k=1}^{n} \frac{k^3 + 6k^2 + 11k + 6 - 1}{(k+3)!}$$

$$= \lim_{n\to\infty} \sum_{k=1}^{n} \frac{(k+1)(k+2)(k+3) - 1}{(k+3)(k+2)(k+1)(k)!}$$

$$= \lim_{n\to\infty} \sum_{k=1}^{n} \left(\frac{1}{k!} - \frac{1}{(k+3)!}\right)$$

$$= \left(\frac{1}{1!} + \frac{1}{2!} + \frac{1}{3!} + \ldots\ldots + \infty\right) - \left(\frac{1}{4!} + \frac{1}{5!} + \frac{1}{6!} + \ldots + \infty\right)$$

(as $n \to \infty$) $\to$

$$= (e-1) - \left(e - 1 - \frac{1}{1!} - \frac{1}{2!} - \frac{1}{3!}\right)$$

$$\left(\because e = 1 + 1 + \frac{1}{2!} + \frac{1}{3!} + \frac{1}{4!} + \ldots + \infty\right)$$

$$= 1 + \frac{1}{2} + \frac{1}{6} = \frac{10}{6} = \frac{5}{3}$$

21. *(1405)* Case (I) When single letter is used, then number of words $= 5$

Case (II) When two distinct letters are used, then number of words $=$

$${}^5C_2 \left(\frac{6!}{2!4!} \times 2 + \frac{6!}{3!3!}\right)$$

$$= 10\left(\frac{6 \times 5 \times 4!}{2 \times 4!}\right) \times 2 + \frac{6 \times 5 \times 4 \times 3!}{3 \times 2 \times 1 \times 3!}$$

$$= 10\,(30 + 20) = 500$$

Case (III) When three distinct letters are used, then number of words $=$

$${}^5C_3 \times \frac{6!}{2!\,2!\,2!} = \frac{5!}{3!\,2!} \times \frac{6!}{2!\,2!\,2!}$$

$$= \frac{5 \times 4 \times 3!}{3! \times 2} \times \frac{6 \times 5 \times 4 \times 3 \times 2!}{2 \times 2 \times 2}$$

$$= 10 \times 90 = 900$$

Hence, total number of words

$$= 5 + 500 + 900 = 1405$$

22. *(24)*

Key idea

For $x \in R$, $x = [x] + \{x\}$, where $[\cdot]$ is GIF and $\{x\}$ be the fractional part of x.

Given,

$$\lim_{x\to 0^+} \left(\frac{x\left(\left[\frac{1}{x}\right] + \left[\frac{2}{x}\right] + \left[\frac{3}{x}\right] + \ldots + \left[\frac{p}{x}\right]\right) - x^2\left(\left[\frac{1}{x^2}\right] + \left[\frac{2^2}{x^2}\right] + \left[\frac{3^2}{x^2}\right] + \ldots + \left[\frac{9^2}{x^2}\right]\right)}{} \right) \geq 1$$

We can write above limit in sigma (Σ) notation as

$$\lim_{x\to 0^+} \left(\sum_{r=1}^{p} x\left[\frac{r}{x}\right] - \sum_{k=1}^{9} x^2\left[\frac{k^2}{x^2}\right]\right) \geq 1$$

$$\Rightarrow \sum_{r=1}^{p} \lim_{x\to 0^+} x\left[\frac{r}{x}\right] - \sum_{k=1}^{9} \lim_{x\to 0^+} x^2\left[\frac{k^2}{x^2}\right] \geq 1 \quad \ldots \text{(i)}$$

Now,

$$\lim_{x\to 0^+} x\left[\frac{r}{x}\right] = \lim_{x\to 0^+} x \cdot \left(\frac{r}{x} - \left\{\frac{r}{x}\right\}\right)$$

where $\{\cdot\}$ denotes FPF

$$= \lim_{x\to 0^+} \left(r - x\left\{\frac{r}{x}\right\}\right)$$

$$= \lim_{x\to 0^+} r - \lim_{x\to 0^+} x\left\{\frac{r}{x}\right\}$$

$$= r - 0 \left(\because 0 \leq \left\{\frac{r}{x}\right\} < 1\right)$$

$$= r = 0$$

Similarly, we can find

$$= \lim_{x\to 0^+} x^2\left[\frac{k^2}{x^2}\right]$$

$$= \lim_{x\to 0^+} x^2\left(\frac{k^2}{x^2} - \left\{\frac{k^2}{x^2}\right\}\right)$$

$$= k^2 - 0 = k^2$$

From Eq. (i), we get

$$\sum_{r=1}^{p} r - \sum_{k=1}^{9} k^2 \geq 1$$

$$\Rightarrow \frac{p(p+1)}{2} - \frac{9 \times (9+1)(2 \times 9 + 1)}{6} \geq 1$$

$$\Rightarrow \frac{p(p+1)}{2} - 285 \geq 1$$

$$\Rightarrow p(p+1) \geq 2 \times 286$$

$$\Rightarrow p(p+1) \geq 572$$

$\therefore$ Least natural value of $p = 24$

$(\because 24^2 = 576 > 572.)$

23. *(5)* Given,

$$\cos^{-1} x = \pi + \sin^{-1} x + \sin^{-1}(2x+1) \quad \ldots \text{(i)}$$

$$\Rightarrow \cos^{-1} x + \cos^{-1} x = \pi + \cos^{-1} x + \sin^{-1} x + \sin^{-1}(2x+1)$$

$$\Rightarrow 2\cos^{-1} x = \pi + \frac{\pi}{2} + \sin^{-1}(2x+1)$$

$$= \frac{3\pi}{2} + \sin^{-1}(2x+1)$$

$$\Rightarrow \quad 2m = \frac{3\pi}{2} + n$$

where $\quad m = \cos^{-1} x$

and $\quad n = \sin^{-1}(2x+1)$

$$\Rightarrow \cos(2m) = \cos\left(\frac{3\pi}{2} + n\right) = \sin n$$

$$\Rightarrow 2\cos^2 m - 1 = \sin n$$

$$\Rightarrow 2x^2 - 1 = 2x + 1$$

$$(\because m = \cos^{-1} x,\ n = \sin^{-1}(2x+1))$$

$$\therefore x^2 - x - 1 = 0$$

$$\Rightarrow \quad x = \frac{1 \pm \sqrt{5}}{2}$$

But $x = \dfrac{1+\sqrt{5}}{2}$ (rejected) because at this value of x, Eq. (i) will be undefined.

$$\therefore x = \frac{1-\sqrt{5}}{2}$$

So, $\sum\limits_{x \in s} (2x-1)^2 = \left[2 \times \left(\dfrac{1-\sqrt{5}}{2}\right) - 1\right]^2$

$$= [1 - \sqrt{5} - 1]^2$$

$$= 5$$

24. *(2)* Given, $A = \begin{bmatrix} 2 & -1 \\ 1 & 0 \end{bmatrix}$

$$\Rightarrow A^2 = A \cdot A \cdot = \begin{bmatrix} 2 & -1 \\ 1 & 0 \end{bmatrix}\begin{bmatrix} 2 & -1 \\ 1 & 0 \end{bmatrix}$$

$$= \begin{bmatrix} 3 & -2 \\ 2 & -1 \end{bmatrix}$$

$$\Rightarrow A^3 = A^2 \cdot A = \begin{bmatrix} 3 & -2 \\ 2 & -1 \end{bmatrix}\begin{bmatrix} 2 & -1 \\ 1 & 0 \end{bmatrix}$$

$$= \begin{bmatrix} 4 & -3 \\ 3 & -2 \end{bmatrix}$$

$$\Rightarrow A^4 = A^3 \cdot A = \begin{bmatrix} 4 & -3 \\ 3 & -2 \end{bmatrix}\begin{bmatrix} 2 & -1 \\ 1 & 0 \end{bmatrix}$$

$$= \begin{bmatrix} 5 & -4 \\ 4 & -3 \end{bmatrix}$$

$$\Rightarrow A^5 = \begin{bmatrix} 6 & -5 \\ 5 & -4 \end{bmatrix}, A^6 = \begin{bmatrix} 7 & -6 \\ 6 & -5 \end{bmatrix}$$

and so on

$$\Rightarrow \quad A^m = \begin{bmatrix} m+1 & -m \\ m & -(m-1) \end{bmatrix}$$

$$\text{and } A^{m^2} = \begin{bmatrix} m^2+1 & -m^2 \\ m^2 & -(m^2-1) \end{bmatrix}$$

Now, $A^{m^2} + A^m = 3I - A^{-6}$

$$\Rightarrow \begin{bmatrix} m^2+1 & -m^2 \\ m^2 & -(m^2-1) \end{bmatrix} + \begin{bmatrix} m+1 & -m \\ m & -(m-1) \end{bmatrix}$$

$$= 3\begin{bmatrix} 1 & 0 \\ 0 & 1 \end{bmatrix} - \begin{bmatrix} -5 & 6 \\ -6 & 7 \end{bmatrix}$$

$[\because A^{-6} = (A^6)^{-1}]$

$$= \begin{bmatrix} 8 & -6 \\ 6 & -4 \end{bmatrix}$$

$$\Rightarrow m^2 + 1 + m + 1 = 8 \Rightarrow m^2 + m - 6 = 0$$

$$\Rightarrow m = -3, 2$$

$\therefore n(S) = 2$

25. *(112)* Given $\int_0^1 f(\lambda x)\, d\lambda = af(x)$ ...(i)

In LHS of Eq. (i), we get

Put $\lambda x = t$

$$\Rightarrow d\lambda = \frac{1}{x}\, dt$$

λ	0	1
t	0	x

So, Eq. (i) becomes,

$$\frac{1}{x}\int_0^x f(t)\, dt = af(x)$$

$$\Rightarrow \quad \int_0^x f(t)\, dt = ax\, f(x)$$

Differentiating both sides, we get

$$(f(x) \times 1 - 0) = a\,(xf'(x) + f(x) \times 1)$$

$$\Rightarrow f(x)(1-a) = a \cdot x \cdot f'(x)$$

$$\Rightarrow \quad \frac{f'(x)}{f(x)} = \frac{1-a}{a} \times \frac{1}{x}$$

$$\Rightarrow \quad \int \frac{f'(x)}{f(x)}\, dx = \int\left(\frac{1-a}{a}\right) \times \frac{1}{x}\, dx$$

$$\Rightarrow \quad \ln f(x) = \frac{1-a}{a} \ln x + C \quad \text{...(ii)}$$

where C is constant of integration.

Given initial conditions as

$$f(1) = 1$$

So, put $x = 1$ and $f(1) = 1$ in Eq. (ii) we get,

$$\ln f(1) = \frac{1-a}{a} \ln 1 + C$$

$$\Rightarrow \quad 0 = 0 + C$$

$$\Rightarrow \quad C = 0 \qquad [\because \ln f(1) = \ln 1 = 0]$$

For Eq. (ii), we get

$$\Rightarrow \ln f(x) = \frac{1-a}{a} \ln x + 0 = \frac{1-a}{a} \ln x \quad \text{...(iii)}$$

Again, putting $x = 16$ and $f(16) = \frac{1}{8}$ [as data given]

$$\ln f(16) = \frac{1-a}{a} \ln 16$$

$$\Rightarrow \ln \frac{1}{8} = \ln(16)^{\frac{1-a}{a}}$$

$$\Rightarrow \quad \frac{1}{8} = (16)^{\frac{1-a}{a}} \Rightarrow -3 = \frac{4-4a}{a}$$

$$\Rightarrow \quad a = 4$$

Hence, $f(x) = x^{-\frac{3}{4}}$

$$\Rightarrow \quad f'(x) = \frac{-3}{4} \cdot x^{-7/4}$$

Thus,

$$16 - f'\left(\frac{1}{16}\right) = 16 - \left(\frac{-3}{4} \times (2^{-4})^{-\frac{7}{4}}\right)$$

$$= 16 - \left(\frac{-3}{4} \times 2^7\right)$$

$$= 16 + \frac{3}{2^2} \times 2^7$$

$$= 16 + 96 = 112$$

Physics

26. *(b)* A sufficiently negative potential which stops the photoelectric effect is called as stopping potential. So, assertion is correct.

Stopping potential varies linearly with frequency as given by the equation.

$$V_0 = \frac{h\nu}{e} - \frac{\phi_0}{e}$$

Hence, reason is also correct.

27. *(d)* As u increases, v decreases, while u is –ve and v is +ve.

Further, relationship between u and v is not a straight line as seen from the following equation.

$$\frac{1}{v} - \frac{1}{u} = \frac{1}{f}$$

Thus

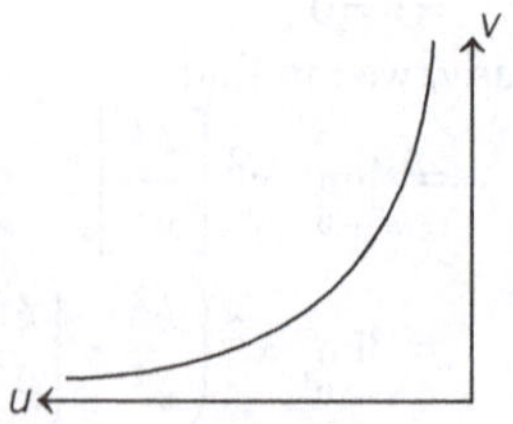

28. *(b)* Flux, $\phi = NBA \cos\theta$

For $\quad \theta = 90°$

$\phi = 0$

Induced emf,

$$e = \frac{-d\phi}{dt}$$

$$e = -\frac{d}{dt}(NBA \cos\theta)$$

Here, $\quad \theta = \omega t$

$$\Rightarrow \quad e = \frac{-d}{dt}(NBA \cos \omega t)$$

$$= +NBA \sin \omega t \times \omega$$

$$e = NBA\omega \sin \omega t$$

For $\theta = 90° = \omega t \Rightarrow e = NBA\omega$

29. *(b)* Let the dipole is slightly rotated from its mean position

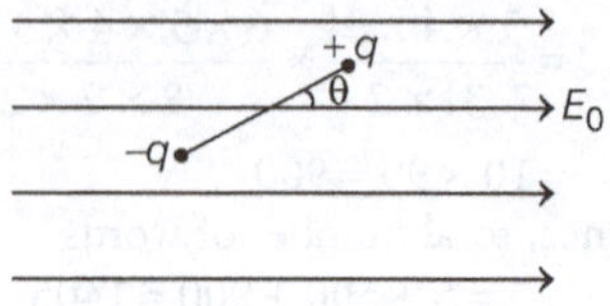

$$\tau = I\alpha$$

$$pE_0 \sin\theta = 2 \times m\left(\frac{l}{2}\right)^2\left(\frac{-d^2\theta}{dt^2}\right)$$

$$\frac{ml^2}{2}\frac{d^2\theta}{dt^2} + pE_0 \sin\theta = 0$$

$$\frac{d^2\theta}{dt^2} + \frac{2pE_0 \sin\theta}{ml^2} = 0$$

For θ to be small, $\sin\theta \simeq \theta$

$$\Rightarrow \frac{d^2\theta}{dt^2} + \frac{2pE_0\theta}{ml^2} = 0$$

$$\Rightarrow \omega = \sqrt{\frac{2pE_0}{ml^2}} = \sqrt{\frac{2qlE_0}{ml^2}} = \sqrt{\frac{2qE_0}{ml}}$$

$$T = 2\pi\sqrt{\frac{ml}{2qE_0}}$$

30. *(*)*

Key idea

Whenever the difference in the refractive index is large, the value of critical angle is small. In the question, difference between n_3 and n_1 is more as compared to n_2 and n_1 therefore $\theta_{2c} < \theta_{1c}$.

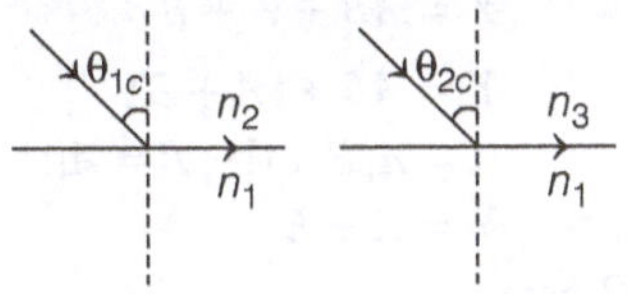

$$\sin\theta_{1C} = \frac{n_1}{n_2} \quad \text{...(i)}$$

$$\sin\theta_{2C} = \frac{n_1}{n_3} \quad \text{...(ii)}$$

Divide Eq. (ii) by Eq. (i)

$$\frac{\sin\theta_{2C}}{\sin\theta_{1C}} = \frac{n_1}{n_3} \times \frac{n_2}{n_1} = \frac{n_2}{n_3} = \frac{2}{5}$$

$$\Rightarrow \sin\theta_{2C} = \frac{2}{5}\sin\theta_{1C}$$

As given in the question

$$\sin\theta_{2C} - \sin\theta_{1C} = \frac{1}{2}$$

$$\Rightarrow \frac{2}{5}\sin\theta_{1C} - \sin\theta_{1C} = \frac{1}{2}$$

$$\Rightarrow \frac{-3}{5}\sin\theta_{1C} = \frac{1}{2}$$

$$\Rightarrow \sin\theta_{1C} = \frac{-5}{6}$$

$$\Rightarrow \theta_{1C} = \sin^{-1}\left(\frac{-5}{6}\right)$$

31. *(b)* Given,

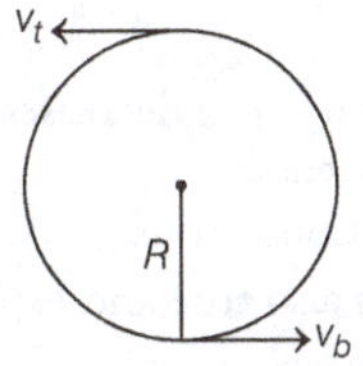

$$v_t = n\sqrt{gR}$$

Applying conservation of mechanical energy between bottom and top points

$$KE_i + PE_i = KE_f + PE_f$$

$$\frac{1}{2}mv_b^2 + 0 = \frac{1}{2}m(n\sqrt{gR})^2 + mg \times 2R$$

$$\frac{1}{2}v_b^2 = \frac{1}{2}n^2 gR + 2gR$$

$$v_b^2 = n^2 gR + 4gR$$

$$v_b^2 = gR(n^2 + 4)$$

$$v_b = \sqrt{gR(n^2 + 4)}$$

$$\text{Now, } \frac{\text{KE at top}}{\text{KE at bottom}} = \frac{\frac{1}{2}mv_t^2}{\frac{1}{2}mv_b^2} = \left(\frac{v_t}{v_b}\right)^2$$

$$= \left(\frac{n\sqrt{gR}}{\sqrt{gR(n^2 + 4)}}\right)^2$$

$$= \frac{n^2 gR}{gR(n^2 + 4)} = \frac{n^2}{n^2 + 4}$$

Alternate Solution

If v_b is the speed at the bottom, then speed at any other point is given by

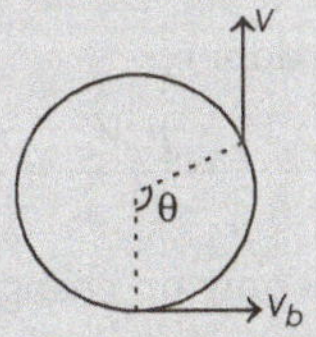

$$v = \sqrt{v_0^2 - 2gR(1 - \cos\theta)}$$

So, according to question

$$v = n\sqrt{gR}, \; v_0 = v_b, \; \theta = 180^\circ$$

Using the formula

$$v = \sqrt{v_0^2 - 2gR(1 - \cos\theta)}$$

$$n\sqrt{gR} = \sqrt{v_b^2 - 2gR[1 - \cos 180^\circ]}$$

$$\Rightarrow v_b = \sqrt{gR(n^2 + 4)}$$

$$\text{Now, } \frac{\text{KE at top}}{\text{KE at bottom}} = \frac{\frac{1}{2}mv_t^2}{\frac{1}{2}mv_b^2} = \frac{n^2}{n^2 + 4}$$

32. *(b)* To calculate the speed of

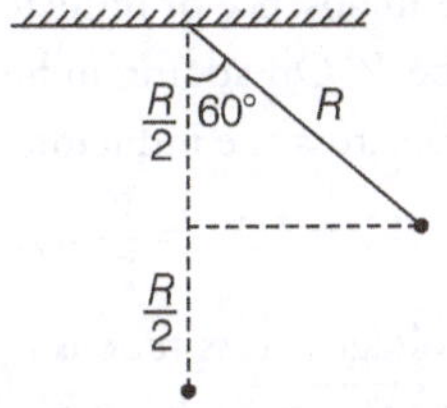

A just before the collision, apply conservation of mechanical energy.

$$KE_i + PE_i = KE_f + PE_f$$

$$0 + \frac{mgR}{2} = \frac{1}{2}mv_1^2 + 0 \Rightarrow v_1 = \sqrt{gR}$$

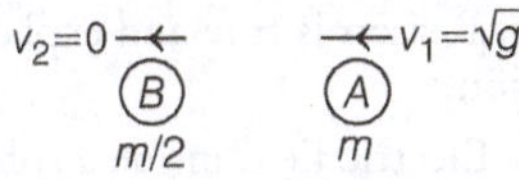

Let final velocities of *A* and *B* after collisions be v_1' and v_2'

By applying conservation of momentum

$$mv_1 + \frac{mv_2}{2} = mv_1' + \frac{mv_2'}{2}$$

$$\sqrt{gR} + 0 = v_1' + \frac{v_2'}{2} \quad \text{...(i)}$$

By using coefficient of restitution

$$e = \frac{v_2' - v_1'}{v_1 - v_2} = 1 \Rightarrow v_2' - v_1' = v_1 - v_2$$

$$v_2' - v_1' = v$$

$$v_2' = v + v_1' \quad \text{...(ii)}$$

Put value of v_2' in Eq. (i)

$$\sqrt{gR} = v_1' + \frac{v}{2} + \frac{v_1'}{2} \Rightarrow \sqrt{gR} - \frac{v}{2} = \frac{3v_1'}{2}$$

$$\sqrt{gR} - \frac{\sqrt{gR}}{2} = \frac{3v_1'}{2}$$

$$\frac{\sqrt{gR}}{2} = \frac{3v_1'}{2} \Rightarrow v_1' = \frac{\sqrt{gR}}{3}$$

33. *(b)* $\phi_1 = L_1 I_1 + M_{12} I_2$

$$e_1 = -\frac{d\phi_1}{dt} = -\frac{d}{dt}[L_1 I_1 + M_{12} I_2]$$

$$e_1 = -L_1\frac{dI_1}{dt} - M_{12}\frac{dI_2}{dt}$$

34. *(b)* Inductor in an AC circuit simply provides potential difference without consuming any energy. So, an inductor in the form of choke coil is connected in series with fluoroscent - mercury tube fitting to reduce the voltage across it and protect it. So, Assertion is true.

Voltage across tube is same as source voltage in absence of inductor.
Let it be V. On adding inductor,
Voltage across the inductor,

$$V_L = I\,X_L = \frac{V}{Z}X_L$$

New voltage across tube is

$$V_R' = \sqrt{V^2 - V_L^2}$$

$$= \sqrt{V^2 - \frac{V^2X_L^2}{Z^2}} = V\sqrt{\frac{Z^2 - X_L^2}{Z^2}}$$

$$V_R' = \frac{VR}{Z} \Rightarrow \frac{V_R'}{V} = \frac{R}{Z} = \frac{R}{\sqrt{R^2 + X_L^2}}$$

Hence, Reason is true and explains the Assertion.

35. *(*)* A. Electric field inside a spherical shell is always 0.
B. Electric field due to infinite plane sheet is $\sigma / 2\varepsilon_0$
C. Electric field outside a spherical shell is

$$\frac{q}{4\pi\varepsilon_0 r^2} \text{ or } \frac{\sigma}{\varepsilon_0}\frac{R^2}{r^2}$$

D. Electric field between two oppositely charged infinite sheet is σ / ε_0.

36. *(d)* For a particle, de-Broglie wavelength is given as

$$\lambda = \frac{h}{\sqrt{2mK}}$$

$$2mK = \frac{h^2}{\lambda^2} \Rightarrow \frac{1}{2mK} = \frac{\lambda^2}{h^2}$$

The graph between $\frac{1}{K}$ and λ is upward parabola.

37. *(a)* a. Surface tension = $[MT^{-2}]$
Impulse = $[MLT^{-1}]$
b. Angular momentum = $[ML^2T^{-1}]$
Planck's constant = $[ML^2T^{-1}]$
c. Pressure = $[ML^{-1}T^{-2}]$
Young's modulus = $[ML^{-1}T^{-2}]$
d. Torque = $[ML^2T^{-2}]$
Energy = $[ML^2T^{-2}]$

38. *(*)* Electromagnetic waves carry both energy and momentum.
So, Assertion is false.
Rest mass of the photon is zero but mass of the photon is not zero.
So, Reason is also false.

39. *(c)* Time period = $2\pi\sqrt{\frac{l}{g_{\text{eff}}}}$

As we go up, g_{eff} decreases, thus time period increases.
So, Assertion is true.
Reason is also true and explain the Assertion.

40. *(c)* $v = At^2 + \frac{Bt}{C+t}$

By principle of homogeneity
1. $C = [T]$
2. $v = At^2$

$$A = \frac{v}{t^2} = \frac{[LT^{-1}]}{[T^2]} = [LT^{-3}]$$

3. $v = \frac{Bt}{t}$

$$B = [LT^{-1}]$$

Now $ABC = [LT^{-3}][LT^{-1}][T]$

$$= [L^2T^{-3}] = [M^0L^2T^{-3}]$$

41. *(d)* $\frac{\Delta V}{V} = \frac{-\Delta p}{B} = -\frac{(p_0 + \rho gh)}{B}$

$$= -\left(\frac{10^5 + 1000\times 10\times 2.5\times 10^3}{2\times 10^9}\right)$$

$$= -\left(\frac{10^5 + 250 + 10^5}{2\times 10^9}\right)$$

$$= -\frac{251\times 10^5}{2\times 10^9} = -125.5\times 10^{-4}$$

$$\frac{\Delta V}{V} = -125.5\times 10^{-4}$$

$$\frac{\Delta V}{V}\times 100 = -125.5\times 10^{-2}\ \% = -1.25\%$$

42. *(b)* (Work done)$_a$

$$= \frac{p_1V_1 - p_2V_2}{\gamma - 1} = \frac{nR(T_1 - T_2)}{\gamma - 1}$$

So, depends only on change of temperature.

43. *(d)* Maximum height = $\frac{u^2 \sin^2\theta}{2g}$

For $\theta = 45° - \alpha$

$$H_1 = \frac{u^2\sin^2(45° - \alpha)}{2g}$$

for $\theta = 45° + \alpha$

$$H_2 = \frac{u^2\sin(45° + \alpha)}{2g}$$

$$\frac{H_1}{H_2} = \frac{\sin^2(45° - \alpha)}{\sin^2(45° + \alpha)}$$

$$= \frac{1 - \cos(90° - 2\alpha)}{1 - \cos(90° + 2\alpha)} = \frac{1 - \sin 2\alpha}{1 + \sin 2\alpha}$$

44. *(d)*

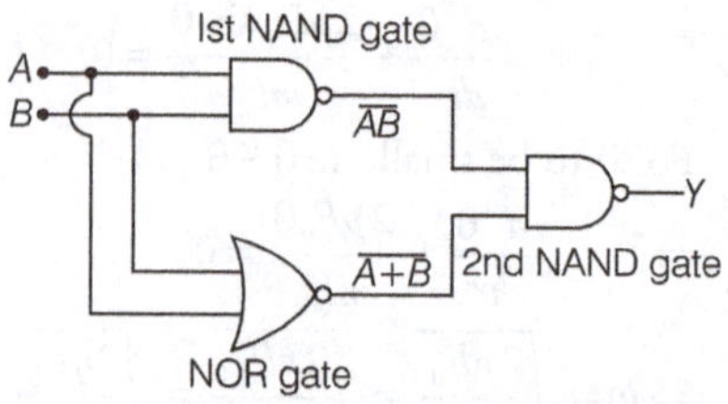

For 1st NAND gate, inputs are A and B while output is $\overline{AB}$.
For NOR gate, inputs are A and B, while output is $\overline{A+B}$.
For IInd NAND gate, inputs are $\overline{AB}$ and $\overline{A+B}$, so output

$$Y = \overline{\overline{AB}\cdot\overline{A+B}}$$

Applying de-Morgan's first law

$$Y = \overline{\overline{AB}} + \overline{\overline{A+B}}$$

$$Y = AB + (A+B)$$

$$= A[B+1] + B = A[1] + B$$

$$Y = A + B$$

OR gate

45. *(d)*

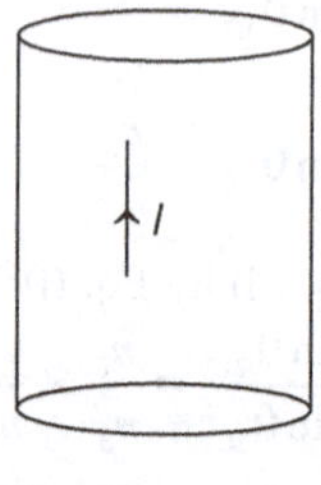

Due to a thick wire, maximum magnetic

Field is at its surface, $B_{\max} = \frac{\mu_0 I}{2\pi a}$

Internal magnetic field is, $B_i = \frac{\mu_0 Ir}{2\pi a^2}$

For $\quad B_i = \frac{1}{2}B_{\max}$

$$\frac{\mu_0 Ir}{2\pi a^2} = \frac{1}{2}\times\frac{\mu_0 I}{2\pi a} \Rightarrow r = \frac{a}{2}$$

External magnetic field is

$$B_0 = \frac{\mu_0 I}{2\pi r}$$

For $B_0 = \frac{1}{2} \times B_{max}$

$$\frac{\mu_0 I}{2\pi r} = \frac{1}{2} \times \frac{\mu_0 I}{2\pi a}$$

$$r = 2a$$

46. *(2)* $\tau = \mathbf{r} \times \mathbf{F}$

Point of force (1, 1, 1)

Point about which torque needed to calculate (0, 0, 0)

Thus $\mathbf{r} = (1-0)\hat{i} + (1-0)\hat{j} + (1-0)\hat{k}$

$$= \hat{i} + \hat{j} + \hat{k}$$

$$\mathbf{F} = \hat{i} - \hat{j} + \hat{k}$$

Now $\tau = \mathbf{r} \times \mathbf{F}$

$$= (\hat{i} + \hat{j} + \hat{k}) \times (\hat{i} - \hat{j} + \hat{k})$$

$$\begin{vmatrix} \hat{i} & \hat{j} & \hat{k} \\ 1 & 1 & 1 \\ 1 & -1 & 1 \end{vmatrix}$$

$$\tau = 2\hat{i} + 0\hat{j} - 2\hat{k}$$

Magnitude of torque in *z*-direction is 2.

47. *(327)* Since volume in fixed

$$\Rightarrow \frac{p_1}{T_1} = \frac{p_2}{T_2} \Rightarrow \frac{p}{300} = \frac{2p}{T_2}$$

$$T_2 = 600\text{ K or } 327°\text{C}$$

48. *(6)* Given

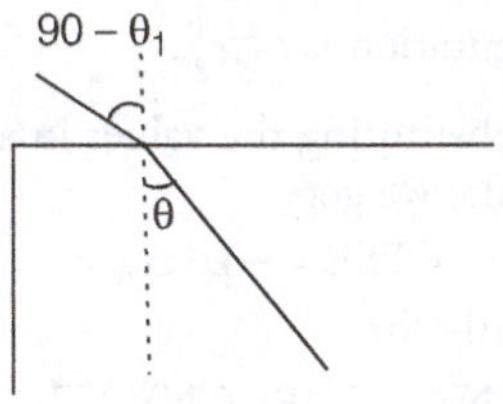

$$\theta_1 = \cos^{-1}\left(\frac{n_2}{2n_1}\right) \Rightarrow \cos\theta_1 = \frac{n_2}{2n_1}$$

By Snell's law

$$\frac{\sin(90° - \theta_1)}{\sin\theta} = \frac{n_2}{n_1}$$

$$\frac{\cos\theta_1}{\sin\theta} = \frac{n_2}{n_1}$$

$$\frac{n_2}{2n_1 \times \sin\theta} = \frac{n_2}{n_1} \Rightarrow \sin\theta = \frac{1}{2}$$

or $\theta = 30°$

Now, $\tan\theta = \frac{d}{2 \times t}$

$$t = \frac{d}{2\tan\theta} = \frac{4\sqrt{3}}{2 \times \tan 30°} = 6\text{ cm}$$

49. *(2000)* $v_r = 9\text{ km/h} = \frac{9 \times 5}{18} = 2.5\text{ m/s}$

v_b in still water = 27 km/h

$$= 27 \times \frac{5}{18} = 7.5\text{m/s}$$

Velocity of boat in moving river as observed from ground

$$= 2.5 + 7.5 = 10\text{ m/s}$$

This becomes horizontal speed of ball i.e. 10 m/s

Vertical speed of ball is 10 m/s

Net speed of ball

$$= \sqrt{10^2 + 10^2} = 10\sqrt{2}\text{ m/s}$$

Angle of throw as observed from ground, $\tan\theta = \frac{u_y}{u_x} = \frac{10}{10} = 1$

$$\theta = 45°$$

$$\text{Range} = \frac{u^2 \sin 2\theta}{g}$$

$$= \frac{(10\sqrt{2})^2 \sin 2 \times 45°}{10} = \frac{200}{10} = 20\text{ m}$$

$$= 2000\text{ cm}$$

50. *(5)* $F_2 = \frac{F_1 A_2}{A_1} = \frac{100 \times 1500}{6}$

$$= \frac{15}{6} \times 10^4\text{ N}$$

Work done = Force × Displacement

$$= \frac{15}{6} \times 10^4 \times 0.2$$

$$= \frac{3}{6} \times 10^4 = 5000\text{ J} = 5\text{ kJ}$$

Chemistry

51. *(b)* The correct match is

A-IV, B-I, C-II, D-III,

52. *(b)*

> **Key idea**
> To solve this question, first of all write the balanced chemical equation, then using this find the value of change in number of e^-. After that use Nernst-equation to obtain the correct option.

According to Nernst equation,

$$E_{cell} = E°_{cell} - \frac{RT}{nF} \ln Q$$

The cell reaction given in question is

$$Mg(s) + 2Ag^+(aq) \rightleftharpoons 2Ag(s) + Mg^{2+}(aq)$$

$$Q = \frac{[Mg^{2+}]}{[Ag^+]^2}$$

Also, as there is change in 2 number of electrons. So, $n = 2$

$$\therefore \quad E = E°_{Cell} - \frac{RT}{2F} \ln\left[\frac{[Mg^{2+}]}{[Ag^+]^2}\right]$$

$$\text{or} \quad E = E°_{Cell} + \frac{RT}{2F} \ln\left[\frac{[Ag^+]^2}{[Mg^{2+}]}\right]$$

53. *(a)* The given reaction is an example of nucleophilic aromatic substitution reaction and gives the following product '*P*'.

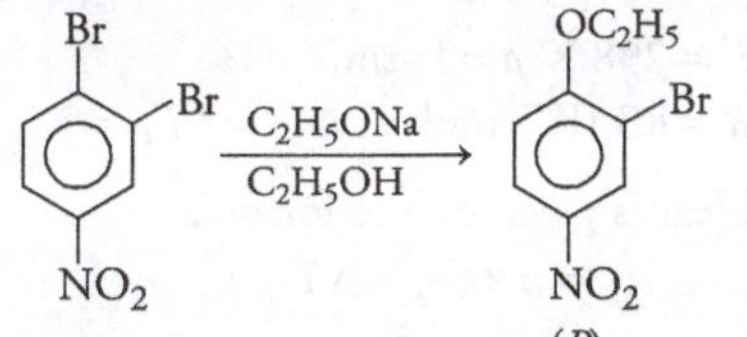

(*P*)

54. *(a)* The relation between molar conductivity and concentration is as follows

$$\text{Molar conductivity} \propto \frac{1}{\sqrt{C}}$$

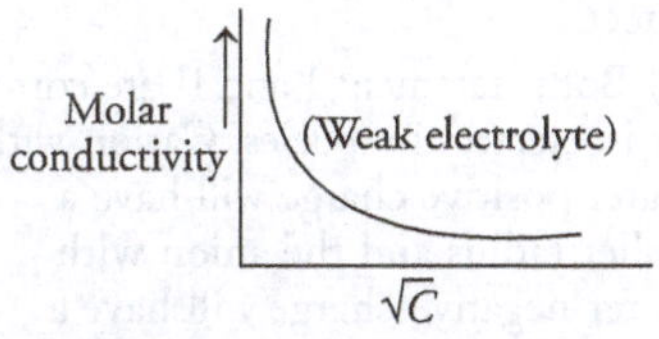

Thus, for weak electrolyte molar conductivity decreases sharply as there is an increases in concentration. Thus, option (a) is the correct answer.

55. *(b)* Given, Bohr radius = a_0

de-Broglie wavelength = λ

Radius of *n*th orbit; $r_n = n^2 a_0 = (2)^2 a_0$

$$= 4a_0 \quad [\because n = 2]$$

Using the formula,

$$2\pi r_n = n\lambda \Rightarrow 2\pi(4a_0) = n\lambda$$

$$\Rightarrow \quad \lambda = \frac{8\pi a_0}{n}$$

56. *(a)* As the difference between ionisation enthalpy and electron gain enthalpy increases, ionic characteristic increases. Thus, the correct order of the ionic characteristic is

$$EB > EA > EC > ED$$

57. *(b)* The rate of reaction is given by

$$\text{Rate} = k_2[A][B_2] \quad \ldots(\text{i})$$

Also $\left(\frac{k_1}{k_{-1}}\right) = \left(\frac{[A]^2}{[A_2]}\right)$

Or $[A] = \sqrt{\frac{k_1}{k_{-1}}}\sqrt{A_2}$

Substituting the value of $[A]$ in Eq (i),

$$r_{\text{ate}} = k_2\sqrt{\frac{k_1}{k_{-1}}}.[A_2]^{1/2}[B_2]$$

So, the order will be given by $\frac{1}{2}+1=\frac{3}{2}$

$\Rightarrow$ Order = 1.5

58. *(a)* Among the given statements, statement given in C, D and E are correct while A and B are incorrect.

The correct forms are

Weight of substance is the force applied by gravity on body.

Mass of a substance is the amount of matter present in it.

59. *(c)* The correct match is

A-II, B-III, C-IV, D-I

60. *(a)* Melting point of the given pair is in order (increasing)

Mn < Fe, Tc < Ru, Os < Re

So, the order given in option (a) is correct.

61. *(b)* Both statement I and II are correct. For isoelectronic species. Cation with greater positive charge will have a smaller radius and the anion with greater negative charge will have a larger radius

$$Mg^{2+} < Na^{+} < F^{-} < O^{2-}$$

Magnitude of E.G.E.

Cl > F > Br > I

62. *(a)* The less the value of CFSE the less stable will be the compound, so using the formula,

$$\text{CFSE} = -0.4\,\Delta_0 \text{ (number of } e^- \text{ in } t_{2g} \text{ orbital)} + 0.6\,\Delta_0 \text{(number of } e^- \text{ in } e_g \text{ orbital)}$$

For $[Mn(CN)_6]^{3-}$ the electronic configuration is $t_{2g}^4 e_g^0$.

On substituting the values in above formula, we get

$$\text{CFSE} = -1.6\,\Delta_0$$

Similarly for

$[Co(CN)_6]^{4-}$, $[Fe(CN)_6]^{4-}$, $[Fe(CN)_6]^{3-}$ the value of CFSE are $-1.8\Delta_0, -2.4\Delta_0, -2.0\Delta_0$ respectively

Thus, the correct increasing order of stability is

$$[Mn(CN)_6]^{3-} < [Co(CN)_6]^{4-} < [Fe(CN)_6]^{3-} < [Fe(CN)_6]^{4-}$$

i.e., I < II < IV < III

63. *(b)* As the standard reduction potential value (positive) increases, oxidising capacity increases.

Hence, option (b) has highest standard reduction potential value.

64. *(a)* Among the given options, there are five nucleophiles i.e,

NH_3, PhSH, $(H_3C)_2S$, $\overset{\ominus}{O}H$ and $CH_2{=}CH_2$

65. *(c)* The given reaction proceeds via Clemmensen reduction in which carbonyl group is reduced to methylene group.

The complete reaction involved is as follow

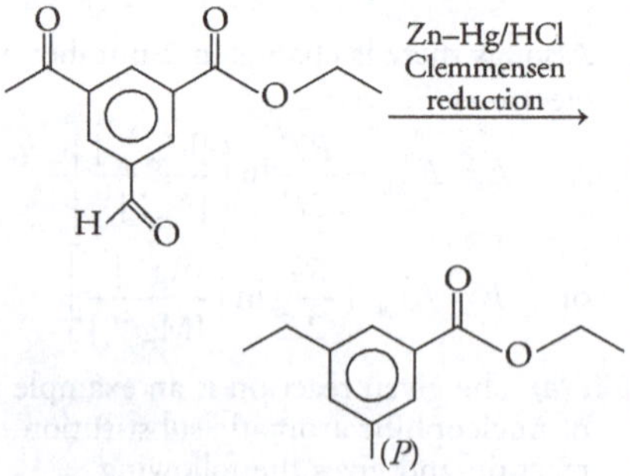

66. *(d)* Given, $\Delta q = 500$ J, $n = 0.5$ mol

$T = 298$ K $p = 1$ atm,

$R = 8.3\ \text{JK}^{-1}\text{mol}^{-1}$, $\Delta U = ?\ T_f = ?$

Heat is given by the formula,

$$q_p = n \times C_p \times \Delta T$$

$$500 = 0.5 \times \frac{5}{2} \times 8.3(T_f - 298) \quad \left[\because C_p = \frac{5R}{2}\right]$$

On solving, we get,

$$T_f = 346.2 \text{ K} \simeq 348 \text{ K}$$

Also, we know that

$$\frac{\Delta H}{\Delta U} = \frac{C_p}{C_V} = \frac{5/2}{3/2} = \frac{5}{3}$$

$$\Delta U = \frac{3}{5} \times 500 = 300 \text{ J}$$

So, the final temperature and change in internal energy, respectively are 348 K and 300 J.

67. *(b)* The correct statement is AX_2 is fully ionised while AY_2 is completly unionised.

For AX_2, $\Delta T_b = K_b \times m \times i$

Given, $T_b = 100.0156°\text{C} = 373.0156$ K

$$\therefore \Delta T_b = T_b - T_b^\circ = (373.0156 - 373)\text{ K} = 0.0156 \text{ K}$$

$$K_b = 0.52 \text{ K kg mol}^{-1},$$

$$\text{Molality } (m) = \frac{1.24\text{ g}/124\text{ g mol}^{-1}}{1\text{ kg}}$$

$$0.0156 = 0.52 \times \frac{0.01}{1} \times i_{AX_2}$$

$\Rightarrow$ $i_{AX_3} = 3$

$\Rightarrow$ Completely ionised

For AY_2

$\therefore$ $\Delta T_b = K_b \times m \times i$

Given, $T_b = 100.0260°\text{C} = 373.0260$ K

$$\therefore \Delta T_b = T_b - T_b^\circ = (373.0260 - 373)\text{ K} = 0.0260 \text{ K}$$

$$\text{Molality } (m) = \frac{25.4\text{ g}/250\text{ g mol}^{-1}}{2\text{ kg}} = 0.0508 \text{ mol kg}^{-1}$$

$$0.026 = 0.52 \times 0.0508 \times i_{AY_2}$$

$$= i_{AY_2} \approx 1$$

$\Rightarrow$ Completely unionised.

68. *(c)* The given reaction is as follows

$$AB_2(g) \rightleftharpoons AB(g) + \frac{1}{2}B_2(g)$$

	$AB_2(g)$	$AB(g)$	$\frac{1}{2}B_2(g)$
At $t = 0$	1	0	0
At $t = t_{eq}$	$1-x$	x	$\frac{x}{2}$

$$\therefore K_p = \frac{\left(\frac{x}{1+x/2}\right)p\left(\frac{x/2}{1+x/2}\right)p}{\left(\frac{1-x}{1+x/2}\right)p}$$

$\because \quad x << 1 \Rightarrow \frac{x}{2} + 1$ and $1 - x \simeq 1$

$$\therefore \quad K_p = \frac{(xp)\left(\frac{xp}{2}\right)^{1/2}}{p}$$

So, the value of x is terms of K_p and p

is $x = \sqrt[3]{\frac{2K_p^2}{p}}$

69. *(a)* The correct match is

A-IV, B-II, C-I, D-III.

A. $[MnBr_4]^{2-}$

In the given complex Mn is in +2 oxidation state. So, electronic configuration of Mn^{2+} is

$$Mn^{2+} \rightarrow [Ar]\, 3d^5$$

In presence of weak field ligand, Br

[Ar]	1	1	1	1	1	××	××	××	××
	3d					4s	4p		

$= sp^3$ hybridisation, paragmagnetic

B. $[FeF_6]^{3-}$

In the complex Fe is in +3 oxidation state. So, electronic configuration of Fe^{3+} is

$$Fe^{3+} = [Ar]\, 3d^5$$

In presence of weak field ligand, F

[Ar]	1	1	1	1	1	××	××	××	××	××	××			
	3d					4s	4p			4d				

sp^3d^2 hybridisation , paramagnetic

C. $[Co(C_2O_4)_3]^{3-}$

In the complex Co is in +3 oxidation state. So, electronic configuration of Co^{3+} is

$$Co^{3+} = [Ar]3d^6$$

In presence of strong field ligand, C_2O_4

[Ar]	⇅	⇅	⇅	××	××	××	××	××	××
	3d					4s	4p		

d^2sp^3 hybridisation, diamagnetic

D. $[Ni(CO)_4]$

In this complex Ni is present in zero oxidation state. So, electronic configuration of Ni^0 is

$$Ni^0 = [Ar]\, 3d^8 4s^2$$

In presence of strong fied ligand, CO

⇅	⇅	⇅	⇅	⇅	××	××	××	××
3d					4s	4p		

sp^3 hybridisation, diamagnetic

70. *(a)* Among the given compounds, compound present in, A and B are steam volatile.

This is due to intermolecular hydrogen bonding among them.

So, option (a) is the correct answer.

71. *(160)* The reaction involved in the given question is as follows.

$$4FeCr_2O_4 + 8\,Na_2CO_3 + 7O_2 \downarrow 8Na_2CrO_4 + 2Fe_2O_3 + 8\,CO_2$$

Fe_2O_3 is water insoluble, so it's molar mass is

$$[2 \times 56 + 3 \times 16] = 160 \text{ g/mol}$$

Concept Enhancer

Chromite ore is an oxide of iron chromium. It is an oxide mineral belonging to the spinel group. Substitution of the element aluminium can also occur leading to formation of hercynite ($FeAl_2O_4$).

72. *(341)* Among the given compounds benzyl amine is most basic due to localised lone pair. The reaction involved as per question is as follow.

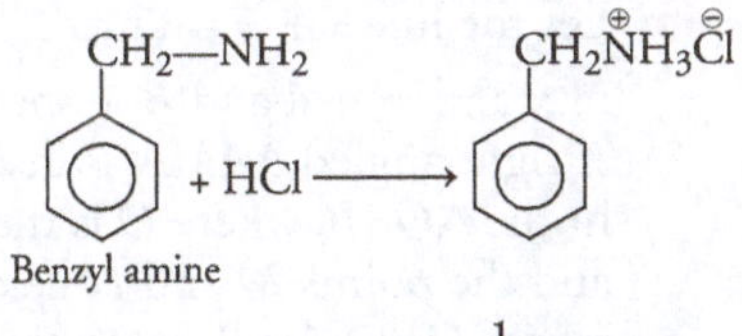

$$\text{Mole of benzyl amine} = \frac{1}{107} = 0.00934 \text{ mol}$$

1 mol of benzyl amine consumed 1 mole of HCl.

So, mole of HCl consumed

$= 0.00934$ mol.

mass of HCl consumed

$= 0.00934 \times$ molar mass of HCl

$= 0.00934 \times 36.5 = 0.341$ g $= 341$ mg

73. *(15)* The structure of hex-1, 3-dien-5-yne is

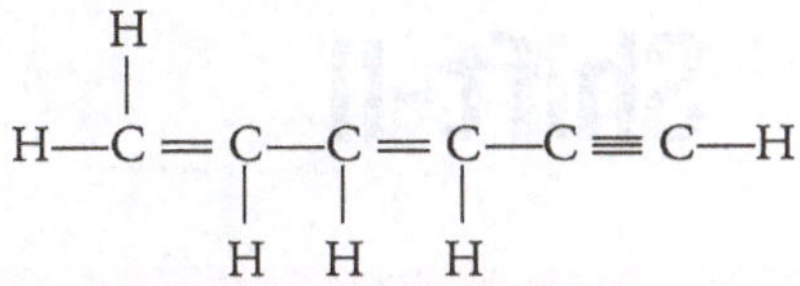

So, σ bond $= 11$, π bond $= 4$

Thus, total $\sigma + \pi = 11 + 4 = 15$

74. *(16)* The reaction involved is as follows

$$A_2B \longrightarrow 2A^+ + B^{2-}, y = 3$$

$$\alpha = 0.3$$

$$i = 1 + (y-1)\alpha$$

On substituting the value of y and α in above equation, we get

$i = 1 + (3-1)(0.3) = 1.6$ or 16×10^{-1}

75. *(13)* The complete reaction is as follows

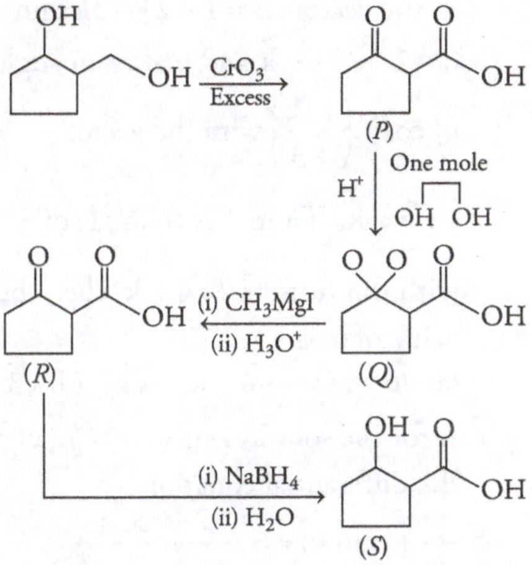

0.1 mol of compound (*S*) weigh in g

$= 0.1 \times$ molar mass of compound (*S*)

$= 0.1 \times 130 = 13$ g

Time 3:00-6:00 | **MM** : 300 Marks

29 January
Shift-II

Online Question Paper

JEE Main 2025

Mathematics

Section A
Objective Type Questions

01/ Let $f(x)=\int_0^x t(t^2-9t+20)\,dt$, $1\le x\le 5$. If the range of f is $[\alpha,\beta]$, then $4(\alpha+\beta)$ equals

(a) 253 (b) 154 (c) 125 (d) 157

02/ Let **a** be a unit vector perpendicular to the vectors $\mathbf{b}=\hat{i}-2\hat{j}+3\hat{k}$ and $\hat{c}=2\hat{i}+3\hat{j}-\hat{k}$ and makes an angle of $\cos^{-1}\left(-\frac{1}{3}\right)$ with the vector $\hat{i}+\hat{j}+\hat{k}$. If $\hat{a}$ makes an angle of $\frac{\pi}{3}$ with the vector $\hat{i}+\alpha\hat{j}+\hat{k}$, then the value of α is

(a) $\sqrt{6}$ (b) $-\sqrt{6}$ (c) $-\sqrt{3}$ (d) $\sqrt{3}$

03/ If for the solution curve $y=f(x)$ of the differential equation $\frac{dy}{dx}+(\tan x)y=\frac{2+\sec x}{(1+2\sec x)^2}$, $x\in\left(\frac{-\pi}{2},\frac{\pi}{2}\right)$, $f\left(\frac{\pi}{3}\right)=\frac{\sqrt{3}}{10}$, then $f\left(\frac{\pi}{4}\right)$ is equal to

(a) $\frac{\sqrt{3}+1}{10(4+\sqrt{3})}$ (b) $\frac{5-\sqrt{3}}{2\sqrt{2}}$

(c) $\frac{9\sqrt{3}+3}{10(4+\sqrt{3})}$ (d) $\frac{4-\sqrt{2}}{14}$

04/ Let P be the foot of the perpendicular from the point $(1,2,2)$ on the line $L:\frac{x-1}{1}=\frac{y+1}{-1}=\frac{z-2}{2}$. Let the line $\hat{r}=(-\hat{i}+\hat{j}-2\hat{k})+\lambda(\hat{i}-\hat{j}+\hat{k}),\lambda\in R$, intersect the line L at Q. Then, $2(PQ)^2$ is equal to.

(a) 25 (b) 19

(c) 29 (d) 27

05/ Let $A=[a_{ij}]$ be a matrix of order 3×3 with $a_{ij}=(\sqrt{2})^{i+j}$. If the sum of all the elements in the third row of A^2 is $\alpha+\beta\sqrt{2}$, $\alpha,\beta\in Z$, then $\alpha+\beta$ is equal to

(a) 280 (b) 224

(c) 210 (d) 168

06/ Let the line $x+y=1$ meet the axes of x and y at A and B, respectively. A right angled ΔAMN is inscribed in the ΔOAB, where O is the origin and the points M and N lie on the where OB and AB, respectively. If the area of the ΔAMN is $\frac{4}{9}$ of the area of the ΔOAB and $AN:NB=\lambda:1$, then the sum of all possible values(s) of is λ.

(a) 2 (b) $\frac{5}{2}$

(c) $\frac{1}{2}$ (d) $\frac{13}{6}$

07/ If all the words with or without meaning made using all the letters of the word "KANPUR" are arranged as in a dictionary, then the word at 440th position in this arrangement is.

(a) PRNAUK (b) PRKANU
(c) PRKAUN (d) PRNAKU

08/ If the set of all $a \in R$ for which the equation $2x^2 + (a-5)x + 15 = 3a$ has no real roots is the interval (α, β) and $X = (x \in Z : \alpha < x < \beta)$, then $\sum_{x \in X} x^2$ is equal to

(a) 2109 (b) 2129
(c) 2119 (d) 2139

09/ Let $A = [a_{ij}]$ be a 2×2 matrix such that $a_{ij} \in \{0, 1\}$ for all i and j. Let the random variable X denote the possible values of the determinant of the matrix A. Then, the variance of X is

(a) $\frac{3}{4}$ (b) $\frac{5}{8}$
(c) $\frac{3}{8}$ (d) $\frac{1}{4}$

Imp.

10/ Let the function $f(x) = (x^2 - 1)|x^2 - ax + 2| + \cos|x|$ be not differentiable at the two points $x = \alpha = 2$ and $x = \beta$. Then, the distance of the point (α, β) from the line $12x + 5y + 10 = 0$ is equal to

(a) 5 (b) 4 (c) 3 (d) 2

Differentiability of Function

11/ Let the area enclosed between the curves $|y| = 1 - x^2$ and $x^2 + y^2 = 1$ be α. If $9\alpha = \beta\pi + \gamma; \beta, \gamma$ are integers, then the value of $|\beta - \gamma|$ equals

(a) 27 (b) 33
(c) 15 (d) 18

12/ The remainder, when 7^{103} is divided by 23 is equal to

(a) 6 (b) 17
(c) 9 (d) 14

13/ If $\alpha x + \beta y = 109$ is the equation of the chord of the ellipse $\frac{x^2}{9} + \frac{y^2}{4} = 1$, whose mid-point is $\left(\frac{5}{2}, \frac{1}{2}\right)$, then $\alpha + \beta$ is equal to

(a) 58 (b) 46
(c) 37 (d) 72

14/ If the domain of the function $\log_5(18x - x^2 - 77)$ is (α, β) and the domain of the function $\log_{(x-1)}\left(\frac{2x^2 + 3x - 2}{x^2 - 3x - 4}\right)$ is (γ, δ), then $\alpha^2 + \beta^2 + \gamma^2$ is equal to

(a) 195 (b) 179 (c) 186 (d) 174

15/ Let a circle C pass through the points (4, 2) and (0, 2) and its centre lie on $3x + 2y + 2 = 0$. Then, the length of the chord of the circle C, whose mid-point is (1, 2) is

(a) $\sqrt{3}$ (b) $2\sqrt{2}$
(c) $2\sqrt{3}$ (d) $4\sqrt{2}$

16/ Let a straight line L pass through the point $P(2, -1, 3)$ and be perpendicular to the lines $\frac{x-1}{2} = \frac{y+1}{1} = \frac{z-3}{-2}$ and $\frac{x-3}{1} = \frac{y-2}{3} = \frac{z+2}{4}$. If the line L intersects the YZ-plane at the point Q. Then, the distance between the points P and Q is

(a) $\sqrt{10}$ (b) $2\sqrt{3}$
(c) 2 (d) 3

17/ Bag 1 contain 4 white balls and 5 black balls and Bag 2 contains n white balls and 3 black balls. One ball is drawn randomly from Bag 1 and transferred to Bag 2. A ball is then drawn randomly from Bag 2. If the probability that the ball drawn is white is 29/45, then n is equal to

(a) 6 (b) 3
(c) 5 (d) 4

18/ Let $\alpha, \beta \ (\alpha \neq \beta)$ be the values of m for which the equations $x + y + z = 1; x + 2y + 4z = m$ and $x + 4y + 10z = m^2$ have infinitely many solutions. Then, the value of $\sum_{n=1}^{10} (n^\alpha + n^\beta)$ is equal to

(a) 3080 (b) 560
(c) 3410 (d) 440

19/ Let $S = N \cup \{0\}$. Define a relation R from S to R by $R = \left\{(x, y) : \log_e y = x \log_e\left(\frac{2}{5}\right), x \in S, y \in R\right\}$

Then, the sum of all the elements in the range of R is equal to

(a) $\frac{10}{9}$ (b) $\frac{3}{2}$
(c) $\frac{5}{2}$ (d) $\frac{5}{3}$

20/ If $\sin x + \sin^2 x = 1, x \in \left(0, \frac{\pi}{2}\right)$, then
$(\cos^{12} x + \tan^{12} x) + 3(\cos^{10} x + \tan^{10} x + \cos^8 x + \tan^8 x) + (\cos^6 x + \tan^6 x)$ is equal to

(a) 4 (b) 1 (c) 3 (d) 2

Section B
Numerical Type Questions

Imp.

21/ If $24\int_0^{\frac{\pi}{4}} \left(\sin\left|4x - \frac{\pi}{12}\right| + [2\sin x] \right) dx = 2\pi + \alpha$, where [.] denotes the greatest integer function, then α is equal to...........

Properties of Definite Integral

22/ Let $a_1, a_2, \ldots\ldots, a_{2024}$ be an arithmetic progression such that $a_1 + (a_5 + a_{10} + a_{15} + \ldots\ldots\ldots\ldots + a_{2020}) + a_{2024} = 2233$. Then, $a_1 + a_2 + a_3 + \ldots\ldots\ldots + a_{2024}$ is equal to..........

23/ If $\lim_{t \to 0} \left(\int_0^1 (3x+5)^t \, dx \right)^{\frac{1}{t}} = \frac{\alpha}{5e}\left(\frac{8}{5}\right)^{\frac{2}{3}}$, then α is equal to......

24/ Let $y^2 = 12x$ be the parabola and S be its focus. Let PQ be a focal chord of the parabola such that $(SP)(SQ) = 147/4$. Let C be the circle described taking PQ as a diameter. If the equation of a circle C is $64x^2 + 64y^2 - \alpha x - 64\sqrt{3}y = \beta$, then $\beta - \alpha$ is equal to.........

25/ Let integers $a, b \in [-3, 3]$ be such that $a + b \neq 0$. Then, the number of all possible ordered pairs (a, b) for which $\left|\frac{z-a}{z+b}\right| = 1$ and
$$\begin{vmatrix} z+1 & \omega & \omega^2 \\ \omega & z+\omega^2 & 1 \\ \omega^2 & 1 & z+\omega \end{vmatrix} = 1, z \in C,$$
where ω and ω^2 are the roots of $x^2 + x + 1 = 0$ is equal to.......

Physics

Section A
Objective Type Questions

26/ Given below are two statements. One is labelled as Assertion (A) and the other is labelled as Reason (R).

Assertion (A)

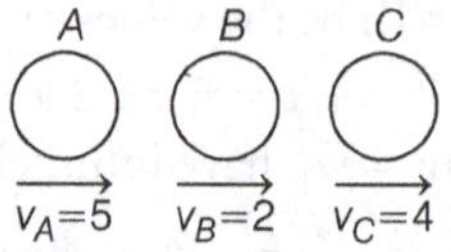

Three identical spheres of same mass undergo one dimensional motion as shown in figure with initial velocities $v_A = 5$ m/s, $v_B = 2$ m/s, $v_C = 4$ m/s. If we wait sufficiently long for elastic collision to happen, then $v_A = 4$ m/s, $v_B = 2$ m/s, $v_C = 5$ m/s will be the final velocities.

Reason (R) In an elastic collision between identical masses, two objects exchange their velocities.

In the light of the above statements, choose the correct answer from the options given below.

(a) A is false but R is true.
(b) Both A and R are true and R is not the correct explanation of A.
(c) Both A and R are true and R is the correct explanation of A.
(d) A is true but R is false.

Imp.

27/ Two identical symmetric double convex lenses of focal length f are cut into two equal parts L_1, L_2 by AB plane and L_3, L_4 by XY-plane as shown in figure, respectively. The ratio of focal lengths of lenses L_1 and L_3 is

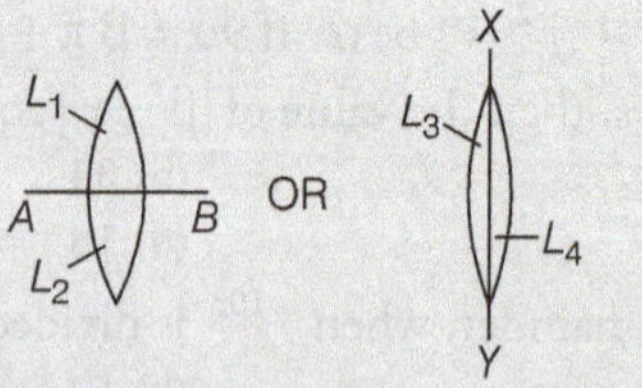

(a) 1 : 1 (b) 1 : 2 (c) 1 : 4 (d) 2 : 1

Lens

28/ Two bodies A and B of equal mass are suspended from two massless springs of spring constant k_1 and k_2, respectively. If the bodies oscillate vertically such that the their amplitudes are equal, the ratio of the maximum velocity of A to the maximum velocity of B is

(a) $\frac{k_1}{k_2}$ (b) $\sqrt{\frac{k_1}{k_2}}$ (c) $\sqrt{\frac{k_2}{k_1}}$ (d) $\frac{k_2}{k_1}$

29/ Given below are two statements. One labelled as Assertion (A) and the other is labelled as Reason (R).

Assertion (A) With the increase in the pressure of an ideal gas, the volume falls off more rapidly in an isothermal process in comparison to adiabatic process.

Reason (R) In isothermal process, pV = constant, while in adiabatic process pV^{γ} = constant. Here γ is the ratio of specific heats, p is the pressure and V is the volume of the ideal gas.

In the light of the above statements, choose the correct answer from the options given below.

(a) Both A and R are true and R is the correct explanation of A.
(b) Both A and R are true and R is not the correct explanation of A.
(c) A is true but R is false.
(d) A is false but R is true.

30/ The truth table for the circuit given below is

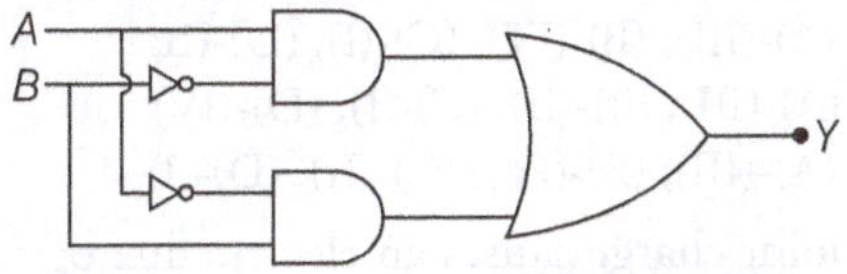

(a)

A	B	Y
0	0	0
1	0	0
1	1	0
0	1	1

(b)

A	B	Y
0	0	0
1	1	1
1	0	1
0	1	1

(c)

A	B	Y
0	0	0
1	0	1
0	1	0
1	1	0

(d)

A	B	Y
0	0	0
0	1	1
1	0	1
1	1	0

31/ The difference of temperature in a material can convert heat energy into electrical energy. To harvest the heat energy, the material should have

(a) high thermal conductivity and high electrical conductivity
(b) low thermal conductivity and low electrical conductivity
(c) high thermal conductivity and low electrical conductivity
(d) low thermal conductivity and high electrical conductivity

32/ Match List-I with List-II

	List-I		List-II
(A)	Young's modulus	(I)	$[ML^{-1}T^{-1}]$
(B)	Torque	(II)	$[ML^{-1}T^{-2}]$
(C)	Coefficient of viscosity	(III)	$[M^{-1}L^{3}T^{-2}]$
(D)	Gravitational constant	(IV)	$[ML^{2}T^{-2}]$

Choose the correct answer from the options given below.

(a) (A)-(I), (B)-(III), (C)-(II), (D)-(IV)
(b) (A)-(IV), (B)-(II), (C)-(III), (D)-(I)
(c) (A)-(II), (B)-(IV), (C)-(I), (D)-(III)
(d) (A)-(II), (B)-(I), (C)-(IV), (D)-(III)

33/ A sand dropper drops sand of mass $m(t)$ on a conveyer belt at a rate proportional to the square root of speed (v) of the belt, i.e $\frac{dm}{dt} \propto \sqrt{v}$. If P is the power delivered to run the belt at constant speed, then which of the following relationship is true?

(a) $P \propto \sqrt{v}$ (b) $P \propto v$ (c) $P^2 \propto v^5$ (d) $P^2 \propto v^3$

34/

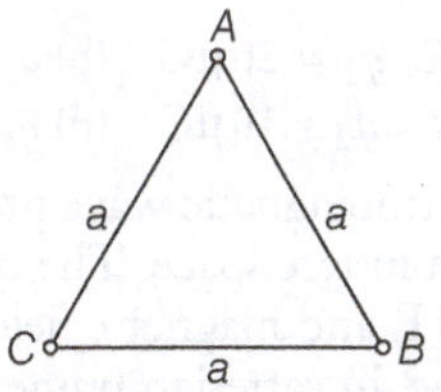

Three equal masses m are kept at vertices (A, B, C) of an equilateral triangle of side a in free space. At $t = 0$, they are given an initial velocity $\mathbf{v}_A = v_0\mathbf{AC}$, $\mathbf{v}_B = v_0\mathbf{BA}$ and $\mathbf{v}_C = v_0\mathbf{CB}$. Here, **AC**, **CB** and **BA** are unit vectors along the edges of the triangle. If the three masses interact gravitationally, then the magnitude of the net angular momentum of the system at the point of collision is

(a) $3\,a m v_0$ (b) $\frac{3}{2}\,a m v_0$

(c) $\frac{\sqrt{3}}{2}\,a m v_0$ (d) $\frac{1}{2}\,a m v_0$

35/ The number of spectral lines emitted by atomic hydrogen that is in the 4th energy level, is

(a) 3 (b) 1 (c) 6 (d) 0

36/ A convex lens made of glass (refractive index = 1.5) has focal length 24 cm in air. When it it totally immersed in water (refractive index = 1.33), its focal length changes to

(a) 24 cm (b) 96 cm (c) 48 cm (d) 72 cm

37/ In an experiment with photoelectric effect, the stopping potential,

(a) increases with increase in the intensity of the incident light
(b) decreases with increase in the intensity of the incident light
(c) increases with increase in the wavelength of the incident light
(d) is $(1/e)$ times the maximum kinetic energy of the emitted photoelectrons

38/ A capacitor $C_1 = 6\,\mu F$ is charged to a potential difference of $V_0 = 5$ V using a 5 V battery. The battery is removed and another capacitor $C_2 = 12\,\mu F$ is inserted in place of the battery. When the switch S is closed, the charge flows between the capacitors for some time until equilibrium condition is reached. What are the charges (q_1 and q_2) on the capacitors C_1 and C_2 when equilibrium condition is reached ?

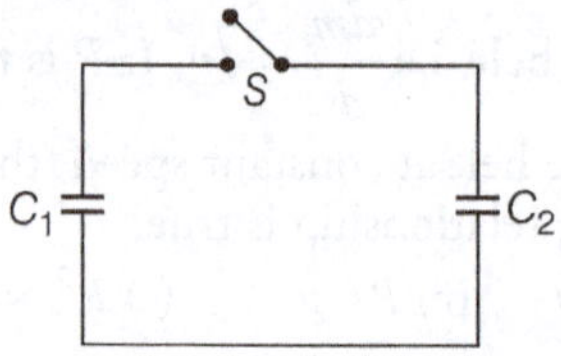

(a) $q_1 = 10\,\mu C, q_2 = 20\,\mu C$ (b) $q_1 = 30\,\mu C, q_2 = 15\,\mu C$
(c) $q_1 = 20\,\mu C, q_2 = 10\,\mu C$ (d) $q_1 = 15\,\mu C, q_2 = 30\,\mu C$

39/ A plane electromagnetic wave propagates along the +x-direction in free space. The components of the electric filed **E** and magnetic field, **B** vectors associated with the wave in cartesian frame are

(a) E_x, B_y (b) E_y, B_z (c) E_z, B_y (d) E_y, B_x

40/ A cup of coffee cools from 90°C to 80°C in t minutes when the room temperature is 20°C. The time taken by the similar cup of coffee to cool from 80°C to 60°C at the same room temperature is

(a) $\frac{13}{10}t$ (b) $\frac{10}{13}t$ (c) $\frac{5}{13}t$ (d) $\frac{13}{5}t$

41/

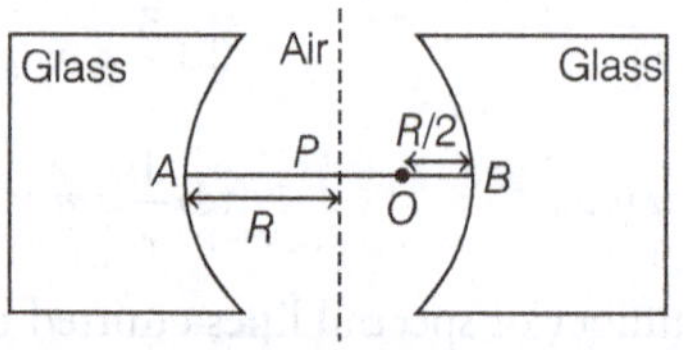

Two concave refracting surfaces of equal radii of curvature and refractive index 1.5 face each other in air as shown in figure. A point object O is placed mid-way, between P and B. The separation between the images of O, formed by each refracting surface is

(a) $0.214R$ (b) $0.411R$ (c) $0.124R$ (d) $0.114R$

42/

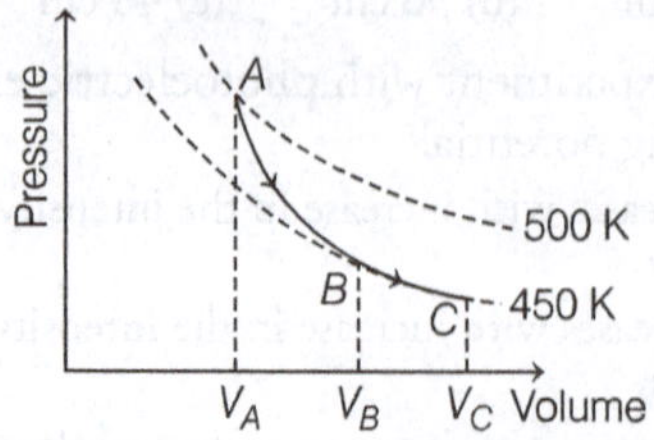

A poly-atomic molecule ($C_V = 3R, C_p = 4R$, where R is gas constant) goes from phase space point A ($p_A = 10^5$ Pa, $V_A = 4 \times 10^{-6}\,m^3$) to point B ($p_B = 5 \times 10^4$ Pa, $V_B = 6 \times 10^{-6}\,m^3$) to point C($p_C = 10^4$ Pa, $V_C = 8 \times 10^{-6}\,m^3$). A to B is an adiabatic path and B to C is an isothermal path. The net heat absorbed per unit mole by the system is

(a) $500R$ (In 3 + In 4) (b) $450R$ (In 4-In 3)
(c) $500R$ In2 (d) $400R$ In 4

43/ Match List-I with List-II

	List-I		List-II
(A)	Magnetic induction	(I)	Ampere metre^2
(B)	Magnetic intensity	(II)	Weber
(C)	Magnetic flux	(III)	Gauss
(D)	Magnetic moment	(IV)	Ampere metre

Choose the correct answer from the options given below.

(a) (A)-(I), (B)-(II), (C)-(III), (D)-(IV)
(b) (A)-(III), (B)-(IV), (C)-(I), (D)-(II)
(c) (A)-(III), (B)-(II), (C)-(I), (D)-(IV)
(d) (A)-(III), (B)-(IV), (C)-(II), (D)-(I)

44/ A point charge causes an electric flux of $-2 \times 10^4\ Nm^2C^{-1}$ to pass through a spherical gaussian surface of 8.0 cm radius, centred on the charge. The value of the point charge is (Given, $\varepsilon_0 = 8.85 \times 10^{-12} C^2N^{-1}m^{-2}$)

(a) 15.7×10^{-8} C (b) 17.7×10^{-8} C
(c) -15.7×10^{-8} C (d) -17.7×10^{-8} C

45/ An electric dipole is placed at a distance of 2 cm from an infinite plane sheet having positive charge density σ_0. Choose the correct option from the following

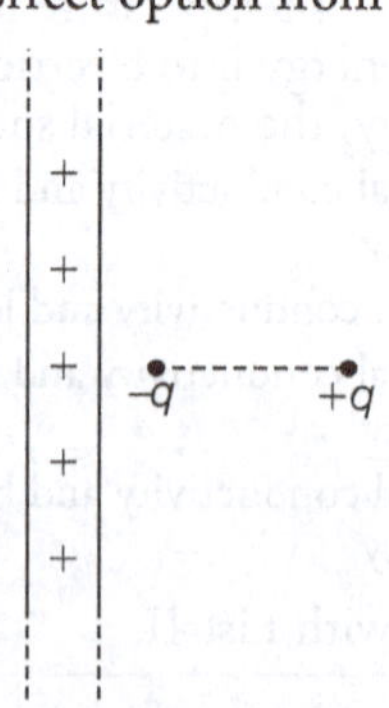

(a) Potential energy and torque both are maximum.
(b) Torque on dipole is zero and net force is directed away from the sheet.
(c) Torque on dipole is zero and net force acts towards the sheet.
(d) Potential energy of dipole is minimum and torque is zero.

Section B
Numerical Type Questions

46/ The magnetic field inside a 200 turns solenoid of radius 10 cm is 2.9×10^{-4} T.
If the solenoid carries a current of 0.29 A, then the length of the solenoid is.............π cm

Imp.

47/ A parallel plate capacitor consisting of two circular plate of radius 10 cm is being charged by a constant current of 0.15 A. If the rate of change of potential difference between the plates is 7×10^8 V/s, then the integer value of the distance between the parallel plates is $\left(\text{Take, } \varepsilon_0 = 9 \times 10^{12} \frac{\text{F}}{\text{m}}, \pi = \frac{22}{7}\right)$........ μm.

Displacement Current

48/ Two planets, A and B are orbiting a common star in circular orbits of radii R_A and R_B, respectively, with $R_B = 2R_A$. The planet B is $4\sqrt{2}$ times more massive the planet A. The ratio (L_B / L_A) of angular momentum L_B of planet B to that of planet $A(L_A)$ is closest to integer...........

49/ Two cars P and Q are moving on a road in the same direction. Acceleration of car P increases linearly with time, whereas car Q moves with a constant acceleration. Both cars cross each other at time $t = 0$, for the first time. The maximum possible number of crossing(s) (including the crossing at $t = 0$) is.........

50/ A physical quantity Q is related to four observables a, b, c, d as follows.
$Q = ab^4 / cd$, where $a = (60 \pm 3)$ Pa, $b = (20 \pm 0.1)$ m, $c = (40 \pm 0.2)$ Nsm^{-2} and $d = (50 \pm 0.1)$m, then the percentage error in Q is $\frac{x}{1000}$, where $x =$........

Chemistry

Section A
Objective Type Questions

51/ Consider the equilibrium,

$$CO(g) + 3H_2(g) \rightleftharpoons CH_4(g) + H_2O(g)$$

If the pressure applied over the system increases by two fold at constant temperature then
(A) Concentration of reactants and products increases.
(B) Equilibrium will shift in forward direction.
(C) Equilibrium constant increases since concentration of products increases.
(D) Equilibrium constant remains unchanged as concentration of reactants and products remain same.
Choose the correct answer from the options given below
(a) (A), (B) and (C) only (b) (A) and (B) only
(c) (A),(B) and (D) only (d) (B) and (C) only

52/ Drug X becomes ineffective after 50% decomposition. The original concentration of drug in a bottle was 16 mg/ mL which becomes 4 mg/mL in 12 months. The expiry time of the drug in months is ....
Assume that the decomposition of the drug follows first order kinetics.
(a) 2 (b) 6
(c) 12 (d) 3

Imp.

53/ O_2 gas will be evolved as a product of electrolysis of
(A) an aqueous solution of $AgNO_3$ using silver electrodes.
(B) an aqueous solution of $AgNO_3$ using platinum electrodes.
(C) a dilute solution of H_2SO_4 using platinum electrodes.
(D) a high concentration solution of H_2SO_4 using platinum electrodes.
Choose the correct answer from the options given below
(a) (A) and (C) only (b) (B) and (C) only
(c) (A) and (D) only (d) (B) and (D) only

Product of Electrolysis

54/ Identify the homoleptic complexes with odd number of d-electrons in the central metal
(A) $[FeO_4]^{2-}$ (B) $[Fe(CN)_6]^{3-}$
(C) $[Fe(CN)_5NO]^{2-}$ (D) $[CoCl_4]^{2-}$
(E) $[Co(H_2O)_3F_3]$
Choose the correct answer from the options given below
(a) (A), (B) and (D) only (b) (C) and (E) only
(c) (B) and (D) only (d) (A), (C) and (E) only

55/ Which one of the following, with HBr will give a phenol?

(a) 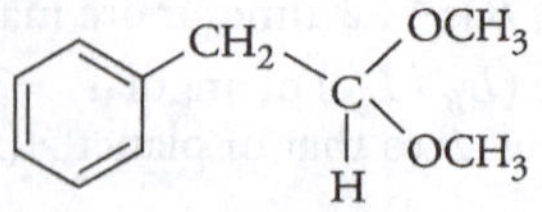

(b) 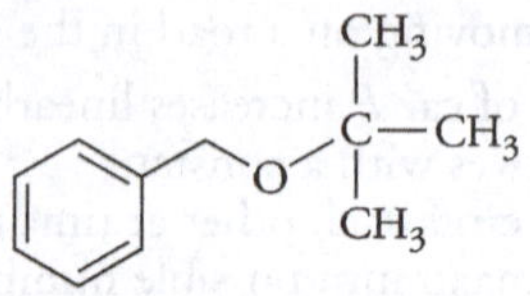

(c) 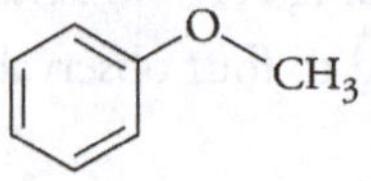

(d) 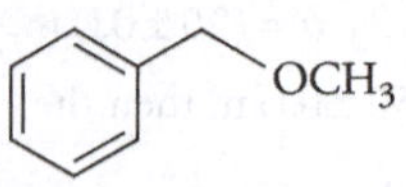

56/ Given below are two statements.

Statement I NaCl is added to the ice at 0°C, present in the ice cream box to prevent the melting of ice cream.

Statement II On addition of NaCl to ice at 0°C, there is a depression in freezing point.

In the light of the above statements, choose the correct answer from the options given below:

(a) Both Statement I and Statement II are false.
(b) Statement I is false but Statement II is true.
(c) Statement I is true but Statement II is false.
(d) Both Statement I and Statement II are true.

57/ Match List-I with List-II.

	List-I Applications		List-II Batteries/Cell
(A)	Transistors	(I)	Anode – Zn/Hg; Cathode –HgO+C
(B)	Hearing aids	(II)	Hydrogen fuel cell
(C)	Invertors	(III)	Anode-Zn;Cathode-Carbon
(D)	Apollo space ship	(IV)	Anode-Pb;Cathode-Pb|PbO_2

Choose the correct answer from the options given below

(a) (A)-(III), (B)-(II), (C)-(IV), (D)-(I)
(b) (A)-(II), (B)-(III), (C)-(IV), (D)-(I)
(c) (A)-(III), (B)-(I), (C)-(IV), (D)-(II)
(d) (A)-(IV), (B)-(III), (C)-(II), (D)-(I)

58/ For hydrogen like species, which of the following graphs provides the most appropriate representation of *E vs Z* plot for a constant *n*?

[*E* = energy of the stationary state, *Z* = atomic number, *n* = principal quantum number]

(a) 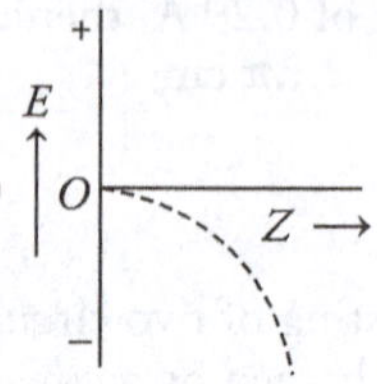

(b) 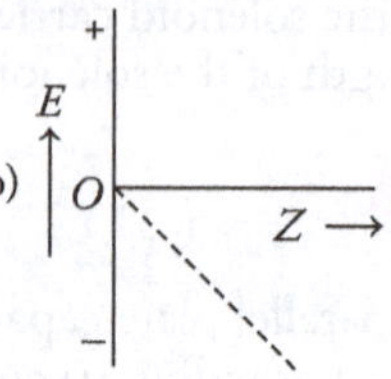

(c) 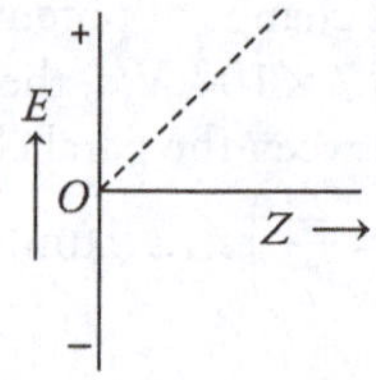

(d) 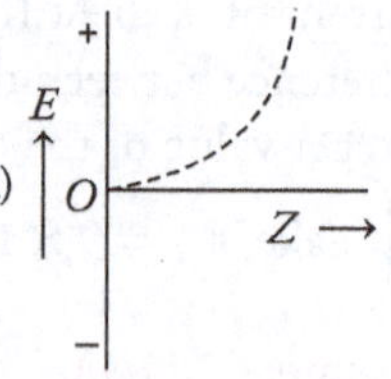

59/ Which one of the following reaction sequences will give an azo dye?

(a) 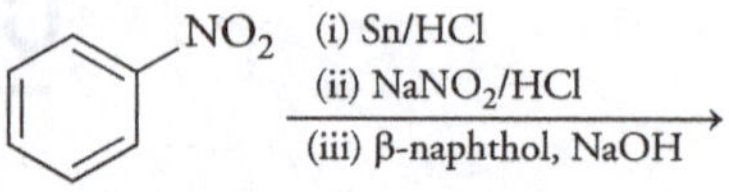

(b) 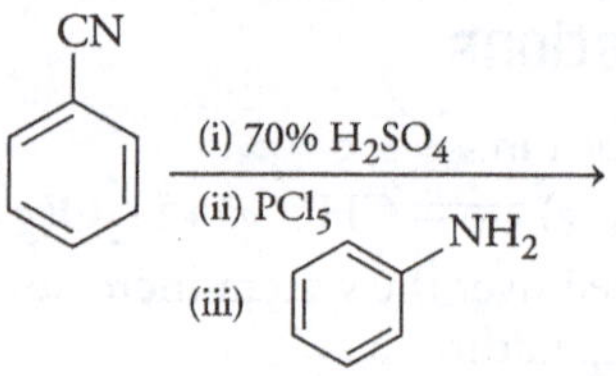

(c) 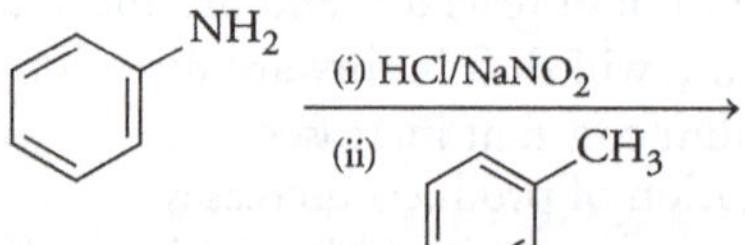

(d) 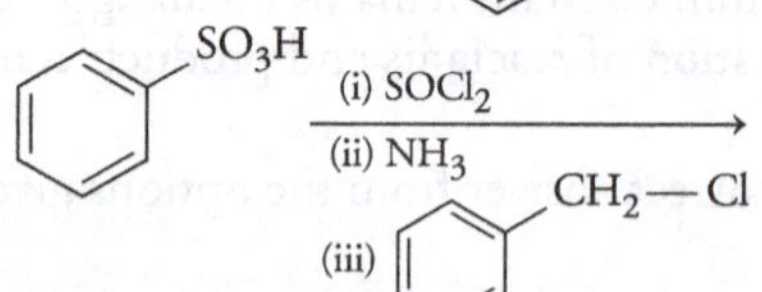

60/ Identify the essential amino acids from below

(A) Valine (B) Proline
(C) Lysine (D) Threonine
(E) Tyrosine

Choose the correct answer from the options given below

(a) (A), (C) and (E) only
(b) (A), (C) and (D) only
(c) (C), (D) and (E) only
(d) (B), (C) and (E) only

61/ The calculated spin-only magnetic moments of $K_3[Fe(OH)_6]$ and $K_4[Fe(OH)_6]$ respectively are
(a) 3.87 and 4.90 BM
(b) 4.90 and 5.92 BM
(c) 4.90 and 4.90 BM
(d) 5.92 and 4.90 BM

Imp.

62/ The type of oxide formed by the element among Li, Na, Be, Mg, B and Al that has the least atomic radius is
(a) A_2O (b) A_2O_3 (c) AO_2 (d) AO

Types of Oxides in Blend with Atomic Radius

63/ Total number of sigma (σ) ....and pi (π).....bonds respectively present in hex-1-en–4-yne are
(a) 3 and 13 (b) 11 and 3 (c) 13 and 3 (d) 14 and 3

64/ First ionisation enthalpy values of first four group 15 elements are given below. Choose the correct value for the element that is a main component of apatite family.
(a) 1402 kJ mol^{-1} (b) 834 kJ mol^{-1}
(c) 1012 kJ mol^{-1} (d) 947 kJ mol^{-1}

65/ Given below are two statements.
Statement I On nitration of *m*-xylene with HNO_3, H_2SO_4 followed by oxidation, 4-nitrobenzene-1,3-dicarboxylic acid is obtained as the major product.
Statement II —CH_3 group is *o*/*p*-directing while —NO_2 group is *m*-directing group.
In the light of the above statements, choose the correct answer from the options given below:
(a) Both Statement I and Statement II are false.
(b) Statement I is false but Statement II is true.
(c) Statement I is true but Statement II is false.
(d) Both Statement I and Statement II are true.

66/ Which among the following halides will generate the most stable carbocation in the nucleophilic substitution reaction?

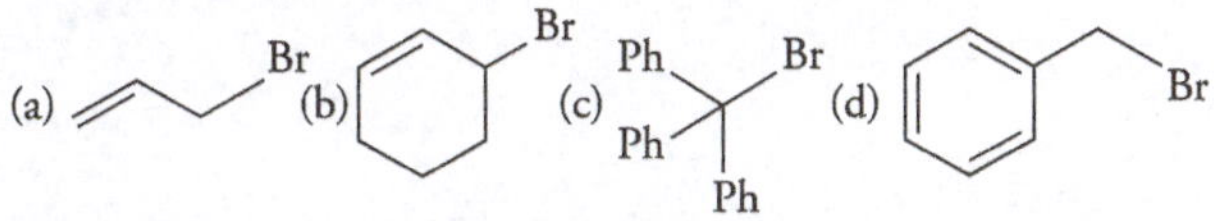

67/ Given below are two statements.
Statement I It is impossible to specify simultaneously with arbitrary precision, both the linear momentum and the position of a particle.
Statement II If the uncertainty in the measurement of position and uncertainty in measurement of momentum are equal for an electron, then the uncertainty in the measurement of velocity is

$$\geq \sqrt{\frac{h}{\pi}} \times \frac{1}{2m}$$

In the light of the above statements, choose the correct answer from the options given below
(a) Statement I is false but Statement II is true.
(b) Both Statement I and Statement II are false.
(c) Both Statement I and Statement II are true.
(d) Statement I is true but Statement II is false.

68/ 0.1 M solution of KI reacts with excess of H_2SO_4 and KIO_3 solutions. According to equation

$$5I^- + IO_3^- + 6H^+ \longrightarrow 3I_2 + 3H_2O$$

Identify the correct statements.
(A) 200 mL of KI solution reacts with 0.004 mol of KIO_3.
(B) 200 mL of KI solution reacts with 0.006 mol of H_2SO_4.
(C) 0.5 L of KI solution produced 0.005 mol of I_2.
(D) Equivalent weight of KIO_3 is equal to $\left(\frac{\text{Molecular weight}}{5}\right)$

Choose the correct answer from the options given below
(a) (A) and (D) only
(b) (C) and (D) only
(c) (B) and (C) only
(d) (A) and (B) only

69/ Given below are two statements.
Statement I In partition chromatography, stationary phase is thin film of liquid present in the inert support.
Statement II In paper chromatography, the material of paper acts as a stationary phase.
In the light of the above statements, choose the correct answer from the options given below:
(a) Statement I is true but Statement II is false.
(b) Statement I is false but Statement II is true.
(c) Both Statement I and Statement II are false.
(d) Both Statement I and Statement II are true.

70/ If C (diamond) $\longrightarrow$ C(graphite) $+X$ kJ mol^{-1}

C (diamond) + $O_2(g) \longrightarrow CO_2(g) + Y$ kJ mol^{-1}

C (graphite) + $O_2(g) \longrightarrow CO_2(g) + Z$ kJ mol^{-1}

at constant temperature.

Then

(a) $X = -Y + Z$ (b) $-X = Y + Z$

(c) $X = Y + Z$ (d) $X = Y - Z$

Section B

Numerical Type Questions

71/ In the sulphur estimation, 0.20 g of a pure organic compound gave 0.40 g of barium sulphate. The percentage of sulphur in the compound is ……… $\times 10^{-1}$%.

(Molar mass : O = 16, S = 32, Ba = 137 in g mol^{-1})

72/ Isomeric hydrocarbons $\longrightarrow$ negative Baeyer's test (Molecular formula C_9H_{12})

The total number of isomers from above with four different non-aliphatic substitution sites is

Imp.

73/ Consider the following low-spin complexes $K_3[Co(NO_2)_6]$, $K_4[Fe(CN)_6]$, $K_3[Fe(CN)_6]$, $Cu_2[Fe(CN)_6]$ and $Zn_2[Fe(CN)_6]$.

The sum of the spin-only magnetic moment values of complexes having yellow colour is …… BM (answer in nearest integer).

Spin Only Magnetic Moment

74/ In the Claisen-Schmidt reaction to prepare, dibenzalacetone from 5.3g of benzaldehyde, a total of 3.51 g of product was obtained. The percentage yield in this reaction was………%.

75/ Total number of non-bonded electrons present in NO_2^- ion based on Lewis theory is………

Answer Key

Mathematics

1 (d)	2 (b)	3 (d)	4 (d)	5 (b)	6 (a)	7 (c)	8 (d)	9 (c)	10 (c)
11 (b)	12 (d)	13 (a)	14 (c)	15 (c)	16 (d)	17 (a)	18 (d)	19 (d)	20 (d)
21 (12)	22 (11132)	23 (64)	24 (1328)	25 (10)					

Physics

26 (a)	27 (b)	28 (b)	29 (a)	30 (d)	31 (d)	32 (c)	33 (c)	34 (c)	35 (c)
36 (b)	37 (d)	38 (a)	39 (b)	40 (d)	41 (d)	42 (b)	43 (*)	44 (d)	45 (d)
46 (8)	47 (1320)	48 (8)	49 (2)	50 (77)					

Chemistry

51 (b)	52 (b)	53 (b)	54 (c)	55 (c)	56 (d)	57 (a)	58 (a)	59 (a)	60 (b)
61 (d)	62 (b)	63 (c)	64 (c)	65 (d)	66 (c)	67 (c)	68 (a)	69 (a)	70 (d)
71 (275)	72 (2)	73 (0)	74 (60)	75 (12)					

***Note** (*) None of the option is correct.*

Detailed Explanations

29 January
Shift-II

Mathematics

1. *(d)* We have,

$$f(x) = \int_0^x t(t^2 - 9t + 20)dt, 1 \le x \le 5$$

$$\Rightarrow f'(x) = x^3 - 9x^2 + 20x$$

$$= x(x-4)(x-5)$$

$-$ (below 0) $+$ (0 to 4) $-$ (4 to 5) $+$ (above 5) on the number line with points 0, 4, 5

$f(x)$ is increasing in (0, 4)

$\therefore$ Minimum value of $f(x)$ at $x = 1$ and maximum value of $x = 4$

$$\therefore \; f(x) = \frac{x^4}{4} - \frac{9x^3}{3} + \frac{20x^2}{2}$$

$$\Rightarrow f(1) = \frac{1}{4} - 3 + 10 = \frac{29}{4} = \alpha$$

$$\Rightarrow f(4) = \frac{256}{4} - 3(64) + 10 \times (16)$$

$$= 32 = \beta$$

$$\therefore 4(\alpha + \beta) = 4\left(\frac{29}{4} + 32\right) = 157$$

2. *(b)*

Key idea

The angle θ between **a** and **b** is given by

$$\cos\theta = \frac{\mathbf{a}\cdot\mathbf{b}}{|\mathbf{a}||\mathbf{b}|}$$

Let $\mathbf{P} = \hat{i} + \hat{j} + \hat{k}$

$$\text{and } \mathbf{b}\times\mathbf{c} = \begin{vmatrix} \hat{i} & \hat{j} & \hat{k} \\ 1 & -2 & 3 \\ 2 & 3 & -1 \end{vmatrix}$$

$$= -7\hat{i} + 7\hat{j} + 7\hat{k}$$

$$= -7(\hat{i} - \hat{j} - \hat{k})$$

$$\text{Now, } \mathbf{a} = \frac{\hat{i} - \hat{j} - \hat{k}}{\sqrt{3}} \text{ or } \frac{-\hat{i} + \hat{j} + \hat{k}}{\sqrt{3}}$$

$$\cos\theta = \frac{\mathbf{a}\cdot\mathbf{p}}{|\mathbf{p}|} = \frac{1-1-1}{\sqrt{3}\sqrt{3}} = -\frac{1}{3}$$

$$\text{or} \quad \cos\theta = \frac{-1+1+1}{3} = \frac{-1}{3} = \frac{1}{3}$$

(rejected)

$$\Rightarrow \quad \mathbf{a} = \frac{\hat{i} - \hat{j} - \hat{k}}{\sqrt{3}}$$

$$\text{Now,} \quad \cos\frac{\pi}{3} = \frac{\mathbf{a}\cdot(\hat{i} + \alpha\hat{j} + \hat{k})}{\sqrt{1 + \alpha^2 + 1}}$$

$$\Rightarrow \quad \frac{1}{2} = \frac{1 - \alpha - 1}{\sqrt{3}\sqrt{\alpha^2 + 2}}$$

$$\Rightarrow \frac{\sqrt{3}}{2}\sqrt{\alpha^2 + 2} = -\alpha \qquad [\because \alpha < 0]$$

On squaring both sides, we get

$$3\alpha^2 + 6 = 4\alpha^2$$

$$\Rightarrow \quad \alpha = \pm\sqrt{6}$$

$$\Rightarrow \quad \alpha = -\sqrt{6}$$

3. *(d)* $\text{IF} = e^{\int \tan x \, dx} = e^{\ln(\sec x)} = \sec x$

$$\therefore y.\sec x = \int \left\{\frac{2 + \sec x}{(1 + 2\sec x)^2}\right\} \sec x \, dx$$

$$= \int \frac{2\cos x + 1}{(\cos x + 2)^2} dx$$

Let $\quad t = \tan x$

$$\therefore \quad \cos x = \frac{1 - t^2}{1 + t^2}$$

Now, differentiating both sides w.r.t x we get

$$-\sin x dx = \frac{(1+t^2)(-2t) - (1-t^2)(2t)}{(1+t^2)^2} dt$$

$$\Rightarrow -\sin x dx = \frac{-2t(1 + t^2 + 1 - t^2)}{(1+t^2)^2} dt$$

$$\Rightarrow -\sin x dx = -\frac{4t dt}{(1+t^2)^2}$$

$$\Rightarrow dx = \frac{4t dt}{(1+t^2)^2 \sin x}$$

If $\cos x = \dfrac{1-t^2}{1+t^2}$, then

$$\sin x = \frac{2t}{1+t^2}$$

$$\therefore \; dx = \frac{2}{1+t^2} dt$$

$$= \int \frac{2\left(\frac{1-t^2}{1+t^2}\right) + 1}{\left(\frac{1-t^2}{1+t^2} + 2\right)^2} \cdot \frac{2}{(1+t^2)} dt$$

$$= \int \frac{\frac{(2 - 2t^2 + 1 + t^2)}{(1 - t^2 + 2 + 2t^2)^2} \cdot \frac{2}{(1+t^2)^2}}{(1+t^2)^2} dt$$

$$= \int \frac{(3 - t^2)}{(t^2 + 3)} \cdot 2dt \Rightarrow 2\int \left(\frac{3 - t^2}{t^2 + 3}\right) dt$$

Let $t + \dfrac{3}{t} = u$

$$\Rightarrow \left(1 - \frac{3}{t^2}\right) dt = du = -2\int \frac{du}{u^2}$$

$$y.(\sec x) = \frac{2}{u} + C$$

$$y \cdot \sec x = \frac{2}{t + \frac{3}{t}} + C \qquad \ldots \text{(i)}$$

$$\text{at } x = \frac{\pi}{3}, t = \tan\frac{x}{2} = \frac{1}{\sqrt{3}}$$

$$2 \cdot \frac{\sqrt{3}}{10} = \frac{2}{\frac{1}{\sqrt{3}} + 3\sqrt{3}} + C$$

$$\frac{2\sqrt{3}}{10} = \frac{2\sqrt{3}}{10} + C$$

$$\Rightarrow \quad C = 0$$

$$\text{at } x = \frac{\pi}{4}, t = \tan\frac{x}{2} = \sqrt{2} - 1$$

$$\therefore \quad y.\sqrt{2} = \frac{2}{\sqrt{2} - 1 + \frac{3}{\sqrt{2} - 1}}$$

$$\Rightarrow \; y.\sqrt{2} = \frac{2(\sqrt{2} - 1)}{6 - 2\sqrt{2}}$$

$$\Rightarrow \quad y = \frac{\sqrt{2}(\sqrt{2} - 1)}{2(3 - \sqrt{2})}$$

$$= \frac{1}{\sqrt{2}} \times \frac{(2\sqrt{2} - 1)}{7} = \frac{4 - \sqrt{2}}{14}$$

4. *(d)*

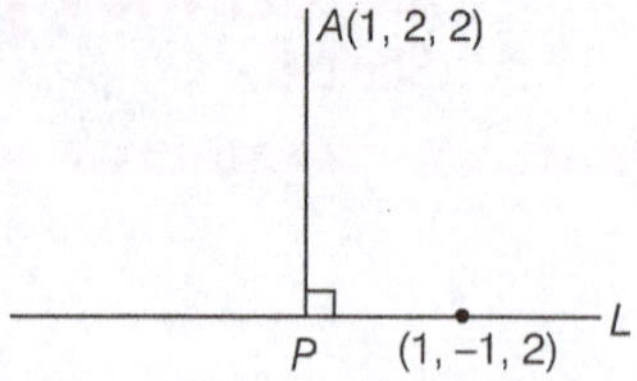

(1, −1, 2) = for of given line

$L: \frac{x-1}{1} = \frac{y+1}{-1} = \frac{z-2}{2} = \mu(\text{say})$

Put $(\mu+1, -\mu-1, 2\mu+2)$

$\mathbf{AP}\cdot\mathbf{d} = 0$

$\Rightarrow (\mu, -\mu-3, 2\mu)\cdot(1, -1, 2) = 0$

$\Rightarrow \quad \mu + \mu + 3 + 4\mu = 0$

$\Rightarrow \quad \mu = \frac{-1}{2}$

$\therefore \; P\left(\frac{-1}{2}+1, +\frac{1}{2}-1, 2\left(\frac{-1}{2}\right)+2\right)$

$P\left(\frac{1}{2}, -\frac{1}{2}, 1\right)$

Now, general point on L' is

$a(-1+\lambda, 1-\lambda, -2+\lambda)$

For point of intersection of line L and L'

$\mu + 1 = -1 + \lambda \;\big|\; -\mu - 1 = 1 - \lambda$

$\mu = \lambda - 2 \;\big|\; \mu = \lambda - 2$

$2\mu + 2 = -2 + \lambda$

$2(\lambda - 2) + 2 = -2 + \lambda$

$\therefore \quad \mu = -2, \lambda = 0$

$\Rightarrow \quad Q \equiv (-1, 1, -2)$

So,

$PQ = \sqrt{\left(\frac{1}{2}+1\right)^2 + \left(\frac{-1}{2}-1\right)^2 + (1+2)^2}$

$= \sqrt{\frac{9}{4}+\frac{9}{4}+9} = \sqrt{\frac{54}{4}}$

$\therefore 2(PQ)^2 = 2\left(\frac{54}{4}\right) = 27$

5. *(b)* We have,

$A = [a_{ij}]$, a 3×3 matrix,

where $a_{ij} = (\sqrt{2})^{i+j}$

$$\Rightarrow A = \begin{bmatrix} (\sqrt{2})^2 & (\sqrt{2})^3 & (\sqrt{2})^4 \\ (\sqrt{2})^3 & (\sqrt{2})^4 & (\sqrt{2})^5 \\ (\sqrt{2})^4 & (\sqrt{2})^5 & (\sqrt{2})^6 \end{bmatrix}$$

$$\Rightarrow A = \begin{bmatrix} 2 & 2\sqrt{2} & 4 \\ 2\sqrt{2} & 4 & 4\sqrt{2} \\ 4 & 4\sqrt{2} & 8 \end{bmatrix}$$

$$A^2 = 2^2\begin{bmatrix} 1 & \sqrt{2} & 2 \\ \sqrt{2} & 2 & 2\sqrt{2} \\ 2 & 2\sqrt{2} & 4 \end{bmatrix}\begin{bmatrix} 1 & \sqrt{2} & 2 \\ \sqrt{2} & 2 & 2\sqrt{2} \\ 2 & 2\sqrt{2} & 4 \end{bmatrix}$$

$$= 4\begin{bmatrix} - & & - \\ - & & - \\ (2+4+8) & (2\sqrt{2}+4\sqrt{2}+8\sqrt{2} & \\ & - & \\ & - & \\ & (4+8+16) & \end{bmatrix}$$

Sum of elements of 3rd row

$= 4(14 + 14\sqrt{2} + 28)$

$= 4(42 + 14\sqrt{2})$

$= 168 + 56\sqrt{2}$

$= \alpha + \beta\sqrt{2}$

$\therefore \; \alpha + \beta = 168 + 56 = 224$

6. *(a)*

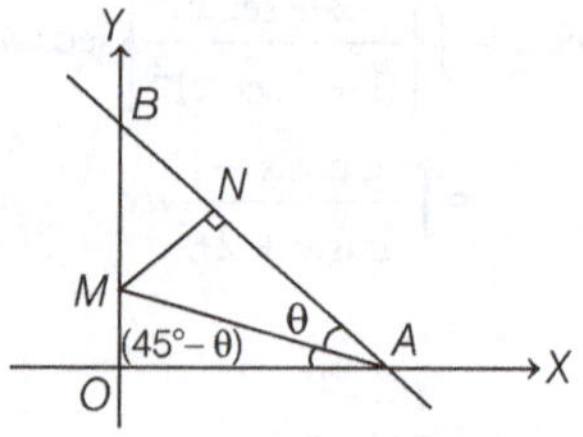

ar $(AOB) = \frac{1}{2}$

ar $(\Delta AMN) = \frac{4}{9}\times\frac{1}{2} = \frac{2}{9}$

Equation of AB is $x + y = 1$

$OA = 1, AM = \sec(45° - \theta)$

$AN = \sec(45° - \theta)\cos\theta$

$MN = \sec(45° - \theta)\sin\theta$

Now, ar (ΔAMN)

$= \frac{1}{2}\sec^2(45° - \theta)\sin\theta\cdot\cos\theta = \frac{2}{9}$

$\Rightarrow \tan\theta = 2, \frac{1}{2}$

$\tan\theta = 2$ is rejected

So, $\frac{AN}{NB} = \frac{\lambda}{1} = \cot\theta = 2$

7. *(c)* We have,

A, K, N, P, R, U

[A] .... = 5! = 120

[K] .... = 5! = 120

[N] .... = 5! = 120

[P] [A] .... = 4! = 24

[P] [K] .... = 4! = 24

[P] [N] .... = 4! = 24

[P] [R] [A] .... = 3! = 6

[P] [R] [K] [A] [N] [U] .... = 1

[P] [R] [K] [A] [U] [N] .... = 1

Total = 440

$\Rightarrow$ 440th word = PRKAUN

8. *(d)* We have,

$2x^2 + (a-5)x + 15 = 3a$

has no real root

So, $(a-5)^2 - 8(15-3a) < 0$

$= a^2 + 14a + 25 - 120 < 0$

$= a^2 + 14a - 95 < 0$

$= (a+19)(a-5) < 0$

$= a \in (-19, 5)$

$\therefore \sum_{x\in X} x^2 = (1^2 + 2^2 + \ldots.. + 4^2)$

$+ (1^2 + 2^2 + \ldots.. + 18^2)$

$= \frac{4\times5\times9}{6} + \frac{18\times19\times37}{6}$

$= 30 + 2109 = 2139$

Concept Enhancer

$\sum_{r=1}^{n} r^2 = 1^2 + 2^2 + 3^2 + \ldots + n^2$

$= \frac{n(n+1)(2n+1)}{6}$

9. *(c)* We have,

$A = \begin{bmatrix} a_{11} & a_{12} \\ a_{21} & a_{22} \end{bmatrix}$

$|A| = a_{11}a_{22} - a_{12}a_{21} = \{-1, 0, 1\}$

X	p_i	P_iX_i	$P_iX_i^2$
−1	$\frac{3}{16}$	$\frac{-3}{16}$	$\frac{3}{16}$
0	$\frac{10}{16}$	0	0
1	$\frac{3}{16}$	$\frac{3}{16}$	$\frac{3}{16}$
		$\Sigma P_iX_i = 0$	$\Sigma P_iX_i^2 = \frac{3}{8}$

$\therefore \text{Var}(x) = \Sigma P_i X_i^2 - (\Sigma P_i X_i)^2$

$= \frac{3}{8} - 0 = \frac{3}{8}$

10. *(c)* $\because \cos|x|$ is always differentiable

$\therefore$ We have to check only for

$|x^2 - ax + 2|$

$\therefore$ Not differentiable at

$x^2 - ax + 2 = 0$

One root is given $\alpha = 2$

$\therefore \quad 4 - 2a + 2 = 0$

$a = 3$

$\Rightarrow$ Other root $\beta = 1$

But for $x = 1$, $f(x)$ is differentiable

So, it is bonus

$\therefore$ Distance from (α, β) i.e. (2, 1) to line $12x + 5y + 10 = 0$

$= \left|\frac{12(2) + 5(1) + 10}{\sqrt{144 + 25}}\right| = \frac{39}{13} = 3$

11. *(b)* We have,

$c_1 : |y| = 1 - x^2$

$c_2 : x^2 + y^2 = 1$

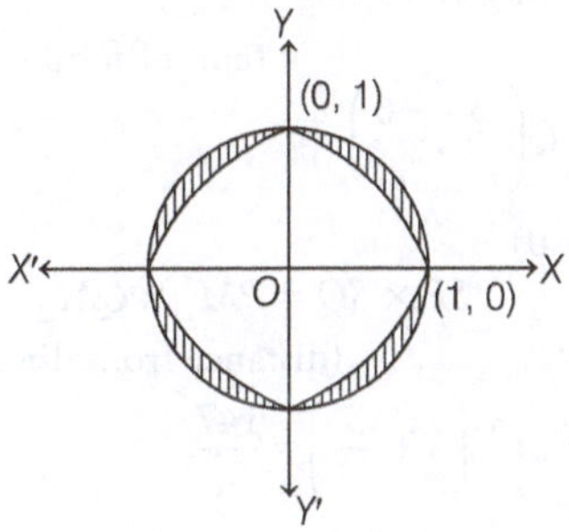

$\because$ Required area

$= \alpha = 4$ [area of circle is 1st quadrant $- \int_0^1 (1 - x^2) dx$]

$= 4\left[\frac{\pi}{4} - \left[x - \frac{x^3}{3}\right]_0^1\right]$

$\alpha = \pi - \frac{8}{3}$

$\therefore \; 3\alpha = 3\pi - 8$

$\Rightarrow 9\alpha = 9\pi - 24$

$\therefore \quad \beta = 9, \gamma = -24$

$\therefore |\beta - \gamma| = 33$

12. *(d)* $7^{103} = 7(7)^{102} = 7(343)^{34}$

$= 7(345 - 2)^{34}$

$\Rightarrow \; 7^{103} = 23k_1 + 7 \cdot 2^{34}$

Now, $7 \cdot 2^{34} = 7 \cdot 2^2 \cdot 2^{32}$

$= 28 \cdot (256)^4$

$= 28(253 + 3)^4$

$\therefore 28 \times 81 \Rightarrow (23 + 5)(69 + 12)$

$\equiv 23k_2 + 60$

$\therefore$ Remainder $= 14$

13. *(a)* Given, equation of ellipase

$\frac{x^2}{9} + \frac{y^2}{4} = 1$

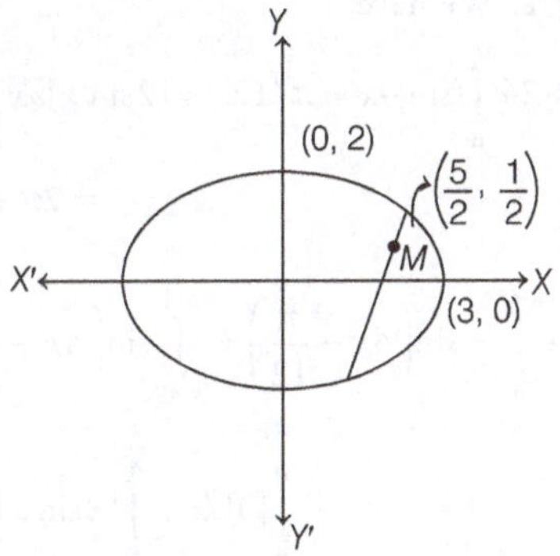

Equation of chord

$\frac{5}{2}\left(\frac{x}{9}\right) + \frac{1}{2}\left(\frac{y}{4}\right) = \frac{25}{36} + \frac{1}{16}$

$\Rightarrow \frac{5x}{18} + \frac{y}{8} = \frac{100 + 9}{144} = \frac{109}{144}$

$\Rightarrow 40x + 18y = 109$

Compare with $\alpha x + \beta y = 109$, we get

$\Rightarrow \alpha = 40, \beta = 18$

$\Rightarrow \alpha + \beta = 58$

> **Concept Enhancer**
>
> The equation of the chord of the ellipse $\frac{x^2}{a^2} + \frac{y^2}{b^2} = 1$ bisected at the point (x_1, y_1) is $T = S_1$

14. *(c)* Let $f_1(x) = \log_5(18x - x^2 - 77)$

$\therefore 18x - x^2 - 77 > 0$

$x^2 - 18x + 77 < 0$

$x \in (7, 11)$

$\Rightarrow \alpha = 7, \beta = 11$

and $f_2(x) = \log_{(x-1)}\left(\frac{2x^2 + 3x - 2}{x^2 - 3x - 4}\right)$

$\therefore x - 1 > 0, \; x - 1 \neq 1, \; \frac{2x^2 + 3x - 2}{x^2 - 3x - 4} > 0$

$x > 1, \; x \neq 2 \; \frac{(2x-1)(x+2)}{(x-4)(x+1)} > 0$

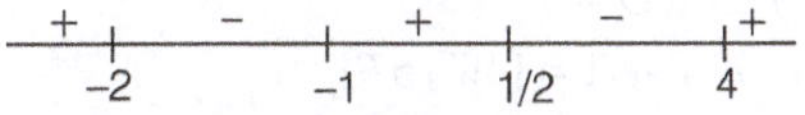

$\therefore \qquad x \in [4, \infty)$

$\therefore \qquad \gamma = 4$

$\therefore \alpha^2 + \beta^2 + \gamma^2 = 49 + 121 + 16$

$= 186$

15. *(c)*

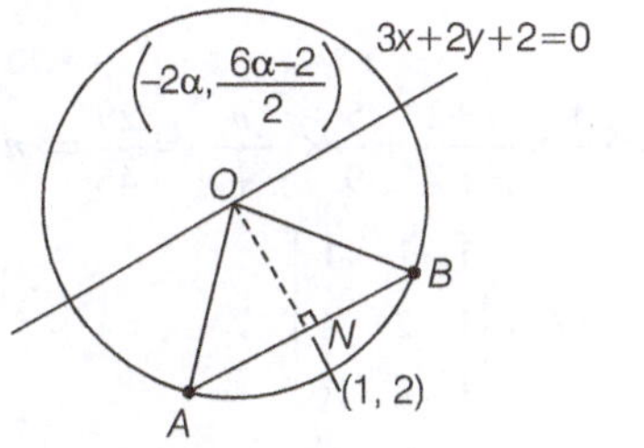

Let the centre be

$\left(-2\alpha, \frac{6\alpha - 2}{2}\right) = (-2\alpha, 3\alpha - 1)$

Centre is at equal distance from (4, 2) and (0, 2)

$= \sqrt{(4 + 2\alpha)^2 + (3\alpha - 3)^2}$

$= \sqrt{(-2\alpha - 0)^2 + (3\alpha - 3)^2}$

$= 4\alpha^2 + 16 + 16\alpha = 4\alpha^2$

$\Rightarrow \qquad \alpha = -1$

$\therefore$ Centre $O = (2, -4)$

$r = OA = \sqrt{(2-4)^2 + (2+4)^2} = \sqrt{40}$

Mid-point of chord is $N \equiv (1, 2)$

$\therefore \qquad ON = \sqrt{37}$

$\therefore$ Length of chord $= 2\sqrt{r^2 - (ON)^2}$

$= 2\sqrt{40 - 37} = 2\sqrt{3}$

16. *(d)* Vector parallel to 'L'

$= \begin{vmatrix} \hat{i} & \hat{j} & \hat{k} \\ 2 & 1 & -2 \\ 1 & 3 & 4 \end{vmatrix} = 10\hat{i} - 10\hat{j} + 5\hat{k}$

$= 5(2\hat{i} - 2\hat{j} + \hat{k})$

Equation of 'L'

$\frac{x-2}{2} = \frac{y+1}{-2} = \frac{z-3}{1} = \lambda$ (say)

Let $Q(2\lambda + 2, -2\lambda - 1, \lambda + 3)$

$\Rightarrow 2\lambda + 2 = 0$

$\Rightarrow \qquad \lambda = -1$

$\Rightarrow \quad Q(0, 1, 2)$

$\therefore$ Perpendicular distance between P and $Q = 3$

17. *(a)* Bag $1 = \{4W, 5B\}$

Bag $2 = \{nW, 3B\}$

$$P\left(\frac{W}{\text{Bag2}}\right) = \frac{29}{45}$$

$$\Rightarrow P\left(\frac{W}{B_1}\right) \times P\left(\frac{W}{B_2}\right) + P\left(\frac{W}{B_1}\right) \times P\left(\frac{W}{B_2}\right) = \frac{29}{45}$$

$$\Rightarrow \frac{4}{9} \times \frac{n+1}{n+4} + \frac{5}{9} \times \frac{n}{n+4} = \frac{29}{45} \Rightarrow n = 6$$

18. *(d)* $\Delta = \begin{vmatrix} 1 & 1 & 1 \\ 1 & 2 & 4 \\ 1 & 4 & 10 \end{vmatrix}$

$$= 1(20-16) - 1(10-4) + 1(4-2)$$
$$= 4 - 6 + 2 = 0$$

For infinite solutions

$\Delta_x = \Delta_y = \Delta_z = 0$

$m^2 - 3m + 2 = 0$

$m = 1, 2$

$\alpha = 1, \beta = 2$

$$\therefore \sum_{n=1}^{10}(n^\alpha + n^\beta) = \sum_{n=1}^{10} n^1 + \sum_{n=1}^{10} n^2$$

$$= \frac{10(11)}{2} + \frac{10(11)(21)}{6}$$

$$= 55 + 385 = 440$$

19. *(d)* $S = \{0, 1, 2, 3, \ldots\}$

$$\log_e y = x \log_e\left(\frac{2}{5}\right) \Rightarrow y = \left(\frac{2}{5}\right)^x$$

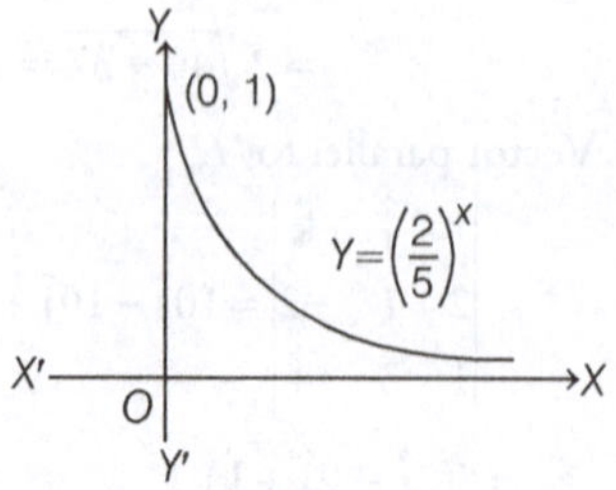

$\therefore$ Required sum

$$= 1 + \left(\frac{2}{5}\right)^1 + \left(\frac{2}{5}\right)^2 + \left(\frac{2}{5}\right)^3 + \ldots$$

$$= \frac{1}{1 - \frac{2}{5}} = \frac{5}{3}$$

20. *(d)* We have, $\sin x + \sin^2 x = 1$

$\Rightarrow \sin x = \cos^2 x$

$\Rightarrow \tan x = \cos x$

$\therefore$ Given expression

$$= 2\cos^{12} x + 6[\cos^{10} x + \cos^8 x] + 2\cos^6 x$$

$$= 2[\sin^6 x + 3\sin^5 x + 3\sin^4 x + \sin^3 x]$$

$$= 2\sin^3 x[(\sin x + 1)^3]$$

$$= 2[\sin^2 x + \sin x]^3 = 2$$

21. *(12)* We have,

$$\Rightarrow 24\int_0^{\pi/4} (\sin|4x - \pi/12|) + [2\sin x]dx = 2\pi + \alpha$$

$$\Rightarrow 24\int_0^{\pi/48} -\sin\left(4x - \frac{\pi}{12}\right) + \int_{\pi/48}^{\pi/4} \sin\left(4x - \frac{\pi}{12}\right) + \int_0^{\frac{\pi}{6}}[0]dx + \int_{\frac{\pi}{6}}^{\frac{\pi}{4}}[2\sin x]dx$$

$$= 24\left[\frac{\left(1 - \cos\frac{\pi}{12}\right)}{4} - \frac{\left(-\cos\frac{\pi}{12} - 1\right)}{4}\right] + \frac{\pi}{4} - \frac{\pi}{6}$$

$$= 24\left(\frac{1}{2} + \frac{\pi}{12}\right) = 2\pi + 12$$

Compare with $2\pi + \alpha$, we get

$\Rightarrow \alpha = 12$

22. *(11132)* Given,

$a_1, a_2, \ldots\ldots, a_{2024}$ be arithmetic progression such that

$$a_1 + (a_5 + a_{10} + \ldots\ldots + a_{2020}) + a_{2024} = 2233$$

In an AP the sum of terms equidistant from ends is equal

$$\Rightarrow a_1 + a_{2024} = a_5 + a_{2020} = a_{10} + a_{2015}$$

$\Rightarrow 203$ pairs

$\Rightarrow 203\ (a_1 + a_{2024}) = 2233$

$$\text{Hence, } S_{2024} = \frac{2024}{2}(a_1 + a_{2024})$$
$$= 1012 \times 11 = 11132$$

23. *(64)* Given limit is 1^∞ form

$$\text{Now, } L = e^{\lim\limits_{t \to 0} \frac{\left(\left.\frac{(3x+5)^{t+1}}{3(t+1)}\right|_0^{1}\ -1\right)}{t}}$$

$$= e^{\lim\limits_{t \to 0} \frac{8^{t+1} - 5^{t+1} - 3t - 3}{3t(t+1)}}$$

$$= e^{\frac{8\ln 8 - 5\ln 5 - 3}{3}}$$

$$= \left(\frac{8}{5}\right)^{\frac{2}{3}}\left(\frac{64}{5}\right) = \frac{\alpha}{5e}\left(\frac{8}{5}\right)^{\frac{2}{3}}$$

On comparing, we get $\alpha = 64$

24. *(1328)*

> **Key idea**
> Equation of the circle, when the coordinates of end points of a diameter are (x_1, y_1) and (x_2, y_2) is
> $(x - x_1)(x - x_2) + (y - y_1)(y - y_2) = 0$

We have, $y^2 = 12x$

$\Rightarrow \quad a = 31$

and $SP \times SQ = \frac{147}{4}$

Let $P(3t^2, 6t)$ and $t_1 t_2 = -1$

(end of focal chord)

So, $Q\left(\frac{3}{t^2}, \frac{-6}{t}\right)$

$S(3, 0)$

$\therefore \quad SP \times SQ = PM_1 \times QM_2$

(distance from directrix)

$$(3 + 3t^2)\left(3 + \frac{3}{t^2}\right) = \frac{147}{4}$$

$$\Rightarrow \quad \frac{(1+t^2)^2}{t^2} = \frac{49}{12}$$

$$\Rightarrow \quad t^2 = \frac{3}{4}, \frac{4}{3} \Rightarrow t = \pm\frac{\sqrt{3}}{2}, \pm\frac{2}{\sqrt{3}}$$

Considering $t = \frac{\sqrt{3}}{2}$

$P\left(\frac{9}{4}, -3\sqrt{3}\right)$ and $Q(4, 4\sqrt{3})$

Hence, diametric circle

$$(x - 4)\left(x - \frac{9}{4}\right) + (y + 3\sqrt{3})(y - 4\sqrt{3}) = 0$$

$$\Rightarrow x^2 + y^2 - \frac{25}{4}x - \sqrt{3}y - 27 = 0$$

$\Rightarrow \alpha = 400, \beta = 1728$

$\beta - \alpha = 1328$

25. *(10)* We have,

$a, b \in I, -3 \leq a, b \leq 3$

$a + b \neq 0$

$\Rightarrow \quad |z - a| = |z + b|$

$$\Rightarrow \begin{vmatrix} z+1 & \omega & \omega^2 \\ \omega & z+\omega^2 & 1 \\ \omega^2 & 1 & z+\omega \end{vmatrix} = 1$$

$$\Rightarrow \begin{vmatrix} z & z & z \\ \omega & z+\omega^2 & 1 \\ \omega^2 & 1 & z+\omega \end{vmatrix} = 1$$

$$\Rightarrow z\begin{vmatrix} 1 & 1 & 1 \\ \omega & z+\omega^2 & 1 \\ \omega^2 & 1 & z+\omega \end{vmatrix} = 1$$

$$\Rightarrow z\begin{vmatrix} 1 & 0 & 0 \\ \omega & z+\omega^2-\omega & 1-\omega \\ \omega^2 & 1-\omega^2 & z+\omega-\omega^2 \end{vmatrix} = 1$$

$\Rightarrow \quad z^3 = 1 \Rightarrow z = w, w^2, 1$

Now, $|1 - a| = |1 + b|$

$\Rightarrow$ 10 pairs

Physics

26. *(a)*

Key idea
In an elastic collision both linear momentum and total kinetic energy of system remains conserved.

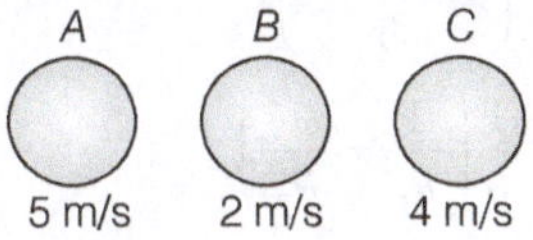

Reason is true as on collision between identical masses, two objects exchange their velocities. Now, in assertion, first A and B collide and exchange their velocities.

v_A becomes 2 m/s.

v_B becomes 5 m/s.

After that balls B and C collide and exchange their velocities.

v_B becomes 4 m/s.

v_C becomes 5 m/s.

So, final velocities are

$v_A = 2$ m/s, $v_B = 4$ m/s, $v_C = 5$ m/s

Thus, assertion is wrong.

27. *(b)* Since, lens is cut horizontally, hence, focal length remains same. So, focal length of L_1 is f

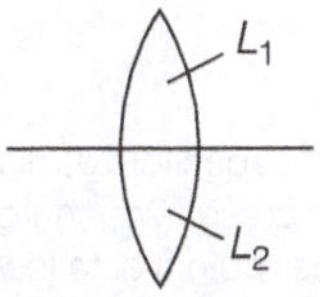

In 2nd lens, it is cut vertically, hence, focal length becomes $2f$.

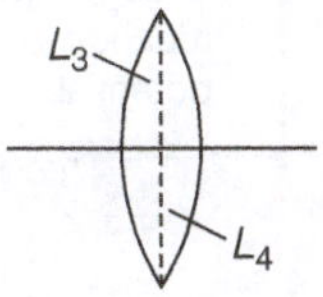

$$\Rightarrow \quad \frac{f_{L_1}}{f_{L_3}} = \frac{f}{2f} = \frac{1}{2}$$

28. *(b)*

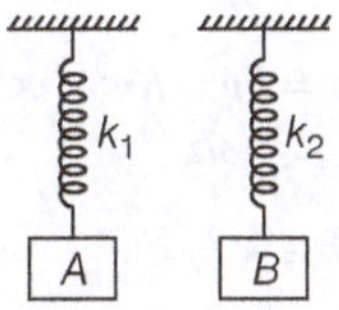

Angular velocity of SHM, $\omega = \sqrt{\frac{k}{m}}$

Angular velocity of 1st SHM,

$$\omega_1 = \sqrt{\frac{k_1}{m}}$$

Angular velocity of 2nd SHM,

$$\omega_2 = \sqrt{\frac{k_2}{m}}$$

Maximum velocity of SHM, $v = a\omega$

Maximum velocity of 1st SHM,

$$v_1 = a\omega_1$$

Maximum velocity of 2nd SHM,

$$v_2 = a\omega_2$$

According to question,

$$\frac{v_1}{v_2} = \frac{a\omega_1}{a\omega_2} = \sqrt{\frac{k_1}{m}} \times \sqrt{\frac{m}{k_2}}$$

$$\Rightarrow \quad \frac{v_1}{v_2} = \sqrt{\frac{k_1}{k_2}}$$

29. *(a)* For isothermal process, $pV = C$, which for adiabatic process $pV^\gamma = C$

The graphs are as shown.

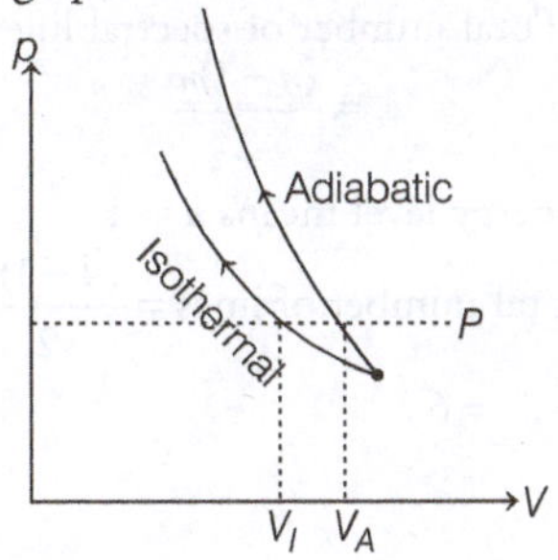

As seen from the graph, for the same pressure p,

$$V_I < V_A$$

i.e.

Volume falls more rapidly for isothermal process.

So, assertion is correct.

Reason is also true and explains the assertion.

30. *(d)*

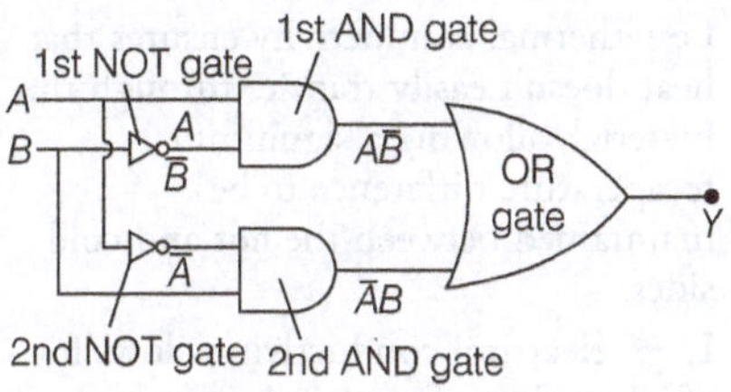

For 1st NOT gate, B is input,

So output is $\bar{B}$.

For 2nd NOT gate, A is input,

So output is $\bar{A}$.

For 1st AND gate, A and $\bar{B}$ are inputs.

So output is $A\bar{B}$.

For 2nd AND gate, $\bar{A}$ and B are input.

So, output is $\bar{A}B$.

Ultimately for OR gate, $A\bar{B}$ and $\bar{A}B$ are the inputs, so output is

$Y = A\bar{B} + \bar{A}B$

This is the XOR gate, whose truth table is

A	B	Y
0	0	0
0	1	1
1	0	1
1	1	0

Alternate Solution

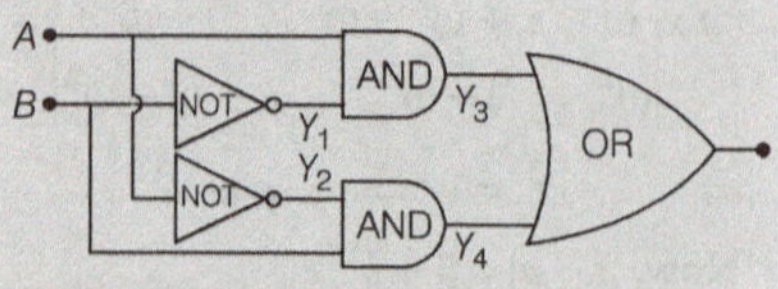

Truth table as per charactersitics of different gates is as follows.

A	B	$y_1 = \bar{B}$	$y_2 = \bar{A}$	y_3	y_4	y
0	0	1	1	0	0	0
1	0	1	0	1	0	1
0	1	0	1	0	1	1
1	1	0	0	0	0	0

y_3 is output of $A + y_1$ by AND gate

$y_4 = B$ and y_2

31. *(d)*

Key idea

This phenomenon is called the seeback effect, and materials with these properties are called thermoelectric materials.

Low thermal conductivity ensures that heat doesn't easily transfer through the material, allowing a significant temperature difference to be maintained between the hot and cold sides.

High electrical conductivity allows for efficient flow of electrical current generated by the temperature difference.

32. *(c)* A. Young's modulus = Pressure

$$= [ML^{-1}T^{-2}]$$

B. Torque = Work = $[ML^2T^{-2}]$

C. Coefficient of viscosity (η)

$$F = 6\pi r\eta v$$

$$\Rightarrow \eta = \frac{F}{rv} = \frac{[MLT^{-2}]}{[L][LT^{-1}]}$$

$$\therefore \eta = [ML^{-1}T^{-1}]$$

D. Gravitational constant (G)

$$F = \frac{Gm_1m_2}{r^2}$$

$$\Rightarrow G = \frac{Fr^2}{m_1m_2} = \frac{[MLT^{-2}][L^2]}{[M][M]}$$

$$\therefore G = [M^{-1}L^3T^{-2}]$$

33. *(c)* Given, $\frac{dm}{dt} \propto \sqrt{v}$

$$\Rightarrow \frac{dm}{dt} = K\sqrt{v}$$

By Newton's 2nd law

$$F_{ext} = \frac{mdv}{dt} + v\frac{dm}{dt}$$

As speed is constant

$$\Rightarrow \frac{dv}{dt} = 0$$

$$\Rightarrow F_{ext} = v\frac{dm}{dt} = v \times K\sqrt{v} = Kv^{3/2}$$

Now power $= Fv = Kv^{3/2} \times v$

$$P = Kv^{5/2}$$

$$P^2 = K^2v^5$$

$$\therefore P^2 \propto v^5$$

34. *(c)* Since only internal forces are acting, hence, net external torque is zero on the system. Thus angular momentum is conserved.

$$\Rightarrow L_i = L_f$$

By symmetry, the masses collide at the centroid of the triangle. So, total angular momentum at point C.

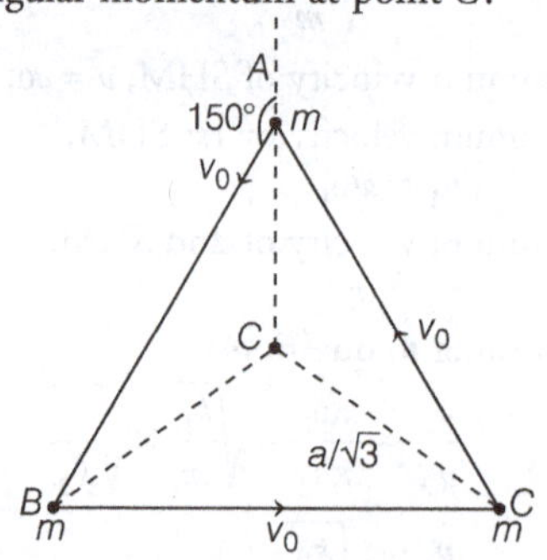

$$L_C = 3 \times mv_0 \times \frac{a}{\sqrt{3}} \sin 150°$$

$$= \frac{\sqrt{3}}{2} mv_0 a$$

35. *(c)* Total number of spectral line

$$= \frac{(n-1)n}{2}$$

4th energy level means $n = 4$

$$\Rightarrow \text{Total number of lines} = \frac{(4-1)4}{2}$$

$$= 6$$

36. *(b)* Focal length of a lens

$$\frac{1}{f} = \left(\frac{n_2}{n_1} - 1\right)\left(\frac{1}{R_1} - \frac{1}{R_2}\right)$$

For lens in air

$$n_1 = 1, n_2 = 1.5, f = f_a$$

$$\frac{1}{f_a} = \left(\frac{1.5}{1} - 1\right)\left(\frac{1}{R_1} - \frac{1}{R_2}\right)$$

$$= \frac{1}{2}\left(\frac{1}{R_1} - \frac{1}{R_2}\right) \quad \text{...(i)}$$

For lens in water

$$n_1 = 1.33 = \frac{4}{3}, n_2 = 1.5 = \frac{3}{2}, f = f_w$$

$$\frac{1}{f_w} = \left(\frac{3}{2 \times \frac{4}{3}} - 1\right)\left(\frac{1}{R_1} - \frac{1}{R_2}\right)$$

$$= \frac{1}{8}\left(\frac{1}{R_1} - \frac{1}{R_2}\right) \quad \text{...(ii)}$$

Divide Eq. (i) by Eq. (ii)

$$\frac{\frac{1}{f_a}}{\frac{1}{f_w}} = \frac{\frac{1}{2}\left(\frac{1}{R_1} - \frac{1}{R_2}\right)}{\frac{1}{8}\left(\frac{1}{R_1} - \frac{1}{R_2}\right)}$$

$$\Rightarrow f_w = 4f_a$$

$$= 4 \times 24 = 96 \text{ cm}$$

37. *(d)* In photo electric effect, stopping potential is independent of the intensity of light while it decreases with the increase of wavelength. Hence, options (a), (b) and (c) are wrong.

Now,

$$V_0 = \frac{KE_{max}}{e}$$

Hence, option (d) is correct.

38. *(a)*

Key idea

When two capacitors at different potentials are joined, charge will flow from higher potential to lower potential. Charge flow occurs till both capacitors reaches to a common (same) potential.

$$V_{Common} = \frac{C_1V_1 + C_2V_2}{C_1 + C_2}$$

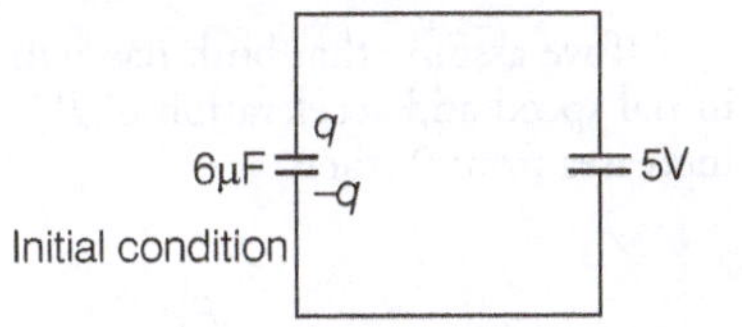

$$q = CV = 6 \times 5 = 30\,\mu C$$

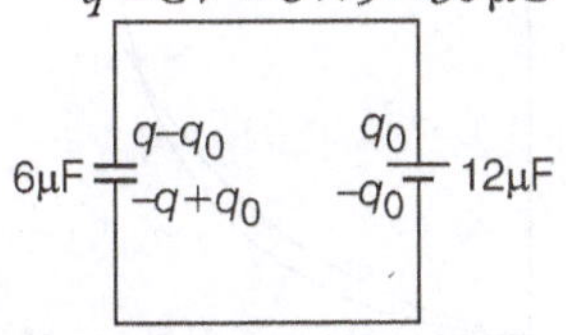

Let charge q_0 flows from C_1 to C_2.
Now apply KVL in the loop

$$\frac{q_0}{12} = \frac{q - q_0}{6}$$

$$6q_0 = 12q - 12q_0$$

$$18q_0 = 12q$$

$$q_0 = \frac{2}{3}q$$

According to the question,

$$q_1 = q - q_0$$

$$= q - \frac{2}{3}q = \frac{q}{3} = \frac{30}{3} = 10\,\mu C$$

$$q_2 = q_0 = \frac{2}{3}q = \frac{2}{3} \times 30 = 20\,\mu C$$

39. *(b)* $\hat{E} \times \hat{B}$ gives the direction of the wave propagation.

Since, wave is propagating in positive x-direction.

$\Rightarrow$ Electric field is in y-direction (E_y).

and magnetic field is in z-direction (B_z).

40. *(d)* According to Newton's law of cooling

$$-\frac{(T_f - T_i)}{t} = C\left[\frac{T_f + T_i}{2} - T_0\right]$$

For $T_i = 90°C$, $T_f = 80°C$, $T_0 = 20°C$, $t = t$

$$\Rightarrow \frac{-(80-90)}{t} = C\left[\frac{90+80}{2} - 20\right]$$

$$\Rightarrow \quad \frac{10}{t} = 65\,C \quad \text{...(i)}$$

or $T_i = 80°\,C$, $T_f = 60°C$, $T_0 = 20°C$, $t = t_1$

$$\Rightarrow -\frac{60-80}{t_1} = C\left[\frac{80+60}{2} - 20\right]$$

$$\Rightarrow \quad \frac{20}{t_1} = 50C \quad \text{...(ii)}$$

Divide Eq. (i) by Eq. (ii)

$$\frac{\frac{10}{t}}{\frac{20}{t_1}} = \frac{65C}{50C}$$

$$\Rightarrow \quad \frac{t_1}{2t} = \frac{13}{10} \Rightarrow t_1 = \frac{13}{5}t$$

41. *(d)*

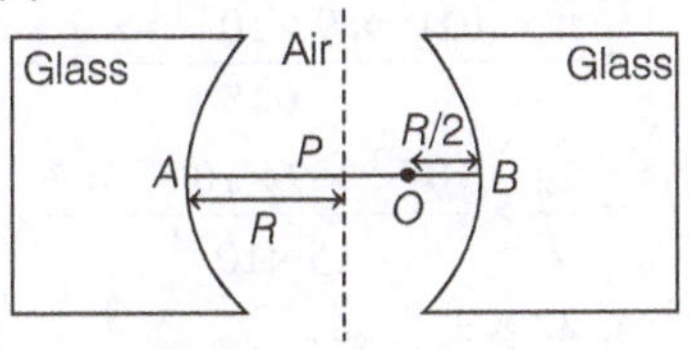

Image formed by a single refracting surface is given by

$$\frac{n_2}{v} - \frac{n_1}{u} = \frac{n_2 - n_1}{R} \quad \text{...(i)}$$

For surface A,

$$n_1 = 1,\ n_2 = 1.5$$

$$u = \frac{-3R}{2},\ R = -R$$

Applying Eq. (i)

$$\frac{1.5}{v_A} - \frac{1}{\frac{-3R}{2}} = \frac{1.5-1}{-R}$$

$$\Rightarrow \frac{1.5}{v_A} = \frac{-0.5}{R} - \frac{2}{3R}$$

$$\Rightarrow \frac{1.5}{v_A} = \frac{-1.5-2}{3R}$$

$$\Rightarrow \frac{1.5}{v_A} = \frac{-3.5}{3R}$$

$$\Rightarrow v_A = \frac{-4.5R}{3.5} = \frac{-9R}{7}$$

Similarly for surface B,

$$n_1 = 1,\ n_2 = 1.5$$

$$u = \frac{-R}{2},\ R = -R$$

Applying Eq. (i)

$$\frac{1.5}{v_B} - \frac{1}{\frac{-R}{2}} = \frac{1.5-1}{-R}$$

$$\Rightarrow \quad \frac{1.5}{v_B} = \frac{-0.5}{R} - \frac{2}{R}$$

$$\Rightarrow \quad \frac{1.5}{v_B} = \frac{-2.5}{R}$$

$$\Rightarrow \quad v_B = \frac{-1.5R}{2.5} = \frac{-3R}{5}$$

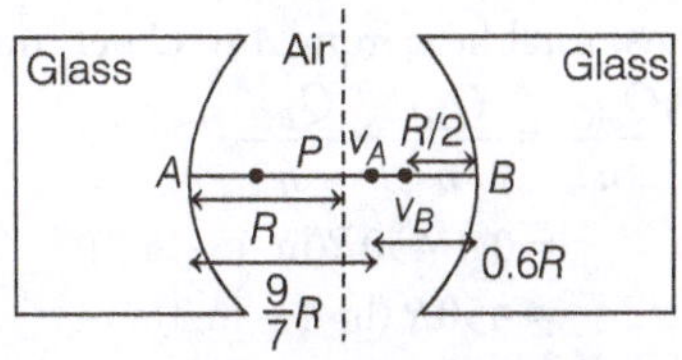

Distance between v_A and v_B is given by

$$d = 2R - (|v_A| + |v_B|)$$

$$= 2R - \left(\frac{9R}{7} + \frac{3R}{5}\right)$$

$$= 2R - \frac{9R}{7} - \frac{3R}{5}$$

$$= \left(\frac{70-45-21}{35}\right)R = \frac{4R}{35}$$

$$= 0.114R$$

42. *(b)*

Key idea

Heat absorbed in an adiabatic process $\Delta Q = 0$. Heat absorbed in an isothermal process,

$$\Delta Q_{\text{isothermal}} = W_{\text{isothermal}}$$

$$= nRT\ln\left(\frac{V_f}{V_i}\right)$$

$$\therefore \quad \Delta U_{\text{isothermal}} = 0$$

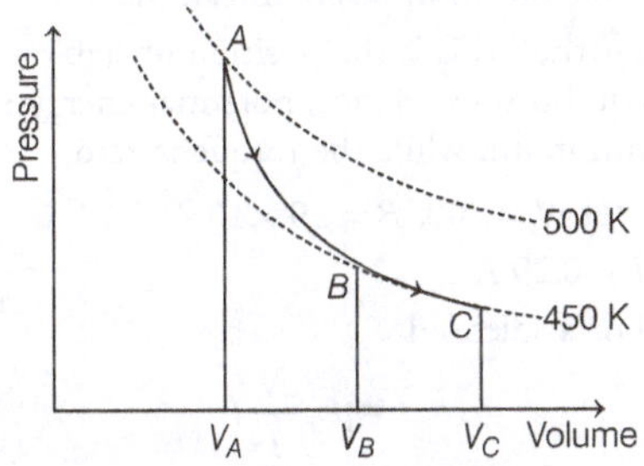

Since, process A to B is adiabatic, hence no heat get absorbed.

$$Q_{AB} = 0$$

or B to C, process is isothermal.
Temperature is 450 K.

$$\Delta Q = \Delta W = nRT \ln \frac{V_C}{V_B}$$

$$\frac{\Delta Q_{BC}}{n} = R \times 450 \times \ln\left(\frac{8 \times 10^{-6}}{6 \times 10^{-6}}\right)$$

$$= 450R \ln\left(\frac{4}{3}\right)$$

$$\frac{\Delta Q_{BC}}{n} = 450R(\ln 4 - \ln 3)$$

Thus, total heat from A to C per mole

$$\frac{Q_{AC}}{n} = \frac{Q_{AB}}{n} + \frac{Q_{BC}}{n}$$

$$= 0 + 450R(\ln 4 - \ln 3)$$

$$= 450R(\ln 4 - \ln 3)$$

43. *(*)* A. Magnetic induction has unit Tesla or Gauss.

B. Magnetic intensity

$$H = \frac{m}{\text{Volume}} = \frac{\text{Ampere metre}^2}{\text{metre}^3}$$

$$= \text{Ampere/metre}$$

C. Magnetic flux has unit Weber.

D. Magnetic moment has unit

$$m = IA = \text{Ampere metre}^2$$

44. *(d)* By Gauss law

$$\phi = \frac{q_{in}}{\varepsilon_0}$$

$$\Rightarrow -2\times10^4 = \frac{q_{in}}{8.85\times10^{-12}}$$

$$\Rightarrow \quad q_{in} = -17.70\times10^{-8}\text{ C}$$

45. *(d)* Infinite plane sheet produces uniform electric field. Hence, both the charges of the dipole experience same force but in opposite direction.

Further, this is the position of stable equilibrium. Hence, potential energy is minimum while the torque is zero.

46. *(8)* $N = 200$, $B = 2.9\times10^{-4}$ T, $I = 0.29$ A

For a solenoid

$$B = \mu_0 \frac{N}{l} I$$

$$\Rightarrow 2.9\times10^{-4} = \frac{4\pi\times10^{-7}\times200\times0.29}{l}$$

$$\Rightarrow l = \frac{4\pi\times10^{-7}\times2\times29}{2.9\times10^{-4}}$$

$$l = 80\pi\times10^{-3}\text{ m} = 8\pi\text{ cm}$$

47. *(1320)* $\frac{dV}{dt} = 7\times10^8$ V/s, $i = 0.15$ A, $r = 10$ cm $= 0.1$ m

From displacement current,

$$i_d = C\frac{dV}{dt}$$

$$\Rightarrow 0.15 = \frac{A\varepsilon_0}{d}\times7\times10^8$$

$$\Rightarrow d = \frac{A\varepsilon_0\times7\times10^8}{0.15}$$

$$= \frac{\pi\times(0.1)^2\times9\times10^{-12}\times7\times10^8}{0.15}$$

$$= \frac{22}{7}\times\frac{10^{-2}\times9\times10^{-4}\times7}{15\times10^{-2}}$$

$$= \frac{22\times9}{15}\times10^{-4}$$

$$= 13.2\times10^{-4}\text{ m}$$

$$= 1320\,\mu\text{m}$$

48. *(8)* Angular momentum, $L = mvr$

For planet A,

$$v_A = \sqrt{\frac{GM}{R_A}}$$

$$r_A = R_A$$

$$m_A = m$$

$$L_A = m_A v_A r_A$$

$$= m\sqrt{\frac{GM}{R_A}}\times R_A$$

$$= m\sqrt{GMR_A} \quad \text{...(i)}$$

For planet B

$$v_B = \sqrt{\frac{GM}{2R_A}}$$

$$r_B = 2R_A$$

$$m_B = 4\sqrt{2}\text{ m}$$

$$L_B = m_B r_B v_B$$

$$= 4\sqrt{2}m\times2R_A\times\sqrt{\frac{GM}{2R_A}}$$

$$= 8m\sqrt{GMR_A} \quad \text{...(ii)}$$

Divide Eq. (ii) by Eq. (i)

$$\frac{L_B}{L_A} = \frac{8m\sqrt{GMR_A}}{m\sqrt{GMR_A}} = 8$$

49. *(2)* If we assume that both has zero initial speed and acceleration of P increases from 0, then

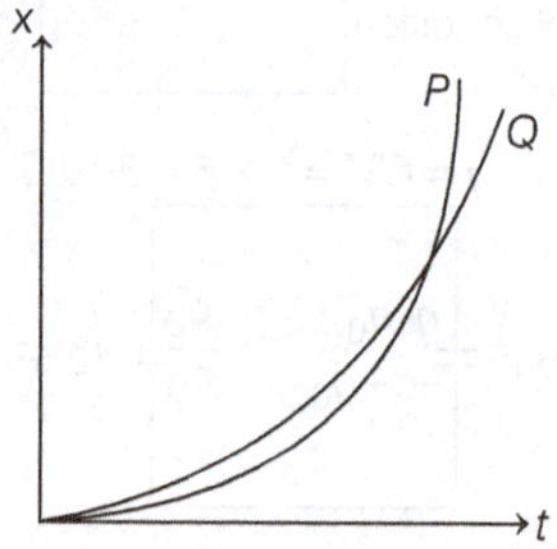

Q will be ahead initially due to higher acceleration and thus higher speed.

Now with time acceleration of P increases which increases its speed and crosses Q.

Now Q having lesser speed and acceleration than P won't able to cross it again.

50. *(77)* $a = 60 \pm 3$

$$\frac{\Delta a}{a}\times100 = \frac{3}{60}\times100 = 5\%$$

$$b = 20\pm0.1$$

$$\frac{\Delta b}{b}\times100 = \frac{0.1}{20}\times100 = 0.5\%$$

$$c = 40\pm0.2$$

$$\frac{\Delta c}{c}\times100 = \frac{0.2}{40}\times100 = 0.5\%$$

$$d = 50\pm0.1$$

$$\frac{\Delta d}{d}\times100 = \frac{0.1}{50}\times100 = 0.2\%$$

Now, $Q = \frac{ab^4}{cd}$

Percentage error in Q is given by

$$\frac{\Delta Q}{Q}\times100 = \frac{\Delta a}{a}\times100 + 4\frac{\Delta b}{b}\times100 + \frac{\Delta c}{c}\times100 + \frac{\Delta d}{d}\times100$$

$$= 5\% + 4\times0.5\% + 0.5\% + 0.2\%$$

$$= 7.7\%$$

Now, $\frac{7.7}{100} = \frac{x}{1000}$

$$\Rightarrow \quad x = 77$$

Chemistry

51. *(b)* Among the given statements, A and B are correct, while C and D are incorrect.

It is because equilibrium constant remains unchanged as the temperature is constant.

52. *(b)* Given, $[A] = 16$ mg/mL

$[A_0] = 4$ mg/mL, $T = 12$ months

Using the formula of rate constant for 1st order reaction.

$$k = \frac{2.303}{t} \log \frac{A}{[A_0]} \quad \ldots(i)$$

$$= \frac{2.303}{12} \log \frac{16}{4}$$

On solving, we get

$$k = 0.1155$$

Now, since 50% decomposition takes place. So,

$$[A] = [A_0] - \frac{50}{100}[A_0]$$

$$[A] = 0.5[A_0]$$

Now, substitute the obtained values in Eq. (i) to calculate the expiry time

$$0.1155 = \frac{2.303}{t} \log \frac{[A_0]}{[0.5A_0]}$$

$$t = 6 \text{ months}$$

53. *(b)* Among the given statements only B and C are correct while A and D are incorrect.

Silver (Ag) itself will oxidise to Ag^+ instead of water, so O_2 will not be evolved.

At very high concentrations, sulphate ions might undergo oxidation instead of water, forming peroxydisulphate.

54. *(c)*

Key idea

Homoleptic complexes are those in which all the ligand which are connected with central atom are same. So, look for the complex in which same ligands are present in central atom.

Among the given complexes, the one which is homoleptic complex with odd number of *d*-electrons in the central metal are $[Fe(CN)_6]^{3-}$ with d^5 and $[CoCl_4]^{2-}$ with d^7 configuration i.e., B and D only.

55. *(c)* Among the given compounds, compound given in option (c) will react with HBr to give a phenol.

The complete reaction is as follows,

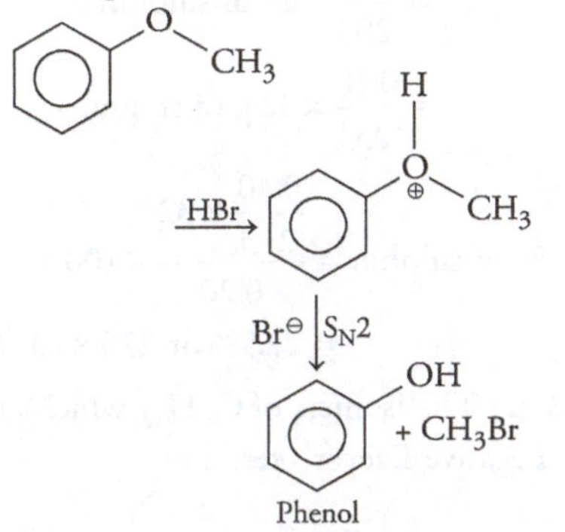

56. *(d)* Both statement I and statement II are true.

57. *(a)* The correct match is

A-III, B-II, C-IV, D-I

58. *(a)* The relation between energy (*E*) and atomic number (*Z*) is

$$E = -\frac{13.6Z^2}{n^2} \text{ eV}$$

$$E \propto -Z^2$$

So, the correct graph is

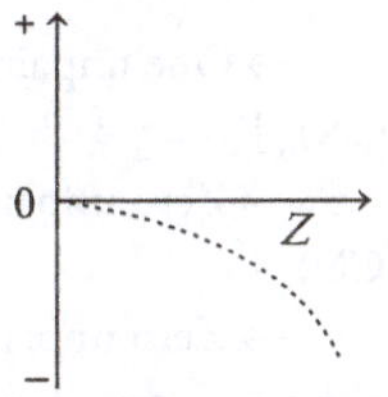

59. *(a)* Among the given reactions, nitrobenzene will give an azo dye.

The complete reaction is as follows :

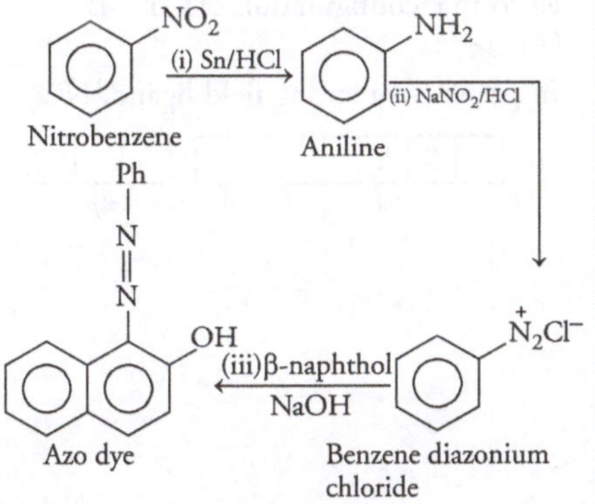

60. *(b)* The essential amino acids from the given options are valine, lysine and threonine. i.e., A, C and D only.

61. *(d)* In $K_3[Fe(OH)_6]$, Fe is in +3 oxidation state. So, it has d^5 electronic configuration. Thus, the number of unpaired electrons = 5

So, $\mu = \sqrt{5(5+2)} = 5.92$ BM

Similarly, in $K_4[Fe(OH)_6]$, Fe is +2 oxidation state. So, it has d^4 electronic configuration. Thus, the number of unpaired electrons = 4

So, $\mu = \sqrt{4(4+2)} = 4.90$ BM

Thus, the calculated spin-only magnetic moment of $K_3[Fe(OH)_6]$ and $K_4[Fe(OH)_6]$, respectively are 5.92 and 4.90 BM.

62. *(b)* The type of oxide formed by elements like Li, Na, Be, B, Mg and Al that has least atomic radii is A_2O_3.

63. *(c)* The structure of hex-1-en-4-yne is

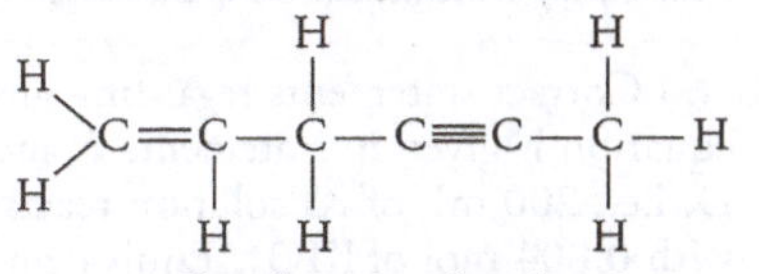

σ-bonds = 13, π-bonds = 3

64. *(c)* The apatite family consists mainly of calcium phosphates, meaning phosphorus (P) is the main component.

The first ionisation enthalpy values are

Nitrogen (N) $\rightarrow$ 1402 kJ/mol

Phosphorus (P) $\rightarrow$ 1012 kJ/mol

Arsenic (As) $\rightarrow$ 947 kJ/mol

Antimony (Sb) $\rightarrow$ 834 kJ/mol

Since, phosphorus is the main component of apatite, the correct answer is option c.

65. *(d)* Both statement I and II are true.

Reaction for statement I is

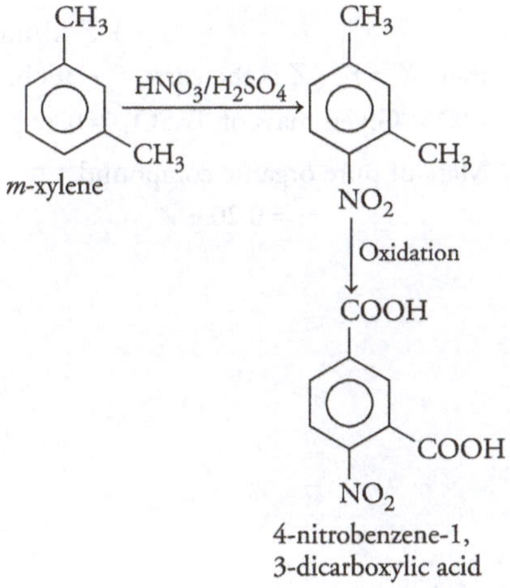

— CH_3 group is *o/p* directing while —NO_2 group is *meta* directing.

66. *(c)* Stability order of carbocation is

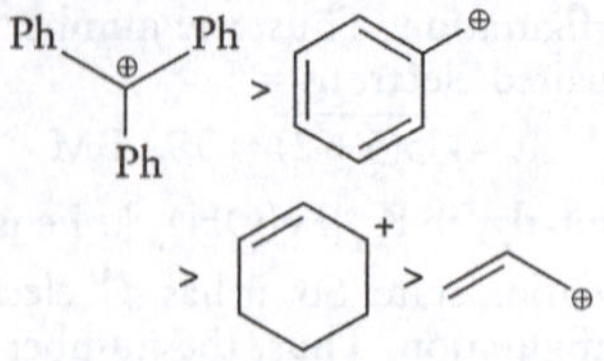

So, the most stable carbocation is given in option (c).

67. *(c)* Both statement I and statement II are true.

> **Concept Enhancer**
> Heisenberg's uncertainty principle is applicable only when the location and momentum are along the same axis. i.e if Δx is uncertainty in position along the *x*-axis then Δp must also be uncertainty in momentum along the *x*-axis.

68. *(a)* Correct statements regarding given equation is given in statements A and D. i.e., 200 mL of KI solution reacts with 0.004 mol of KIO_3. Equivalent weight of KIO_3 is equal to $\left(\frac{\text{molecular weight}}{5}\right)$.

69. *(a)* Statement I is true but statement II is false. The correct form of statement II is :

In paper chromatography, stationary phase is water.

70. *(d)* C (diamond) $\longrightarrow$ C(graphite) $+ X$ kJ/mol

C(diamond) + $O_2(g) \longrightarrow CO_2(g)$ $+ Y$ kJ/mol

C(graphite) + $O_2(g) \longrightarrow CO_2(g)$ $+ Z$ kJ/mol

then $X = Y - Z$ is the correct relation.

71. *(275)* Given mass of $BaSO_4 = 0.40$ g

Mass of pure organic compound $= 0.20$ g

0.20 g of organic compound gives 0.40 g of $BaSO_4$.

$$\text{Moles of } BaSO_4 = \frac{0.40}{233} \text{ mol of } BaSO_4$$

$$= \frac{0.40}{233} \text{ mol of sulphur}$$

$$= \frac{0.40}{233} \times 32 \text{ g of sulphur}$$

$$\% \text{ of sulphur} = \frac{\frac{0.40}{233} \times 32}{0.20} \times 100$$

$$= 27.5\% \text{ or } 275 \times 10^{-1}\%$$

72. *(2)* The isomers of C_9H_{12} which give negative Baeyer's test are

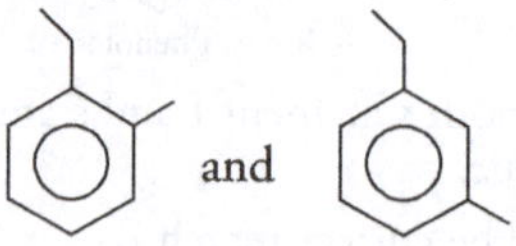

Above two isomers of C_9H_{12} have four different sites for aromatic electrophilic substitution reaction.

73. *(0)* $K_3[Co(NO_2)_6]$

$\rightarrow$ Zero unpaired electron.

$K_4[Fe(CN)_6]$

$\rightarrow$ Zero unpaired electron.

$K_3[Fe(CN)_6]$

$\rightarrow$ One unpaired electron.

$Cu_2[Fe(CN)_6]$

$\rightarrow$ Zero unpaired electron.

$Zn_2[Fe(CN)_6]$

$\rightarrow$ Zero unpaired electron.

Yellow colour compounds are $K_3[Co(NO_2)_6]$ and $K_4[Fe(CN)_6]$

Now, in $K_3[Co(NO_2)_6]$, the outer electronic configuration of Co^{3+} is $[Ar]3d^6$

In presence of strong field ligand, NO_2

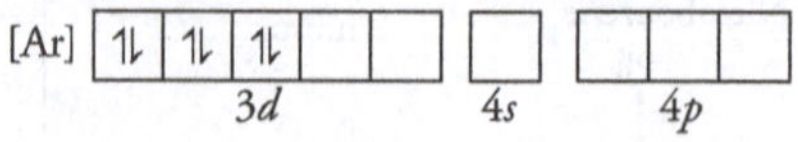

Magnetic moment $(\mu) = \sqrt{n(n+2)}$ BM $= 0$ BM

In $K_4[Fe(CN)_6]$, Fe^{2+} is $[Ar]3d^6$

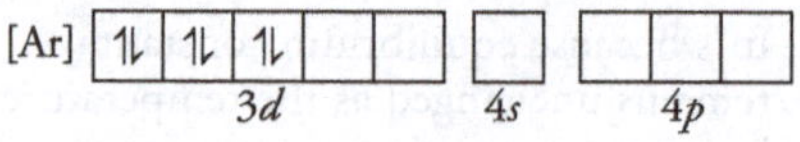

$$\mu = \sqrt{n(n+2)} \text{ BM}$$
$$= 0 \text{ BM}$$

So, total spin only magnetic moment value of complex with yellow colour is zero.

74. *(60)* The complete reaction is as follows

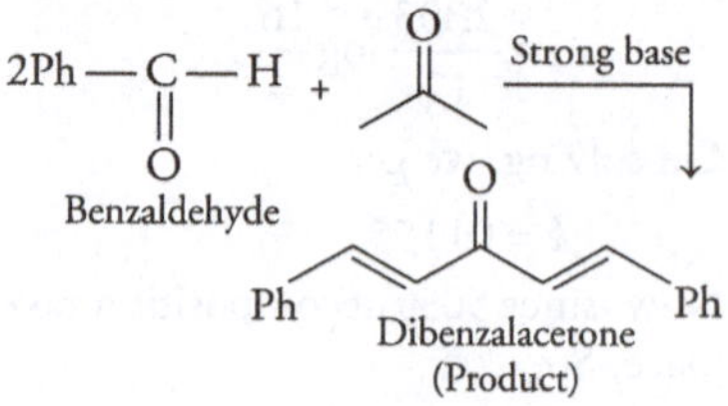

Now, 5.3 g of benzaldehyde $= 3.51$ g of dibenzalacetone

$\therefore \frac{5.3}{106}$ mol of benzaldehyde

$= \frac{3.51}{234}$ mol of dibenzalacetone

or, $\frac{1}{20}$ mol (Actual) of benzaldehyde

$= 0.015$ mol of dibenzalacetone

Theoretical $= \frac{1}{40}$ mol

[$\because$ 2 mole sof benzaldehyde give 1 mol]

$$\% \text{ yield} = \frac{0.015}{1/40} \times 100 = 60\%$$

75. *(12)* The structure of NO_2^- according to Lewis theory is as follows

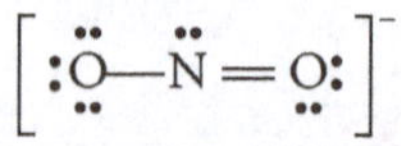

Thus, it has 12 non-bonded electrons.